Feature Papers in *BioChem*, 2nd Edition

Feature Papers in *BioChem*, 2nd Edition

Guest Editors

Buyong Ma
Manuel Aureliano

MDPI

Basel • Beijing • Wuhan • Barcelona • Belgrade • Novi Sad • Cluj • Manchester

Guest Editors
Buyong Ma
School of Pharmacy
Shanghai Jiaotong University
Shanghai
China

Manuel Aureliano
Department of Biological
Sciences and Biotechnology
University of Algarve
Faro
Portugal

Editorial Office
MDPI AG
Grosspeteranlage 5
4052 Basel, Switzerland

This is a reprint of the Special Issue, published open access by the journal *BioChem* (ISSN 2673-6411), freely accessible at: https://www.mdpi.com/journal/biochem/special_issues/46HUT2S6AP.

For citation purposes, cite each article independently as indicated on the article page online and as indicated below:

Lastname, A.A.; Lastname, B.B. Article Title. *Journal Name* **Year**, *Volume Number*, Page Range.

ISBN 978-3-7258-7739-3 (Hbk)
ISBN 978-3-7258-7740-9 (PDF)
https://doi.org/10.3390/books978-3-7258-7740-9

Contents

About the Editors

Buyong Ma

Buyong Ma is a professor in the school of pharmacy, Shanghai Jiao Tong University. Dr. Ma received his Ph.D. in physical chemistry from the University of Georgia at Athens in 1995. In 1998, he joined the National Cancer Institute (USA) as a Senior Scientist. He moved to his current position in 2020. He has made outstanding contributions to computer-aided antibody drug design, biomacromolecular structure and interactions, and protein–protein interactions and functions. He contributed to the development of the conformational selection theory, which has become a model alongside the classic 'Lock–Key' and 'Induced Fit' models. He was the first in the world to reveal the structure of the Aβ amyloid associated with Alzheimer's disease. His research findings have been extensively cited in areas such as the characterization of protein structure dynamics, enzyme catalysis, protein–protein interactions and drug design, and neurodegenerative diseases, thereby advancing progress in these fields.

Manuel Aureliano

Manuel Aureliano is a Full Professor of Biochemistry, University of Algarve, Portugal; he is also the Director of the Biochemistry degree (1998–2013; 2021–2025) and the PhD in Biological Sciences (2025–). He investigates the role and applications of decavanadate and other polyoxometalates in the environment and biomedicine. He was awarded the title of "Outstanding Reviewer" for *Metallomics* (2017, 2018, 2019). He has reviewed more than 950 papers from about 150 journals and has been an Editor, Associate Editor, and/or Guest Editor for several journals. From 2021 to 2025, he was included in the 'World's Top 2% Scientists List' (impact-career and year). Recently, he was awarded the third edition of the "UAlg Researcher Award".

BioChem

MDPI

Editorial

Feature Papers in *BioChem*, 2nd Edition

Manuel Aureliano [1,2,*,†] and Buyong Ma [3,*,†]

1 Faculdade de Ciências e Tecnologia, Universidade do Algarve, Campus de Gambelas, 8005-139 Faro, Portugal
2 Centro de Ciências do Mar do Algarve (CCMAR/CIMAR LA), Universidade do Algarve, 8005-139 Faro, Portugal
3 School of Pharmacy, Shanghai Jiaotong University, Shanghai 200240, China
* Correspondence: maalves@ualg.pt (M.A.); mabuyong@sjtu.edu.cn (B.M.)
† These authors contributed equally to this work.

1. Introduction and Scope

As mentioned in the Editorial of the first edition of the Special Issue on "Feature Papers in *BioChem*", biochemistry acts as a key cog in the "clock of the knowledge, permitting that wheels from several science areas move each other" [1]. Moreover, the understanding of life on earth is a perpetual topic in biochemistry, with boundaries dependent on the specificities of each time period and scientific environment [1]. Thus, following the first edition, this Special Issue on "Feature Papers in *BioChem*, 2nd Edition" still emphases interdisciplinary research in diverse and interconnected scientific areas, with a majority from molecular medicine, cell biology and of course biochemistry (Table 1). Among these scientific fields, a wide range of topics were addressed, as expressed by the diversity of the titles and keywords that can be found in each contribution (Table 1).

Table 1. Type of contribution, scientific area, title and keywords of the 16 contributions of the SI, according to the order of publication.

Contribution/ Type/Scientific Area	Title	Keywords
Contribution 1/ review/cell biology	Advances in Personalized Cancer Vaccine Development: AI Applications from Neoantigen Discovery to mRNA Formulation	personalized cancer vaccines; neoantigen prediction; codon optimization; 5′ UTR; mRNA design; deep learning; immunotherapy
Contribution 2/ review/microbiology	From Microbes to Molecules: Synthetic Biology Approaches for Advanced Materials Design	synthetic biology; biopolymer; nanotechnology; biomineralization; MICP
Contribution 3/ review/medicine	The Mini-TRH Test, Dopamine Transmission, and Schizophrenia Symptoms	schizophrenia; dopamine; TRH; positive symptoms
Contribution 4/ review/cell biology	Single-Cell Transcriptomics in Spinal Cord Studies: Progress and Perspectives	single-cell RNA sequencing; spinal cord; transcriptomics; spinal cord injury; neurodegeneration; developmental biology; spatial transcriptomics; multi-omics integration; cell atlas; regenerative medicine
Contribution 5/ article/medicine	HCC in MASLD and ALD: Biochemical Pathways, Epidemiology, Diagnosis, and Treatment	hepatocellular carcinoma; HCC; liver cancer; MASLD; MASH; alcohol-related liver disease; ALD

Table 1. *Cont.*

Contribution/ Type/Scientific Area	Title	Keywords
Contribution 6/ review/medicine	The Occult Cascade That Leads to CTEPH	pulmonary hypertension; pulmonary emboli; microthrombi
Contribution 7/ review/cell biology	Enhancing Mitochondrial Maturation in iPSC-DerivedCardiomyocytes: Strategies for Metabolic Optimization	mitochondrial maturation; iPSC-derived cardiomyocytes; oxidative phosphorylation; mitochondrial biogenesis; fatty acid oxidation; energy metabolism; metabolic preconditioning; cardiac regeneration; regenerative medicine
Contribution 8/ review/medicine	Cancer Therapy-Related Left Ventricular Dysfunction: Are There New Gatekeepers?	cardio-oncology; cancer therapy-related cardiac dysfunction; tailored approach; left ventricular impairment
Contribution 9/ review/biotechnology	Therapeutic Potential of Essential Oils and Their Bioactive Compounds Against Colon Cancer: Focus on Colon-Specific Micro- and Nanocarriers	anticancer essential oils; bioactive compounds; colon cancer; colon-targeted drug delivery; micro- and nanoapproaches
Contribution 10/ opinion/medicine	An Opinion on the Supplementation of Folic Acid 1 mg + Iron (Ferrous Sulfate) 90 mg in the Prevention and Treatment of Anemia	anemia; iron deficiency; folic acid; supplementation; ferrous sulfate
Contribution 11/ review/medicine	Osteoporosis: Focus on Bone Remodeling and Disease Types	osteoporosis; pathogenesis; secondary osteoporosis
Contribution 12/ article/biochemistry	AI-Assisted Identification of the Functional Residues of Ginsenoside Biosynthesis-Associated UGTs	UGT; ginsenoside biosynthesis; functional residue; artificial intelligence; 3D structure
Contribution 13/ review/medicine	RASopathy and Sudden Cardiac Death: A Literature Review	RASopathy; RASopathies; sudden cardiac death; sudden death
Contribution 14/ feature paper/biochemistry	Polyoxometalates' Progress for the Treatment of Alzheimer's Disease	polyoxometalates; Alzheimer's disease; β-amyloid aggregation; neurotoxicity; blood–brain barrier; reactive oxygen species production;β-amyloid disaggregation; acetylcholinesterase inhibitors; transition metal chelators; nanoparticles
Contribution 15/ article/medicine	Drug Allergy in Hospitalized Patients: Three Years of Consultation Experience in a Tertiary Care Setting	allergy consultation; drug allergy; adverse drug reaction; epidemiology; inpatient allergology; anaphylaxis; scar; drug allergy diagnosis; desensitization; allergy stewardship
Contribution 16/ review/biotecnhology	Lactic Acid Bacteria: From Bioprocessing to Nanomedicine	lactic acid bacteria; nanomedicine; *Lactobacillus*; postbiotics; paraprobiotics; probiotics; encapsulation; functional foods; nanoparticles

Altogether, this Special Issue shared three articles, one opinion, one feature paper and eleven reviews, coming to a total of 16 contributions. A total of 53 authors from 9 countries, namely, Italy (5), USA (4), Portugal (2), Korea (2), Japan, Finland, Bulgaria, Indonesia and Iran, were involved, with the majority of topics being related to medicine, biochemistry and biotechnology. Whereas the eleven contributions of the first edition have gathered 32 citations and 50,114 views as of 20 March 2026, indicating an average of 2.9 citations and 4555 views per publication, these 16 contributions have already gathered 31 citations and 52,707 views, indicating an average of 1.4 citations and 3294 views per publication.

https://doi.org/10.3390/biochem6020009

2. Contributions

This Special Issue contains three articles (contributions 5, 12, 15), one opinion (10), one feature paper (14) and eleven reviews (1–4, 6–9, 11, 13, 16), for a total of 16 contributions. The first article contribution published in the present Special Issue (SI) is entitled "HCC in MASLD and ALD: Biochemical Pathways, Epidemiology, Diagnosis, and Treatment" (contribution 5) and it deals with hepatocellular carcinoma (HCC) with metabolic-dysfunction-associated steatohepatitis (MASH) and alcohol-related liver disease (ALD). The authors emphasize the consensus regarding the screening of HCC in cirrhosis patients, and the lack of consensus on screening strategies for non-cirrhotic MASH patients who are also at risk for HCC. It was concluded that multidisciplinary and personalized approaches to prevention, diagnosis, and treatment are crucial for improving outcomes in HCC, especially for groups affected by MASH and ALD, thus preventing healthcare disparities. A recent review from Dr Patel analyzed glucagon-like peptide-1 receptor agonists (GLP-1 RAs) in the reduction of alcohol use and in the treating of alcohol use disorder [2]. The second article (contribution 12) is entitled "AI-Assisted Identification of the Functional Residues of Ginsenoside Biosynthesis-Associated UGTs" and the authors identified amino acid residues in UGTs (UDP-dependent glycosyltransferases) that may alter their reaction kinetics. They also identified, using AI, several other residues that potentially change ligand interactivity of ginsenoside biosynthesis-associated UDP-dependent glycosyltransferases (UGTs) in plant enzymes. As referenced in the article, the conversion of ginsenosides into various molecules presents numerous pharmaceutical applications such as anti-cancer, anti-diabetic, cardio-protective, anti-osteoporosis, antiviral, and anti-Parkinson's disease, among many others, highlighting how modulating the sugar moiety of ginsenoside molecules may alter their pharmaceutical activity. Recent studies from Professor Kim's research group, which also used structure-based AI models, described four promising amyloid-β amino acid residues to be targeted which might treat Alzheimer's disease [3]. The third article (contribution 15) is about "Drug Allergy in Hospitalized Patients: Three Years of Consultation Experience in a Tertiary Care Setting". The study aims to characterize the clinical features of patients referred for allergology consultation due to suspected drug allergy, since drug hypersensitivity reactions (DHRs) are an important cause of morbidity in hospitalized patients.

Also based on clinical experience, and from another hospital medical research team, we received an opinion paper (contribution 10) about "An Opinion on the Supplementation of Folic Acid 1 mg + Iron (Ferrous Sulfate) 90 mg in the Prevention and Treatment of Anemia". The authors discussed the recommendation of using folic acid 1 mg + iron (ferrous sulfate) 90 mg supplementation in the prevention and treatment of anemia in adults. Following a similar concern, Professor Hugo and collaborators' recent publications include a review about the appropriate prescription of non-steroidal anti-inflammatory drugs (NSAIDs) in geriatric patients [4]. This SI also contains a feature paper about "Polyoxometalates' Progress for the Treatment of Alzheimer's Disease" (contribution 14). Behind the inorganic biochemistry perspective of the evolution of the utilization of polyoxometalates' in Alzheimer's disease (AD), focusing on the diverse types of POM structures, the POMs' effects and future directions in the field, the authors also aim to bring together AD caregivers, doctors, psychologists, nurses and family members. Recent studies from Professor Aureliano's research group include new polyoxometalates with anti-cancer properties and Ca^{2+}-ATPase inhibition potential [5,6].

Besides the above contributions, this SI also contains eleven review papers. The first review published is about "Advances in Personalized Cancer Vaccine Development: AI Applications from Neoantigen Discovery to mRNA Formulation" (contribution 1). This contribution is a review about the recent developments using AI-driven tools to enhance

https://doi.org/10.3390/biochem6020009

personalized cancer vaccines, although this is still in its early stages. The second review (contribution 2) is "From Microbes to Molecules: Synthetic Biology Approaches for Advanced Materials Design". This article seeks to present a review of recent developments in microbial-mediated synthesis of inorganic and organic materials, and it explores the mechanisms of microbial biomineralization, particularly how microorganisms influence the formation of the mineral deposits and participate in key biogeochemical cycles. The next review is entitled "The Mini-TRH Test, Dopamine Transmission, and Schizophrenia Symptoms". The author mentions that in psychotic patients, care must be taken to reduce the effect of stress on the thyrotropin-releasing hormone (TRH) analysis and the mini-TRH test, including the use of local anesthesia to control stress induced by blood extraction. In fact, in psychotic patients, the assessment of unstimulated stress is not easy. It was concluded that correlative analyses of symptoms may provide information on symptom-specific alterations in dopamine transmission in non-affective psychotic symptoms. Following a similar concern, recent studies from the same author stated that the most common symptoms of schizophrenia, paranoid delusions and hallucinations, were not significantly associated with the mini-TRH test. However, if individual symptoms of schizophrenia could be associated with basal blood prolactin level, the mini-TRH could help discover whether it could be associated with dopamine inhibition of prolactin secretion [7]. The following review entitled "Single-Cell Transcriptomics in Spinal Cord Studies: Progress and Perspectives", related to spinal cord research and single-cell RNA sequencing (scRNA-seq) analysis, summarizes recent progress in the application of scRNA-seq to spinal cord development, injury, and neurodegenerative diseases. scRNA-seq platforms and their contributions to spinal cord research are described along with current challenges and future directions. The 5th review is about "The Occult Cascade That Leads to CTEPH". Chronic thromboembolic pulmonary hypertension (CTEPH) is a rare, progressive form of pre-capillary pulmonary hypertension. The authors concluded that the ongoing research into molecular mechanisms and biomarker-driven diagnostics promises earlier identification and more personalized management.

The 6th review is entitled "Enhancing Mitochondrial Maturation in iPSC-Derived Cardiomyocytes: Strategies for Metabolic Optimization" and highlights the potential of using pluripotent stem cell-derived cardiomyocytes (iPSC-CMs) for cardiovascular regenerative medicine, although their clinical application is hindered by suboptimal mitochondrial maturation and metabolic inefficiencies. The authors point out that metabolic preconditioning with fatty acids and oxidative phosphorylation enhancers improves mitochondrial architecture, cristae density, and ATP production. Recent studies from Professor Sugiura and collaborators include the exploration of extracellular vesicles (EVs) as powerful mediators of mitochondrial transfer and cardiomyocyte repair of mitochondrial dysfunction in cardiac disease [8]. The 7th review, also associated with cardiovascular diseases, is about "Cancer Therapy-Related Left Ventricular Dysfunction: Are There New Gatekeepers?". The authors mention that although the growing success of oncologic therapies has led to a significant improvement in patient survival, this has been accompanied by an increasing incidence of cardiovascular adverse events, particularly cancer therapy-related cardiac dysfunction (CTRCD). Specifically, the authors summarized the molecular mechanism of damage of the most dangerous drugs that cause left ventricular (LV) impairment, with the molecular mechanisms of anthracycline-induced cardiotoxicity highlighted, as well as the "new gatekeepers", that is, the drugs preventing ventricular dysfunction such as angiotensin-converting enzyme inhibitors, angiotensin receptor blockers and β-blockers. Recent studies from authors of this research group include the need to identify and implement more sensitive and specific biomarkers for cardiovascular disorders in clinical practice [9].

https://doi.org/10.3390/biochem6020009

The next review is about "Therapeutic Potential of Essential Oils and Their Bioactive Compounds Against Colon Cancer: Focus on Colon-Specific Micro- and Nanocarriers". Lifestyle and dietary factors, namely high consumption of processed foods, red meat, and alcohol, coupled with sedentary behavior, are well known contributors of the development of colon cancer, ranking among the most prevalent and lethal cancers worldwide. Although essential oils cannot replace conventional chemotherapy, they can mitigate some of its adverse effects and improve the efficacy of associated chemotherapy drugs. Current advancements in micro- and nanoencapsulation techniques for essential-oil-targeted delivery to the colon are also explored. The authors from this research group have also published recent studies about humic substances as therapeutic agents and the use of essential oils in enhancing sports performance and recovery [10,11]. Another review (9th) in the field of molecular medicine is about "Osteoporosis: Focus on Bone Remodeling and Disease Types". It is also known that aging, hormonal changes, comorbidities, and drugs influence the process that leads to osteoporosis. Osteoporosis is a skeletal disease that leads to increased bone fragility, associated with increased risk of fracture. The review describes the main pathways involved in osteoclasts differentiation and bone resorption, and osteoblasts differentiation and proliferation, since osteoporosis results from a chronic imbalance in bone remodeling, where bone resorption by osteoclasts exceeds bone formation. As referenced in this review, the multidisciplinary nature of osteoporosis intersects with many fields of molecular medicine, making it a common challenge across specialties.

The 10th review paper is entitled "RASopathy and Sudden Cardiac Death: A Literature Review". It discusses how the increased risk of sudden cardiac death, especially in relation to hypertrophic cardiomyopathy (HCM) and ventricular arrhythmias, is associated with RASopathies, genetic syndromes caused by germline mutations in gene-encoding proteins of the RAS/MAPK pathway. The review provides an updated overview of the main RASopathies with cardiac involvement. Genotype–phenotype correlations, pathogenic mechanisms underlying sudden cardiac death, and current diagnosis, monitoring, and prevention strategies are explored. Exploring similar concerns, recent studies from the authors have analyzed the causes of youth sudden cardiac death [12]. Finally, the 11th and last review paper is entitled "Lactic Acid Bacteria: From Bioprocessing to Nanomedicine", which aimed to explore a new horizon for future research and development to exploit the capabilities of lactic acid bacteria across biotechnology, food science, and nanomedicine. The authors concluded that although the biological origin of biosynthesized nanoparticles (NPs) may confer greater biocompatibility and biodegradability than physically or chemically synthesized nanoparticles (NPs) using lactic acid bacteria, thereby reducing the risk of in vivo toxicity, the safety of nanomaterials remains a crucial challenge, especially concerning their long-term use. In fact, despite using safe substances and synthesis methods, the bioaccumulation of NPs in tissues, along with their potential cytotoxicity and genotoxicity, are among the critical concerns, which can arise from repeated administration and the interactions between NPs and cells. Recent studies from the authors included the applications of bioactive glass nanoparticles in dentistry [13].

The understanding of human health is an endless topic in biochemistry. As can be observed in Table 1, the majority of the keywords are associated with medicine, such as drug allergy, Alzheimer's disease, osteoporosis, cardiac dysfunction, sudden cardiac death, pulmonary hypertension, anemia, schizophrenia as well as hepatocellular carcinoma, colon cancer and cancer vaccines. Besides human diseases, others complementary keywords such as dopamine, mRNA design, nanotechnology, mitochondria biogenesis, bioactive compounds, essentials oils, iron deficiency, ginsenoside biosynthesis, polyoxometalates, ß-amyloid aggregation, acetylcholinaesterase inhibitors, transition metal chelators and nanoparticles can be found, highlighting the cellular and molecular basis of biochemistry

https://doi.org/10.3390/biochem6020009

in the understanding of diseases. Allegorically, biochemistry represents a fundamental knowledge base, being a scientific safe harbor that provides molecular understanding for the treatment of several human diseases such as Alzheimer's disease, osteoporosis, cardiac diseases, pulmonary hypertension, anemia, drug allergy, schizophrenia, and cancer, among others.

In addition, this Special Issue also reflects on the emerging direction of biochemistry, i.e., from "Understanding Life" to "Designing Life". The first notable feature of this is the growing role of data-driven methodologies, especially deep learning and artificial intelligence, in elucidating biomolecular structure–function relationships and enabling rational therapeutic design. These advances are not only accelerating the identification of clinically actionable targets, such as neoantigens in cancer immunotherapy, but are also reshaping our ability to engineer biomolecules with enhanced precision, as exemplified by innovations in mRNA optimization and enzyme functional annotation.

Equally significant is the expansion of biochemistry into multi-scale and systems-level investigations. The integration of single-cell and spatial transcriptomics, together with multi-omics approaches, is redefining our understanding of cellular heterogeneity, tissue organization, and disease progression. Such approaches are particularly transformative in complex biological contexts, including neurodegeneration, spinal cord injury, and metabolic and cardiovascular disorders, where the interplay between molecular pathways and cellular environments is critical.

Taken together, the contents of the two Special Issues in *BioChem* follow overall global trends in the field of biochemistry, namely, structural biology revolution in AI protein folding prediction, RNA therapeutics explosion, single-cell and spatial omics, and translational and clinical focus. Finally, while chemical biology and small molecule mechanisms are less emphasized in this issue, interested readers can find them in the first Special Issue.

Again, we hope that this Special Issue will serve as both a valuable resource and a source of inspiration for researchers across disciplines, fostering continued advances at the frontiers of biochemistry.

Funding: M.A. thanks to Algarve University, and to Portuguese national funds from FCT—Foundation for Science and Technology, through contracts UID/04326/2025, UID/PRR/04326/2025 and LA/P/0101/2020 (https://doi.org/10.54499/LA/P/0101/2020), B.M. thanks the Natural Science Foundation of China (Grant No. 32171246) for their support.

Data Availability Statement: The data are available in the original research papers.

Acknowledgments: The authors would like to acknowledge to all the contributing authors and reviewers.

Conflicts of Interest: The authors declare no conflicts of interest.

List of Contributions:

1. Kong, H. Advances in Personalized Cancer Vaccine Development: AI Applications from Neoantigen Discovery to mRNA Formulation. *BioChem* **2025**, *5*, 5. https://doi.org/10.3390/biochem5020005.
2. Ramachandran, R.; Macabenta, F.; Bettencourt, G.; Feng, S. From Microbes to Molecules: Synthetic Biology Approaches for Advanced Materials Design. *BioChem* **2025**, *5*, 12. https://doi.org/10.3390/biochem5020012.
3. Spoov, J. The Mini-TRH Test, Dopamine Transmission, and Schizophrenia Symptoms. *BioChem* **2025**, *5*, 15. https://doi.org/10.3390/biochem5020015.
4. Maihemuti, M.; Mimi, M.A.; Sohag, S.M.; Hasan, M.M. Single-Cell Transcriptomics in Spinal Cord Studies: Progress and Perspectives. *BioChem* **2025**, *5*, 16. https://doi.org/10.3390/biochem5020016.

https://doi.org/10.3390/biochem6020009

5. Patel, S.; Kasem, F.; Flaherty, D.; Barve, A. HCC in MASLD and ALD: Biochemical Pathways, Epidemiology, Diagnosis, and Treatment. *BioChem* **2025**, *5*, 19. https://doi.org/10.3390/biochem5030019.
6. Fox, C.; Pandit, L.M. The Occult Cascade That Leads to CTEPH. *BioChem* **2025**, *5*, 22. https://doi.org/10.3390/biochem5030022.
7. Shahannaz, D.C.; Sugiura, T.; Ferrell, B.E. Enhancing Mitochondrial Maturation in iPSC-DerivedCardiomyocytes: Strategies for Metabolic Optimization. *BioChem* **2025**, *5*, 23. https://doi.org/10.3390/biochem5030023.
8. Piscione, M.; Di Marcantonio, M.C.; Pala, B.; Mincione, G. Cancer Therapy-Related Left Ventricular Dysfunction: Are There New Gatekeepers? *BioChem* **2025**, *5*, 25. https://doi.org/10.3390/biochem5030025.
9. Gvozdeva, Y.; Georgieva, P. Therapeutic Potential of Essential Oils and Their Bioactive Compounds Against Colon Cancer: Focus on Colon-Specific Micro- and Nanocarriers. *BioChem* **2025**, *5*, 26. https://doi.org/10.3390/biochem5030026.
10. Gomes, J.; Silva, J.B.; José, C.V.; Ribeiro, H. An Opinion on the Supplementation of Folic Acid 1 mg + Iron (Ferrous Sulfate) 90 mg in the Prevention and Treatment of Anemia. *BioChem* **2025**, *5*, 30. https://doi.org/10.3390/biochem5030030.
11. Castellani, C.; De Martino, E.; Scapato, P. Osteoporosis: Focus on Bone Remodeling and Disease Types. *BioChem* **2025**, *5*, 31. https://doi.org/10.3390/biochem5030031.
12. Jung, K.; Kim, N.; Park, C.; Kim, J. AI-Assisted Identification of the Functional Residues of Ginsenoside Biosynthesis-Associated UGTs. *BioChem* **2025**, *5*, 36. https://doi.org/10.3390/biochem5040036.
13. Salzillo, C.; Marzullo, A. RASopathy and Sudden Cardiac Death: A Literature Review. *BioChem* **2025**, *5*, 38. https://doi.org/10.3390/biochem5040038.
14. Aureliano, M.; Mateus, J.; Rijo, D.M. Polyoxometalates' Progress for the Treatment of Alzheimer's Disease. *BioChem* **2025**, *5*, 41. https://doi.org/10.3390/biochem5040041.
15. Ratti, C.P.; Chiei Gallo, A.; Barei, F.; Botta, A.; Cavara, M.; Bono, E.; Caron, L.; Ortolani, V.G.R.; Iemoli, E. Drug Allergy in Hospitalized Patients: Three Years of Consultation Experience in a Tertiary Care Setting. *BioChem* **2025**, *5*, 42. https://doi.org/10.3390/biochem5040042.
16. Rezvani, M.; Manconi, M.; Düzgüneş, N. Lactic Acid Bacteria: From Bioprocessing to Nanomedicine. *BioChem* **2026**, *6*, 3. https://doi.org/10.3390/biochem6010003.

References

1. Aureliano, M.; Ma, B. Feature Papers in *BioChem*. *BioChem* **2025**, *5*, 17. [CrossRef]
2. Patel, S.; Blaney, H.; Nassar, S.; Singal, A.K. GLP-1 receptor agonists and alcohol use disorder: A systematic review. *Alcohol Alcohol.* **2026**, *61*, agaf069. [CrossRef]
3. Hwang, H.; Park, D.; Kim, J. Structural Assessment of Fibrillated Beta Amyloid to Reveal the Potential Epitopes for Alzheimer's Disease. *Cell Biochem. Biophys.* **2026**. [CrossRef] [PubMed]
4. Costa, C.; Soares, D.; Borges, A.; Gonçalves, A.; Andrade, J.P.; Ribeiro, H. Appropriate Prescription of Non-Steroidal Anti-Inflammatory Drugs in Geriatric Patients—A Systematic Review. *BioChem* **2024**, *4*, 300–312. [CrossRef]
5. Brito, B.R.; Camilo, H.d.S.; Cruz, A.F.d.; Ribeiro, R.R.; de Sá, E.L.; Camargo de Oliveira, C.; Fraqueza, G.; Klassen, G.; Aureliano, M.; Nunes, G.G. Mixed-Valence Pentadecavanadate with Ca^{2+}-ATPase Inhibition Potential and Anti-Breast Cancer Activity. *Inorganics* **2025**, *13*, 306. [CrossRef]
6. Meskini, I.; Capet, F.; Fraqueza, G.; Dege, N.; Tahir, M.N.; Ayed, B.; Aureliano, M. Dawson- and Lindqvist-Type Hybrid Polyoxometalates: Synthesis, Characterization and Ca^{2+}-ATPase Inhibition Potential. *Molecules* **2025**, *30*, 4334. [CrossRef]
7. Spoov, J. Plasma prolactin and symptoms of schizophrenia. *J. Neural Transm.* **2025**, *132*, 897–898. [CrossRef] [PubMed]
8. Shahannaz, D.C.; Sugiura, T.; Yoshida, T. Mitochondria-Enriched Extracellular Vesicles (EVs) for Cardiac Bioenergetics Restoration: A Scoping Review of Preclinical Mechanisms and Source-Specific Strategies. *Int. J. Mol. Sci.* **2025**, *26*, 11052. [CrossRef] [PubMed]
9. Celeski, M.; Segreti, A.; Crisci, F.; Cricco, R.; Piscione, M.; Di Gioia, G.; Nusca, A.; Fossati, C.; Pigozzi, F.; Ussia, G.P.; et al. The Role of Cardiac Troponin and Other Emerging Biomarkers Among Athletes and Beyond: Underlying Mechanisms, Differential Diagnosis, and Guide for Interpretation. *Biomolecules* **2024**, *14*, 1630. [CrossRef]
10. Gvozdeva, Y.; Peneva, P.; Katsarov, P. Biomedical Applications of Humic Substances: From Natural Biopolymers to Therapeutic Agents. *Antioxidants* **2025**, *14*, 1139. [CrossRef]

https://doi.org/10.3390/biochem6020009

11. Ivanova, S.; Dzhakova, Z.; Gvozdeva, Y.; Petkova, G.; Ivanova, A.; Dzhambazova, E. The Role of Essential Oils in Sports Recovery and Performance. *Molecules* **2025**, *30*, 3771. [CrossRef] [PubMed]
12. Salzillo, C.; Marzullo, A. Hereditary Aortopathies as Cause of Sudden Cardiac Death in the Young: State-of-the-Art Review in Molecular Medicine. *Diseases* **2024**, *12*, 264. [CrossRef] [PubMed]
13. Rezvani, M.; Düzgüneş, N. Chapter 16—Multifaceted application of bioactive glass nanoparticles in dentistry: The state of the art. In *Micro and Nano Technologies, Applications of Nanomaterials in Dentistry*; Kesharwani, P., Ed.; Elsevier: Amsterdam, The Netherlands, 2026; pp. 353–374. [CrossRef]

https://doi.org/10.3390/biochem6020009

BioChem

MDPI

Review

Advances in Personalized Cancer Vaccine Development: AI Applications from Neoantigen Discovery to mRNA Formulation

Hyunseung Kong

Interdisciplinary Program in Bioinformatics, Seoul National University, Seoul 08826, Republic of Korea; hskong@snu.ac.kr

Abstract: Personalized cancer vaccines are a promising immunotherapy targeting patient-specific tumor neoantigens, yet their design and efficacy remain challenging. Recent advances in artificial intelligence (AI) provide powerful tools to enhance multiple stages of cancer vaccine development. This review systematically evaluates AI applications in personalized cancer vaccine research over the past five years, focusing on four key areas: neoantigen discovery, codon optimization, untranslated region (UTR) sequence generation, and mRNA vaccine design. We examine AI model architectures (e.g., neural networks), datasets (from omics to high-throughput assays), and outcomes in improving vaccine development. In neoantigen discovery, machine learning and deep learning models integrate peptide–MHC binding, antigen processing, and T cell receptor recognition to enhance immunogenic neoantigen identification. For sequence optimization, deep learning models for codon and UTR design improve protein expression and mRNA stability beyond traditional methods. AI-driven strategies also optimize mRNA vaccine constructs and formulations, including secondary structures and nanoparticle delivery systems. We discuss how these AI approaches converge to streamline effective personalized vaccine development, while addressing challenges such as data scarcity, tumor heterogeneity, and model interpretability. By leveraging AI innovations, the future of personalized cancer immunotherapy may see unprecedented improvements in both design efficiency and clinical effectiveness.

Keywords: personalized cancer vaccines; neoantigen prediction; codon optimization; 5′ UTR; mRNA design; deep learning; immunotherapy

1. Introduction

Cancer immunotherapy has been revolutionized by approaches that direct the immune system to target tumors. Personalized cancer vaccines, which stimulate an immune attack against patient-specific tumor antigens, exemplify this paradigm shift. Unlike prophylactic cancer vaccines targeting viral oncogenes, personalized therapeutic vaccines aim to induce T cells against neoantigens, which are novel peptides arising from somatic mutations in the tumor [1]. Two seminal 2017 studies demonstrated the feasibility and immunogenicity of personalized neoantigen vaccines in melanoma patients [2,3]. In these trials, vaccines composed of patient-specific mutant peptides or mRNAs elicited T cell responses against multiple neoantigens without serious toxicity. These results established a proof of concept that vaccinating against neoantigens can induce antitumor immunity in humans.

Despite this promise, designing an effective personalized vaccine is complex. Each patient's tumor presents a unique set of mutations and human leukocyte antigen (HLA) types, requiring identification of suitable neoantigen targets [4]. The peptide–MHC binding process is a critical first step in this cascade, where tumor-derived peptides must fit precisely into the binding groove of the patient's specific HLA molecules, forming a stable complex that can be recognized by T cells. This interaction is governed by anchor residues within the peptide sequence that determine binding affinity and stability, directly influencing immunogenicity. T cell recognition then occurs through the T cell receptor (TCR), which engages with the peptide–MHC complex in a highly specific manner, with the CDR3 loops of the TCR making direct contact with the presented neoantigen peptide, enabling discrimination between self and tumor-specific antigens.

Once neoantigens are identified, vaccine constructs (such as long peptide pools or mRNA encoding multiple epitopes) must be engineered for high expression and immunogenicity [5]. The typical pipeline involves neoantigen discovery by sequencing the tumor to find mutations, predicting which mutant peptides bind strongly to the patient's HLA molecules, and assessing their likelihood of being immunogenic. Vaccine sequence optimization follows, particularly for DNA or mRNA vaccines, optimizing the coding sequence for efficient production of the neoantigen peptides, including codon optimization to maximize protein expression in the chosen system [6] and designing regulatory regions such as 5′ and 3′ UTRs for optimal translation and stability. The engineering of these untranslated regions is particularly important for mRNA vaccines, as they contain elements that interact with cellular RNA-binding proteins to modulate mRNA stability, localization, and translation efficiency, with optimal 5′ UTRs enhancing ribosome loading and 3′ UTRs controlling poly(A) tail length and protection from exonucleases.

Lastly, mRNA vaccine design ensures the mRNA molecule has an appropriate structure (cap, UTRs, coding region, poly(A) tail) and includes the design of delivery mechanisms (e.g., lipid nanoparticles) to efficiently elicit an immune response [7]. The self-assembly principles of lipid nanoparticles involve precise formulation of ionizable cationic lipids, helper lipids, cholesterol, and PEG lipids at specific molar ratios. These components spontaneously form bilayer structures in aqueous environments, with the ionizable lipids becoming positively charged at acidic pH to complex with negatively charged mRNA. The resulting particles protect the mRNA from degradation while facilitating cellular uptake via endocytosis and subsequent endosomal escape through pH-dependent membrane disruption, allowing the mRNA to reach ribosomes in the cytoplasm.

Personalized cancer vaccines rely on multiple stages of optimization, from neoantigen selection to mRNA sequence engineering and delivery formulation (Figure 1). Each step plays a crucial role in ensuring effective antigen presentation and immune activation, and recent AI-driven approaches have significantly improved these processes, enhancing both vaccine efficacy and efficiency.

Traditional neoantigen prediction pipelines often produce many candidates, but only a small fraction (~2%) prove to elicit T cells experimentally [8]. Similarly, maximizing protein yield from an mRNA vaccine through sequence changes is a high-dimensional problem, as there is an enormous number of possible synonymous coding sequences and UTR variants for any given peptide [9]. AI techniques have rapidly become indispensable in tackling these complexities [10].

Over the past five years, researchers have developed a range of AI-driven tools to enhance personalized cancer vaccine development [11]. These include deep learning models such as convolutional neural networks, recurrent networks, and transformers for predicting peptide–MHC binding and T cell recognition, generative models for creating optimized genetic sequences, and multitask frameworks that integrate diverse biological features [12]. Early results show that such models can dramatically improve the accuracy of neoantigen identification and yield of vaccine antigens relative to the conventional methods [13].

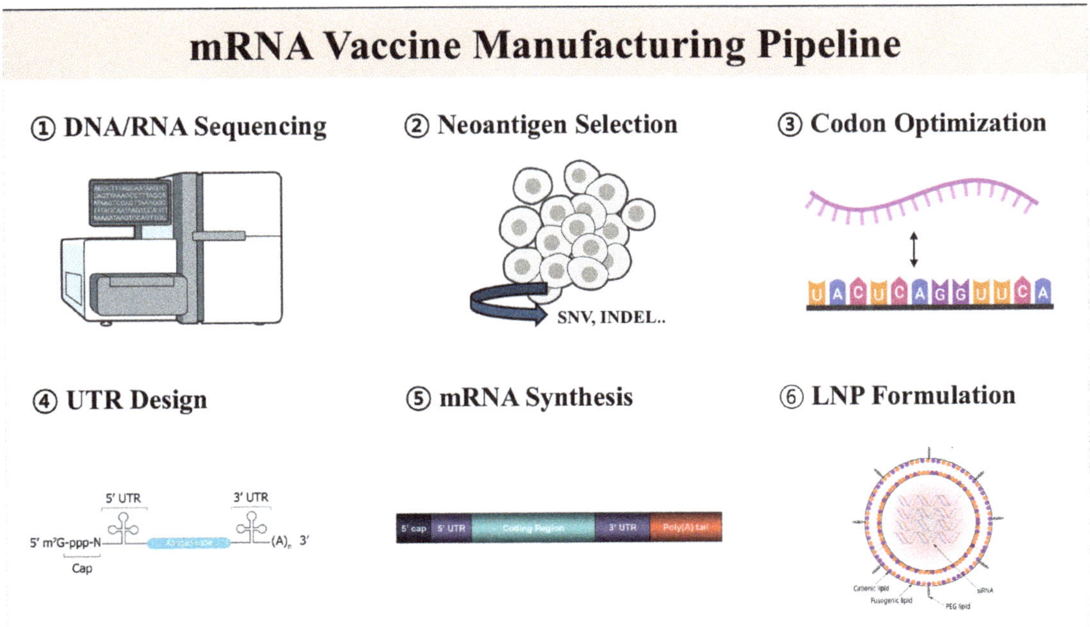

Figure 1. Schematic representation of the mRNA vaccine manufacturing process, including neoantigen selection, codon optimization, untranslated region (UTR) design, mRNA synthesis, and lipid nanoparticle (LNP) formulation.

This review provides a comprehensive overview of these developments. We focus on four major aspects where AI is applied: neoantigen discovery, identifying and prioritizing tumor-specific antigen targets using AI; codon optimization, using AI to modify the coding sequence for improved expression; UTR sequence generation, designing 5′ and 3′ UTRs to enhance translation and stability of mRNA vaccines; and mRNA vaccine design, with AI-driven improvements in vaccine construct design and formulation. Recent studies in these areas are summarized in Table 1, highlighting key advancements and their implications for AI-driven mRNA vaccine development.

Table 1. Representative AI applications in personalized cancer vaccine development.

Aspect	Reference	AI Model	Data and Inputs	Key Outcome
Neoantigen discovery	Springer et al. (2020) [14]	LSTM neural network for TCR–peptide binding prediction	>170,000 TCR–peptide pairs	Learned TCR–peptide specificity; achieved performance on par with state-of-the-art methods
	Montemurro et al. (2021) [15]	CNN model for paired TCRα and TCRβ sequences	Known TCRαβ sequences and cognate peptides (HLA-A*02:01, 9-mer)	Improved prediction of TCR–peptide binding (79% specificity at a 2% false-positive rate)
	Lu et al. (2021) [16]	Transfer learning deep learning model for pMHC–TCR binding	Sequence data: mutated peptide, patient's MHC class I, and TCR sequence	Achieved high accuracy (AUC 0.827 on independent testing) in predicting TCR binding

Table 1. *Cont.*

Aspect	Reference	AI Model	Data and Inputs	Key Outcome
Codon optimization	Fu et al. (2020) [17]	BiLSTM-CRF deep learning model using "codon boxes"	*E. coli* expression of multiple genes (e.g., vaccine antigen FALVAC-1, PTP4A3)	AI-optimized genes showed a higher protein expression than industry-optimized sequences
	Costa/Absci (2024) [18]	Transformer-based language model ("CO-BERT") for codon choice	Large-scale coding sequence dataset (multiple organisms)	Predicted optimal synonymous codons for maximal protein expression
	Wang et al. (2023) [19]	Multi-species transformer ("CodonTransformer") for codon optimization	>1 million DNA–protein sequence pairs from 164 species	Learned codon usage across species; enabled cross-species gene optimization
UTR sequence generation	Song et al. (2024) [20]	Deep generative model ("Smart5UTR")—multitask autoencoder with a CNN encoder	5′ UTR library with >200,000 sequences tested via MPRA	Generated optimized 5′ UTRs for N1-methyl-pseudouridine mRNA, improving vaccine efficacy
	Castillo-Hair et al. (2024) [21]	Deep learning regression + generative design for 5′UTR (CNN models and optimization)	Polysome profiling data from random 5′UTR libraries in 3 human cell types	Designed synthetic 5′ UTRs that enhanced translation of a gene editor enzyme
	Chu et al. (2024) [22]	Pretrained transformer language model ("UTR-LM") for 5′UTRs	Endogenous 5′ UTR sequences from multiple species (unsupervised pretraining)	Learned a language of 5′ UTRs enabling improved translation initiation efficiency
mRNA vaccine design	Cafri et al. (2020) [23]	Personalized neoantigen mRNA vaccine (clinical trial)	13 patients with gastrointestinal tumors; 5–20 neoantigens per mRNA vaccine	First-in-human phase I trial of an individualized mRNA neoantigen vaccine
	Mekki-Berrada et al. (2021) [24]	Two-step ML for lipid nanoparticle formulation optimization	Data from combinatorial synthesis of polymeric nanoparticles for mRNA delivery	AI-guided nanoparticle formulation improved mRNA delivery efficiency
	He et al. (2023) [25]	CNN with self-attention ("RNAdeformer") for mRNA degradation prediction	Public mRNA stability datasets (e.g., OpenVaccine COVID-19 mRNA data)	Achieved state-of-the-art accuracy in predicting mRNA half-life and degradation sites

Abbreviations: MHC—major histocompatibility complex; TCR—T cell receptor; LSTM—long short-term memory; CNN—convolutional neural network; HLA—human leukocyte antigen; MPRA—massively parallel reporter assay; UTR—untranslated region.

2. AI in Neoantigen Discovery

Identifying suitable neoantigens is the first and arguably most critical step in personalized cancer vaccine design [26]. Neoantigens result from non-synonymous mutations (or other alterations such as fusions) that create novel peptides not present in normal human proteins [27]. When a neoantigen peptide is presented on the cell surface by major histocompatibility complex (MHC) molecules, it can be recognized as "non-self" by T cells, triggering an immune attack on the tumor cell [28]. However, not all tumor mutations produce immunogenic neoantigens [29]. The challenge is to predict which mutated peptides will bind to the patient's HLA alleles and elicit a T cell response [29].

Classical in silico pipelines use sequential filters: identify tumor-specific mutations from exome sequencing, predict peptide–MHC binding (often with such tools as NetMHC), filter for those with sufficient tumor expression and processing, and sometimes predict T cell recognition or immunogenicity [30]. Many pipeline implementations exist (e.g., pVACseq,

MuPeXI, Vaxrank), typically relying on motif-based or machine learning predictors for MHC binding and applying heuristic filters for expression and antigen processing steps [31]. While these pipelines can shortlist candidate neoantigens, their precision is limited—only about 1–5% of the predicted high-affinity binders tend to be true immunogenic neoantigens in practice [32]. This low precision leads to many false positives, requiring labor-intensive experimental validation and risking missing the truly effective targets [33].

2.1. Enhancing Neoantigen Prediction with Machine Learning

AI has been introduced into neoantigen discovery to improve each stage of the prediction pipeline. One major focus has been on improving peptide–MHC binding prediction, as peptide immunogenicity is contingent on stable MHC presentation. Earlier algorithms (e.g., NetMHCpan) were already machine learning-based, using neural networks trained on binding affinity datasets, and were fairly accurate for many HLA alleles [34]. Recent advances use deep learning to capture more complex sequence patterns or even structural features [35].

For example, NeoaPred is a deep learning framework that constructs peptide–HLA class I complex structures in silico and evaluates binding and surface features to predict immunogenicity [36]. NeoaPred achieved approximately 82% accuracy in modeling peptide–MHC complex structures and used those structures to improve the identification of peptides likely to be immunogenic. Such structure-based approaches go beyond sequence motifs by considering how a peptide sits in the MHC binding groove and is exposed for T cell recognition. Similarly, DeepHLApan (2020) and other CNN/LSTM-based predictors incorporate flanking residues or the entire proteasomal processing context to better predict which peptides will be generated and presented by MHC [37]. These deep models, trained on large datasets of peptide–MHC interactions (including mass spectrometry data of naturally presented peptides), often outperform older tools in binding prediction benchmarks [38].

Another critical area is predicting T cell recognition of peptide–MHC complexes. Even a peptide that binds to MHC well may not trigger a T cell response if the T cell repertoire does not recognize it or if it resembles self-peptides. AI models bridging peptides and T cell receptors (TCRs) are being developed for this purpose. For instance, Springer et al. developed ERGO, an LSTM recurrent neural network that predicts whether a given TCR sequence will bind a given peptide–MHC complex [14]. ERGO was trained on a large dataset of known TCR–peptide pairs and outputs a binding probability score.

Building on that, NetTCR-2.0 (2021) [15] incorporated both TCR α- and β-chain sequences, using a convolutional neural network architecture. In this model, the amino acid sequences of the peptide and complementarity-determining regions (CDR3) of TCRα and TCRβ were one-hot encoded and passed through CNN and pooling layers, then merged in fully connected layers. NetTCR-2.0 achieved high specificity in identifying true peptide–TCR pairs, improving accuracy over previous models.

A notable advancement came with pMTnet [16], a transfer learning-based model that first learns numerical embeddings of TCRs and peptide–MHC complexes from large general-purpose datasets, then trains a final predictive network on the known TCR–pMHC pairs. This strategy allows it to transfer knowledge from extensively studied viral epitopes to predict TCR binding of tumor-specific neoantigens.

Another AI approach to predicting peptide immunogenicity incorporates multiple features. Certain sequence properties correlate with immunogenicity: for instance, immunogenic epitopes often have an intermediate length, certain hydrophobicity patterns, and minimal similarity to self-proteins [39]. AI models, including deep neural networks and gradient-boosted classifiers, integrate these features to improve immunogenicity prediction.

Zhang et al. (2022) introduced an integrated pipeline where separate deep models predict MHC binding, antigen processing, and immunogenicity, then combine these outputs to rank neoantigens [20].

Kai Xin et al. (2024) developed NUCC, a multi-branch deep neural network integrating multiple biological features to predict neoantigen immunogenicity [40]. NUCC takes multiple inputs for each peptide, including its sequence, the patient's HLA allele, and numeric features such as predicted presentation probability and peptide–MHC binding stability. NUCC outperformed NetMHCpan and identified experimentally validated neoantigens in a gastric cancer dataset.

2.2. Improved Pipelines and Web Services

To facilitate the clinical application of AI-driven neoantigen prediction, models are being integrated into user-friendly pipelines and web platforms. DeepNeo is a web server that employs deep learning to predict neoantigen immunogenicity based on structural and sequence properties [41]. Similarly, pVACtools offers a framework that integrates multiple prediction methods, expression filtering, and optional experimental validation steps to generate a ranked list of candidate neoantigens. Another notable tool, DeepVACPred, utilizes an autoencoder-based deep learning model to predict and design optimal epitope sets for vaccines [42]. The implementation of these AI-driven tools has led to improved outcomes in recent clinical trials of personalized neoantigen vaccines, demonstrating higher validation rates of predictions [43]. This advancement reduces wasted resources on non-immunogenic peptides and accelerates the transition from sequencing to vaccine formulation, ultimately enhancing the efficiency of personalized cancer vaccine development.

3. AI in Codon Optimization for Vaccine Antigens

Once neoantigen peptides are selected, the next step—particularly for DNA or mRNA-based vaccines—is to encode them in a genetic sequence that can be efficiently translated into proteins. The efficacy of personalized cancer vaccines largely depends on how well the selected antigens are expressed and presented by antigen-presenting cells. Codon optimization plays a crucial role in ensuring high expression levels of vaccine antigens by modifying the coding sequence without altering the amino acid sequence.

Different organisms prefer different synonymous codons due to variations in tRNA abundances and regulatory mechanisms. For instance, a human gene may not be optimally expressed in *E. coli* due to differences in codon usage biases. Even within human cells, certain codon choices can influence translation speed, mRNA stability, and protein folding efficiency. Kudla et al. demonstrated that synonymous variants of GFP showed up to 250-fold differences in expression levels in *E. coli*, primarily due to mRNA secondary structure differences near the translation start site [44]. The MDR1 gene provides another compelling example, where a synonymous C3435T mutation alters cotranslational folding dynamics, resulting in a protein with an identical amino acid sequence but modified substrate specificity. Similarly, studies have shown that a "slow ramp" of rare codons at the beginning of genes can enhance overall translation by preventing ribosomal traffic jams, while the strategic reuse of the same tRNA through codon autocorrelation can increase translation efficiency. Traditional codon optimization methods rely on heuristics such as the codon adaptation index (CAI), GC content balancing, or elimination of problematic motifs (e.g., RNA secondary structures, cryptic splicing signals) [45]. However, these approaches may overlook complex, nonlinear interactions between codon usage and translation efficiency. AI-driven codon optimization aims to overcome these limitations by learning optimal codon usage patterns from large biological datasets and predicting the best synonymous codon substitutions for a given antigen.

3.1. Deep Learning Models for Codon Optimization

One of the pioneering AI approaches in codon optimization was introduced by Fu et al. (2020), who developed a BiLSTM-CRF (bidirectional LSTM with a conditional random field) model for selecting optimal codons [17]. This model leverages a concept called "codon boxes," which groups codons by their nucleotide composition rather than by their sequence order. By training on highly expressed genes, the BiLSTM-CRF model captured preferred codon patterns that yield enhanced protein production. When experimentally validated, codon sequences optimized by this model resulted in significantly higher protein expression compared to sequences optimized by traditional rule-based methods.

Another major advancement is the CO-BERT model developed by Absci in 2024, which applies a transformer-based language model to codon selection [28]. Inspired by natural language processing (NLP), CO-BERT treats codons as tokens and learns context-aware representations of coding sequences. This approach enables the model to predict the most likely (or optimal) codon in a given genetic context. By training on large-scale genomic datasets, CO-BERT can suggest codon substitutions that maximize protein expression in specific cellular environments, automating what was traditionally an experimental trial-and-error process.

Similarly, CodonBERT, an AI model introduced by Li et al. (2024), was trained on over 10 million mRNA sequences from diverse organisms to learn codon-level sequence preferences [19]. Unlike traditional optimization strategies that optimize codons in isolation, CodonBERT captures long-range dependencies between codons, ensuring that synonymous substitutions do not disrupt local regulatory motifs or RNA secondary structures. The model was tested on vaccine-relevant antigen genes, where its optimized sequences led to increased protein yields compared to industry-standard codon optimization methods.

In addition to supervised learning models, generative models have also been explored for codon optimization. RiboCode, a deep generative optimization framework, integrates ribosome profiling data to design codon sequences that enhance ribosome loading and elongation efficiency [46]. This model uses reinforcement learning to iteratively improve codon sequences based on an objective function that maximizes translation efficiency while minimizing structural constraints. Early results suggest that RiboCode-designed sequences outperform heuristic-based optimization in driving higher protein production.

3.2. Experimental Validation of AI-Optimized Codon Sequences

The success of AI-driven codon optimization has been demonstrated across multiple experimental studies. Fu et al. (2020) tested AI-optimized versions of several genes, including a malaria vaccine antigen (FALVAC-1) and the human phosphatase PTP4A3, in *E. coli* expression systems. Compared to native sequences and traditionally optimized counterparts, the AI-optimized genes consistently exhibited higher protein expression levels [17]. Similarly, CodonBERT-optimized sequences of influenza vaccine antigens resulted in higher antigen yields in mammalian cell systems, supporting their potential use in vaccine manufacturing.

These improvements in protein expression have direct implications for personalized cancer vaccines. AI-guided codon optimization can ensure that mRNA vaccine constructs encoding patient-specific neoantigens produce robust antigen levels in dendritic cells, leading to stronger immune activation. This is particularly relevant when designing multivalent mRNA vaccines, where ensuring balanced expression of multiple antigens is crucial. AI models can optimize codon choices in a way that prevents translation competition between different antigenic segments, thereby ensuring each neoantigen is adequately produced.

4. AI in UTR Sequence Generation and mRNA Design

Messenger RNA (mRNA) stability and translational efficiency are critical factors in determining the potency of an mRNA-based cancer vaccine. While codon optimization enhances the efficiency of the coding sequence, additional regulatory elements, such as untranslated regions (UTRs), significantly influence the overall expression of vaccine antigens. The 5′ UTR plays a crucial role in translation initiation by regulating ribosome scanning and start codon recognition [47], while the 3′ UTR affects mRNA stability, localization, and translation efficiency through interactions with RNA-binding proteins and microRNAs [48]. Optimizing these regulatory regions has traditionally been based on the empirical selection of UTRs from highly expressed genes. However, AI-driven approaches now allow for de novo UTR design, creating synthetic sequences with optimized translation and stability characteristics.

4.1. AI-Driven 5′ UTR Optimization

The 5′ UTR influences how efficiently ribosomes initiate translation, particularly through its nucleotide composition, secondary structure, and upstream open reading frames (uORFs). AI models have been developed to design 5′ UTRs that enhance translation initiation while avoiding inhibitory structures.

One of the most prominent AI approaches is Smart5UTR, a deep generative model introduced by Tang et al. (2024) that designs synthetic 5′ UTRs optimized for translation efficiency in mRNA vaccines [49]. The model was trained on a massively parallel reporter assay (MPRA) dataset containing over 200,000 randomized 5′ UTR sequences, each experimentally tested for translation efficiency. Smart5UTR employs a multitask autoencoder with a convolutional neural network (CNN) encoder, allowing it to learn hidden sequence patterns that contribute to high translation initiation. When applied to vaccine mRNAs encoding the SARS-CoV-2 spike protein, Smart5UTR-generated UTRs significantly improved antigen expression in mice compared to natural UTRs from housekeeping genes. In vivo validation further demonstrated that Smart5UTR-designed 5′ UTRs enhanced antibody titers against SARS-CoV-2 variants by up to 120 times compared to the conventional designs, while maintaining excellent safety profiles.

Another AI-driven approach, developed by Castillo-Hair et al. (2024), focused on optimizing 5′ UTRs for gene-editing enzymes delivered as mRNA [21]. They performed polysome profiling on large libraries of randomized 5′ UTRs to measure translation efficiency across different cell types. A deep learning regression model was then trained on these data to predict which UTR sequences would maximize ribosome loading. When applied to megaTAL nucleases (a class of gene-editing proteins), the AI-designed UTRs significantly enhanced the editing activity in human cells.

Chu et al. (2024) developed UTR-LM, a pretrained transformer language model that learns sequence representations of endogenous 5′ UTRs from diverse species [22]. This unsupervised model was fine-tuned to predict translation initiation efficiency (TIE) based on sequence features. UTR-LM successfully identified previously unknown synthetic UTR sequences that outperformed natural UTRs in driving protein expression. The ability of UTR-LM to generalize across species suggests it may be useful for optimizing vaccine mRNAs across different expression systems.

4.2. AI-Driven 3′ UTR Optimization

The 3′ UTR primarily influences mRNA stability, degradation, and localization. While many vaccines borrow 3′ UTRs from highly stable endogenous transcripts (such as α-globin), AI-driven approaches can design synthetic 3′ UTRs with improved stability and translational properties.

Morrow et al. (2024) developed a machine learning model for 3′ UTR design using high-throughput RNA stability assays [50]. Their dataset included thousands of synthetic 3′ UTR sequences, each measured for mRNA half-life in human cells. The model, trained on these data, predicted mRNA stability based on sequence features and identified de novo 3′ UTR designs that extended mRNA half-life beyond the commonly used endogenous UTRs. When experimentally tested, AI-optimized 3′ UTRs led to prolonged antigen expression in mammalian cells, a crucial advantage for cancer vaccines requiring sustained immune stimulation.

Similarly, Liu et al. (2024) introduced LinearDesign2, an AI framework that co-optimizes both the 5′ UTR and the coding sequence to enhance translation initiation and stability simultaneously [51]. Traditional UTR and codon optimization methods have often been applied separately, but LinearDesign2 integrates them into a multi-objective optimization pipeline, balancing translation efficiency with mRNA folding stability. Their computational analysis suggested that jointly optimizing the 5′ UTR and coding sequence leads to greater protein expression gains than optimizing either component alone.

4.3. Implications for mRNA Vaccine Design

AI-driven UTR and mRNA design has several significant implications for personalized cancer vaccines. One major advantage is the increased vaccine potency, as AI-optimized UTRs can double or even triple antigen expression, enhancing vaccine immunogenicity without requiring a higher mRNA dose. This improvement can lead to more efficient T cell priming while reducing the necessary vaccine dosage. Additionally, AI enables circumvention of biological constraints, addressing challenges where certain neoantigen sequences are difficult to express due to RNA secondary structures or cryptic motifs. By designing custom 5′ and 3′ UTRs, AI ensures robust antigen production despite these inherent limitations.

Another critical aspect is context-specific optimization, where AI models tailor UTRs to different cell types or conditions. For instance, a UTR optimized for dendritic cell translation—where vaccine antigens undergo immune processing—may differ significantly from one designed for general mammalian cells [52]. Furthermore, AI-driven approaches facilitate the integration of codon optimization, allowing simultaneous optimization of coding regions, UTRs, and secondary structures. This comprehensive strategy represents the next frontier in computational vaccine design, potentially yielding even greater improvements in mRNA vaccine performance.

5. AI in mRNA Vaccine Formulation and Delivery

Optimizing the sequence of mRNA vaccines is crucial, but ensuring effective delivery and stability within the human body is equally important. Particularly, mRNA vaccines must be efficiently translated into proteins inside cells while avoiding premature degradation by extracellular RNases and immune system overactivation. Lipid nanoparticles (LNPs) are the most widely used delivery system for mRNA vaccines, as they protect mRNA and facilitate its uptake by cells [53]. Additionally, immune-stimulatory adjuvants are often included to enhance vaccine efficacy [54]. AI is playing an increasing role in optimizing both mRNA delivery vehicles and adjuvant formulations to maximize vaccine performance.

5.1. AI-Guided Lipid Nanoparticle (LNP) Optimization

LNPs are composed of ionizable lipids, phospholipids, cholesterol, and polyethylene glycol (PEG) lipids [55]. The specific composition and ratio of these components influence mRNA encapsulation efficiency, stability, and cellular uptake. Traditionally, LNP formulations have been developed through trial and error, but AI is now accelerating the discovery of optimized nanoparticle formulations.

Mekki-Berrada et al. (2021) developed a two-step machine learning approach to optimize polymeric nanoparticles for mRNA delivery [24]. First, they applied unsupervised clustering on a large combinatorial dataset of nanoparticle formulations to identify promising formulation families. Then, they trained a supervised regression model to predict mRNA delivery efficiency based on nanoparticle composition. This approach significantly reduced the experimental burden, enabling the identification of highly effective LNP formulations with improved cellular uptake and transfection efficiency.

Similarly, AGILE—a deep learning-powered platform introduced in 2024—leverages a graph neural network in combination with high-throughput combinatorial lipid synthesis to predict the mRNA transfection potency of lipid nanoparticle (LNP) formulations [56]. Trained on thousands of LNP compositions evaluated for stability, delivery efficiency, and immune activation, AGILE accurately identifies optimal ionizable lipid combinations for mRNA vaccine delivery, outperforming traditional heuristic-based screening methods.

Another AI-driven approach, TransLNP, is a transformer-based model designed to accelerate the screening of ionizable lipid nanoparticles for mRNA delivery [57]. By integrating coarse-grained atomic sequence information with fine-grained spatial relationships, it captures the key structure–efficiency correlations. Utilizing pretraining and a BalMol block for label and feature smoothing, TransLNP effectively overcomes data scarcity and imbalance, while also identifying transfection cliffs where similar structures exhibit drastically different efficiencies.

5.2. AI in Adjuvant and Immune-Stimulatory Element Design

In addition to delivery optimization, AI is being used to design novel adjuvants and immune-modulating elements that enhance vaccine efficacy [58]. Adjuvants help stimulate the immune system and ensure a robust response to the vaccine antigen. Traditional adjuvants include aluminum salts, toll-like receptor (TLR) agonists, and cytokine-based stimulants. However, the discovery of novel adjuvant molecules with improved safety and efficacy profiles is a growing area of AI-driven research.

According to Chaudhury et al. (2018), distinct adjuvant formulations produce unique immune signatures in non-human primates [59]. By integrating serological, cellular, and cytokine data from multiple compartments and applying machine learning techniques (such as random forest models), the study achieved up to 92% accuracy in predicting adjuvant conditions from the immune profile, highlighting a quantitative approach for optimal vaccine–adjuvant pairing.

Leveraging computer-aided design and machine learning, a library of 46 AuNP-based adjuvants with TLR-targeting ligands was developed, leading to the identification of AuNP27 and AuNP35 as potent dendritic cell activators. These nano-adjuvants enhance antigen presentation and T cell responses via multiple TLR pathways, thereby improving antitumor immunity in preclinical models [60].

5.3. AI in Personalized mRNA Vaccine Formulation

AI is not only optimizing general vaccine delivery strategies, but also enabling personalized vaccine formulation. Each patient's immune system and tumor microenvironment are unique, meaning a one-size-fits-all vaccine formulation may not be optimal. AI models integrating genomic, proteomic, and immunological data can help tailor vaccine formulations to individual patients [61].

For example, AI frameworks that analyze patient-specific immune signatures can predict which LNP formulation will work best for a given individual. Patients with high levels of inflammation might benefit from a formulation with lower reactogenicity, while patients with immunosuppressive tumor environments may require stronger adjuvants. AI can rapidly generate these predictions, guiding personalized vaccine formulation in real time.

Moreover, AI-powered models can predict how a patient's immune system will respond to different adjuvant combinations, allowing clinicians to select the best formulation for each patient. This is particularly important for cancer patients, where immune evasion mechanisms often hinder vaccine effectiveness. By using AI-driven formulation tools, personalized cancer vaccines can be fine-tuned to elicit the strongest possible immune response.

6. Challenges and Future Perspectives

The integration of AI into personalized cancer vaccine development is still in its early stages, and several challenges must be addressed to fully realize its potential. One of the most pressing issues is data limitations, as high-quality data are essential for training AI models. In neoantigen prediction, determining which mutations elicit T cell responses at scale remains difficult, requiring patient samples or labor-intensive experiments. Current models rely heavily on small datasets, typically containing only tens to hundreds of immunogenic peptides, which can introduce bias. Similarly, in mRNA design, the sequence space is vast, yet only a small fraction has been experimentally tested. While techniques such as transfer learning and active learning are improving model performance by prioritizing the collection of the most informative data, continued accumulation of experimental data—such as neoantigen immunogenicity, epitope–HLA–TCR structures, and mRNA translation outcomes—will be crucial. Initiatives such as TESLA for neoantigens and efforts to publish mRNA design datasets, such as the MPRA data from Smart5UTR, are valuable. A promising future direction is the integration of multi-omics data, combining tumor proteomics and immunopeptidomics to better identify which mutations generate peptides that are naturally presented on MHC molecules.

Another key challenge is tumor heterogeneity and immune evasion [62]. Tumors evolve, and a neoantigen present in all cancer cells at the time of vaccine design may be lost or downregulated before the immune system can target it due to immune editing. AI could potentially predict which neoantigens are less likely to be lost, such as those arising from truncal mutations essential to tumor survival. However, vaccines themselves could impose selective pressure on tumors, necessitating future AI models that simulate this evolutionary game. By selecting a combination of neoantigens that tumors cannot easily evade—perhaps by simultaneously targeting multiple essential neoantigens—AI could help develop more effective vaccine strategies. This challenge intersects with game theory and may benefit from AI techniques used in strategic planning, such as reinforcement learning, where the "opponent" is the evolving tumor.

Generalizability is another significant concern, as many AI models are trained on data from specific populations or experimental systems. For instance, neoantigen prediction models often rely on data from Caucasian patients with common HLA alleles, which may reduce their accuracy for individuals with rare HLA alleles or cancers with unique mutation profiles [63]. Similarly, codon optimization models trained on one cell type may not generalize well to another. Addressing this issue requires strategies such as training ensemble models that integrate multiple prediction methods, a practice already employed in pipelines such as pVACtools. Additionally, AI can be used to estimate its own uncertainty; if a model is unsure about a particular neoantigen, it can deprioritize the candidate or flag it for experimental validation first.

7. Conclusions

Personalized cancer vaccines sit at the intersection of genomics, immunology, and bioengineering. Artificial intelligence has rapidly emerged as an enabling technology at this intersection, addressing critical bottlenecks in vaccine design. In this review, we surveyed how AI techniques—particularly machine learning and deep learning models—are being applied to enhance neoantigen discovery, optimize vaccine coding sequences, design regulatory regions, and refine vaccine formulation and strategy. Over the last five years, tangible progress has been made: neoantigen prediction models integrating deep learning have improved the precision of target selection, reducing false positives and uncovering cryptic immunogenic mutations that earlier methods overlooked. AI-optimized coding sequences and UTRs have demonstrated superior protein expression, translating to stronger immune responses in preclinical vaccine studies. Furthermore, AI-driven insights are guiding the development of better delivery systems and combination therapies, inching the field closer to the goal of effective, individualized cancer immunotherapy.

The journey is far from over. As discussed, challenges such as data scarcity, model interpretability, and ensuring robust clinical efficacy need continued attention. However, the trajectory is clear: AI techniques will become ever more integrated into the pipeline of personalized vaccine development. In the near future, it is plausible that upon sequencing a patient's tumor, automated AI pipelines will propose a "vaccine design" that experts will then validate and implement, compressing what used to be a multiyear research endeavor into a process that fits within the window of clinical decision-making for the patient. Early adopters of these technologies in clinical trials are already setting the stage, and as successes accumulate, confidence in AI-assisted design will grow.

In conclusion, AI is accelerating the development of personalized cancer vaccines by making the design process smarter, faster, and more tailored to each patient's unique cancer. It exemplifies the power of interdisciplinary innovation—when computational algorithms are guided by biological understanding, the result is more than the sum of its parts. Continued collaboration and data-sharing will be essential to fully unlock AI's potential in this field. The ultimate success will be measured in improved patient outcomes: longer survival, cancer remission, and, perhaps, one day, effective vaccines that can prevent cancer from recurring.

Funding: This research received no external funding.

Data Availability Statement: Data is contained within the article.

Conflicts of Interest: The author declares no conflict of interest.

References

1. Blass, E.; Ott, P.A. Advances in the development of personalized neoantigen-based therapeutic cancer vaccines. *Nat. Rev. Clin. Oncol.* **2021**, *18*, 215–229. [CrossRef] [PubMed]
2. Ott, P.A.; Hu, Z.; Keskin, D.B.; Shukla, S.A.; Sun, J.; Bozym, D.J.; Zhang, W.; Luoma, A.; Giobbie-Hurder, A.; Peter, L.; et al. An immunogenic personal neoantigen vaccine for patients with melanoma. *Nature* **2017**, *547*, 217–221. [CrossRef]
3. Sahin, U.; Derhovanessian, E.; Miller, M.; Kloke, B.P.; Simon, P.; Löwer, M.; Bukur, V.; Tadmor, A.D.; Luxemburger, U.; Schrörs, B. Personalized RNA mutanome vaccines mobilize poly-specific therapeutic immunity against cancer. *Nature* **2017**, *547*, 222–226. [CrossRef] [PubMed]
4. Keskin, D.B.; Anandappa, A.J.; Sun, J.; Tirosh, I.; Mathewson, N.D.; Li, S.; Oliveira, G.; Giobbie-Hurder, A.; Felt, K.; Gjini, E.; et al. Neoantigen vaccine generates intratumoral T cell responses in glioblastoma. *Nature* **2019**, *565*, 234–239. [CrossRef]
5. Sahin, U.; Türeci, Ö. Personalized vaccines for cancer immunotherapy. *Science* **2018**, *359*, 1355–1360. [CrossRef]
6. Liu, M.A. A comparison of plasmid DNA and mRNA as vaccine technologies. *Vaccines* **2019**, *7*, 37. [CrossRef]
7. Pardi, N.; Hogan, M.J.; Porter, F.W.; Weissman, D. mRNA vaccines—A new era in vaccinology. *Nat. Rev. Drug Discov.* **2018**, *17*, 261–279. [CrossRef]

8. Nguyen, T.B.Q.; Pham, Q.T.M.; Tran, L.S. 84P T cell receptor repertoire profiles of tumor-infiltrating lymphocytes improves neoantigen prioritization for personalized cancer immunotherapy. *Ann. Oncol.* **2023**, *34*, S1499. [CrossRef]
9. Hong, W.; Chen, C.; Zhu, Z.; Tang, K. An Elite Archive-Assisted Multi-Objective Evolutionary Algorithm for mRNA Design. In Proceedings of the 2024 IEEE Congress on Evolutionary Computation (CEC), Yokohama, Japan, 30 June–5 July 2024; pp. 1–8.
10. Kaushik, R.; Kant, R.; Christodoulides, M. Artificial intelligence in accelerating vaccine development-current and future perspectives. *Front. Bacteriol.* **2023**, *2*, 1258159. [CrossRef]
11. Imani, S.; Li, X.; Chen, K.; Maghsoudloo, M.; Kaboli, P.J.; Hashemi, M.; Khoushab, S.; Li, X. Computational biology and artificial intelligence in mRNA vaccine design for cancer immunotherapy. *Front. Cell. Infect. Microbiol.* **2025**, *14*, 1501010. [CrossRef]
12. Xin, P.; Yang, D.; Zhou, Y.; Peng, S. TlcMHCpan: A Novel Deep Learning Model for Enhanced Pan-Specific Prediction of Peptide-HLA Binding. *IEEE Access* **2024**, *12*, 184644–184656. [CrossRef]
13. Pu, T.; Peddle, A.; Zhu, J.; Tejpar, S.; Verbandt, S. Neoantigen identification: Technological advances and challenges. *Methods Cell Biol.* **2024**, *183*, 265–302. [PubMed]
14. Springer, I.; Besser, H.; Tickotsky-Moskovitz, N.; Dvorkin, S.; Louzoun, Y. Prediction of specific TCR-peptide binding from large dictionaries of TCR-peptide pairs. *Front. Immunol.* **2020**, *11*, 1803.
15. Montemurro, A.; Schuster, V.; Povlsen, H.R.; Bentzen, A.K.; Jurtz, V.; Chronister, W.D.; Crinklaw, A.; Hadrup, S.R.; Winther, O.; Peters, B.; et al. NetTCR-2.0 enables accurate prediction of TCR-peptide binding by using paired TCRα and β sequence data. *Commun. Biol.* **2021**, *4*, 1060.
16. Lu, T.; Zhang, Z.; Zhu, J.; Wang, Y.; Jiang, P.; Xiao, X.; Bernatchez, C.; Heymach, J.V.; Gibbons, D.L.; Wang, J.; et al. Deep learning-based prediction of the T cell receptor–antigen binding specificity. *Nat. Mach. Intell.* **2021**, *3*, 864–875. [CrossRef]
17. Fu, H.; Liang, Y.; Zhong, X.; Pan, Z.; Huang, L.; Zhang, H.; Xu, Y.; Zhou, W.; Liu, Z. Codon optimization with deep learning to enhance protein expression. *Sci. Rep.* **2020**, *10*, 17617.
18. Kumar, A.; Dixit, S.; Srinivasan, K.; M, D.; Vincent, P.M.D.R. Personalized cancer vaccine design using AI-powered technologies. *Front. Immunol.* **2024**, *15*, 1357217.
19. Li, S.; Moayedpour, S.; Li, R.; Bailey, M.; Riahi, S.; Kogler-Anele, L.; Miladi, M.; Miner, J.; Zheng, D.; Wang, J.; et al. CodonBERT: Large language models for mRNA design and optimization. *bioRxiv* **2023**. [CrossRef]
20. Cai, Y.; Chen, R.; Gao, S.; Li, W.; Liu, Y.; Su, G.; Song, M.; Jiang, M.; Jiang, C.; Zhang, X. Artificial intelligence applied in neoantigen identification facilitates personalized cancer immunotherapy. *Front. Oncol.* **2023**, *12*, 1054231.
21. Castillo-Hair, S.; Fedak, S.; Wang, B.; Linder, J.; Havens, K.; Certo, M.; Seelig, G. Optimizing 5′UTRs for mRNA-delivered gene editing using deep learning. *Nat. Commun.* **2024**, *15*, 5284. [CrossRef]
22. Chu, Y.; Yu, D.; Li, Y.; Huang, K.; Shen, Y.; Cong, L.; Zhang, J.; Wang, M. A 5′UTR Language Model for Decoding Untranslated Regions of mRNA and Function Predictions. *bioRxiv* **2023**. [CrossRef]
23. Cafri, G.; Gartner, J.J.; Zaks, T.; Hopson, K.; Levin, N.; Paria, B.C.; Parkhurst, M.R.; Yossef, R.; Lowery, F.J.; Jafferji, M.S.; et al. mRNA vaccine–induced neoantigen-specific T cell immunity in patients with gastrointestinal cancer. *J. Clin. Investig.* **2020**, *130*, 5976–5988. [PubMed]
24. Mekki-Berrada, F.; Ren, Z.; Huang, T.; Wong, W.K.; Zheng, F.; Xie, J.; Tian, I.P.S.; Jayavelu, S.; Mahfoud, Z.; Bash, D.; et al. Two-step machine learning enables optimized nanoparticle synthesis. *Npj Comput. Mater.* **2021**, *7*, 55.
25. He, S.; Gao, B.; Sabnis, R.; Sun, Q. RNAdegformer: Accurate prediction of mRNA degradation at nucleotide resolution with deep learning. *Brief. Bioinform.* **2023**, *24*, bbac581.
26. Laumont, C.M.; Vincent, K.; Hesnard, L.; Audemard, É.; Bonneil, É.; Laverdure, J.-P.; Gendron, P.; Courcelles, M.; Hardy, M.-P.; Côté, C.; et al. Noncoding regions are the main source of tumor-specific antigens. *Sci. Transl. Med.* **2018**, *10*, eaau5516. [CrossRef]
27. Wells, D.K.; van Buuren, M.M.; Dang, K.K.; Hubbard-Lucey, V.M.; Sheehan, K.C.; Campbell, K.M.; Lamb, A.; Ward, J.P.; Sidney, J.; Blazquez, A.B.; et al. Key parameters of tumor epitope immunogenicity revealed through a consortium approach improve neoantigen prediction. *Cell* **2020**, *183*, 818–834.e13. [CrossRef]
28. Schumacher, T.N.; Schreiber, R.D. Neoantigens in cancer immunotherapy. *Science* **2015**, *348*, 69–74. [CrossRef]
29. Hilf, N.; Kuttruff-Coqui, S.; Frenzel, L.P.; Bukur, T.; Stevanović, S.; Gouttefangeas, C.; Platten, M.; Tabatabai, J.; Dutoit, V.; van der Burg, S.H.; et al. Actively personalized vaccination trial for newly diagnosed glioblastoma. *Nature* **2019**, *565*, 240–245. [CrossRef]
30. Yadav, M.; Jhunjhunwala, S.; Phung, Q.; Lupardus, P.; Tanguay, J.; Bumbaca, S.; Franci, C.; Cheung, T.K.; Fritsche, F.; Weinschenk, T.; et al. Predicting immunogenic tumour mutations by combining mass spectrometry and exome sequencing. *Nature* **2014**, *515*, 572–576. [CrossRef]
31. Hundal, J.; Kiwala, S.; McMichael, J.; Miller, C.A.; Xia, H.; Wollam, A.T.; Liu, C.J.; Zhao, S.; Feng, Y.-Y.; Graubert, A.P.; et al. pVACtools: A computational toolkit to identify and visualize cancer neoantigens. *Cancer Immunol. Res.* **2020**, *8*, 409–420. [CrossRef]
32. Laumont, C.M.; Wouters, M.C.A.; Smazynski, J.; Gierc, N.S. The landscape of tumor antigens and neoantigens in cancer immunotherapy. *Nat. Rev. Cancer* **2022**, *22*, 682–696. [CrossRef]
33. Vitiello, A.; Zanetti, M. Neoantigen prediction and the need for validation. *Nat. Biotech.* **2017**, *35*, 815–817.

34. Jurtz, V.; Paul, S.; Andreatta, M.; Marcatili, P.; Peters, B.; Nielsen, M. NetMHCpan-4.0: Improved peptide–MHC class I interaction predictions integrating eluted ligand and peptide binding affinity data. *J. Immunol.* **2017**, *199*, 3360–3368. [CrossRef] [PubMed]
35. Xu, X.; Liu, X.; Li, X.; Zhang, L.; Wang, Y. Advances in deep learning for neoantigen prediction. *Front. Immunol.* **2021**, *12*, 705096. [CrossRef]
36. Zhao, W.; Li, S.; Liu, Y.; Yang, H. NeoaPred: A structure-based deep learning approach for predicting peptide-MHC binding and immunogenicity. *Bioinformatics* **2022**, *38*, 3792–3800. [CrossRef]
37. Abella, J.R.; Brown, J.R.; Pal, M.; Silvestri, G. DeepHLApan: A convolutional neural network for MHC-peptide binding prediction. *PLoS Comput. Biol.* **2020**, *16*, e1007869. [CrossRef]
38. Racle, J.; Michaux, J. Machine learning-based approaches for neoantigen prediction. *Trends Cancer* **2022**, *8*, 45–57. [CrossRef]
39. Carri, I.; Schwab, E.; Podaza, E.; Alvarez, H.M.G.; Mordoh, J.; Nielsen, M.; Barrio, M.M. Beyond MHC binding: Immunogenicity prediction tools to refine neoantigen selection in cancer patients. *Explor. Immunol.* **2023**, *3*, 82–103.
40. Xin, K.; Wei, X.; Shao, J.; Chen, F.; Liu, Q.; Liu, B. Establishment of a novel tumor neoantigen prediction tool for personalized vaccine design. *Hum. Vaccines Immunother.* **2024**, *20*, 2300881.
41. Kim, J.Y.; Bang, H.; Noh, S.J.; Choi, J.K. DeepNeo: A webserver for predicting immunogenic neoantigens. *Nucleic Acids Res.* **2023**, *51*, W134–W140.
42. Yang, Z.; Bogdan, P.; Nazarian, S. An in silico deep learning approach to multi-epitope vaccine design: A SARS-CoV-2 case study. *Sci. Rep.* **2021**, *11*, 3238.
43. Bulashevska, A.; Nacsa, Z.; Lang, F.; Braun, M.; Machyna, M.; Diken, M.; Childs, L.; König, R. Artificial intelligence and neoantigens: Paving the path for precision cancer immunotherapy. *Front. Immunol.* **2024**, *15*, 1394003.
44. Plotkin, J.B.; Kudla, G. Synonymous but not the same: The causes and consequences of codon bias. *Nat. Rev. Genet.* **2011**, *12*, 32–42. [PubMed]
45. Sharp, P.M.; Li, W.H. The codon adaptation index—A measure of directional synonymous codon usage bias, and its potential applications. *Nucleic Acids Res.* **1987**, *15*, 1281–1295.
46. Li, Y.; Wang, F.; Yang, J.; Han, Z.; Chen, L.; Jiang, W.; Zhou, H.; Li, T.; Tang, Z.; Deng, J.; et al. Deep Generative Optimization of mRNA Codon Sequences for Enhanced Protein Production and Therapeutic Efficacy. *bioRxiv* **2024**. [CrossRef]
47. Joshi, M.; Wang, A.; Myong, S. 5′UTR G-quadruplex structure enhances translation in size dependent manner. *Nat. Commun.* **2024**, *15*, 3963. [CrossRef]
48. Hong, D.; Jeong, S. 3′UTR Diversity: Expanding Repertoire of RNA Alterations in Human mRNAs. *Mol. Cells* **2023**, *46*, 48–56. [CrossRef]
49. Tang, X.; Huo, M.; Chen, Y.; Huang, H.; Qin, S.; Luo, J.; Qin, Z.; Jiang, X.; Liu, Y.; Duan, X.; et al. A novel deep generative model for mRNA vaccine development: Designing 5′ UTRs with N1-methyl-pseudouridine modification. *Acta Pharm. Sin. B* **2024**, *14*, 1814–1826.
50. Morrow, A.; Thornal, A.; Flynn, E.D.; Hoelzli, E.; Shan, M.; Garipler, G.; Kirchner, R.; Reddy, A.J.; Tabchouri, S.; Gupta, A.S.; et al. ML-driven design of 3′ UTRs for mRNA stability. *bioRxiv* **2024**. [CrossRef]
51. Liu, Y.; Gao, J.; Zhang, X.; Fang, X. Joint Design of 5′Untranslated Region and Coding Sequence of mRNA. *arXiv* **2024**, arXiv:2410.20781.
52. Castillo-Hair, S.M.; Seelig, G. Machine learning for designing next-generation mRNA therapeutics. *Acc. Chem. Res.* **2021**, *55*, 24–34. [PubMed]
53. Shi, R.; Liu, X.; Wang, Y.; Pan, M.; Wang, S.; Shi, L.; Ni, B. Long-term stability and immunogenicity of lipid nanoparticle COVID-19 mRNA vaccine is affected by particle size. *Hum. Vaccines Immunother.* **2024**, *20*, 2342592.
54. Xie, C.; Yao, R.; Xia, X. The advances of adjuvants in mRNA vaccines. *npj Vaccines* **2023**, *8*, 162. [PubMed]
55. Li, X.; Qi, J.; Wang, J.; Hu, W.; Zhou, W.; Wang, Y.; Li, T. Nanoparticle technology for mRNA: Delivery strategy, clinical application and developmental landscape. *Theranostics* **2024**, *14*, 738.
56. Xu, Y.; Ma, S.; Cui, H.; Chen, J.; Xu, S.; Gong, F.; Golubovic, A.; Zhou, M.; Wang, K.C.; Varley, A.; et al. AGILE platform: A deep learning powered approach to accelerate LNP development for mRNA delivery. *Nat. Commun.* **2024**, *15*, 6305.
57. Wu, K.; Yang, X.; Wang, Z.; Li, N.; Zhang, J.; Liu, L. Data-balanced transformer for accelerated ionizable lipid nanoparticles screening in mRNA delivery. *Brief. Bioinform.* **2024**, *25*, bbae186.
58. Zhang, W.Y.; Zheng, X.L.; Coghi, P.S.; Chen, J.H.; Dong, B.J.; Fan, X.X. Revolutionizing adjuvant development: Harnessing AI for next-generation cancer vaccines. *Front. Immunol.* **2024**, *15*, 1438030.
59. Chaudhury, S.; Duncan, E.H.; Atre, T.; Storme, C.K.; Beck, K.; Kaba, S.A.; Lanar, D.E.; Bergmann-Leitner, E.S. Identification of immune signatures of novel adjuvant formulations using machine learning. *Sci. Rep.* **2018**, *8*, 17508.
60. Ma, J.; Wang, S.; Zhao, C.; Yan, X.; Ren, Q.; Dong, Z.; Qiu, J.; Liu, Y.; Shan, Q.; Xu, M.; et al. Computer-Aided Discovery of Potent Broad-Spectrum Vaccine Adjuvants. *Angew. Chem. Int. Ed.* **2023**, *62*, e202301059.
61. Gude, S.; Abburi, S.K.; Gali, P.K.; Gorlagunta, S. Advancing single-shot vaccine design through AI and computational models. *Transl. Regul. Sci.* **2025**. [CrossRef]

62. McGranahan, N.; Swanton, C. Clonal heterogeneity and tumor evolution: Past, present, and the future. *Cell* **2017**, *168*, 613–628.
63. Sugiyama, N.; Terry, F.E.; Gutierrez, A.H.; Hirano, T.; Hoshi, M.; Mizuno, Y.; Martin, W.; Yasunaga, S.; Niiro, H.; Fujio, K.; et al. Individual and population-level variability in HLA-DR associated immunogenicity risk of biologics used for the treatment of rheumatoid arthritis. *Front. Immunol.* **2024**, *15*, 1377911. [CrossRef]

BioChem

MDPI

Review

From Microbes to Molecules: Synthetic Biology Approaches for Advanced Materials Design

Roshini Ramachandran *, Frank Macabenta, Grace Bettencourt and Shulammite Feng

Department of Biology and Chemistry, California State University Monterey Bay, Seaside, CA 93955, USA; sfeng@csumb.edu (S.F.)
* Correspondence: rramachandran@csumb.edu

Abstract: Traditional materials synthesis often involves energy-intensive processes with significant waste generation and limited control over material properties. This review examines synthetic biology as a sustainable alternative for designing advanced materials with enhanced precision and versatility. It explores microbial biomineralization, detailing how microorganisms influence the formation of mineral deposits and participate in key biogeochemical cycles. It highlights recent research advancements in using a wide variety of microorganisms for the synthesis of inorganic materials such as metal and metal oxide nanoparticles, quantum dots, magnetic nanoparticles, and thin films. The review also discusses the production and properties of various biopolymers. Important factors that can influence the size, morphology, and uniformity of these biomaterials are covered in detail. Emphasis is placed on advancements utilizing synthetic biology tools, such as protein engineering and genome editing, and recent research for creating smart and responsive materials. Considering the present limitations of synthetic biology, challenges related to scale-up, yield, and uniformity are discussed, and suggestions for future research are detailed.

Keywords: synthetic biology; biopolymer; nanotechnology; biomineralization; MICP

1. Introduction

The ongoing quest for materials of improved performance, designed for particular functionality and applications, has been achieved traditionally through synthetic routes that are high in energy demand, extreme in chemical conditions, and detrimental in environmental impact [1–5]. From the elevated temperatures of ceramic processing to the multi-step, intricate chemical syntheses of polymers, traditional material synthesis is geared to generate large amounts of waste, greenhouse gas emissions, and copious use of solvents and catalysts [2,6,7]. Furthermore, the degree of control over the properties of materials, including nanoscale architecture and precise chemical composition, may be lost because of the intrinsic limitations of these top–down approaches. These challenges necessitate a revolutionary change toward sustainable and specific material synthesis approaches. Synthetic biology is a new discipline that takes advantage of the natural capacity of living systems to construct biological entities with specific functions [8]. This strategy has the potential to transform materials science through the facilitation of a green, bottom–up route to the production of new materials with higher versatility and precision [5]. Microorganisms are very efficient biofactories that have the capacity to synthesize an incredibly large number of organic and inorganic materials under mild conditions and from renewable feedstocks. The potential to apply biological processes in the fabrication of materials presents appealing avenues of inquiry culminating in the development of sustainable alternatives for conventional paradigms of manufacture and addressing acute concerns in the area of environmental sustainability and depletion of resources [2,9,10]. Theoretical underpinnings of biofabricated materials are often directly inspired by nature. Living systems have evolved sophisticated methodologies for the synthesis of intricate architectures with extremely high precision and efficiency. From the complex silica structure of diatoms to the durable nacre of mollusc shells, nature provides the template for developing materials with specific properties [11]. Through imitation of these biological processes, researchers are developing bio-inspired materials with performance characteristics that exceed their traditional synthetic counterparts [12,13]. The intersection of bio-inspired design and synthetic biology is therefore catalyzing an exponential growth in research interest in the area of biofabrication and elevating it to a forefront position in materials science research [14].

The "biofabrication paradigm" is a fundamental departure from traditional materials synthesis, initiating an era where living organisms, particularly microorganisms, are conceptualized as programmable factories [12]. This paradigm shift moves away from conventional, often environmentally expensive, chemical processes and towards biologically mediated manufacturing, taking advantage of the inherent efficiency and diversity of living systems [15–17]. It envisions a future in which microbes, with their genetic and metabolic pathways exactly controlled, produce all kinds of advanced materials, from complex nanomaterials to elaborate biopolymers, in a sustainable manner [18–20]. Inspired by the efficiencies of nature's solutions and driven by the accuracy of synthetic biology, the biofabrication paradigm will transform materials science, creating a future where design and fabrication are inseparably intertwined with biological processes. Nature has created processes for the synthesis of materials with complex properties, as in the case of bone's capacity for self-healing and spider silk's combination of rigidity and lightness [21]. Researchers are working on the creation of materials that can mimic natural processes to exhibit unprecedented properties. Microorganisms are at the forefront of this effort since they can be utilized to create bio-inspired materials with desired properties. The growing emphasis on naturally derived materials highlights the promise of synthetic biology to help advance materials science.

This article seeks to present a review of recent developments in microbial-mediated synthesis of inorganic and organic materials—in particular, how recent advancements in synthetic biology tools that utilize genetic and metabolic engineering have led to robust methods for industrial use. We will explore the ways in which microorganisms create molecules through biomineralization and focus on the microbial synthesis of inorganic nanostructures and biopolymers. We will discuss pathways microbes undertake to fabricate materials and highlight how microbes can be engineered to synthesize materials with control over their composition. We will discuss the prospects of microbes being used to engineer enhanced properties, improved biocompatibility, and biodegradability. We will deliberate the challenges that currently hinder the widespread adoption of these technologies, including issues related to scale-up, yield, product uniformity, and environmental considerations, and postulate future directions for research and development, highlighting the potential of emerging synthetic biology tools and synergistic approaches to overcome existing limitations.

2. Biomineralization

2.1. Overview

Biomineralization is the process by which living organisms produce minerals. Microbial pathways contribute to the formation of mineral deposits on the earth, such as carbonates, sulfides, and oxides [22]. Microbes are the key drivers of biogeochemical cycles, influencing the transformation of elements such as carbon, nitrogen, sulfur, and metals. Microbial biomineralization plays a vital role in these cycles by affecting the solubility and mobility of various elements [23]. Microbial processes such as respiration, fermentation, and photosynthesis can alter the pH of their surroundings; for example, the hydrolysis of urea by ureolytic bacteria releases ammonia, thus increasing pH and promoting the precipitation of carbonate minerals [22,24–26]. Conversely, the production of organic acids during the degradation of organic matter can decrease the pH, leading to mineral dissolution [27,28]. Microbes can also catalyze redox reactions by changing the oxidation state of metals and other elements, thereby forcing the precipitation or dissolution of minerals. For example, the reduction in ferric iron (Fe^{3+}) to ferrous iron (Fe^{2+}) by iron-reducing bacteria can lead to the formation of Fe^{2+}-based minerals like siderite and magnetite [29,30]. Magnetotactic bacteria (MTB), which are capable of synthesizing magnetic nanoparticles (Fe_3O_4 or Fe_3S_4) using specialized organelles called magnetosomes, are an especially fascinating example of highly stereotyped and robust biomineralization with a lot of potential for industrial applications [31]. Similarly, diatoms, a group of photosynthetic algae, construct elaborate cell walls composed of amorphous silica (SiO_2) called frustules. These frustules display hierarchical porosity and nanoscale patterning—features that have inspired the development of biomimetic materials for filtration, optics, and catalysis [11]. Furthermore, the microbial production of CO_2 or H_2S gases can influence mineral formation; CO_2 production can increase carbonate saturation, while H_2S production by sulfate-reducing bacteria (SRB) can lead to metal sulfide precipitation [32,33]. This influences sulfur cycling in nature and the availability of metals in different environments.

2.2. Bioremediation

A well-studied application of biomineralization is bioremediation, i.e., the use of microbes to remove pollutants from contaminated environments [34,35]. The origins of human-caused heavy metal contamination stem from both industrial and domestic sources. The metal ion contaminants that disperse into the environment experience difficulty degrading, introducing potential issues for human health and physical systems. In an effort to combat the bioaccumulation of these contaminants, many technological advancements, such as induced chemical precipitation, physical adsorption, ion exchange, and membrane filtration, have been in production but experience difficulties with progression due to financial constraints. Therefore, there is a focus on the advancement of biomineralization technology towards the goal of providing a stable and sustainable sequestration of heavy metals. One method of bioremediation is through microbial-induced carbonate precipitation [22,35]. This process produces the carbonate anion, CO_3^{2-}, through microorganisms such as bacteria, fungi, or enzymes, which then precipitates the heavy metal ions into free, stable carbonates that do not leave harmful impacts on the environment. The precipitation of heavy metals can also be completed through other methods such as urea hydrolysis, denitrification, or sulfate reduction. Urea hydrolysis stands as the most advantageous method due to its time and money efficiency. Within this process, microorganisms decompose urea into compounds, which increases the overall pH, lessening the residual impact the contaminant would originally have on its surroundings [35–37]. Urea hydrolysis is driven largely by urease-producing bacteria (UPB), which can be easily ordered through commercial collection centers, making it an overall more preferable method of remediation.

Another method is phosphate precipitation, which utilizes microbial activity during the dissolution of inorganic phosphate and organic phosphorus to create an enzyme-induced phosphate precipitation reaction [38–40]. The process enables the precipitation of free metal ions into stable heavy metal phosphates, mitigating their toxic effects. Among the most well-studied methods of bioremediation, phosphate-solubilizing bacteria (PSB) play a crucial role in making organic phosphorus and inorganic phosphate soluble through the usage of enzymes and organic acids, which speed up the process [35]. Moreover, scientists have discovered that the addition of an increased amount of organic acids and microorganisms (such as *Bacillus*, *Ochrobactrum*, *Pseudomonas*, *Marcescens*, and *Rahnella*) improves the efficiency of the reaction, as they demonstrate increased decomposition rates [34,35]. The method of phosphate-based remediation introduces many advantages over carbonate precipitation, such as its greater thermodynamic stability, tolerance to lower pH levels, and lack of emission of harmful gases [38,41–43]. However, it remains an unattainable method to practice, as the cost is out of range for sustainable, long-term use. The efficiency of microbial precipitation is largely influenced by multiple conditions concerning its environment and the bacterial species involved in the heavy-metal deterioration process. In regard to phosphate precipitation-based bioremediation, differing species of bacteria exhibit varying degrees of tolerance to heavy metal ions [35]. As for carbonate precipitation, certain fungal species inherit higher capabilities of urea hydrolysis [44]. The situational differences in heavy-metal removal, lack of funding towards research, provision of materials, and overall high costs inhibit large steps towards technological advancements.

While bioremediation represents an important application of harnessing microbe-driven mineral formation, the processes involved in biomineralization also underpin their ability to synthesize a wider range of inorganic materials. Building on the understanding of how microorganisms control mineral precipitation for bioremediation, researchers have explored their potential in the synthesis of novel inorganic materials. Figure 1 highlights two mechanisms of biomineralization-induced precipitation.

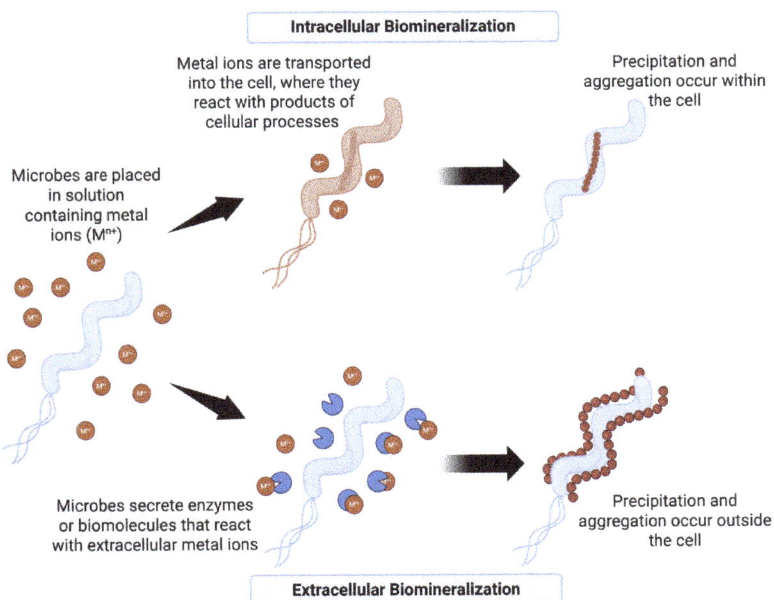

Figure 1. Mechanism of biomineralization using microbial-induced precipitation.

3. Synthesis of Advanced Materials

Biomineralization has long inspired materials scientists. By using synthetic biology to design materials, researchers are moving beyond traditional top–down approaches toward bottom–up, self-organizing systems with new functionalities. By accessing the metabolic potential of microorganisms, scientists have engineered biological systems that synthesize materials under less harsh conditions, using renewable biomass feedstocks and reducing toxic waste [1,3,45,46]. Another significant advantage of microbial-mediated synthesis is its versatility. Microorganisms are able to readily adjust to changing environments and can metabolize substrates with broad diversity, thereby enabling the production of a wide range of products [47].

3.1. Inorganic Nanostructures

Microorganisms can be genetically designed to synthesize inorganic nanostructures with defined size and shape, eliminating the necessity for toxic chemical reductants and high-temperature ovens. Naturally occurring microbial metabolic processes like photosynthesis and respiration can alter the chemical environment and lead to mineral or metal precipitation.

3.1.1. Metal Nanoparticles

Metal nanoparticles, especially those involving noble metals, are ubiquitous in catalytic converters [45]. Some metal nanoparticles are also utilized in drug delivery and therapeutics, making biocompatibility an essential requirement [48]. Due to their inherent ability to reduce metal ions, microorganisms have the ability to synthesize a range of metal nanoparticles [49]. By controlling the biological systems and reaction conditions, the size, shape, and composition of the nanoparticles can be precisely tuned [49,50].

The reduction in metal ions by microorganisms can occur both intracellularly and extracellularly (Figure 1). In intracellular synthesis, metal ions are transported into the microbial cell, where they are reduced and assembled into nanoparticles (NPs). This process involves specific enzymes (e.g., reductases) that act as bioreductants and stabilizing agents [3,51]. NPs are typically formed within the cell cytoplasm or periplasmic space. Bacteria like *Shewanella oneidensis* can transport gold (Au^{3+}) and silver (Ag^{+}) ions into the cell where they are reduced by intracellular reductases to their metallic form (Au^0 and Ag^0), leading to the accumulation of metal NPs [1,52]. It also utilizes nitrate reductase on its cell surface and in the periplasmic space to catalyze the reduction of gold and silver ions [53,54].

Extracellular synthesis occurs outside the cell, where metal ions are reduced by enzymes or biomolecules secreted by the microbe. Extracellular synthesis simplifies the purification process for obtaining the final product when compared to intracellular synthesis. *Shewanella oneidensis* employs extracellular electron shuttles, such as riboflavin, to transfer electrons to gold and silver ions in the surrounding environment [55,56]. *Pseudomonas aeruginosa* secretes enzymes such as NADH-dependent reductases extracellularly [49,52]. The bacterium also releases organic molecules, such as polysaccharides and proteins, which prevent aggregation of the resulting nanoparticles [54]. Similarly, certain fungi secrete a high concentration of enzymes and proteins from their mycelia, which act as primary reducing agents [57,58]. The proteins and polysaccharides in the fungal filtrate can act as capping agents and stabilize the synthesized NPs.

3.1.2. Metal Oxide Nanostructures

Metal oxide materials have diverse applications in fields such as catalysis, electronics, and medicine [46]. The mechanisms involved in microbial synthesis of metal oxide NPs often overlap with those in the synthesis of metal NPs. However, a key difference is that the metal ions are not reduced to their zero-valent state and instead form complexes with oxygen. Bio-precipitation is the most common mechanism for metal oxide formation within the cell [23,32,59]. Enzymes such as phosphatases or oxidases that are produced can alter the local chemical environment, leading to the precipitation of metal oxides. Similarly, metabolic activities can alter the pH of their surroundings, leading to the precipitation of metal ions as oxides or hydroxides. One example is the precipitation of zinc oxide (ZnO) NPs under alkaline conditions using fungal biomass (*Aspergillus niger*) [60,61]. Phytochemicals present in Azadirachta indica (neem) leaves were also used to synthesize metal oxide nanoparticles [8,62].

Microbial cell surfaces can also bind to metal ions through various functional groups (e.g., carboxyl, phosphate, hydroxyl). This creates a nucleation site for nanoparticle formation. Algae like *Spirulina platensis* have been successfully used as templates for the formation of titanium dioxide (TiO_2) NPs through binding of titanium ions to the algal cell wall and its subsequent oxidation [54,63]. Here, the microbes do not actively mediate synthesis via metabolic processes but serve as structural scaffolds for abiotic mineral formation. This highlights the distinction between true microbial biosynthesis, where the organism plays an active biochemical role, and templated synthesis, where the organism's surface facilitates material deposition. Some bacteria produce siderophores, which chelate with iron ions and facilitate their transport into the cell [64]. In certain cases, the iron ions are further processed to form iron oxide NPs, such as goethite (α-FeOOH) or hematite (α-Fe_2O_3) [23,65]. Microbial enzymes, such as reductases, can also support the synthesis of iron oxide (Fe_2O_3) nanoparticles [66].

3.1.3. Magnetic Nanoparticles

Microbial synthesis of magnetic nanoparticles, especially magnetite (Fe_3O_4), has been researched for producing biocompatible nanomaterials with targeted applications in drug delivery, magnetic hyperthermia therapy, and MRI contrast agents [29,67]. Magnetotactic bacteria (MTB) can naturally synthesize magnetic nanoparticles within specialized organelles called magnetosomes that contain magnetite or greigite (Fe_3S_4) crystals. The magnetosomes can control the size, shape, and crystal structure of the magnetic nanoparticles with precision [68]. *Magnetospirillum magnetotacticum* and *Magnetospirillum gryphiswaldense* are well-studied MTBs that produce uniform magnetite crystals within magnetosomes [69,70]. The process of magnetosome formation involves several steps: uptake of iron, formation of a membrane vesicle, regulation of iron transport and redox state, nucleation and growth of the magnetic crystal, and arrangement of magnetosomes in chains within the cell [68,71,72]. Specific pH conditions, oxygen availability, and the presence of other minerals have an influence on the transformation process [29]. The ability to control the properties of synthesized magnetic nanoparticles via genetics has allowed MTBs to play an outsized role in engineering chimeric magnetic nanoparticles necessary for novel research applications. Magnetogenetics is an emerging technology in biomedical research, allowing for noninvasive manipulation of cell signaling pathways via magnetic fields; an enduring challenge is developing a viable fusion of proteins and magnetic probes tailored to individual research needs. A recent study described a robust method for generating engineered ferritin fused to green fluorescent protein (GFP) as an efficient and biostable in vivo technique to manipulate cellular processes remotely [73].

Some bacteria and fungi can also produce magnetic nanoparticles extracellularly. These microorganisms secrete organic molecules that can reduce Fe^{3+} to Fe^{2+}, leading to the formation of magnetite nanoparticles. Elblbesy et al. explored the extracellular production of nanoparticles by microbes, specifically using *Magnetospirillum* strain AMB-1 to produce magnetic nanoparticles [67]. The experiment showed that this nitrate-reducing Fe (II) oxidizing bacteria produced paramagnetic magnetite nanoparticles of ~50 nm size that possessed a low coercivity. Interestingly, even minute changes to incubation conditions under which biosynthesis was conducted caused small variations in the magnetic characteristics of the produced nanoparticles [67].

Shewanella oneidensis does not possess magnetosomes, but it can perform an extracellular reduction of iron ions [30]. It forms magnetite nanoparticles during the process of anaerobic respiration, where it uses iron oxides as electron acceptors. Sulfate-reducing bacteria are capable of producing magnetite nanoparticles during the process of sulfate reduction [74]. The mechanism involves the production of sulfide, which can react with iron ions to form iron sulfides. These iron sulfide minerals undergo transformation to form magnetite through processes like dissolution, recrystallization, and oxidation–reduction [74].

3.1.4. Semiconductor Nanocrystals

Quantum dots (QDs) are semiconductor nanocrystals ranging in size from 2 to 10 nanometers. Their unique optical and electronic properties arise from quantum confinement effects due to their small size, making them suitable for photocatalysis, optoelectronics, and bioimaging. Both intracellular and extracellular processes have been explored for the microbial synthesis of semiconductor nanocrystals [52,75–79].

Sulfate-reducing bacteria can react with metal ions to create nanocrystals of cadmium sulfide and zinc sulfide [4,80]. Through an intracellular process, they take up metal ions (Cd^{2+}, Zn^{2+}) into their cells. Sulfide ions (S^{2-}), which are generated through sulfate reduction, then react with metal ions within the cell. This leads to the formation of CdS or ZnS QDs within the microbial cell. As expected, the size and properties of the QDs are influenced by the metal ion concentration, sulfide ion availability, temperature, pH, and enzymatic activity [78]. *Escherichia coli* has been engineered to synthesize CdS QDs intracellularly by introducing genes encoding gamma-glutamylcysteine synthetase (gamma-GCS) and glutathione synthetase (GS), which are involved in sulfate reduction [81]. Mi et al. reported the production of sulfide ions within the cells, leading to efficient CdS QD formation [77]. *E. coli* was also reported to synthesize CdTe quantum dots through an extracellular mechanism [76]. The fungus *Fusarium oxysporum* produces proteins and enzymes when exposed to a zinc ion-containing solution [82,83]. These secreted biomolecules convert Zn^{2+} to ZnS. The Gram-negative bacterium *Pseudomonas putida* can perform both intracellular and extracellular synthesis methods to make CdS QDs [75,84].

It is important to recognize that quantum dots are environmentally harmful due to the toxicity of their constituent materials [85]. Quantum dots can degrade and release heavy metals (e.g., Cd, Zn, Te) when improperly disposed of into soil and water [86]. This makes techniques like microbial bioremediation increasingly relevant; however, some studies have shown that quantum dots can be toxic to bacteria and algae. Nonetheless, microbial synthesis is a promising avenue for producing less toxic QDs. For example, a study investigating the effects of biogenic CdSQDs generated using *F. oxysporum* on bacterial viability and lettuce germination found that concentrations had to be at their highest to display antibacterial activity, while biogenic CdSQDs did not inhibit seed germination [87].

3.2. Thin Films Through Microbial Deposition

Microbial deposition leverages the ability of microorganisms to precisely control the nucleation and growth of inorganic materials [55]. It relies purely on biomineralization to precipitate the desired inorganic materials. Unlike traditional chemical vapor deposition or sputtering techniques, microbial deposition occurs under mild conditions, often requiring only room temperatures and neutral pH. For example, *Bacillus subtilis* precipitates $CaCO_3$ through urea hydrolysis [25]; this process can be employed to create thin films of calcium carbonate for concrete reinforcement [88–90].

Some bacteria and fungi can actively transport metal ions into their cells, leading to intracellular precipitation. *Pseudomonas* sp. can accumulate uranium within their cells through active transport mechanisms where uranium ions are taken up and then precipitated as uranium phosphate [91]. *Aspergillus niger* and *Paecilomyces javanicus* also produce uranium phosphates on fungal mycelia in combination with extracellular processes [92]. These insights can be used to create thin biofilms of uranium compounds [93]. Another method is through passive absorption, where the microbial cell surfaces act as a template for nucleation and growth of minerals, owing to the presence of certain functional groups (e.g., carboxyl, phosphate) that bind to ions. A well-known example is diatoms, a type of algae that possess silica cell walls (frustules) with highly ordered nanoporous structures [11]. Research has shown that coating diatom frustules with other materials can fabricate thin films with tailored optical properties [94]. We have summarized the inorganic materials that can be synthesized microbially in Figure 2.

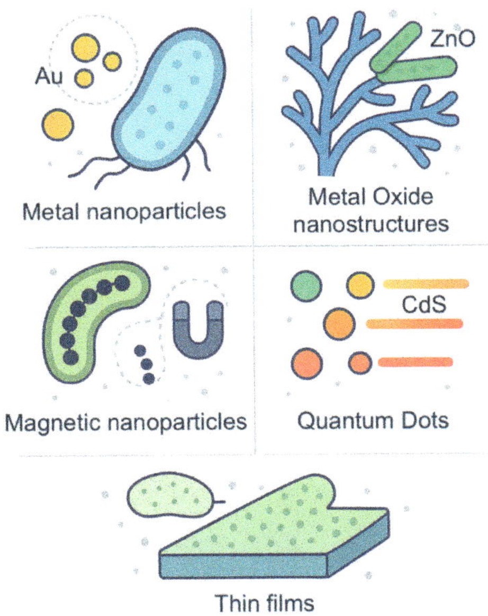

Figure 2. Infographic representing the inorganic nanostructures described in Section 3.

3.3. Biopolymers

Microorganisms offer sustainable alternatives to traditional petroleum-based polymers. These microbial biopolymers are biodegradable, biocompatible, and synthesized from renewable resources. The production of microbial biopolymers involves (i) microbial selection and strain development, (ii) fermentation, i.e., nutrient consumption under controlled conditions, and (iii) extraction of biopolymer from the microbial cells [2,20].

Polysaccharides are made up of sugar monomers, which makes them an ideal avenue for biosynthesis. Some bacteria, such as *Komagataeibacter xylinus*, can synthesize bacterial cellulose with high purity and tensile strength [5,95–97]. *K. xylinus* converts glucose from its surroundings into UDP-glucose, which is then polymerized into glucan chains by cellulose synthase, an enzyme complex located at the bacterial cell membrane. The cellulose nanofibers produced have applications in wound healing, tissue engineering, and high-strength materials [2,20,98,99]. Similarly, xanthan gum, which is widely used in the food and cosmetic industry, can be produced by the bacterium *Xanthomonas campestris* from glucose starting materials [100].

Bacteria from the genera *Azotobacter* and *Pseudomonas* are capable of producing alginates, which are extensively used in wound dressings and drug delivery [5]. The enzymes alginate synthases and epimerases play a key role in determining the final structure and properties of the alginate product. An interesting example to demonstrate the versatility of microbial biopolymers is the production of polyhydroxyalkanoates (PHA) by *Cupriavidus necator* [101,102]. The type of PHA produced can be influenced by the carbon source provided to the bacteria—using glucose leads to polyhydroxybutyrate (PHB) production, while adding valerate leads to the production of polyhydroxybutyrate-co-hydroxyvalerate (PHBV) copolymer [103].

Bacteria are also capable of producing monomers for specific polymers, such as the synthesis of lactic acid by the *Lactobacillus* species [104]. These bacteria ferment carbohydrates like glucose or starch to produce lactic acid, which is then polymerized into polylactic acid. Similarly, succinic acid can be produced by bacteria like *Actinobacillus succinogenes* [105,106]. These bacteria can ferment renewable feedstocks containing glucose, glycerol, and agricultural byproducts to produce succinic acid [104,107,108].

Genetic alterations in microbes can also facilitate the synthesis of biopolymers with designed properties, including enhanced mechanical properties and biodegradability [96]. Researchers can genetically design bacteria to synthesize cellulose from alternative carbon sources, including waste streams [97]. Yeast can be made to secrete lipids with specific kinds of fatty acids, thereby offering a renewable feedstock for biopolymers [109]. Recently, researchers engineered bacteria and yeast to produce spider silk-like proteins [21]. These protein fibers have a high strength and elasticity, with potential to be used in textiles, biomedicine, and composite materials. Table 1 summarizes.

Table 1. Summary of advanced materials discussed in Section 3.

Material	Microbe	Category	Reference
Alginate	*Azotobacter* sp. *Pseudomonas* sp.	Biopolymer	[5]
Bacterial cellulose	*Komagataeibacter xylinus*	Biopolymer	[95–97]
CdS	*Escherichia coli*	QD	[81]
CdS	*Pseudomonas putida*	QD	[75,84]
CdTe	*Escherichia coli*	QD	[76]
Fe_3O_4	*Magnetospirillum strain AMB-1*	NP	[67]
	Magnetospirillum gryphiswaldense		[70]
	Magnetospirillum magnetotacticum		[69]
	Shewanella oneidensis		[30]
Gold	*Shewanella oneidensis*	NP	[52–55]
Polyhydroxyalkanoates	*Cupriavidus necator*	Biopolymer	[101]
Polylactic acid	*Lactobacillus* sp.	Biopolymer	[104]
Silver	*Pseudomonas aeruginosa*	NP	[49,52]
	Fusarium oxysporum	NP	[51,58]
	Shewanella oneidensis	NP	[54,56]
Succinic acid	*Actinobacillus succinogenes*	Monomer	[104–108]
TiO_2	*Bacillariophyceae* (diatoms)	Thin films	[94]
	Spirulina platensis	NP	[54,63]
Uranium phosphate	*Pseudomonas* sp.	Thin films	[93]
Xanthan gum	*Xanthomonas campestris*	Biopolymer	[100]
ZnO	*Aspergillus niger*	NP	[60,61]
ZnS	*Fusarium oxysporum*	QD	[82,83]

4. Factors Affecting Material Properties

By manipulating biological and experimental parameters, we can tailor the properties of the resulting materials for specific applications. There are several factors that influence the morphology and properties of the resulting material (Figure 3), which are discussed in detail below.

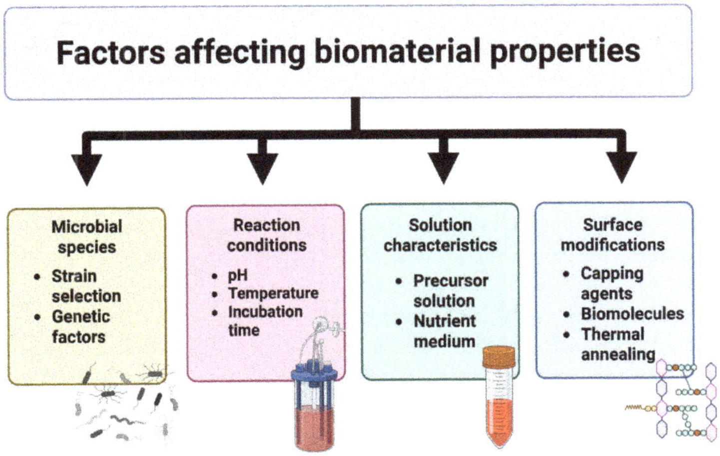

Figure 3. General factors affecting the properties of biosynthesized materials.

4.1. Microbial Species/Strain Selection

Different organisms possess distinct enzyme machinery and metabolic processes, which directly influence the formation of the product. For example, researchers have found that careful selection of the strains of *Fusarium oxysporum* is required to produce monodisperse, spherical silver nanoparticles [15,48,79]. Other fungal species, or even different strains of *Fusarium*, could produce different shapes or a wider size distribution [3,45,49]. Also, *S. xiamenensis* was found to produce a higher yield of gold nanoparticles when compared to *S. oneidensis* [53].

Magnetotactic bacteria (MTB) clearly demonstrate how the choice of strain can affect the desired properties of the material produced. Genetic factors within MTB control iron uptake and crystal growth [71], leading to monodisperse Fe_3O_4 NPs. The size of magnetosomes (typically 35–120 nm) as well as their shape (cuboidal, elongated, and bullet-shaped) is genetically determined [29,68]. Specific proteins within the magnetosome guide crystal growth along preferred crystallographic planes, governing the final shape of the inorganic nanoparticles formed [29,68]. In contrast, sulfate-reducing bacteria (SRB) produce irregularly shaped Fe_3O_4 NPs extracellularly during anaerobic respiration [74].

4.2. Precursor and/or Nutrient Characteristics

As established with solution-based synthesis of inorganic nanoparticles, lower concentrations of metal salts lead to smaller nanoparticles, while higher concentrations of metal salts lead to larger nanoparticles due to agglomeration [62,110]. Carefully controlling the ratio of microbes and cations (e.g., Zn^{2+}) in the growth medium will produce nanoparticles with a narrower size distribution [110,111]. In the synthesis of gold nanoparticles using *Shewanella oneidensis*, increasing the concentration of gold chloride (the metal precursor) in the growth medium generally led to an increase in the average size of the resulting gold nanoparticles [53,112]. Similarly, the size of palladium nanoparticles was controlled by fine-tuning the Pd/cell dry weight [65,112]. Additionally, the presence of other organic matter and salts in the growth medium can play a role in the nanoproduct [113]. Depending on the chemical composition of the product, the presence of specific ligands or additives can guide its morphology.

Biopolymers are also affected by the nutrient medium. In the synthesis of bacterial cellulose (BC) nanofibers, using glucose as the sugar source leads to a high yield of product [96]. In some cases, it has been shown that using glycerol can also lead to high BC yields [96,97]. Use of peptone in the medium improved BC yields from mannitol and sucrose [114]. The addition of polysaccharide pullulan can affect the bundling behavior of BC and improve the mechanical strength of BC [115]. In the biosynthesis of xanthan gum by *Xanthomonas campestris*, the choice of carbon source and composition of the nutrient medium controls the yield and molecular weight of the biopolymer [100,116,117].

4.3. Reaction pH, Temperature, and Time

Reaction pH, temperature, and time influence microbial enzyme activity and the kinetics of nanoparticle nucleation and growth [118]. Optimal ranges for synthesis vary depending on the microorganism and the target material. Therefore, these factors exhibit considerable variability across published research [15]. All factors are interactive, and therefore, they have to be carefully considered together in order to determine the best suitable conditions for a particular microorganism [119].

An initial increase in temperature generally increases the rate of synthesis; however, very high temperatures can denature microbial enzymes and lead to poor-quality products. In general, temperatures of 25–37 °C are ideal for the formation of uniform nanoparticles [46]. Higher temperatures can cause nanoparticle aggregation or loss of capping agent. Similarly, longer incubation times can lead to a higher yield; however, the product may be polydisperse with a larger size distribution due to continued growth or aggregation [111]. For biopolymers, longer fermentation times can lead to increased polymer chain elongation and higher molecular weight [116,120,121], but very long times may lead to degradation.

Synthesis of Ag nanoparticles by *Bacillus* sp. is more efficient at slightly alkaline pH, resulting in smaller, spherical nanoparticles [59,122–124]. This is because acidic pH could lead to delayed nucleation and agglomeration [125]. However, a pH that is too high may cause undesirable precipitation of hydroxides [126]. On the other hand, a slightly acidic pH promotes Fe_3O_4 precipitation by *P. aeruginosa* [72]. The formation of xanthan gum was increased at a pH of 7.0–7.5; however, further increase in the pH decreased the production [127]. For the production of bacterial cellulose, the ideal conditions are a lower pH of 3–7 and a temperature range of 25–30 °C [95].

4.4. Biomolecules and Capping Agents

Proteins and polysaccharides secreted by extracellular processes can act as capping and stabilizing agents, influencing product morphology. For example, specific proteins can act as templates, directing the growth of Ag NPs into different shapes like spherical and triangular [128,129]. Altering the reaction pH can affect the charge of these biomolecules, influencing their interaction with Ag^+ ions and the final particle shape. In the biosynthesis of Au nanorods, some proteins bind to gold ions in a way that directs their assembly into rod-shaped structures rather than spheres [130]. Post-synthesis modifications such as thermal annealing or capping with additional ligands can be employed to fine-tune nanostructure properties.

5. Synergistic Approaches

The field of synthetic biology is naturally interdisciplinary, and its utility in material science has increased by its convergence with other technologies. The integration of synthetic biology with existing tools enables researchers to develop a robust approach to design and fabricate novel materials with increased complexity and functionality.

5.1. Protein Engineering and Genome Editing

One interesting approach is the combination of protein engineering with biomineralization. This entails creating proteins with specific shapes and chemical functionalities, acting as templates for mineral deposition [131–133]. These protein scaffolds can bind specific ions, thereby controlling the nucleation and growth of minerals with desired size, shape, and crystalline phases. For example, genetically engineered proteins, such as silk proteins or peptides, can be used to template the formation of calcium carbonate, silica, or metal nanoparticles [18,134–139]. The resulting materials exhibited enhanced mechanical strength, optical properties, and catalytic activity. Studies have shown that silver nanoparticles prepared using genetically engineered bacteria demonstrated precise control over nanoparticle size and distribution [140].

Similarly, using genome editing such as CRISPR-Cas9, researchers can modify microorganisms to enhance the production of specific building blocks for polymers. This involves deleting competing pathways, introducing new genes, or altering regulatory elements to fine-tune gene expression. An example is the engineering of bacteria to overproduce monomers like lactic acid (for polylactic acid production) or isoprene (for rubber synthesis) [19,104,141,142]. Additionally, new advancements in Cas9-based genetic engineering have allowed for the development of synthetic gene circuits that (1) greatly reduce the leakiness of a system (i.e., the basal expression of a gene in the absence of an inducer) and (2) achieve high maximum expression of a target product. CASwitch is one such system that combines low leakiness with high maximum expression, resulting in high fold induction (i.e., the ratio between maximum expression and leakiness) by combining two established network motifs: the Coherent Feed-Forward Loop (CFFL) and Mutual Inhibition (MI) [143].

5.2. Smart and Responsive Living Materials

Living materials offer unique advantages such as the ability to respond dynamically to their environment [144]. Engineered living tissues with tailored mechanical and biological properties could be used to repair or replace damaged organs and tissues. For example, researchers are exploring the use of 3D-printed living constructs containing cells and extracellular matrix components to create functional bone, cartilage, or skin [13,145,146]. Self-healing hydrogels can use biosynthesis to repair damaged tissues [147,148]. Bacteria can be programmed to synthesize and release therapeutic proteins at tumor sites [149–151]. While the term 'living materials' encompasses a plethora of technologies that combine living cells with nonliving scaffolds, these are distinct from engineered cells (otherwise known as artificial cells), which are nonliving and designed to mimic biological processes [152].

In the field of engineering, an interesting example is bacteria-containing composite materials that are programmed to secrete a mineralizing agent in response to damage. For example, concrete can be engineered to incorporate bacteria that secrete calcium carbonate to repair cracks and restore structural integrity [25,89,90]. Living cells can be engineered to sense specific environmental cues, such as changes in pH, temperature, or the presence of pollutants, and then trigger a visual response or release a signaling molecule [147,153,154]. Microorganisms can be engineered to detect specific pollutants in water and emit a fluorescent signal [154–157].

5.3. Resource Management and Recycling

Environmental chemistry is another area of synergy with synthetic biology. Microorganisms can be engineered to produce biofuels from renewable feedstocks, reducing our reliance on fossil fuels. For example, bacteria can produce high-energy-density biofuels from agricultural waste [10,158,159], while yeast can be programmed to synthesize biodegradable polymers for packaging applications [109,160,161]. Living filters composed of microorganisms can be used to remove heavy metals or organic pollutants from contaminated water [1,16,162]. Synthetic biology can play a key role toward a circular economy by facilitating the recycling of materials [109,163]. For example, enzymes can be engineered to break down complex polymers into their constituent monomers, which can then be used to synthesize new materials [164,165]. Microorganisms can be engineered to convert waste streams, such as agricultural waste or CO_2 gas, into bioplastics or biofuels, closing the loop and reducing our reliance on non-renewable resources [1,22,109].

6. Challenges and Future Directions

Microbial synthesis has the potential to be a greener alternative to material synthesis by avoiding the use of toxic chemicals, generating less waste, and operating at mild conditions [14,16,33,54]. Microorganisms could be genetically engineered to produce substances with specific characteristics, such as enhanced strength, unique optical properties, or catalytic function, through the controlled synthesis mechanisms in the nanoscale. The process of microbial production supports the production of complex hierarchical structures through the self-assembly process, mimicking the biological formation processes for numerous materials. A variety of inorganic materials, ranging from metal nanoparticles, metal oxides, and sulfides to carbonates and phosphates, can be synthesized with the assistance of microorganisms [1,67]. However, the discipline also possesses numerous challenges that must be resolved in order to attain its full potential (Figure 4).

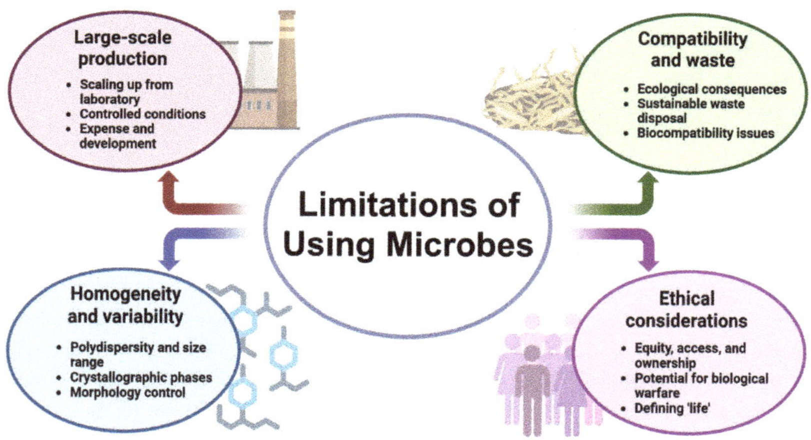

Figure 4. Flowchart depicting the limitations of using microbes for materials synthesis.

6.1. Large-Scale Production

A key challenge is the transition from laboratory experiments to large-scale production [22]. While microorganisms are capable of producing inorganic materials quite well under controlled laboratory conditions, scaling up this process is fraught with numerous issues. Large bioreactors must have extremely controlled temperature, pH, nutrient availability, and oxygen conditions to promote constant microbial activity and material synthesis. Such conditions could be hard to achieve uniformly in a large volume. Process scale-up could increase the expense of bioreactor construction and operation, media preparation, and product purification. Moreover, growth conditions for microbes and synthesis of materials can be very different at the laboratory and production levels. A lot of process development is typically required to attain large scales and high yields and rates of production [14,54]. This can be achieved by creating novel bioreactors in which we can maintain control over important factors at large scales, for example, continuous-flow bioreactors, microfluidic bioreactors, or three-dimensional perfusion systems. In addition to technical scale-up challenges, regulatory frameworks and cost structures remain barriers to commercial adoption of microbial materials. Biomanufacturing of new nanomaterials must comply with biosafety and environmental guidelines, which vary across regions.

6.2. Polydispersity and Uniformity

Microbial growth and deposition rates can be slow compared to conventional techniques. Another problem is to regulate the properties and homogeneity of the produced materials. Microbe-produced materials, particularly nanoparticles, may be of varying sizes and shapes, which can alter their properties and performance [3,16,54]. It is desirable to have nanomaterials with a smaller size range for most applications in semiconductors and biology. Furthermore, regulating the crystal structure and shape of inorganic materials is extremely important. Microbial synthesis may sometimes produce amorphous or poorly crystalline material of non-uniform shape; batch and run consistency may be hard to obtain because of the inherent variability of biological systems [14,166]. Achieving uniform thin films over large areas can be challenging. Therefore, precise control over microbial metabolic pathways through genetic engineering techniques, optimization of precursor production, and designing microorganisms can aid uniform material synthesis [16,54]. Developing in situ methods of analyzing the synthesis process helps in controlling material properties and making them homogeneous.

6.3. Compatibility and Waste Production

For medical uses, the materials produced by the microorganisms must not cause adverse reactions in the host organism [14]. Certain microorganisms or their products can be harmful to human beings or other organisms. Non-toxic microorganisms must be chosen, and methods must be developed to remove any toxic byproducts [8,147]. Both in vivo and in vitro tests for biocompatibility must be conducted to ensure that the synthesized materials are safe and effective [5,98,99]. Finally, the synthesis of large amounts of material using microbes will generate enormous amounts of waste and biomass [1]. Disposal of these wastes must be conducted in a sustainable manner so as not to damage the environment. Even though microbial synthesis is generally greener than chemical synthesis, keen scrutiny of the degree to which the end materials are compatible with living things and their possible impact on the environment must be conducted. The release of engineered living materials in the environment, whether intentional or accidental, may have unexpected effects on ecosystems, biodiversity, and human health [167,168]. Strong risk assessments and control strategies are therefore crucial, but it can be difficult to predict long-term effects. Further research in the form of life-cycle assessments must be conducted to ascertain the effects of microbial synthesis on the environment.

6.4. Ethical Considerations

The emergence of technologies that enable the creation of living or biologically created materials brings forth ethical and societal considerations that require scrutiny. At the forefront is the question of defining "life" itself and the moral status of these novel entities. As we engineer organisms to produce materials with tailored properties, we blur the lines between living and non-living, potentially challenging traditional ethical frameworks. The potential does arise for these techniques to be used in dual applications, such as the development of biological weapons, raising serious security concerns [17,168–171]. The ability to manipulate biological systems with increasing accuracy requires the establishment of strict regulatory structures and international cooperation to prevent abuse [47,169]. In addition, the social implications of large-scale adoption of live materials should be considered. There is a need to address issues of equity, access, and ownership of technologies to ensure that the benefits are widespread and existing inequalities are not increased [8,47,168]. Public commitment and transparent communication are important for promoting informed dialog in these new and emerging areas of research. It is necessary to move forward with a sense of responsibility and acknowledge that synthetic biology has the potential to reshape our world in profound ways, for better or for worse.

7. Conclusions

This review highlights the potential of synthetic biology for materials design, shifting from traditional, environmentally burdensome methods to sustainable, biologically driven approaches. In particular, this review describes a synthesis of natural mechanisms for material production and technological advancements that enhance endogenous mechanisms, either by scaling up production, reducing the toxicity of products, or developing new applications. By harnessing microbial processes, researchers can create advanced materials with precision and versatility, moving beyond conventional synthesis. Microorganisms can act as programmable biofactories to facilitate the production of inorganic nanostructures and biopolymers with tailored properties. For instance, we have described how magnetotactic bacteria (MTBs) have the ability to synthesize magnetic nanoparticles via both intracellular and extracellular mechanisms. This endogenous ability has further been enhanced by synergistic approaches in genetic engineering, culminating in applications in the field of magnetogenetics. The ability to manipulate microbial species, precursor concentrations, and environmental parameters allows for fine-tuning material characteristics for specific applications. However, challenges related to scale-up, yield, uniformity, and environmental impact remain. We have summarized these considerations and conclusions in Table 2. Future research should prioritize the development of robust biomanufacturing platforms, the expansion of genetic engineering tools, and the encouragement of synergistic approaches to create multifunctional biomaterials.

Table 2. Comparison of traditional and microbial methods of materials synthesis.

Factor	Traditional Synthesis	Microbial Synthesis
Energy Requirements	High temperatures and pressures	Ambient or mild conditions
Environmental Impact	High waste generation and toxic byproducts	Generally low waste and greener processes
Precision/Control	Moderate control over structure/composition	Polydispersity and uniformity remain ongoing challenges; high tunability via genetic/metabolic engineering
Raw Material Source	Petroleum-based and non-renewable	Renewable feedstocks (e.g., sugars and agricultural waste)
Scalability	Established large-scale industrial processes	Still challenging; limited by bioreactor design and yield

Table 2. *Cont.*

Factor	Traditional Synthesis	Microbial Synthesis
Cost	Higher due to being energy and purification intensive	Potentially lower, but requires optimization
Speed of Production	Fast, depending on the method	Slower and dependent on microbial growth and metabolic rates
Material Diversity	Broad, well-studied, and chemically intensive	Growing area of research, influenced by microbe strain
Biocompatibility	Often poor; requires surface modification	Often high, especially for medical/biological applications

Author Contributions: Conceptualization, R.R. and F.M.; writing—original draft preparation, R.R., G.B., and S.F.; writing—review and editing, R.R. and F.M.; visualization, R.R. and F.M.; supervision, R.R. All authors have read and agreed to the published version of the manuscript.

Funding: This research received no external funding.

Data Availability Statement: No new data were created or analyzed in this study. Data sharing is not applicable to this article.

Acknowledgments: We thank our department for their continued support and our undergraduate students for their enthusiasm. R.R. gives gratitude to Pradeep Kumar Ragu Chanthar for his support. Authors used Grammarly software 14.1237.0 for editing/proofreading, BioRender 2025 for figures, and utilized the Paperpile extension as a reference management tool.

Conflicts of Interest: The authors declare no conflicts of interest.

References

1. Annamalai, J.; Ummalyma, S.B.; Pandey, A.; Bhaskar, T. Recent Trends in Microbial Nanoparticle Synthesis and Potential Application in Environmental Technology: A Comprehensive Review. *Environ. Sci. Pollut. Res. Int.* **2021**, *28*, 49362–49382. [PubMed]
2. Getahun, M.J.; Kassie, B.B.; Alemu, T.S. Recent Advances in Biopolymer Synthesis, Properties, & Commercial Applications: A Review. *Process Biochem.* **2024**, *145*, 261–287.
3. Hulkoti, N.I.; Taranath, T.C. Biosynthesis of Nanoparticles Using Microbes—A Review. *Colloids Surf. B Biointerfaces* **2014**, *121*, 474–483.
4. Raouf Hosseini, M.; Nasiri Sarvi, M. Recent Achievements in the Microbial Synthesis of Semiconductor Metal Sulfide Nanoparticles. *Mater. Sci. Semicond. Process.* **2015**, *40*, 293–301.
5. Hernández-Arriaga, A.M.; Campano, C.; Rivero-Buceta, V.; Prieto, M.A. When Microbial Biotechnology Meets Material Engineering. *Microb. Biotechnol.* **2022**, *15*, 149–163. [CrossRef]
6. Abid, N.; Khan, A.M.; Shujait, S.; Chaudhary, K.; Ikram, M.; Imran, M.; Haider, J.; Khan, M.; Khan, Q.; Maqbool, M. Synthesis of Nanomaterials Using Various Top-down and Bottom-up Approaches, Influencing Factors, Advantages, and Disadvantages: A Review. *Adv. Colloid Interface Sci.* **2022**, *300*, 102597. [CrossRef]
7. Arole, D.V.; Munde, S.V. Fabrication of Nanomaterials by Top-down and Bottom-up Approaches—An Overview. *J. Mater. Sci* **2014**, *1*, 89–93.
8. Jain, A.; Bhatia, P.; Chugh, A. Microbial Synthetic Biology for Human Therapeutics. *Syst. Synth. Biol.* **2012**, *6*, 9–22.
9. Mironov, V.; Trusk, T.; Kasyanov, V.; Little, S.; Swaja, R.; Markwald, R. Biofabrication: A 21st Century Manufacturing Paradigm. *Biofabrication* **2009**, *1*, 022001.
10. Chen, H.-G.; Zhang, Y.-H.P. New Biorefineries and Sustainable Agriculture: Increased Food, Biofuels, and Ecosystem Security. *Renew. Sustain. Energy Rev.* **2015**, *47*, 117–132. [CrossRef]
11. Reid, A.; Buchanan, F.; Julius, M.; Walsh, P.J. A Review on Diatom Biosilicification and Their Adaptive Ability to Uptake Other Metals into Their Frustules for Potential Application in Bone Repair. *J. Mater. Chem. B Mater. Biol. Med.* **2021**, *9*, 6728–6737. [CrossRef] [PubMed]
12. Raman, R.; Bashir, R. Biomimicry, Biofabrication, and Biohybrid Systems: The Emergence and Evolution of Biological Design. *Adv. Healthc. Mater.* **2017**, *6* 1700496. [CrossRef] [PubMed]

13. Li, T.; Chang, J.; Zhu, Y.; Wu, C. 3D Printing of Bioinspired Biomaterials for Tissue Regeneration. *Adv. Healthc. Mater.* **2020**, *9*, e2000208. [CrossRef]
14. Ghosh, S.; Ahmad, R.; Zeyaullah, M.; Khare, S.K. Microbial Nano-Factories: Synthesis and Biomedical Applications. *Front. Chem.* **2021**, *9*, 626834. [CrossRef] [PubMed]
15. Dhillon, G.S.; Brar, S.K.; Kaur, S.; Verma, M. Green Approach for Nanoparticle Biosynthesis by Fungi: Current Trends and Applications. *Crit. Rev. Biotechnol.* **2012**, *32*, 49–73. [CrossRef]
16. Yang, Y.; Waterhouse, G.I.N.; Chen, Y.; Sun-Waterhouse, D.; Li, D. Microbial-Enabled Green Biosynthesis of Nanomaterials: Current Status and Future Prospects. *Biotechnol. Adv.* **2022**, *55*, 107914. [CrossRef]
17. Rodrigues, R.C.; Pereira, H.S.; Senra, R.L.; Ribon, A.d.O.B.; Mendes, T.A.d.O. Understanding the Emerging Potential of Synthetic Biology for Food Science: Achievements, Applications and Safety Considerations. *Food Chem. Adv.* **2023**, *3*, 100476. [CrossRef]
18. Tamerler, C.; Sarikaya, M. Molecular Biomimetics: Nanotechnology and Bionanotechnology Using Genetically Engineered Peptides. *Philos. Trans. A Math. Phys. Eng. Sci.* **2009**, *367*, 1705–1726. [CrossRef]
19. Isar, J.; Jain, D.; Joshi, H.; Dhoot, S.; Rangaswamy, V. MICROBIAL Isoprene Production: An Overview. *World J. Microbiol. Biotechnol.* **2022**, *38*, 122. [CrossRef]
20. Kreyenschulte, D.; Krull, R.; Margaritis, A. Recent Advances in Microbial Biopolymer Production and Purification. *Crit. Rev. Biotechnol.* **2014**, *34*, 1–15. [CrossRef]
21. Bittencourt, D.M.d.C.; Oliveira, P.; Michalczechen-Lacerda, V.A.; Rosinha, G.M.S.; Jones, J.A.; Rech, E.L. Bioengineering of Spider Silks for the Production of Biomedical Materials. *Front. Bioeng. Biotechnol.* **2022**, *10*, 958486. [CrossRef] [PubMed]
22. Dhami, N.K.; Reddy, M.S.; Mukherjee, A. Biomineralization of Calcium Carbonates and Their Engineered Applications: A Review. *Front. Microbiol.* **2013**, *4*, 314. [CrossRef] [PubMed]
23. Benzerara, K.; Miot, J.; Morin, G.; Ona-Nguema, G.; Skouri-Panet, F.; Férard, C. Significance, Mechanisms and Environmental Implications of Microbial Biomineralization. *Comptes Rendus Geosci.* **2010**, *343*, 160–167. [CrossRef]
24. Keykha, H.A.; Zangani, A.; Romiani, H.M.; Asadi, A.; Kawasaki, S.; Radmanesh, N. Characterizing Microbial and CO2-Induced Carbonate Minerals: Implications for Soil Stabilization in Sandy Environments. *Minerals* **2023**, *13*, 976. [CrossRef]
25. Elgendy, I.M.; Elkaliny, N.E.; Saleh, H.M.; Darwish, G.O.; Almostafa, M.M.; Metwally, K.; Yahya, G.; Mahmoud, Y.A.-G. Bacteria-Powered Self-Healing Concrete: Breakthroughs, Challenges, and Future Prospects. *J. Ind. Microbiol. Biotechnol.* **2024**, *52*, kuae051. [CrossRef]
26. Rahman, M.M.; Hora, R.N.; Ahenkorah, I.; Beecham, S.; Karim, M.R.; Iqbal, A. State-of-the-Art Review of Microbial-Induced Calcite Precipitation and Its Sustainability in Engineering Applications. *Sustainability* **2020**, *12*, 6281. [CrossRef]
27. Lazo, D.E.; Dyer, L.G.; Alorro, R.D. Silicate, Phosphate and Carbonate Mineral Dissolution Behaviour in the Presence of Organic Acids: A Review. *Miner. Eng.* **2017**, *100*, 115–123. [CrossRef]
28. Tan, K.H. Degradation of Soil Minerals by Organic Acids. In *SSSA Special Publications*; Soil Science Society of America: Madison, WI, USA, 2015; pp. 1–27, ISBN 9780891189121.
29. Moisescu, C.; Ardelean, I.I.; Benning, L.G. The Effect and Role of Environmental Conditions on Magnetosome Synthesis. *Front. Microbiol.* **2014**, *5*, 49. [CrossRef]
30. Perez-Gonzalez, T.; Jimenez-Lopez, C.; Neal, A.L.; Rull-Perez, F.; Rodriguez-Navarro, A.; Fernandez-Vivas, A.; Iañez-Pareja, E. Magnetite Biomineralization Induced by Shewanella Oneidensis. *Geochim. Cosmochim. Acta* **2010**, *74*, 967–979. [CrossRef]
31. Wan, J.; Ji, R.; Liu, J.; Ma, K.; Pan, Y.; Lin, W. Biomineralization in Magnetotactic Bacteria: From Diversity to Molecular Discovery-Based Applications. *Cell Rep.* **2024**, *43*, 114995. [CrossRef]
32. Hoffmann, T.D.; Reeksting, B.J.; Gebhard, S. Bacteria-Induced Mineral Precipitation: A Mechanistic Review. *Microbiology* **2021**, *167*, 001049. [CrossRef] [PubMed]
33. Qin, W.; Wang, C.-Y.; Ma, Y.-X.; Shen, M.-J.; Li, J.; Jiao, K.; Tay, F.R.; Niu, L.-N. Microbe-Mediated Extracellular and Intracellular Mineralization: Environmental, Industrial, and Biotechnological Applications. *Adv. Mater.* **2020**, *32*, e1907833. [CrossRef] [PubMed]
34. Lai Huat, L.; Swee Pin, Y.; Yao, H.; Huangpu, J.; Shao Feng, S.; Swee Sen, T. Environmental Bioremediation: Microremediation, Mycoremediation, and Phytoremediation. *Environ. Claims J.* **2024**, *36*, 186–205. [CrossRef]
35. Lai, H.; Ding, X.; Cui, M.; Zheng, J.; Chen, Z.; Pei, J.; Zhang, J. Mechanisms and Influencing Factors of Biomineralization Based Heavy Metal Remediation: A Review. *Biogeotechnics* **2023**, *1*, 100039. [CrossRef]
36. Cui, M.-J.; Lai, H.-J.; Wu, S.-F.; Chu, J. Comparison of Soil Improvement Methods Using Crude Soybean Enzyme, Bacterial Enzyme or Bacteria Induced Carbonate Precipitation. *Géotechnique* **2022**, *74*, 18–26. [CrossRef]
37. Qabany, A.A.L.; Soga, K. Effect of Chemical Treatment Used in MICP on Engineering Properties of Cemented Soils. In *Bio- and Chemo-Mechanical Processes in Geotechnical Engineering*; ICE Publishing: London, UK, 2014; pp. 107–115, ISBN 9780727760531.
38. Zhou, Y.; Zhao, X.; Jiang, Y.; Ding, C.; Liu, J.; Zhu, C. Synergistic Remediation of Lead Pollution by Biochar Combined with Phosphate Solubilizing Bacteria. *Sci. Total Environ.* **2023**, *861*, 160649. [CrossRef]

39. Yu, X.; Jiang, J. Phosphate Microbial Mineralization Consolidation of Waste Incineration Fly Ash and Removal of Lead Ions. *Ecotoxicol. Environ. Saf.* **2020**, *191*, 110224. [CrossRef]
40. Coelho, E.; Reis, T.A.; Cotrim, M.; Mullan, T.K.; Renshaw, J.; Rizzutto, M.; Corrêa, B. Talaromyces Amestolkiae Uses Organic Phosphate Sources for the Treatment of Uranium-Contaminated Water. *Biometals* **2022**, *35*, 335–348. [CrossRef]
41. Zhang, K.; Xue, Y.; Xu, H.; Yao, Y. Lead Removal by Phosphate Solubilizing Bacteria Isolated from Soil through Biomineralization. *Chemosphere* **2019**, *224*, 272–279. [CrossRef]
42. Zeng, G.; Qiao, S.; Wang, X.; Sheng, M.; Wei, M.; Chen, Q.; Xu, H.; Xu, F. Immobilization of Cadmium by Burkholderia Sp. QY14 through Modified Microbially Induced Phosphate Precipitation. *J. Hazard. Mater.* **2021**, *412*, 125156. [CrossRef]
43. Zeng, G.; Wan, J.; Huang, D.; Hu, L.; Huang, C.; Cheng, M.; Xue, W.; Gong, X.; Wang, R.; Jiang, D. Precipitation, Adsorption and Rhizosphere Effect: The Mechanisms for Phosphate-Induced Pb Immobilization in Soils-A Review. *J. Hazard. Mater.* **2017**, *339*, 354–367. [CrossRef] [PubMed]
44. Yin, T.; Lin, H.; Dong, Y.; Li, B.; He, Y.; Liu, C.; Chen, X. A Novel Constructed Carbonate-Mineralized Functional Bacterial Consortium for High-Efficiency Cadmium Biomineralization. *J. Hazard. Mater.* **2021**, *401*, 123269. [CrossRef] [PubMed]
45. Narayanan, K.B.; Sakthivel, N. Biological Synthesis of Metal Nanoparticles by Microbes. *Adv. Colloid Interface Sci.* **2010**, *156*, 1–13.
46. Vanlalveni, C.; Ralte, V.; Zohmingliana, H.; Das, S.; Anal, J.M.H.; Lallianrawna, S.; Rokhum, S.L. A Review of Microbes Mediated Biosynthesis of Silver Nanoparticles and Their Enhanced Antimicrobial Activities. *Heliyon* **2024**, *10*, e32333. [CrossRef]
47. Sridhar, S. *Microbial Genetics*; Wisdom Press: New Delhi, India, 2013; ISBN 9789382006213.
48. Quester, K.; Avalos-Borja, M.; Castro-Longoria, E. Biosynthesis and Microscopic Study of Metallic Nanoparticles. *Micron* **2013**, *54–55*, 1–27. [CrossRef]
49. Siddiqi, K.S.; Husen, A. Fabrication of Metal Nanoparticles from Fungi and Metal Salts: Scope and Application. *Nanoscale Res. Lett.* **2016**, *11*, 98.
50. Korbekandi, H.; Ashari, Z.; Iravani, S.; Abbasi, S. Optimization of Biological Synthesis of Silver Nanoparticles Using Fusarium Oxysporum. *Iran. J. Pharm. Res.* **2013**, *12*, 289–298.
51. Durán, N.; Marcato, P.D.; Alves, O.L.; Souza, G.I.H.D.; Esposito, E. Mechanistic Aspects of Biosynthesis of Silver Nanoparticles by Several Fusarium Oxysporum Strains. *J. Nanobiotechnology* **2005**, *3*, 8. [CrossRef]
52. Tauseef, A.; Hisam, F.; Hussain, T.; Caruso, A.; Hussain, K.; Châtel, A.; Chénais, B. Nanomicrobiology: Emerging Trends in Microbial Synthesis of Nanomaterials and Their Applications. *J. Clust. Sci.* **2023**, *34*, 639–664.
53. Wu, J.-W.; Ng, I.-S. Biofabrication of Gold Nanoparticles by Shewanella Species. *Bioresour. Bioprocess.* **2017**, *4*, 50. [CrossRef]
54. Grasso, G.; Zane, D.; Dragone, R. Microbial Nanotechnology: Challenges and Prospects for Green Biocatalytic Synthesis of Nanoscale Materials for Sensoristic and Biomedical Applications. *Nanomaterials* **2019**, *10*, 11. [CrossRef] [PubMed]
55. Konishi, Y.; Tsukiyama, T.; Tachimi, T.; Saitoh, N.; Nomura, T.; Nagamine, S. Microbial Deposition of Gold Nanoparticles by the Metal-Reducing Bacterium Shewanella Algae. *Electrochim. Acta* **2007**, *53*, 186–192. [CrossRef]
56. Kitching, M.; Ramani, M.; Marsili, E. Fungal Biosynthesis of Gold Nanoparticles: Mechanism and Scale up: Fungal Biosynthesis of AuNPs. *Microb. Biotechnol.* **2015**, *8*, 904–917. [CrossRef]
57. Hammad, S.E.; El-Rouby, M.N.; Abdel-Aziz, M.M.; El-Sayyad, G.S.; Elshikh, H.H. Endophytic Fungi–assisted Biomass Synthesis of Gold, and Zinc Oxide Nanoparticles for Increasing Antibacterial, and Anticancer Activities. *Biomass Convers. Biorefin.* **2025**, *15*, 2285–2302. [CrossRef]
58. Ahmad, A.; Mukherjee, P.; Senapati, S.; Mandal, D.; Khan, M.I.; Kumar, R.; Sastry, M. Extracellular Biosynthesis of Silver Nanoparticles Using the Fungus Fusarium Oxysporum. *Colloids Surf. B Biointerfaces* **2003**, *28*, 313–318. [CrossRef]
59. Gowramma, B.; Keerthi, U.; Rafi, M.; Muralidhara Rao, D. Biogenic Silver Nanoparticles Production and Characterization from Native Stain of Corynebacterium Species and Its Antimicrobial Activity. *3 Biotech* **2015**, *5*, 195–201. [CrossRef]
60. Rilda, Y.; Rinaldi, R.; Syukri, S.; Armaini, A.; Refinel, R.; Agustien, A.; Pardi, H. Biosynthesis of Zinc Oxide (ZnO) Using the Biomass of Aspergillus Niger to Impart Cotton Fabric with Antimicrobial Properties. *ChemistrySelect* **2022**, *7*, e202103824. [CrossRef]
61. Sharma, J.L.; Dhayal, V.; Sharma, R.K. White-Rot Fungus Mediated Green Synthesis of Zinc Oxide Nanoparticles and Their Impregnation on Cellulose to Develop Environmental Friendly Antimicrobial Fibers. *3 Biotech* **2021**, *11*, 269. [CrossRef]
62. Hajiali, S.; Daneshjou, S.; Daneshjoo, S.; Khajeh, K. Biosynthesis Optimization of Antibacterial-Magnetic Iron Oxide Nanoparticles from Bacillus Megaterium. *Biol. Trace Elem. Res.* **2025**, *203*, 467–484. [CrossRef]
63. Vasanth, V.; Murugesh, K.A.; Susikaran, S. Synthesis of Titanium Dioxide Nanoparticles Using Spirulina Platensis Algae Extract. *Pharma Innov. J.* **2022**, *11*, 266–269.
64. Saha, M.; Sarkar, S.; Sarkar, B.; Sharma, B.K.; Bhattacharjee, S.; Tribedi, P. Microbial Siderophores and Their Potential Applications: A Review. *Environ. Sci. Pollut. Res. Int.* **2016**, *23*, 3984–3999. [CrossRef] [PubMed]
65. De Windt, W.; Boon, N.; Van den Bulcke, J.; Rubberecht, L.; Prata, F.; Mast, J.; Hennebel, T.; Verstraete, W. Biological Control of the Size and Reactivity of Catalytic Pd(0) Produced by Shewanella Oneidensis. *Antonie Van Leeuwenhoek* **2006**, *90*, 377–389. [CrossRef] [PubMed]

66. ELtarahony, M.; Zaki, S.; Kheiralla, Z.; Abd-El-Haleem, D. Biogenic Synthesis of Iron Oxide Nanoparticles via Optimization of Nitrate Reductase Enzyme Using Statistical Experimental Design. *J. Adv. Biotechnol* **2016**, *5*, 667–684.
67. Elblbesy, M.A.; Madbouly, A.; Hamdan, T. Bio-Synthesis of Magnetite Nanoparticles by Bacteria. *Am. J. Nano Res. Appl.* **2014**, *2*, 98.
68. Hossain, S.; Bahreini, B.; Pasteur, E. Magnetotactic Bacteria and Magnetosomes–an Overview. *Mater. Sci. Eng.* **2024**, *8*, 83–100. [CrossRef]
69. Noguchi, Y.; Fujiwara, T.; Yoshimatsu, K.; Fukumori, Y. Iron Reductase for Magnetite Synthesis in the Magnetotactic Bacterium Magnetospirillum Magnetotacticum. *J. Bacteriol.* **1999**, *181*, 2142–2147. [CrossRef]
70. Fdez-Gubieda, M.L.; Muela, A.; Alonso, J.; García-Prieto, A.; Olivi, L.; Fernández-Pacheco, R.; Barandiarán, J.M. Magnetite Biomineralization in Magnetospirillum Gryphiswaldense: Time-Resolved Magnetic and Structural Studies. *ACS Nano* **2013**, *7*, 3297–3305. [CrossRef]
71. Amor, M.; Ceballos, A.; Wan, J.; Simon, C.P.; Aron, A.T.; Chang, C.J.; Hellman, F.; Komeili, A. Magnetotactic Bacteria Accumulate a Large Pool of Iron Distinct from Their Magnetite Crystals. *Appl. Environ. Microbiol.* **2020**, *86*, e01278-20. [CrossRef]
72. Khan, A.A.; Khan, S.; Khan, S.; Rentschler, S.; Laufer, S.; Deigner, H.-P. Biosynthesis of Iron Oxide Magnetic Nanoparticles Using Clinically Isolated Pseudomonas Aeruginosa. *Sci. Rep.* **2021**, *11*, 20503. [CrossRef]
73. Liße, D.; Monzel, C.; Vicario, C.; Manzi, J.; Maurin, I.; Coppey, M.; Piehler, J.; Dahan, M. Engineered Ferritin for Magnetogenetic Manipulation of Proteins and Organelles inside Living Cells. *Adv. Mater.* **2017**, *29*, 1700189. [CrossRef]
74. Zhou, Y.; Gao, Y.; Xie, Q.; Wang, J.; Yue, Z.; Wei, L.; Yang, Y.; Li, L.; Chen, T. Reduction and Transformation of Nanomagnetite and Nanomaghemite by a Sulfate-Reducing Bacterium. *Geochim. Cosmochim. Acta* **2019**, *256*, 66–81. [CrossRef]
75. Pan, J.; Qian, H.; Sun, Y.; Miao, Y.; Zhang, J.; Li, Y. Microbially Synthesized Nanomaterials: Advances and Applications in Biomedicine. *Precis. Med. Eng.* **2025**, *2*, 100019. [CrossRef]
76. Bao, H.; Lu, Z.; Cui, X.; Qiao, Y.; Guo, J.; Anderson, J.M.; Li, C.M. Extracellular Microbial Synthesis of Biocompatible CdTe Quantum Dots. *Acta Biomater.* **2010**, *6*, 3534–3541. [CrossRef]
77. Mi, C.; Wang, Y.; Zhang, J.; Huang, H.; Xu, L.; Wang, S.; Fang, X.; Fang, J.; Mao, C.; Xu, S. Biosynthesis and Characterization of CdS Quantum Dots in Genetically Engineered Escherichia Coli. *J. Biotechnol.* **2011**, *153*, 125–132. [CrossRef]
78. Tusher, M.M.H. Microbial Synthesis of Cadmium Selenide Quantum Dots (CdSe QDs), Influencing Factors and Applications. *Opt. Quantum Electron.* **2023**, *55*, 332. [CrossRef]
79. Alghuthaymi, M.A.; Almoammar, H.; Rai, M.; Said-Galiev, E.; Abd-Elsalam, K.A. Myconanoparticles: Synthesis and Their Role in Phytopathogens Management. *Biotechnol. Biotechnol. Equip.* **2015**, *29*, 221–236. [CrossRef]
80. Su, Z.; Li, X.; Xi, Y.; Xie, T.; Liu, Y.; Liu, B.; Liu, H.; Xu, W.; Zhang, C. Microbe-Mediated Transformation of Metal Sulfides: Mechanisms and Environmental Significance. *Sci. Total Environ.* **2022**, *825*, 153767. [CrossRef]
81. Chen, Y.-L.; Tuan, H.-Y.; Tien, C.-W.; Lo, W.-H.; Liang, H.-C.; Hu, Y.-C. Augmented Biosynthesis of Cadmium Sulfide Nanoparticles by Genetically Engineered Escherichia Coli. *Biotechnol. Prog.* **2009**, *25*, 1260–1266. [CrossRef]
82. Mirzadeh, S.; Darezereshki, E.; Bakhtiari, F.; Fazaelipoor, M.H.; Hosseini, M.R. Characterization of Zinc Sulfide (ZnS) Nanoparticles Biosynthesized by Fusarium Oxysporum. *Mater. Sci. Semicond. Process.* **2013**, *16*, 374–378. [CrossRef]
83. Senapati, S.; Syed, A.; Khan, S.; Pasricha, R.; Khan, M.; Kumar, R.; Ahmad, A. Extracellular Biosynthesis of Metal Sulfide Nanoparticles Using the Fungus Fusarium Oxysporum. *Curr. Nanosci.* **2014**, *10*, 588–595. [CrossRef]
84. Oliva-Arancibia, B.; Órdenes-Aenishanslins, N.; Bruna, N.; Ibarra, P.S.; Zacconi, F.C.; Pérez-Donoso, J.M.; Poblete-Castro, I. Co-Synthesis of Medium-Chain-Length Polyhydroxyalkanoates and CdS Quantum Dots Nanoparticles in Pseudomonas Putida KT2440. *J. Biotechnol.* **2017**, *264*, 29–37. [CrossRef] [PubMed]
85. Hardman, R. A Toxicologic Review of Quantum Dots: Toxicity Depends on Physicochemical and Environmental Factors. *Environ. Health Perspect.* **2006**, *114*, 165–172. [CrossRef] [PubMed]
86. Sobhanan, J.; Jones, P.; Kohara, R.; Sugino, S.; Vacha, M.; Subrahmanyam, C.; Takano, Y.; Lacy, F.; Biju, V. Toxicity of Nanomaterials due to Photochemical Degradation and the Release of Heavy Metal Ions. *Nanoscale* **2020**, *12*, 22049–22058. [CrossRef]
87. Calvo-Olvera, A.; De Donato-Capote, M.; Pool, H.; Rojas-Avelizapa, N.G. In Vitro Toxicity Assessment of Fungal-Synthesized Cadmium Sulfide Quantum Dots Using Bacteria and Seed Germination Models. *J. Environ. Sci. Health A Tox. Hazard. Subst. Environ. Eng.* **2021**, *56*, 713–722. [CrossRef]
88. Feng, J.; Chen, B.; Sun, W.; Wang, Y. Microbial Induced Calcium Carbonate Precipitation Study Using Bacillus Subtilis with Application to Self-Healing Concrete Preparation and Characterization. *Constr. Build. Mater.* **2021**, *280*, 122460. [CrossRef]
89. Ahmad, S.S.E.; Elmahdy, M.A.R.; ELShami, A.A.; Yousry, E.-S.M. Bacterial Sustainable Concrete for Repair and Rehabilitation of Structural Cracks. *J. Sustain. Cem.-Based Mater.* **2024**, *20*, e03188. [CrossRef]
90. Fan, Q.; Fan, L.; Quach, W.-M.; Zhang, R.; Duan, J.; Sand, W. Application of Microbial Mineralization Technology for Marine Concrete Crack Repair: A Review. *J. Build. Eng.* **2023**, *69*, 106299. [CrossRef]
91. Renninger, N.; Knopp, R.; Nitsche, H.; Clark, D.S.; Keasling, J.D. Uranyl Precipitation by Pseudomonas Aeruginosa via Controlled Polyphosphate Metabolism. *Appl. Environ. Microbiol.* **2004**, *70*, 7404–7412. [CrossRef]

92. Liang, X.; Hillier, S.; Pendlowski, H.; Gray, N.; Ceci, A.; Gadd, G.M. Uranium Phosphate Biomineralization by Fungi. *Environ. Microbiol.* **2015**, *17*, 2064–2075. [CrossRef]
93. Manobala, T.; Shukla, S.K.; Rao, T.S.; Kumar, M.D. Uranium Sequestration by Biofilm-Forming Bacteria Isolated from Marine Sediment Collected from Southern Coastal Region of India. *Int. Biodeterior. Biodegrad.* **2019**, *145*, 104809. [CrossRef]
94. Toster, J.; Harnagea, C.; Iyer, K.; Rosei, F.; Raston, C. Controlling Anatase Coating of Diatom Frustules by Varying the Binding Layer. *CrystEngComm* **2012**, *14*, 3446–3450. [CrossRef]
95. Lahiri, D.; Nag, M.; Dutta, B.; Dey, A.; Sarkar, T.; Pati, S.; Edinur, H.A.; Abdul Kari, Z.; Mohd Noor, N.H.; Ray, R.R. Bacterial Cellulose: Production, Characterization, and Application as Antimicrobial Agent. *Int. J. Mol. Sci.* **2021**, *22*, 12984. [CrossRef] [PubMed]
96. Islam, M.U.; Ullah, M.W.; Khan, S.; Shah, N.; Park, J.K. Strategies for Cost-Effective and Enhanced Production of Bacterial Cellulose. *Int. J. Biol. Macromol.* **2017**, *102*, 1166–1173. [CrossRef]
97. Kadier, A.; Ilyas, R.A.; Huzaifah, M.R.M.; Harihastuti, N.; Sapuan, S.M.; Harussani, M.M.; Azlin, M.N.M.; Yuliasni, R.; Ibrahim, R.; Atikah, M.S.N.; et al. Use of Industrial Wastes as Sustainable Nutrient Sources for Bacterial Cellulose (BC) Production: Mechanism, Advances, and Future Perspectives. *Polymers* **2021**, *13*, 3365. [CrossRef]
98. Song, J.; Winkeljann, B.; Lieleg, O. Biopolymer-based Coatings: Promising Strategies to Improve the Biocompatibility and Functionality of Materials Used in Biomedical Engineering. *Adv. Mater. Interfaces* **2020**, *7*, 2000850. [CrossRef]
99. Rehm, B.H.; Moradali, M.F. (Eds.) *Biopolymers for Biomedical and Biotechnological Applications*; Wiley: Hoboken, NJ, USA, 2021; ISBN 9783527818310.
100. Rosalam, S.; England, R. Review of Xanthan Gum Production from Unmodified Starches by *Xanthomonas Comprestris* sp. *Enzym. Microb. Technol.* **2006**, *39*, 197–207. [CrossRef]
101. Rond'ošová, S.; Legerská, B.; Chmelová, D.; Ondrejovič, M.; Miertus, S. Optimization of Growth Conditions to Enhance PHA Production by Cupriavidus Necator. *Fermentation* **2022**, *8*, 451. [CrossRef]
102. Zytner, P.; Kumar, D.; Elsayed, A.; Mohanty, A.; Ramarao, B.V.; Misra, M. A Review on Polyhydroxyalkanoate (PHA) Production through the Use of Lignocellulosic Biomass. *RSC Sustain.* **2023**, *1*, 2120–2134. [CrossRef]
103. Policastro, G.; Panico, A.; Fabbricino, M. Improving Biological Production of poly(3-Hydroxybutyrate-Co-3-Hydroxyvalerate) (PHBV) Co-Polymer: A Critical Review. *Rev. Environ. Sci. Biotechnol.* **2021**, *20*, 479–513. [CrossRef]
104. Tian, X.; Chen, H.; Liu, H.; Chen, J. Recent Advances in Lactic Acid Production by Lactic Acid Bacteria. *Appl. Biochem. Biotechnol.* **2021**, *193*, 4151–4171. [CrossRef]
105. Dessie, W.; Xin, F.; Zhang, W.; Jiang, Y.; Wu, H.; Ma, J.; Jiang, M. Opportunities, Challenges, and Future Perspectives of Succinic Acid Production by Actinobacillus Succinogenes. *Appl. Microbiol. Biotechnol.* **2018**, *102*, 9893–9910. [PubMed]
106. Pateraki, C.; Patsalou, M.; Vlysidis, A.; Kopsahelis, N.; Webb, C.; Koutinas, A.A.; Koutinas, M. Actinobacillus Succinogenes: Advances on Succinic Acid Production and Prospects for Development of Integrated Biorefineries. *Biochem. Eng. J.* **2016**, *112*, 285–303. [CrossRef]
107. Eiteman, M.A.; Ramalingam, S. Microbial Production of Lactic Acid. *Biotechnol. Lett.* **2015**, *37*, 955–972. [CrossRef]
108. Juturu, V.; Wu, J.C. Microbial Production of Lactic Acid: The Latest Development. *Crit. Rev. Biotechnol.* **2016**, *36*, 967–977. [CrossRef]
109. Amara, N.I.; Chukwuemeka, E.S.; Obiajulu, N.O.; Chukwuma, O.J. Yeast-Driven Valorization of Agro-Industrial Wastewater: An Overview. *Environ. Monit. Assess.* **2023**, *195*, 1252. [CrossRef]
110. Saravanan, M.; Gopinath, V.; Chaurasia, M.K.; Syed, A.; Ameen, F.; Purushothaman, N. Green Synthesis of Anisotropic Zinc Oxide Nanoparticles with Antibacterial and Cytofriendly Properties. *Microb. Pathog.* **2018**, *115*, 57–63. [CrossRef]
111. Jacob, P.J.; Masarudin, M.J.; Hussein, M.Z.; Rahim, R.A. Optimization of Process Parameters Influencing the Sustainable Construction of Iron Oxide Nanoparticles by a *Novel Tropical Wetlands Streptomyces* spp. *J. Clean. Prod.* **2019**, *232*, 193–202. [CrossRef]
112. Rajput, V.D.; Minkina, T.; Kimber, R.L.; Singh, V.K.; Shende, S.; Behal, A.; Sushkova, S.; Mandzhieva, S.; Lloyd, J.R. Insights into the Biosynthesis of Nanoparticles by the Genus *Shewanella*. *Appl. Environ. Microbiol.* **2021**, *87*, e01390-21. [CrossRef]
113. Khort, A.; Brookman-Amissah, M.; Hedberg, J.; Chang, T.; Mei, N.; Lundberg, A.; Sturve, J.; Blomberg, E.; Odnevall, I. Influence of Natural Organic Matter on the Transformation of Metal and Metal Oxide Nanoparticles and Their Ecotoxic Potency In Vitro. *NanoImpact* **2022**, *25*, 100386. [CrossRef]
114. Mohammadkazemi, F.; Azin, M.; Ashori, A. Production of Bacterial Cellulose Using Different Carbon Sources and Culture Media. *Carbohydr. Polym.* **2015**, *117*, 518–523. [CrossRef]
115. Hu, H.; Catchmark, J.M.; Demirci, A. Effects of Pullulan Additive and Co-Culture of Aureobasidium Pullulans on Bacterial Cellulose Produced by Komagataeibacter Hansenii. *Bioprocess Biosyst. Eng.* **2022**, *45*, 573–587.
116. Chavan, S.; Baig, M. Relationship of Biomass and Xanthan Gum Production by Xanthomonas Campestris: Optimization of Parameters. *Br. Biotechnol. J.* **2016**, *11*, 1–8. [CrossRef]

117. Trindade, R.A.; Munhoz, A.P.; Burkert, C.A.V. Impact of a Carbon Source and Stress Conditions on Some Properties of Xanthan Gum Produced by Xanthomonas Campestris Pv. Mangiferaeindicae. *Biocatal. Agric. Biotechnol.* **2018**, *15*, 167–172. [CrossRef]
118. Zulfiqar, Z.; Khan, R.R.M.; Summer, M.; Saeed, Z.; Pervaiz, M.; Rasheed, S.; Shehzad, B.; Kabir, F.; Ishaq, S. Plant-Mediated Green Synthesis of Silver Nanoparticles: Synthesis, Characterization, Biological Applications, and Toxicological Considerations: A Review. *Biocatal. Agric. Biotechnol.* **2024**, *57*, 103121. [CrossRef]
119. Ibrahim, S.; Ahmad, Z.; Manzoor, M.Z.; Mujahid, M.; Faheem, Z.; Adnan, A. Optimization for Biogenic Microbial Synthesis of Silver Nanoparticles through Response Surface Methodology, Characterization, Their Antimicrobial, Antioxidant, and Catalytic Potential. *Sci. Rep.* **2021**, *11*, 770.
120. Nejadmansouri, M.; Shad, E.; Razmjooei, M.; Safdarianghomsheh, R.; Delvigne, F.; Khalesi, M. Production of Xanthan Gum Using Immobilized Xanthomonas Campestris Cells: Effects of Support Type. *Biochem. Eng. J.* **2020**, *157*, 107554. [CrossRef]
121. Li, P.; Li, T.; Zeng, Y.; Li, X.; Jiang, X.; Wang, Y.; Xie, T.; Zhang, Y. Biosynthesis of Xanthan Gum by Xanthomonas Campestris LRELP-1 Using Kitchen Waste as the Sole Substrate. *Carbohydr. Polym.* **2016**, *151*, 684–691. [CrossRef]
122. Baker, S.; Mohan Kumar, K.; Santosh, P.; Rakshith, D.; Satish, S. Extracellular Synthesis of Silver Nanoparticles by Novel Pseudomonas Veronii AS41G Inhabiting Annona Squamosa L. and Their Bactericidal Activity. *Spectrochim. Acta A Mol. Biomol Spectrosc.* **2015**, *136 Pt C*, 1434–1440. [CrossRef]
123. Saravana Kumar, P.; Balachandran, C.; Duraipandiyan, V.; Ramasamy, D.; Ignacimuthu, S.; Al-Dhabi, N.A. Extracellular Biosynthesis of Silver Nanoparticle Using Streptomyces Sp. 09 PBT 005 and Its Antibacterial and Cytotoxic Properties. *Appl. Nanosci.* **2015**, *5*, 169–180. [CrossRef]
124. Guo, E.; Chen, G.; Yu, D.; Qiu, Y.; Li, S.; Yu, Y. Optimization of Dry Anaerobic Co-Fermentation of Sludge and Corn Straw with Magnetite (Fe_3O_4). *J. Environ. Chem. Eng.* **2022**, *10*, 108618.
125. Tyagi, S.; Tyagi, P.K.; Gola, D.; Chauhan, N.; Bharti, R.K. Extracellular Synthesis of Silver Nanoparticles Using Entomopathogenic Fungus: Characterization and Antibacterial Potential. *SN Appl. Sci.* **2019**, *1*, 1545. [CrossRef]
126. Balashanmugam, P.; Balakumaran, M.D.; Murugan, R.; Dhanapal, K.; Kalaichelvan, P.T. Phytogenic Synthesis of Silver Nanoparticles, Optimization and Evaluation of in Vitro Antifungal Activity against Human and Plant Pathogens. *Microbiol. Res.* **2016**, *192*, 52–64. [CrossRef] [PubMed]
127. Krishna Leela, J.; Sharma, G. Studies on Xanthan Production from Xanthomonas Campestris. *Bioprocess Biosyst. Eng.* **2000**, *23*, 687–689.
128. Chakraborty, I.; Feliu, N.; Roy, S.; Dawson, K.; Parak, W.J. Protein-Mediated Shape Control of Silver Nanoparticles. *Bioconjug. Chem.* **2018**, *29*, 1261–1265.
129. Chakraborty, I.; Parak, W.J. Protein-induced Shape Control of Noble Metal Nanoparticles. *Adv. Mater. Interfaces* **2019**, *6*, 1801407.
130. Murphy, C.J.; Chang, H.-H.; Falagan-Lotsch, P.; Gole, M.T.; Hofmann, D.M.; Hoang, K.N.L.; McClain, S.M.; Meyer, S.M.; Turner, J.G.; Unnikrishnan, M.; et al. Virus-Sized Gold Nanorods: Plasmonic Particles for Biology. *Acc. Chem. Res.* **2019**, *52*, 2124–2135.
131. Duman, E.; Şahin Kehribar, E.; Ahan, R.E.; Yuca, E.; Şeker, U.Ö.Ş. Biomineralization of Calcium Phosphate Crystals Controlled by Protein-Protein Interactions. *ACS Biomater. Sci. Eng.* **2019**, *5*, 4750–4763.
132. Mohammadi, P.; Gandier, J.-A.; Nonappa; Wagermaier, W.; Miserez, A.; Penttilä, M. Bioinspired Functionally Graded Composite Assembled Using Cellulose Nanocrystals and Genetically Engineered Proteins with Controlled Biomineralization. *Adv. Mater.* **2021**, *33*, e2102658.
133. Nie, Z.; Zhang, Y.; Tang, R.; Wang, X. Biomimetic Mineralization: An Emerging Organism Engineering Strategy for Biomedical Applications. *J. Inorg. Biochem.* **2022**, *232*, 111815.
134. Dickerson, M.B.; Sandhage, K.H.; Naik, R.R. Protein- and Peptide-Directed Syntheses of Inorganic Materials. *Chem. Rev.* **2008**, *108*, 4935–4978.
135. Tamerler, C.; Kacar, T.; Sahin, D.; Fong, H.; Sarikaya, M. Genetically Engineered Polypeptides for Inorganics: A Utility in Biological Materials Science and Engineering. *Mater. Sci. Eng. C Mater. Biol. Appl.* **2007**, *27*, 558–564.
136. Qin, K.; Zheng, Z.; Wang, J.; Pan, H.; Tang, R. Biomineralization Strategy: From Material Manufacturing to Biological Regulation. *Giant* **2024**, *19*, 100317. [CrossRef]
137. Miserez, A.; Yu, J.; Mohammadi, P. Protein-Based Biological Materials: Molecular Design and Artificial Production. *Chem. Rev.* **2023**, *123*, 2049–2111.
138. Schulz, A.; Wang, H.; Rijn, P.; Böker, A. Synthetic Inorganic Materials by Mimicking Biomineralization Processes Using Native and Non-Native Protein Functions. *J. Mater. Chem.* **2011**, *21*, 18903–18918.
139. Wang, Y.; Guo, J.; Zhou, L.; Ye, C.; Omenetto, F.G.; Kaplan, D.L.; Ling, S. Design, Fabrication, and Function of Silk-Based Nanomaterials. *Adv. Funct. Mater.* **2018**, *28*, 1805305.
140. Chandraker, S.K.; Kumar, R. Biogenic Biocompatible Silver Nanoparticles: A Promising Antibacterial Agent. *Biotechnol. Genet. Eng. Rev.* **2024**, *40*, 3113–3147. [CrossRef]

141. Russo, P.; De Simone, N.; Capozzi, V.; Mohedano, M.L.; Ruiz-Masó, J.Á.; Del Solar, G.; López, P.; Spano, G. Selection of Riboflavin Overproducing Strains of Lactic Acid Bacteria and Riboflavin Direct Quantification by Fluorescence. *Methods Mol. Biol.* **2021**, *2280*, 3–14.
142. Kant, G.; Pandey, A.; Kumari, S.; Bux, F.; Srivastava, S. Enhancing Bio-Isoprene Production in Escherichia Coli through a Combinatorial Optimization Approach. *Process Biochem.* **2024**, *144*, 210–219.
143. De Carluccio, G.; Fusco, V.; di Bernardo, D. Engineering a Synthetic Gene Circuit for High-Performance Inducible Expression in Mammalian Systems. *Nat. Commun.* **2024**, *15*, 3311.
144. Wang, Y.; Liu, Y.; Li, J.; Chen, Y.; Liu, S.; Zhong, C. Engineered Living Materials (ELMs) Design: From Function Allocation to Dynamic Behavior Modulation. *Curr. Opin. Chem. Biol.* **2022**, *70*, 102188.
145. Sadeghianmaryan, A.; Ahmadian, N.; Wheatley, S.; Alizadeh Sardroud, H.; Nasrollah, S.A.S.; Naseri, E.; Ahmadi, A. Advancements in 3D-Printable Polysaccharides, Proteins, and Synthetic Polymers for Wound Dressing and Skin Scaffolding—A Review. *Int. J. Biol. Macromol.* **2024**, *266*, 131207. [CrossRef] [PubMed]
146. O'Shea, D.G.; Curtin, C.; O'Brien, F. Articulation Inspired by Nature: A Review of Biomimetic and Biologically Active 3D Printed Scaffolds for Cartilage Tissue Engineering. *Biomater. Sci.* **2022**, *10*, 2462–2483. [CrossRef] [PubMed]
147. Heveran, C.M.; Williams, S.L.; Qiu, J.; Artier, J.; Hubler, M.H.; Cook, S.M.; Cameron, J.C.; Srubar, W.V., III. Biomineralization and Successive Regeneration of Engineered Living Building Materials. *Matter* **2020**, *2*, 481–494. [CrossRef]
148. Bi, S.; He, C.; Liu, R.; Zhao, X.; Liu, J.; Gu, J.; Liu, W.; Yan, B. On-Demand Dissociable Antibacterial Self-Healing Hydrogel with Rapid Arginine-Related Biosynthesis Capacity to Promote Full-Thickness Bacteria-Infected Wound Healing. *Ind. Eng. Chem. Res.* **2023**, *62*, 7492–7503. [CrossRef]
149. Yang, M.; Yang, F.; Chen, W.; Liu, S.; Qiu, L.; Chen, J. Bacteria-Mediated Cancer Therapies: Opportunities and Challenges. *Biomater. Sci.* **2021**, *9*, 5732–5744. [CrossRef]
150. Fan, J.-X.; Niu, M.-T.; Qin, Y.-T.; Sun, Y.-X.; Zhang, X.-Z. Progress of Engineered Bacteria for Tumor Therapy. *Adv. Drug Deliv. Rev.* **2022**, *185*, 114296. [CrossRef]
151. Lin, D.; Feng, X.; Mai, B.; Li, X.; Wang, F.; Liu, J.; Liu, X.; Zhang, K.; Wang, X. Bacterial-Based Cancer Therapy: An Emerging Toolbox for Targeted Drug/gene Delivery. *Biomaterials* **2021**, *277*, 121124. [CrossRef]
152. Jiang, W.; Wu, Z.; Gao, Z.; Wan, M.; Zhou, M.; Mao, C.; Shen, J. Artificial Cells: Past, Present and Future. *ACS Nano* **2022**, *16*, 15705–15733. [CrossRef]
153. Choi, S. Electrogenic Bacteria Promise New Opportunities for Powering, Sensing, and Synthesizing. *Small* **2022**, *18*, e2107902. [CrossRef]
154. Shin, H.J. Genetically Engineered Microbial Biosensors for in Situ Monitoring of Environmental Pollution. *Appl. Microbiol. Biotechnol.* **2011**, *89*, 867–877. [CrossRef]
155. Girotti, S.; Ferri, E.N.; Fumo, M.G.; Maiolini, E. Monitoring of Environmental Pollutants by Bioluminescent Bacteria. *Anal. Chim. Acta* **2008**, *608*, 2–29. [CrossRef] [PubMed]
156. Belkin, S. Microbial Whole-Cell Sensing Systems of Environmental Pollutants. *Curr. Opin. Microbiol.* **2003**, *6*, 206–212. [CrossRef] [PubMed]
157. Belkin, S. Genetically Engineered Microorganisms for Pollution Monitoring. In *NATO Science Series*; Springer: Dordrecht, The Netherlands, 2007; pp. 147–160, ISBN 9781402047268.
158. Parakh, P.D.; Nanda, S.; Kozinski, J.A. Eco-Friendly Transformation of Waste Biomass to Biofuels. *Curr. Biochem. Eng.* **2020**, *6*, 120–134. [CrossRef]
159. Zeng, J.; Zeng, H.; Wang, Z. Review on Technology of Making Biofuel from Food Waste. *Int. J. Energy Res.* **2022**, *46*, 10301–10319. [CrossRef]
160. Zhang, F.-L.; Zhang, L.; Zeng, D.-W.; Liao, S.; Fan, Y.; Champreda, V.; Runguphan, W.; Zhao, X.-Q. Engineering Yeast Cell Factories to Produce Biodegradable Plastics and Their Monomers: Current Status and Prospects. *Biotechnol. Adv.* **2023**, *68*, 108222. [CrossRef]
161. Kitamoto, H.K.; Shinozaki, Y.; Cao, X.-H.; Morita, T.; Konishi, M.; Tago, K.; Kajiwara, H.; Koitabashi, M.; Yoshida, S.; Watanabe, T.; et al. Phyllosphere Yeasts Rapidly Break down Biodegradable Plastics. *AMB Express* **2011**, *1*, 44. [CrossRef]
162. Naidu, S.; Singh, I.K.; Singh, A. Microbial Synthesis of Magnetic Nanoparticles for Plant Science and Agriculture. *Plant Nano Biol.* **2023**, *4*, 100036. [CrossRef]
163. Sadh, P.K.; Chawla, P.; Kumar, S.; Das, A.; Kumar, R.; Bains, A.; Sridhar, K.; Duhan, J.S.; Sharma, M. Recovery of Agricultural Waste Biomass: A Path for Circular Bioeconomy. *Sci. Total Environ.* **2023**, *870*, 161904. [CrossRef]
164. Chen, C.-C.; Dai, L.; Ma, L.; Guo, R. Enzymatic Degradation of Plant Biomass and Synthetic Polymers. *Nat. Rev. Chem.* **2020**, *4*, 114–126. [CrossRef]
165. Banerjee, A.; Chatterjee, K. Enzymatic Degradation of Polymers: A Brief Review. *Mater. Sci. Technol.* **2014**, *30*, 567–573. [CrossRef]
166. de Jesús Montaño López, J.; Duran, L.A.; Avalos, J. Physiological Limitations and Opportunities in Microbial Metabolic Engineering. *Nat. Rev. Microbiol.* **2021**, *20*, 35–48. [CrossRef] [PubMed]

167. Varjani, S.J.; Parameswaran, B.; Kumar, S.; Khare, S.K. (Eds.) *Biosynthetic Technology and Environmental Challenges*, 1st ed.; Energy, Environment, and Sustainability; Springer: Singapore, 2017; ISBN 9789811074332.
168. Ahmed, I.A. Ethical Issues of Microbial Products for Industrialization. In *Interdisciplinary Biotechnological Advances*; Springer Nature: Singapore, 2023; pp. 393–411, ISBN 9789819917365.
169. Meena, S.S.; Mohanty, A. Ethical, Patent, and Regulatory Issues in Microbial Engineering. In *Engineering of Microbial Biosynthetic Pathways*; Springer: Singapore, 2020; pp. 133–142, ISBN 9789811526039.
170. Casadevall, A. The Future of Biological Warfare: The Future of Biological Warfare. *Microb. Biotechnol.* **2012**, *5*, 584–587. [CrossRef] [PubMed]
171. Granato, E.T.; Meiller-Legrand, T.A.; Foster, K.R. The Evolution and Ecology of Bacterial Warfare. *Curr. Biol.* **2019**, *29*, R521–R537. [CrossRef] [PubMed]

BioChem

MDPI

Review

The Mini-TRH Test, Dopamine Transmission, and Schizophrenia Symptoms

Johan Spoov

Independent Researcher, 00240 Helsinki, Finland; jspoov@gmail.com

Abstract: Studies in animals and humans suggested that the tonic dopamine inhibition of prolactin release may be estimated by submaximal prolactin stimulation by thyrotropin-releasing hormone (TRH), the mini-TRH test. Because patients with schizophrenia may be more vulnerable to stress-induced elevations of prolactin, great care was taken to avoid stress-induced increases in prolactin, including applying local anaesthesia before blood extraction in our psychotic patients. Basal prolactin levels were in the reference range in all psychotic patients studied by us and were not higher in male patients than in normal men. Results of the mini-TRH test suggested that in acute patients with non-affective psychoses, everyday memory problems, non-paranoid delusions, and first-rank symptoms, but not other Comprehensive Psychopathological Rating Scale (CPRS) positive symptoms, could correlate with decreasing dopamine transmission in lactotrophs. In acute patients with first-episode schizophrenia, increasing negative disorganisation symptoms might correlate with increasing dopamine transmission. In first-episode patients, a hypersensitivity of the TRH response was detected, which could indicate that variability in the basal prolactin levels may confound the interpretation of the mini-TRH response. To avoid that, a smaller dose of TRH was recommended in first-episode patients. Studies using other estimates of basal dopamine release suggested that striatal dopamine transmission reflected delusions and hallucinations but not other Positive and Negative Symptom Scale (PANSS) positive symptoms. Including a wide range of symptoms in the PANSS positive scale may reduce its specificity for assessing basal dopamine transmission, although the scale remains useful for tracking treatment response.

Keywords: schizophrenia; dopamine; TRH; positive symptoms

1. Introduction

The dopamine hypothesis of schizophrenia evolved from observations that amphetamine can induce psychotic symptoms and their improvement by dopamine antagonists. Estimating the basal dopamine level could provide more direct evidence for the hypothesis, which has evolved to include both increased and decreased dopamine transmission dependent on the site of the brain, as reviewed in this text. Much of previous work has concentrated on finding a difference in dopamine transmission between acutely stressed patients with schizophrenia and non-stressed controls, which complicates interpretation. Correlative associations of estimates of dopamine transmission with psychotic symptoms could provide further evidence not necessarily revealed in mean differences in patients and controls. In the following text, such correlative analyses of the prolactin responses to submaximal TRH stimulation are reviewed with an emphasis of minimizing confounders to aid future researchers. Other tests estimating basal dopamine transmission

https://doi.org/10.3390/biochem5020015

are reviewed as well. Currently, only a limited set of schizophrenia symptoms have been directly associated with basal dopamine transmission.

2. The Mini-TRH Test, Studies in Animals and Normal Men

Tuberoinfundibular dopamine (TIDA) is the main prolactin-inhibiting factor in the rat and humans. The medial basal hypothalamus secretes TIDA to stimulate dopamine-2 receptors on the pituitary lactotrophs via the pituitary stalk circulation, and in the rat pituitary plasma, dopamine contributed to an estimated 40–70% of the total prolactin-inhibiting activity [1]. In vitro, a small dose of thyrotropin-releasing hormone (TRH) almost completely antagonised dopamine inhibition of prolactin release, while a larger dose only partially antagonised it [2], suggesting that prolactin responses to smaller, more physiologic doses could be more specific than larger doses in reflecting dopamine transmission. Indeed, in stalk-sectioned monkeys, an increased prolactin response to submaximal stimulation by TRH suggested a role for decreased dopamine in the response [1]. The submaximal response at 15 min may estimate, predominantly or entirely, hypophyseal effects of TRH on prolactin release because TRH passes the blood–brain barrier poorly, and the prolactin short feedback is much too slow to increase TIDA release at 15 min [1].

Our hypothesis was that submaximal prolactin responses to TRH could uncover continuing activity of dopamine-mediated inhibition of prolactin release, i.e., tonic dopamine inhibition of prolactin release in humans. We used the basal prolactin level and the increase in plasma prolactin 15 min after the administration of an i.v. bolus of 12.5 mcg (the mini-TRH test) and the maximal prolactin response to TRH to estimate endogenous dopamine transmission in a series of correlative studies in men. Reproducible doses could be administered using a small insulin needle. We compared correlations of the above tests to other dopamine tests: prolactin response to 0.5 mg i.m. haloperidol, which may reflect dopamine receptor sensitivity [3], 24 h urinary excretions of 17-ketogenic steroids (glucocorticoids), and homovanillyl acid (HVA). Blood samples for the combined TRH test assessing both submaximal and maximal prolactin responses to TRH and a separate haloperidol test [4] were drawn on separate weekends after one hour of rest in a quiet laboratory with no interruptions between 11 a.m. and 1 p.m. to reduce stress and the circadian variability of prolactin release [4]. None of the subjects had an endocrinological disease or were on continuous medication [5]. The subjects had not taken any medication for at least four weeks before the study [5]. The laboratory was familiar to most of the subjects, and only one to three subjects were simultaneously present [6]. They were asked to lie down and relax, and conversation was discouraged [6]. In humans, diet had no significant effects on urinary HVA unless it included large amounts of monoamines [5]. The subjects collected their 24 h urine specimens consuming their habitual diets under relaxed conditions during weekends [5,6]. Because the blood and the urine specimens were (with one exception) obtained on separate days [5], it is not likely that stress reactions associated with the prolactin tests could provoke significant correlations with the urinary test results. The number of days between the 24 h urine collection of HVA and the mini-TRH test was 57.4 ± 11.8 (mean $\pm$ SEM), and the number of days between the urine collection and the haloperidol test was 142 ± 23.8 (mean $\pm$ SEM) [5].

Because the United States Food and Drug Administration did not approve the use of i.v. haloperidol, we used prolactin responses to i.m. haloperidol, the relatively specific dopamine antagonist. Studies in normal men using 1 to 1.5 mg i.m. haloperidol produced maximal prolactin responses and, also, consistently provoked various degrees of sedation and/or restlessness [1]. On the other hand, 0.5 mg i.m. haloperidol produced a prolactin increase of at least 50% (mean increase 60%) of the maximal prolactin increase in 90 min with no noticeable side effects [1]. Prolactin-induced increase in TIDA turnover (the short feedback) started to increase TIDA release after 1 to 3 h, reaching significant levels thereafter [1]. At the 90-minute time point, prolactin feedback loops affecting TIDA release were only starting to become operative.

The Mini-TRH Test in Normal Men

The subjects were 28 male volunteers aged less than 50 years. We did not study normal females. After being provided information about the study, two of the subjects refused to participate in the haloperidol test, and in one, the prolactin response to 200 mcg TRH was not obtained. In 25 normal men, the mini-TRH test, but not the basal prolactin level or the prolactin increase to 200 mcg TRH, significantly correlated with prolactin response to i.m. haloperidol (r = +0.57, p = 0.003) [4] and with 24 h urinary excretion of glucocorticoids (r = −0.64, p < 0.001) [5]. The finding of no correlation between the basal prolactin level and plasma prolactin response to 0.5 mg i.m. haloperidol replicated findings from other studies [4]. Consistent with our results, pre-incubation with glucocorticoids decreased prolactin response to submaximal stimulation by TRH in the rat hypophysis [1]. In 28 normal men, the mini-TRH test, but not the basal prolactin level, negatively correlated with 24 h urinary excretion of HVA (r = −0.48, p = 0.010), even after controlling for activity of the sympathetic nervous system as assessed by 24 h urinary excretion of vanillylmandelic acid (r = −0.44, p = 0.019) [6]. In 27 men, there was no correlation between the maximal prolactin response to TRH and urinary HVA [7]. Our results suggested that factors other than dopamine inhibition of prolactin release determined the ceiling of TRH-induced prolactin release. Dopamine inhibition of prolactin release involves an integrated effect of TIDA levels and dopamine-2 receptor sensitivity on lactotrophs. Our results raised the possibility that a generalised dopamine turnover extended to TIDA. In 26 normal men with both the mini-TRH test and the haloperidol test performed in our laboratory, removing the linear effects of 24 h urinary corticosteroids reduced the correlation between the mini-TRH test and 24 h urinary HVA from r = −0.53, p = 0.0054 to r = –0.16, p = 0.43 and between the haloperidol test and 24 h urinary HVA from r = −0.25, p = 0.22 to r = +0.04 [3]. The widespread origin of urinary HVA raised the possibility that glucocorticoids may integrate TIDA release in other structures sensitive to these steroids [3]. In 25 of the normal men, the maximal prolactin response to TRH significantly correlated with the basal prolactin level in the TRH test (r = +0.52, p < 0.01, but not with the basal prolactin level drawn another day in the haloperidol test (r = +0.04) [5], suggesting that the maximal prolactin response to TRH may be associated with fluctuations in the basal prolactin level. These fluctuations may decrease the accuracy of the basal prolactin level and the maximal prolactin response to TRH when estimating the tonic inhibition of dopamine on prolactin secretion. Prolactin secretory bursts occur every 42 to 65 min, independent of prolactin levels and sex, and the bursts contribute at least as much to the prolactin level as the time-invariant mode of prolactin secretion [1]. A single prolactin level may be of limited value as a trait variable. On the other hand, the time-invariant mode could reflect the tonic dopamine inhibition of prolactin release.

3. Controlling Stress-Induced Prolactin Release

In clinical practice, psychic stress at venipuncture and the inherent pulsatory nature of prolactin release may trigger elevations in blood prolactin levels, leading to a false diagnosis of hyperprolactinemia. The Endocrine Society recommends repeat prolactin testing on a different day at 15 to 20 min intervals to rule out pulsatory prolactin levels when in doubt [8]. Even one normal prolactin level in the reference range should rule out hyperprolactinemia [8]. Another way to control stress-induced prolactin pulses is intravenous cannulation followed by a rest period of 30 to 90 min [8]. A third way is to use local anaesthesia followed by a rest period, which was used in studies with the mini-TRH test in psychotic patients. We took great care to avoid not only stress provoked by venipuncture, but also by apprehension [8]. We used local anaesthetic cream on a vein in the morning of the mini-TRH test and allowed no interruptions during the rest periods before the mini-TRH test. The same investigator who had introduced himself the day before applied the anaesthesia on the day of the test and allowed no interruptions during the rest period [8]. In the first study in psychotic patients, the rest period before the mini-TRH test was one and a half to two hours after local anaesthesia [9]. I found only one study to examine physical stress-induced increases in prolactin levels in schizophrenia [10]. This study reported significantly higher prolactin increases in men with schizophrenia than in control subjects after immersion of a hand in ice water for 2 min [10]. The increase was the same duration as in the control subjects, and prolactin levels returned from stress-induced levels to pre-test levels in 30 min. Because of this study and to avoid an unnecessarily long rest period in a test with a potential use in clinical practice, we used a 30 min rest period in our second study [11].

4. The Mini-TRH Test in Psychotic Subjects

We conducted the mini-TRH test from 10:00 to 11:30 a.m. in all our psychotic patients to avoid the nocturnal rise in pulsatory prolactin secretion, which is stabilised in about 2 h after awakening [1]. In every psychotic patient, the basal prolactin level was in the reference range and, in male patients, not higher than in normal men [1]. In exacerbation of schizophrenia, a controlled study could not substantiate a difference between patients and controls in the mean prolactin response to dopamine agonists [1]. The finding of about two times higher mean mini-TRH test in 16 psychotic men than in 28 normal men could thus be associated with a decreased mean TIDA release rather than a different dopamine-2 receptor sensitivity in patients [11]. Because the method used to measure prolactin in normal men was different from the methods used in patients and because only patients received local anaesthesia, a precise comparison was not possible. These findings do not exclude the possibility that dopamine receptor sensitivity on lactotrophs could be associated with some symptoms of psychosis. Because the distribution of the mini-TRH and the basal prolactin level was skewed, we employed logarithmic transformation before calculating Pearson's correlations in psychotic patients.

4.1. The Mini-TRH Test in Acute Non-Affective Psychosis

In our study of 20 patients admitted to the University Hospital of Helsinki with acute. non-affective psychosis, 9 were diagnosed using DSM-IIIR criteria in the schizophrenia spectrum [9]. Using many sources, we obtained a reliable drug history of all the subjects. The drug-free group included 13 patients, 9 of whom were neuroleptic-naïve, 3 had not taken neuroleptic drugs for at least 4 months before the study, and 1 had not taken neuroleptic drugs for 38 days. After admission to the hospital, a washout period of at least four days was executed in all the patients before the mini-TRH test and rating of the Comprehensive Psychopathological Rating Scale (CPRS) psychosis subscale. Removing the effects of age, sex, and recent drug use revealed a significant correlation between the mini-TRH test and the score for the psychosis subscale (rp = +0.71, $p < 0.01$) [9], more precisely $p = 0.001$ [1]. The partial correlation remained the same even when patients in the drug withdrawal group were excluded (r = +0.69) [9]. These results showed an association of decreasing dopamine transmission in regulating prolactin with increasing CPRS psychosis subscale scores. Of the item-specific correlations, the strongest was between the mini-TRH test and the score for other delusions (rp = +0.73, rp < 0.001). This item excludes delusions of control, pessimistic thoughts, hypochondriasis, ideas of persecution, ideas of grandeur, delusional mood, and morbid jealousy [9]. It thus rates non-paranoid, non-affective delusions [9]. The mini-TRH test also marginally correlated with the score for the item for disrupted thoughts, which rated first-rank symptoms (rp = +0.51, $p < 0.05$) but not with the score for ideas of persecution (rp = +0.05, $p > 0.05$) or with the total score for various hallucinations (rp = +0.36, $p > 0.05$) [9]. Removing the effects of age, sex, and recent drug use did not reveal a significant partial correlation between the basal prolactin level and the rating for the psychosis subscale (rp = −0.20, $p > 0.05$) [9]. In the same 20 patients with non-affective psychoses, 40% of the patients demonstrated memory dysfunction on the CPRS item for failing memory, which is a simple test to estimate everyday memory dysfunction [7]. After controlling effects for age, sex, and recent drug use, the mini-TRH test significantly correlated with the CPRS item for failing memory (rp = +0.67, $p = 0.003$) [7]. Increasing memory function with advancing age could not explain this finding because, after controlling for sex and recent drug use only, the correlation remained about the same (rp = +0.60, $p = 0.008$). These findings suggested that the mini-TRH test could uncover age-independent memory dysfunction in association with decreased pituitary dopamine transmission. There was no correlation between the basal plasma prolactin level and ratings for failing memory (rp = +0.07, $p = 0.78$) [7]. The large fluctuation in the basal prolactin can reduce the accuracy of associations between basal prolactin blood levels and symptoms. Consistent with our results, in 91 patients with drug-naïve schizophrenia, the basal serum prolactin level was not associated with the severity of cognitive symptoms, including memory dysfunction [12].

4.2. The Mini-TRH Test in Acute First-Episode Schizophrenia

Over a four-year period, acute psychotic patients with a presentation of first-episode schizophrenia and no previously known treatment were admitted to the Moisio Hospital, which provides all inpatient psychiatric care for the catchment area of approximately 215,000 inhabitants [11]. We reviewed medical records potentially relevant for diagnostic purposes and possible earlier administration of psychotropic medication. We excluded patients who had previously taken any antipsychotic, antidepressant, or mood-stabilising drugs, as well as patients who, based on their history or examination, appeared to have any clinically significant medical condition or who were suspected to have abused alcohol or drugs. Twenty of the subjects met DSM-IV clinical criteria for schizophrenia, one withdrew her participation in the study. The mean age of the remaining 19 participants was 27.2 years (SD ± 7.7 years), of whom 14 were female and 5 were male [11]. The literature on symptom clusters of the defect state and symptomatic worsening after methylphenidate administration had suggested that the items for poverty of content of speech and inattention in the Scale for the Assessment of Negative Symptoms (SANS) could be particularly related to dopamine activity, more closely than negative or positive symptom clusters [11]. The difference in the mini-TRH test in 14 women and 5 men was not significant, and there was no significant correlation between the mini-TRH test and patient age (r = −0.22). In 19 patients, the mini-TRH test correlated negatively with the SANS scores for poverty of content of speech (r = −0.55, $p = 0.01$) and objective scores for inattention (r = −0.52, $p = 0.02$) [11]. The scores for these items significantly correlated with each other (r = +0.74, $p < 0.0001$). There were no significant correlations between the mini-TRH test and the rest of the SANS symptoms, i.e., negative symptoms not associated with disorganisation. In the patients with first-episode schizophrenia, there was a significant correlation between the mini-TRH test and the basal plasma prolactin level (r = +0.61, $p = 0.006$), not discovered in normal men [5] or in patients with non-affective psychosis [9]. This correlation raised the possibility that the mini-TRH test might have resulted in maximal prolactin responses in some patients with first-episode schizophrenia because in normal men, the basal prolactin level correlated with the maximal prolactin response to TRH [5]. This could be a factor limiting the specificity of the mini-TRH test. To reduce the effects of maximal prolactin responses, a dose of 6.25 mcg in the mini-TRH test was recommended in first-episode patients [1].

More studies were needed to study the correlations of the mini-TRH test with CPRS hallucinations (rp = +0.36, $p > 0.05$) and with disorganisation symptoms other than those rated by the SANS. We administered the Positive and Negative Symptom Scale (PANSS) in the last 13 of the 19 patients recruited with first-episode schizophrenia during the 4-year period after the training for the PANSS was completed by the rater [3]. We first correlated ratings of positive and disorganisation symptoms of schizophrenia based on the consensus five-factor model introduced by Wallwork and Fortgang of the PANSS with ratings of the mini-TRH test [3,13]. The requirement that the strongest loading in at least 24 of 29 factor analytic studies ensured that only the most specific symptoms entered the model [13]. In a total of 33 patients (20 with non-affective psychosis and 13 with first-episode schizophrenia), we calculated the difference in correlations in 2 independent samples. Among the specific PANSS positive symptoms, only the correlation between the mini-TRH test and the rating for hallucinatory behaviour ($p < 0.0125$), but not delusions, unusual thought content, or grandiosity (each $p > 0.125$), differed significantly from the correlation between the mini-TRH test and the rating for the CPRS psychosis subscale [3]. Although hallucinations and hallucinatory behaviour are not synonymous, the results may not support a predominant role of hallucinations in reflecting TIDA transmission. The corresponding correlative analysis with specific PANSS disorganisation symptoms showed that correlations of the mini-TRH test with difficulty in abstract thinking and poor attention differed significantly from the correlation between the mini-TRH test and the rating for the CPRS psychosis subscale (r < 0.005 each) [13]. There was no significant difference in the corresponding association with conceptual disorganisation ($p > 0.05$), which is the only PANSS disorganisation symptom that rates positive formal thought disorder [1].

5. Other Tests of Dopamine Transmission in Psychotic Subjects

Amphetamine-induced increase in intrasynaptic dopamine competes with radioligand occupancy to dopamine receptors, and the relative decrease from pre-amphetamine scan in radioligand binding can be quantified. In nonhuman primates, in the frontal cortex, there was a linear association between amphetamine-induced decrease in the dopamine-2/3 antagonist radiotracer [11C]FLB 457 and increase in extracellular dopamine concentration [14], and correlations were found of amphetamine-induced striatal intrasynaptic dopamine release with corresponding decreases in binding of dopamine-2/3 antagonist radiotracers [123I]IBZM and [11C]raclopride [15,16]. In healthy adults, cortisol levels were positively associated with ventral striatal amphetamine-induced decrease in [11C]raclopride binding [17]. As opposed to physiologic dopamine release, amphetamine releases dopamine to a major degree from both cytosolic and vesicular stores [18]. This may limit the accuracy of the amphetamine challenge tests, i.e., amphetamine release capacity for a given dose of amphetamine, in reflecting physiologic processes. Two important studies investigated amphetamine-induced dopamine release in different extrastriatal regions in schizophrenia. In 20 patients, the release was defective in many cortical and extrastriatal regions, including the midbrain, and did not significantly correlate with the basal PANSS positive scale score [19]. The mean PANSS positive score was 15.1 and comparable to the score in patients stabilised with drugs [1]. This could be a limiting factor because in some extrastriatal structures, the defective amphetamine-induced dopamine release may be more apparent in the acute phase of psychosis, as suggested by a strong inverse correlation between the amphetamine tests in the associative striatum and the prefrontal cortex (r = 0.71,

$p = 0.005$) [20]. Dopamine release in these areas could translate into different symptoms. In a mixed population of 34 patients with schizophrenia, including patients both in the active phase and in remission, a significant correlation was found between striatal amphetamine-induced displacement in the radiotracer [123I]IBZM and the change in PANSS positive scale score ($r = 0.54$, $p < 0.001$) [21]. This result was not verified in smaller studies, with less variability in the clinical status utilizing [11C]NPA [22] or [11C]-(+)-PHNO [23], in patients with schizophrenia and first-episode psychosis, respectively. Two major studies showed an abnormally enhanced striatal amphetamine-induced dopamine release [21,23] in acute patients with PANSS positive scale scores of 18.9 ± 6.5 and 21.4 ± 6.8, respectively, but not in another study with a PANSS positive scale score of 19.5 ± 5.3 [22] or in patients in remission with a PANSS positive score of 16.1 ± 5.8 [21]. On the other hand, none of the studies reported a correlation between amphetamine-induced dopamine release and pre-amphetamine PANSS positive scale, with the largest study reporting the coefficient of correlation ($r = 0.01$, $p = 0.98$) [21]. This was the same study that reported the significant correlation between amphetamine-induced striatal dopamine release and the increase in positive symptoms [21]. The magnitude of the change in PANSS positive symptoms was not associated with the baseline severity of these PANSS symptoms ($r = 0.10$, $p = 0.54$) [21]. There was an antipsychotic-free washout period of at least 2 weeks in all of the above studies before the amphetamine challenge.

Neuromelanin gradually accumulates in the form of neuromelanin–iron complexes, which are paramagnetic and can be estimated by magnetic resonance imaging across midbrain dopaminergic nuclei, reflecting long-lasting dopamine turnover [24]. In thirteen patients with schizophrenia, the scores for the Scale for the Assessment of Positive Symptoms (SAPS) showed a negative correlation with ventral tegmental area (VTA) signalling ($r = -0.69$, $p = 0.012$) [25], consistent with involvement of the VTA-prefrontal cortical tracts. The signal intensity at VTA was significantly lower and did not increase at substantia nigra in patients with schizophrenia compared to normal subjects [25]. The patients were on a relatively long-lasting antipsychotic drug, which, according to the authors, might have affected the results. In nine unmedicated patients with schizophrenia and nine healthy subjects, after controlling for diagnosis and age, a positive correlation was reported between neuromelanin signalling in the substantia nigra and amphetamine-induced striatal dopamine release [26], indexing storage and release capacity [27]. Nine patients with a PANSS positive scale scores of more than 19 showed higher neuromelanin signalling than twenty-four patients with scores of 19 or less, suggesting that patients disposed to react with positive symptoms might show more neuromelanin signalling in the substantia nigra [26]. In 42 unmedicated patients with schizophrenic psychoses, after controlling for age, PANSS negative and general psychopathology scores, neuromelanin contrast in the substantia nigra correlated weakly with the PANSS positive scale score (rp = +0.35, $p = 0.03$) driven by delusions and hallucinatory behaviour [28]. In a model controlling only for age, the effect of PANSS delusions on neuromelanin signalling was robust (rhop = +0.42, $p = 0.006$), but the effect of hallucinatory behaviour was not (rhop = +0.07, $p = 0.66$).

Both extrastriatal and striatal structures contribute to cerebrospinal fluid (CSF) HVA levels [29]. There may be a major contribution of prefrontal rather than striatal HVA to CSF HVA in psychotic patients [29]. This could explain the low CSF HVA reported in patients in association with more positive symptoms [30]. In chronic treatment-refractory schizophrenia, only non-paranoid delusions were significantly associated with reduced 24 h urinary excretion of dopamine [31]. In acute schizophrenia, first-rank symptoms were associated with reduced CSF HVA [32]. In patients with chronic schizophrenia with a high Brief Psychiatric Rating Scale score of 46.3 ± 9.5, low CSF HVA correlated with poor ability to recall visuospatial information and poor results in the Wisconsin Card Sorting

Test (WCST) [33]. In all of the above three studies employing 24 h urinary or CSF HVA, the antipsychotic-free washout period lasted at least 2 weeks. I have not found in the literature increased 24 h urinary or CSF HVA levels in association with the above symptoms in patients with schizophrenia.

6. Factor Analytic Studies of the PANSS

In a meta-analysis of 5-factor studies of the PANSS in schizophrenia, excluding one loading close to 0.4, which was likely to be within the range of noise in the data, the positive dimension consisted of delusions, unusual thought content, suspiciousness/persecution, hallucinatory behavior, and grandiosity [34]. This does not allow the conclusion that all these symptoms could be primarily associated with overall dopamine transmission in the brain. The same investigators who introduced the PANSS recognised the incompleteness of the positive–negative model and published a factor-analytic study of the PANSS, which suggested a pyramidal triangular model with seven components [35]. A syndrome at the vertex in the positive corner was composed of the particular positive symptoms of unusual thought content, delusions, and grandiosity [35], the same PANSS symptoms that could, potentially, be associated with the mini-TRH test [3]. Two of the seven symptoms in the original PANSS positive scale were assigned to the PANSS positive vertex component, and unusual thought content was assigned to it from the general psychopathology scale. Suspiciousness/persecution formed a separate component, and the loading of hallucinative behaviour on the positive component was only 0.43, about the same as the loading of 0.39 on the depressive component. In first-episode psychosis, hallucinations loaded only on a positive dimension, and hallucinatory behaviour loaded on both a positive symptom dimension and a symptom dimension with a mixture of negative and disorganisation items [36].

7. Discussion

Comparisons of dopamine transmission in acutely psychotic patients with non-stressed control subjects may be subject to stress-induced bias in results. In psychotic patients, the assessment of unstimulated stress is not easy. We took great care to reduce the effect of stress in psychotic patients on the mini-TRH test, including the use of local anaesthesia to control stress induced by blood extraction. In first-episode psychosis, a disruption was reported of daily cortisol levels with severity of psychotic symptoms, number of stressful life events, and perceived stress [37]. In schizophrenia, a disruption was observed in the association between stress-induced increase in salivary cortisol and increase in frontal dopamine release [38]. Even with careful routines to reduce stress, the glucocorticoid level may have a limited value to assess stress-induced dopamine transmission in schizophrenia.

Given the strong negative correlation between amphetamine-induced dopamine release in the prefrontal cortex and the associative striatum [20], the interpretation of either of the above is not straightforward. Strong correlations of symptoms with dopamine release at one of the sites might be reflected in weak or even moderate correlations in the opposite direction of these symptoms at the other site, even if dopamine release could translate into different symptoms in the prefrontal cortex and the striatum.

In rats, spontaneous dopamine transients may constitute a major component of extracellular dopamine levels in the brain [39]. In primates, a greater increase and a shorter half-life of striatal than cortical response to amphetamine resulted in more fluctuations in amphetamine-induced striatal than cortical extracellular dopamine levels [40]. In psychotic patients, a similar difference might indicate more fluctuations in striatal than cortical dopamine release and contribute to the lack of correlation between amphetamine-induced striatal dopamine release and pre-amphetamine PANSS positive scale. Even the correlation between the more stable neuromelanin signalling and the basal PANSS positive scale was weak. Inclusion of a broad range of positive symptoms may limit the specificity of the PANSS positive scale to assess basal dopamine transmission, although this scale can track drug-induced changes in dopaminergic symptoms.

It is unlikely that TIDA could mediate symptoms of schizophrenia. A more plausible explanation for the findings with the mini-TRH test is a synchronised action of dopamine inhibition of prolactin release and the activity of cortical structures involved in symptoms of schizophrenia. The thalamus can integrate communication between brain networks [1]. Reduced thalamic connections with frontal regions were associated with PANSS delusions, and increased thalamic connections with parietal regions were associated with a negative disorganisation symptom, PANSS difficulty in abstract thinking [1]. These associations might have been reflected in decreasing TIDA transmission in association with CPRS non-paranoid and non-affective delusions and increasing TIDA transmission in association with increasing negative disorganisation symptoms.

In schizophrenia-related disorders and in first-episode psychosis, studies using the mini-TRH test, extrastriatal amphetamine-induced dopamine release, the MRI neuromelanin signal in the VTA, and urinary or CSF HVA suggested that dopamine transmission might be reduced at extrastriatal sites, including in the VTA. In schizophrenia-related disorders, studies using striatal amphetamine-induced dopamine release and the MRI neuromelanin signal in the substantia nigra suggested almost invariably that dopamine transmission may be increased in the striatum and the substantia nigra. A different radiotracer or the small population of patients studied might contribute to the divergent result [22].

In a review of 36 studies observing acute effects of amphetamine or methylphenidate in schizophrenia, 43% of the patients with active psychotic symptoms worsened as opposed to 35% with nonactive symptoms, and 36% showed no significant change. A total of 21% of patients with active psychotic symptoms showed improvement as opposed to only 6% without active symptoms, suggesting that the improvement was greater in patients with positive symptoms [41]. A possible explanation for these findings is that the net effect of psychostimulant drugs could depend on the worsening of positive symptoms mediated by the dorsal striatum and improving them by effects on extrastriatal structures, such as the prefrontal cortex. The improvement in active symptoms in 21% of patients is considerable, considering the diminished responsivity to dopamine agonists in mesocortical neurons compared to nigrostriatal neurons [29], which could lead to classifying some of the change in mesocortical neurons in the group with no significant change. In the largest study using the amphetamine challenge, there was no correlation between dopamine release in the striatum and the basal PANSS positive scale score [21], suggesting that the latter may not be mainly determined by dopamine transmission in the striatum. The finding of no correlation between the magnitude of change in the PANSS positive scale provoked by amphetamine and the baseline severity of the PANSS positive scale [21] could indicate that amphetamine-induced changes in positive symptoms might differ from basal positive symptoms.

In nonhuman primates, chronic blockade of dopamine-2 receptors downregulates dopamine-1 receptors in the prefrontal cortex and produces severe impairments in working memory [42]. In drug-treated patients with schizophrenia, this could weaken recovery from impaired working memory. In patients with schizophrenia on antipsychotic medication during probabilistic decision-making under uncertainty, reduced activation was reported in the prefrontal cortex and other associated networks [43]. Because of the considerable similarity between the extrastriatal networks associated with probabilistic decision-making and working memory [43], it is possible that drug-induced reduction in extrastriatal dopamine activity could weaken the recovery from impaired probabilistic assessment as well.

8. Conclusions

In conclusion, correlative analyses of symptoms may provide information on symptom-specific alterations in dopamine transmission in non-affective psychotic symptoms. Among symptoms of schizophrenia, decreasing effects of endogenous dopamine on prolactin release as estimated by the mini-TRH test may reflect increasing memory problems, non-paranoid delusions, and first-rank symptoms. On the other hand, increasing endogenous striatal dopamine transmission may reflect increasing delusions and hallucinations. Amphetamine administration is known to increase delusions, paranoid ones in particular [8]. The mini-TRH test and MRI neuromelanin signalling in the substantia nigra are affordable and readily available tests to study basal dopamine transmission, not associated with rapid fluctuations in dopamine release. Using them together would provide a broader view of central dopamine transmission.

9. Future Directions

Most of the studies in this review suggested that derangements in dopamine transmission may be most pronounced in patients in the acute phases of psychosis or in patients prone to experiencing them frequently. A study investigating amphetamine-induced dopamine release at major extrastriatal sites and their associations with psychotic symptoms in the acute phase of schizophrenia is needed. This could help localise sites that are reflected in the mini-TRH test. I have not found a study in the literature comparing dopamine transmission in acutely psychotic patients to dopamine transmission in control subjects experiencing stress of a comparable degree to that found in acutely psychotic patients.

The correlation of the mini-TRH test with 24 h urinary excretion of HVA might associate the mini-TRH test with prefrontal dopamine transmission. In patients with prominent non-paranoid delusions or everyday memory problems, strong correlations of the mini-TRH test with these symptoms could be nonspecifically reflected in weak or even moderate correlations in the opposite direction in the striatum, and great care should be exercised in the interpretation of such results.

It remains to be established whether the mini-TRH test could reflect working memory dysfunction and/or impaired probabilistic assessment in antidopaminergic drug-free or drug-treated psychotic patients. In drug-free patients with schizophrenia-related diseases, it has not been studied whether the mini-TRH test could predict nonresponse to antidopaminergic drugs. Even in the framework of lower dopamine transmission in therapy-resistant patients compared to other patients, there could be a correlation between the mini-TRH test and non-paranoid delusions and/or unusual thoughts represented as first-rank symptoms.

Studying basal dopamine release using the SAPS with its items that rate bizarre delusions or the PANSS positive vertex component with its item for unusual thought content instead of the PANSS positive scale could prove beneficial in schizophrenia, the DSM-V diagnosis of which requires either unwanted obsessive or strange thoughts.

Funding: This review received no external funding.

Conflicts of Interest: The author declares no conflicts of interest.

References

1. Spoov, J. The Mini-TRH Test. *Pharmacopsychiatry* **2023**, *56*, 51–56. [CrossRef]
2. Hill-Samli, M.; MacLeod, R.M. Interaction of thyrotropin-releasing hormone and dopamine on the release of prolactin from the rat anterior pituitary in vitro. *Endocrinology* **1974**, *95*, 1189–1192. [CrossRef]
3. Spoov, J.; Bredbacka, P.E.; Stenman, U.H. Separation of hallucinations from other positive symptoms by prolactin response to 12.5 µg i.v. TRH (mini-TRH test). *Schizophr. Res.* **2020**, *220*, 287–288. [CrossRef] [PubMed]

4. Spoov, J. Submaximal plasma prolactin response to TRH and dopamine activity in man. *Pharmacopsychiatry* **1985**, *18*, 330–332. [CrossRef]
5. Spoov, J. Plasma prolactin response to submaximal stimulation by TRH and endogenous corticosteroids in man. *Pharmacopsychiatry* **1987**, *20*, 96–98.6. [CrossRef]
6. Spoov, J.; Karonen, S.L. Urinary HVA reflects central dopamine activity in man. *New Trends Exp. Clin. Psychiat.* **1987**, *3*, 93–99.
7. Spoov, J.; Bredbacka, P.E. Failing Memory in Nonaffective Psychosis and Prolactin Response to 12.5 µg i.v. TRH. *Pharmacopsychiatry* **2021**, *54*, 246–247. [CrossRef] [PubMed]
8. Spoov, J. Artefactual hyperprolactinemia. In *Advances in Health and Disease*; Duncan, L.T., Ed.; Nova Scotia Science: New York, NY, USA, 2023; Volume 60, pp. 162–176.
9. Spoov, J.; Bredbacka, P.E.; Appelberg, B.; Katila, H.; Karonen, S.L.; Rimón, R.; Sarna, S. Prolactin response to submaximal stimulation by TRH in nonaffective psychoses. *Biol. Psychiatry* **1991**, *29*, 204–210. [CrossRef]
10. Tsuchiya, K. Studies on prolactin in major psychoses—With reference to prolactin response to stress in schizophrenia. *Folia Psychiatr. Neurol. Jpn.* **1984**, *38*, 53–56. [CrossRef]
11. Spoov, J.; Bredbacka, P.E.; Stenman, U.H. An abnormal relation between basal prolactin levels and prolactin response to 12.5 microg TRH i.v. in drug-naïve patients with first-episode schizophrenia. *Schizophr. Res.* **2010**, *119*, 41–46. [CrossRef]
12. Luo, G.; Bai, F.; Qu, X.; Jing, Y.; Wang, S.; Xuekelaiti, Z.; Yao, C.; Li, M.; Li, J. The relationship between serum prolactin levels and cognitive function in drug-naïve schizophrenia patients: A cross-sectional study. *J. Neural Trasm.* **2024**, *131*, 385–391. [CrossRef]
13. Spoov, J.; Bredbacka, P.E.; Stenman, U.H. Separation of positive and disorganization symptoms by prolactin response to 12.5 µg intravenous TRH. *Schizophr. Res.* **2020**, *215*, 449–450. [CrossRef]
14. Narendran, R.; Jedema, H.P.; Lopresti, B.J.; Mason, N.S.; Gurnsey, K.; Ruszkiewicz, J.; Chen, C.M.; Deuitch, L.; Frankle, W.G.; Bradberry, C.W. Imaging dopamine transmission in the frontal cortex: A simultaneous microdialysis and [11C]FLB 457 PET study. *Mol. Psychiatry* **2014**, *19*, 302–310. [CrossRef] [PubMed]
15. Laruelle, M.; Iyer, R.N.; al-Tikriti, M.S.; Zea-Ponce, Y.; Malison, R.; Zoghbi, S.S.; Baldwin, R.M.; Kung, H.F.; Charney, D.S.; Hoffer, P.B.; et al. Microdialysis and SPECT measurements of amphetamine-induced dopamine release in nonhuman primates. *Synapse* **1997**, *25*, 1–14. [CrossRef]
16. Breier, A.; Su, T.P.; Saunders, R.; Carson, R.E.; Kolachana, B.S.; de Bartolomeis, A.; Weinberger, D.R.; Weisenfeld, N.; Malhotra, A.K.; Eckelman, W.C.; et al. Schizophrenia is associated with elevated amphetamine-induced synaptic dopamine concentrations: Evidence from a novel positron emission tomography method. *Proc. Natl. Acad. Sci. USA* **1997**, *94*, 2569–2574. [CrossRef]
17. Oswald, L.M.; Wong, D.F.; McCaul, M.; Zhou, Y.; Kuwabara, H.; Choi, L.; Brasic, J.; Wand, G.S. Relationships among ventral striatal dopamine release, cortisol secretion, and subjective responses to amphetamine. *Neuropsychopharmacol.* **2005**, *30*, 821–832. [CrossRef] [PubMed]
18. Sulzer, D. How addictive drugs disrupt presynaptic dopamine neurotransmission. *Neuron* **2011**, *69*, 628–649. [CrossRef] [PubMed]
19. Slifstein, M.; van de Giessen, E.; Van Snellenberg, J.; Thompson, J.L.; Narendran, R.; Gil, R.; Hackett, E.; Girgis, R.; Ojeil, N.; Moore, H.; et al. Deficits in prefrontal cortical and extrastriatal dopamine release in schizophrenia: A positron emission tomographic functional magnetic resonance imaging study. *JAMA Psychiatry* **2015**, *72*, 316–324. [CrossRef]
20. Frankle, W.G.; Himes, M.; Mason, N.S.; Mathis, C.A.; Narendran, R. Prefrontal and Striatal Dopamine Release Are Inversely Correlated in Schizophrenia. *Biol. Psychiatry* **2022**, *92*, 791–799. [CrossRef]
21. Laruelle, M.; Abi-Dargham, A.; Gil, R.; Kegeles, L.; Innis, R. Increased dopamine transmission in schizophrenia: Relationship to illness phases. *Biol. Psychiatry* **1999**, *46*, 56–72. [CrossRef]
22. Frankle, W.G.; Paris, J.; Himes, M.; Mason, N.S.; Mathis, C.A.; Narendran, R. Amphetamine-Induced Striatal Dopamine Release Measured With an Agonist Radiotracer in Schizophrenia. *Biol. Psychiatry* **2018**, *83*, 707–714. [CrossRef] [PubMed]
23. Weidenauer, A.; Bauer, M.; Sauerzopf, U.; Bartova, L.; Nics, L.; Pfaff, S.; Philippe, C.; Berroterán-Infante, N.; Pichler, V.; Meyer, B.M.; et al. On the relationship of first-episode psychosis to the amphetamine-sensitized state: A dopamine $D_{2/3}$ receptor agonist radioligand study. *Transl. Psychiatry* **2020**, *10*, 2. [CrossRef] [PubMed]
24. Schulz, J.; Zimmermann, J.; Sorg, C.; Menegaux, A.; Brandl, F. Magnetic resonance imaging of the dopam, ine system in schizophrenia—A scoping review. *Front. Psychiatry* **2022**, *13*, 925476. [CrossRef]
25. Yamashita, F.; Sasaki, M.; Fukumoto, K.; Otsuka, K.; Uwano, I.; Kameda, H.; Endoh, J.; Sakai, A. Detection of changes in the ventral tegmental area of patients with schizophrenia using neuromelanin-sensitive MRI. *Neuroreport* **2016**, *27*, 289–294. [CrossRef]
26. Cassidy, C.M.; Zucca, F.A.; Girgis, R.R.; Baker, S.C.; Weinstein, J.J.; Sharp, M.E.; Bellei, C.; Valmadre, A.; Vanegas, N.; Kegeles, L.S.; et al. Neuromelanin-sensitive MRI as a noninvasive proxy measure of dopamine function in the human brain. *Proc. Natl. Acad. Sci. USA* **2019**, *116*, 5108–5117. [CrossRef]
27. Horga, G.; Wengler, K.; Cassidy, C.M. Neuromelanin-Sensitive Magnetic Resonance Imaging as a Proxy Marker for Catecholamine Function in Psychiatry. *JAMA Psychiatry* **2021**, *78*, 788–789. [CrossRef] [PubMed]

28. Wengler, K.; Baker, S.C.; Velikovskaya, A.; Fogelson, A.; Girgis, R.R.; Reyes-Madrigal, F.; Lee, S.; de la Fuente-Sandoval, C.; Ojeil, N.; Horga, G. Generalizability and Out-of-Sample Predictive Ability of Associations Between Neuromelanin-Sensitive Magnetic Resonance Imaging and Psychosis in Antipsychotic-Free Individuals. *JAMA Psychiatry* **2024**, *81*, 198–208. [CrossRef]
29. Pickar, D. Perspectives on a time-dependent model of neuroleptic action. *Schizophr. Bull.* **1988**, *14*, 255–268. [CrossRef]
30. Gasnier, M.; Ellul, P.; Plaze, M.; Ahad, P.A. A New Look on an Old Issue: Comprehensive Review of Neurotransmitter Studies in Cerebrospinal Fluid of Patients with Schizophrenia and Antipsychotic Effect on Monoamine's Metabolism. *Clin. Psychopharmacol. Neurosci.* **2021**, *19*, 395–410. [CrossRef]
31. Karoum, F.; Karson, C.N.; Bigelow, L.B.; Lawson, W.B.; Wyatt, R.J. Preliminary evidence of reduced combined output of dopamine and its metabolites in chronic schizophrenia. *Arch. Gen. Psychiatry* **1987**, *44*, 604–607. [CrossRef]
32. Post, R.M.; Fink, E.; Carpenter, W.T., Jr.; Goodwin, F.K. Cerebrospinal fluid amine metabolites in acute schizophrenia. *Arch. Gen. Psychiatry* **1975**, *32*, 1063–1069. [CrossRef] [PubMed]
33. Kahn, R.S.; Harvey, P.D.; Davidson, M.; Keefe, R.S.; Apter, S.; Neale, J.M.; Mohs, R.C.; Davis, K.L. Neuropsychological correlates of central monoamine function in chronic schizophrenia: Relationship between CSF metabolites and cognitive function. *Schizophr. Res.* **1994**, *11*, 217–224. [CrossRef]
34. Lim, K.; Peh, O.H.; Yang, Z.; Rekhi, G.; Rapisarda, A.; See, Y.M.; Rashid, N.A.A.; Ang, M.S.; Lee, S.A.; Sim, K.; et al. Large-scale evaluation of the Positive and Negative Syndrome Scale (PANSS) symptom architecture in schizophrenia. *Asian J. Psychiatr.* **2021**, *62*, 102732. [CrossRef]
35. Kay, S.R.; Sevy, S. Pyramidical model of schizophrenia. *Schizophr. Bull.* **1990**, *16*, 537–545. [CrossRef] [PubMed]
36. McGorry, P.D.; Bell, R.C.; Dudgeon, P.L.; Jackson, H.J. The dimensional structure of first episode psychosis: An exploratory factor analysis. *Psychol. Med.* **1998**, *28*, 935–947. [CrossRef] [PubMed]
37. Mondelli, V.; Dazzan, P.; Hepgul, N.; Di Forti, M.; Aas, M.; D'Albenzio, A.; Di Nicola, M.; Fisher, H.; Handley, R.; Marques, T.R.; et al. Abnormal cortisol levels during the day and cortisol awakening response in first-episode psychosis: The role of stress and of antipsychotic treatment. *Schizophr. Res.* **2010**, *116*, 234–242. [CrossRef]
38. Schifani, C.; Tseng, H.H.; Kenk, M.; Tagore, A.; Kiang, M.; Wilson, A.A.; Houle, S.; Rusjan, P.M.; Mizrahi, R. Cortical stress regulation is disrupted in schizophrenia but not in clinical high risk for psychosis. *Brain* **2018**, *141*, 2213–2224. [CrossRef]
39. Owesson-White, C.A.; Roitman, M.F.; Sombers, L.A.; Belle, A.M.; Keithley, R.B.; Peele, J.L.; Carelli, R.M.; Wightman, R.M. Sources contributing to the average extracellular concentration of dopamine in the nucleus accumbens. *J. Neurochem.* **2012**, *121*, 252–262. [CrossRef]
40. Jedema, H.P.; Narendran, R.; Bradberry, C.W. Amphetamine-induced release of dopamine in primate prefrontal cortex and striatum: Striking differences in magnitude and timecourse. *J. Neurochem.* **2014**, *130*, 490–497. [CrossRef]
41. Lieberman, J.A.; Kane, J.M.; Alvir, J. Provocative tests with psychostimulant drugs in schizophrenia. *Psychopharmacology* **1987**, *91*, 415–433. [CrossRef]
42. Castner, S.A.; Williams, G.V.; Goldman-Rakic, P.S. Reversal of antipsychotic-induced working memory deficits by short-term dopamine D1 receptor stimulation. *Science* **2000**, *287*, 2020–2022. [CrossRef] [PubMed]
43. Krug, A.; Cabanis, M.; Pyka, M.; Pauly, K.; Kellermann, T.; Walter, H.; Wagner, M.; Landsberg, M.; Shah, N.J.; Winterer, G.; et al. Attenuated prefrontal activation during decision-making under uncertainty in schizophrenia: A multi-center fMRI study. *Schizophr. Res.* **2024**, *152*, 176–183. [CrossRef] [PubMed]

BioChem

MDPI

Review

Single-Cell Transcriptomics in Spinal Cord Studies: Progress and Perspectives

Mst. Afsana Mimi [1] and Md. Mahmudul Hasan [2,*]

1 Independent Researcher, 2-41-21 Fujigaoka, Aoba Ward, Yokohama 227-0049, Japan
2 Earth-Life Science Institute, The Institute of Science Tokyo, Tokyo 152-8550, Japan
* Correspondence: hasan.mahmudbio@gmail.com

Abstract: Single-cell RNA sequencing (scRNA-seq) has revolutionized neuroscience by enabling the analysis of cellular heterogeneity and dynamic molecular processes at the single-cell resolution. In spinal cord research, scRNA-seq provides critical insights into cell type diversity, developmental trajectories, and pathological mechanisms. This review summarizes recent progress in the application of scRNA-seq to spinal cord development, injury, and neurodegenerative diseases and discusses the current challenges and future directions. Relevant studies focusing on the key applications of scRNA-seq, including advances in spatial transcriptomics and multi-omics integration, were retrieved from PubMed and the Web of Science. scRNA-seq has enabled the identification of distinct spinal cord cell populations and revealed the gene regulatory networks driving development. Injury models have revealed the temporal dynamics of immune and glial responses, alongside potential regenerative processes. In neurodegenerative conditions, scRNA-seq highlights cell-specific vulnerabilities and molecular changes. The integration of spatial transcriptomics and computational tools, such as machine learning, has further improved the resolution of spinal cord biology. However, challenges remain in terms of data complexity, sample acquisition, and clinical translation. Single-cell transcriptomics is a powerful approach for understanding spinal cord biology. Its integration with emerging technologies will advance both basic research and clinical applications, supporting personalized and regenerative therapy. Addressing these technical and analytical barriers is essential to fully realize the potential of scRNA-seq in spinal cord science.

Keywords: single-cell RNA sequencing; spinal cord; transcriptomics; spinal cord injury; neurodegeneration; developmental biology; spatial transcriptomics; multi-omics integration; cell atlas; regenerative medicine

1. Introduction

The spinal cord, a pivotal component of the central nervous system (CNS), transmits sensory and motor signals between the brain and peripheral tissues and coordinates reflex activities independently of the brain. The structure of the spinal cord is intricately connected to its function; the outer white matter comprises myelinated ascending and descending nerve tracts, whereas the central gray matter contains neuronal cell bodies and local neural circuits [1].

The complexity of the spinal cord arises from its diverse cellular composition, including neurons, glial cells, and other specialized cell types, as well as the precise functional specialization and intricate interactions among these cells [2]. Advances in single-cell transcriptomics have revealed the heterogeneity of neuronal and non-neuronal cell populations,

https://doi.org/10.3390/biochem5020016

offering new insights into their roles in health and disease [3]. For example, motor neurons in the spinal cord have been found to consist of multiple subtypes, each with distinct gene expression profiles and functional roles [4].

Moreover, the response of glial cells to spinal cord injury (SCI) has attracted considerable attention. Research has shown significant alterations in astrocytes, microglia, and oligodendrocytes post-injury, which critically influence neuronal survival and regeneration [5–7]. These findings highlight the importance of exploring the cellular composition and function of the spinal cord, contributing to our understanding of its roles in health and disease and opening new avenues for therapeutic intervention.

Single-cell RNA sequencing (scRNA-seq) has emerged as a powerful tool for elucidating the intricate cellular and molecular mechanisms underlying spinal cord function and disease. This technology allows for a detailed analysis of gene expression at the single-cell level, providing unprecedented insights into cellular heterogeneity and function [8]. Specifically, methods such as SPLiT-seq and Smart-seq3 have been shown to be particularly suitable for spinal cord tissues due to their ability to handle complex and heterogeneous samples with high sensitivity and accuracy. For instance, SPLiT-seq has been successfully applied to study the developing mouse spinal cord, revealing distinct cell types and lineages [9]. Similarly, Smart-seq3 has been highlighted for its ability to generate high-quality, full-length transcriptomes [10], which is crucial for understanding the nuanced gene expression profiles in spinal cord cells. These advancements not only enhance our understanding of spinal cord biology but also pave the way for future therapeutic interventions.

The groundbreaking development of single-cell transcriptomics dates back to 2009, when Tang et al. first successfully sequenced the transcriptome of individual mouse embryonic cells, overcoming the challenge posed by low RNA content in single cells [11]. Subsequently, advancements in microfluidics and laboratory automation have dramatically increased the capacity of single-cell transcriptomic analyses, facilitating the commercialization of platforms capable of processing thousands to tens of thousands of cells per experiment [12]. The rapid evolution of scRNA-seq has emerged as a powerful tool for unraveling cellular heterogeneity within spinal cord tissues.

Spatial transcriptomics has further advanced this field by integrating gene expression profiles with spatial tissue context. Its origins can be traced back to the application of in situ hybridization by Walter Gehring's group in 1986, which localized antennapedia mRNA in Drosophila embryos and imaginal discs [13]. The term "spatial transcriptomics" was coined by Ståhl et al. in 2016, describing a novel method that captures mRNA molecules directly from tissue sections on glass slides, enabling the high-resolution mapping of gene expression integrated with tissue morphology [14]. While single-cell transcriptomics provides genome-wide expression profiles at single-cell resolution, it lacks contextual tissue information, a gap effectively addressed by spatial transcriptomics, which elucidates intercellular interactions and tissue heterogeneity [12,15]. However, spatial transcriptomics still faces challenges, particularly regarding resolution and the detection of low-abundance transcripts [15].

The integration of single-cell and spatial transcriptomics is driving advances in precision medicine, offering robust tools for disease research, drug development, and regenerative medicine [16]. By sequencing transcripts at the single-cell level, researchers can identify distinct cellular subpopulations and clarify their specific roles in spinal cord development, homeostasis, and injury repair processes. For instance, Delile et al. (2019) employed scRNA-seq to profile the developing mouse spinal cord, uncovering spatial and temporal gene expression dynamics essential for neural progenitor differentiation and neuronal maturation, significantly enriching our understanding of spinal cord developmental processes [17]. Blum et al. (2021) utilized single-nucleus transcriptomics to reveal substantial heterogeneity among motor neurons in adult mouse spinal cords, identifying subtype-specific transcriptional signatures closely associated with motor functions and disease mechanisms [4]. Recently, Zhang et al. (2024) [18] combined spatial transcriptomics with single-nucleus RNA sequencing to create a detailed spatial atlas of adult human spinal cord cells and identified and localized 21 neuronal subclusters. This comprehensive mapping linked molecular identities directly to anatomical structures, greatly expanding our knowledge of the cellular composition of the human spinal cord [18]. Collectively, advances in single-cell and spatial transcriptomics offer unparalleled opportunities to elucidate the complex cellular architecture of the spinal cord and open innovative pathways for precise therapeutic interventions in spinal cord injury and associated diseases.

With the widespread application of single-cell and spatial transcriptomics in spinal cord research, a growing number of studies have uncovered the cellular heterogeneity and dynamic changes that occur in the spinal cord during development, physiological function, and pathological conditions [9,19]. Despite these advances, a comprehensive review that systematically integrates these findings and explores their implications for neuroscience and clinical translation is lacking. This review aims to fill this gap by providing a thorough overview of the recent progress in the application of single-cell and spatial transcriptomic technologies in spinal cord research. By synthesizing the current findings, we aim to offer a conceptual framework that deepens the understanding of complex cellular behaviors within the spinal cord and guides future developments in related fields.

This review covers several key aspects, including technological advancements, the construction of comprehensive cellular atlases, an analysis of microenvironmental dynamics, and challenges associated with translational medicine. By tracing the evolution of these technologies and highlighting representative clinical applications, we aim to bridge basic and translational research, offering insights into both the fundamental biology of the spinal cord and its relevance to regenerative medicine and therapeutic development for neurological disorders. Ultimately, this review aspires to serve as a valuable resource for neuroscientists and clinicians alike, providing a systematic reference that promotes a deeper understanding of spinal cord complexity and informs future research directions and treatment strategies.

2. Single-Cell Transcriptomics: Technologies and Methodologies

2.1. Principles and Evolution of Single-Cell RNA Sequencing Technologies

scRNA-seq is a powerful technique that enables transcriptomic analysis at the resolution of individual cells, allowing for an exploration of cellular heterogeneity and complex gene expression patterns. The core workflow of scRNA-seq typically involves the isolation of single cells, cell lysis, the reverse transcription of mRNA into cDNA, library construction, and high-throughput sequencing. These steps allow researchers to capture a comprehensive transcriptional profile of individual cells [20].

Since its initial introduction in 2009 [11], scRNA-seq has undergone rapid technological advancements. In 2012, the development of SMART-seq significantly improved full-length cDNA amplification strategies, resulting in enhanced transcript coverage [21]. One of the most notable breakthroughs in scRNA-seq is the exponential increase in throughput. While early methods such as STRT-seq could only process hundreds of cells [22], the advent of droplet-based microfluidics in 2015—exemplified by Drop-seq and the 10x Genomics Chromium system—enabled the profiling of tens of thousands to millions of cells in a single experiment [23–26]. This high-throughput capacity has made it feasible to investigate cellular heterogeneity in complex tissues, such as the spinal cord.

Moreover, these technical advancements have not only increased scalability but also reduced the cost per cell, facilitating the broad application of scRNA-seq across various disciplines, including developmental biology, oncology, and neuroscience. In summary, the development of scRNA-seq has provided a transformative tool for dissecting cellular functions and tissue architecture, particularly in elucidating the diverse cellular landscape of complex systems, such as the spinal cord.

2.2. Comparison of Major scRNA-Seq Platforms and Their Features

The selection of scRNA-seq technologies should be guided by specific research objectives and experimental conditions, as different platforms vary significantly in terms of throughput, sensitivity, and cost.

Among the currently available scRNA-seq platforms, the 10x Genomics Chromium system is one of the most widely adopted high-throughput technology. The 10x Genomics Chromium platform has significantly advanced scRNA-seq with its high-throughput capabilities and ability to detect rare cell types [24]. Based on droplet microfluidics, it encapsulates thousands of individual cells into nanoliter droplets containing unique barcodes, enabling efficient and parallel transcriptomic profiling [25]. Its high throughput allows for the detection of rare cell populations, and its relatively low cost and scalability make it particularly suitable for complex experimental designs involving multiple conditions or time points [27]. However, it is important to acknowledge the limitations of the platform, particularly in the context of specific applications. For instance, the 10x Genomics Chromium system is known for its 3′ bias, which can limit the detection of full-length transcripts and affect the identification of transcript isoforms [28]. This bias may be particularly problematic when studying alternative splicing events or low-abundance transcripts, as these require comprehensive transcript coverage [27]. Additionally, the platform experiences higher dropout rates for low-expression genes, which can result in the loss of important biological information [27]. These limitations highlight the need for researchers to carefully consider the specific requirements of their studies when selecting an scRNA-seq platform.

In contrast, SMART-seq2 and SMART-seq3 are well-established plate-based scRNA-seq methods that enable full-length mRNA amplification and sequencing. These technologies are particularly advantageous for detecting low-abundance transcripts and transcript isoforms [29,30]. SMART-seq3 further improves the sensitivity over its predecessor and supports quantitative transcript detection at the single-molecule level [30]. However, these methods are generally limited by lower throughput, longer processing times, and higher per-cell costs, making them more suitable for studies focusing on rare cell populations and functional subtypes.

Drop-seq and inDrops, both droplet-based microfluidic approaches, offer cost-effective and scalable solutions for high-throughput scRNA-seq. Drop-seq, developed by Macosko et al., co-encapsulates single cells with barcoded beads in droplets, providing a robust method with high efficiency and relatively low cost [23]. inDrops introduces improvements in barcode design and reaction chemistry, enhancing cell capture rates and data consistency while also exhibiting remarkably low background noise [31]. Furthermore, its flexible architecture facilitates adaptation to other sequencing-based assays [31]. While both methods maintain high throughput, their sensitivity is generally lower than that of SMART-seq-based techniques [32].

SPLiT-seq is a unique combinatorial barcoding approach that eliminates the need for physical cell isolation. By sequentially tagging fixed cells through multiple rounds of barcode delivery, SPLiT-seq significantly reduces the reliance on specialized instrumentation and greatly increases throughput [9]. Using this method, Rosenberg et al. mapped the spatiotemporal dynamics of 30 neuronal subtypes in the developing mouse spinal cord, demonstrating the utility of SPLiT-seq for comprehensive transcriptomic profiling in complex multicellular systems [9].

In recent years, spatial transcriptomics technologies, such as 10x Genomics Visium and BGI's Stereo-seq, have enabled the in situ mapping of gene expression across complex tissues, such as the spinal cord. For instance, Zhang et al. utilized single-nucleus RNA sequencing in combination with Visium spatial transcriptomics to delineate the spatial distribution of spinal cord neurons in nine human donors [18]. Stereo-seq further advances the field by improving spatial resolution and expanding the capture area, offering the potential to generate spatial gene expression maps at subcellular resolution [33]. To aid comparison, Table 1 summarizes key single-cell RNA sequencing platforms, highlighting their technical features, advantages, limitations, and representative applications in spinal cord research.

Table 1. Summary of Single-cell RNA Sequencing Platforms Used in Spinal Cord Research.

Sequencing Method	Key Features	Advantages	Limitations	Applications in Spinal Cord Research	Temporal Evolution
SMART-seq2/SMART-seq3	Plate-based; full-length transcript coverage (SMART-seq3 improves sensitivity)	High sensitivity; isoform detection; full-length transcripts	Low throughput; expensive; time-consuming	Detecting rare cell types or novel isoforms	SMART-seq2 (~2013); SMART-seq3 (~2019)
10x Genomics Chromium	Droplet-based; barcode encapsulation of cells	High throughput; scalable; relatively low cost	Partial transcript coverage; dropout effect	Developmental mapping; SCI response; human/fetal profiling	Mainstream since ~2016
Drop-seq	Droplet-based; barcoded beads in droplets	Cost-effective; scalable; high efficiency	Lower sensitivity than SMART-seq	Identifying glial/inflammatory responses	Introduced ~2015
inDrops	Improved droplet-based; enhanced barcoding and chemistry	Low background noise; flexible; improved capture	Still limited in sensitivity; complex protocol	Flexible transcriptomic profiling with high consistency	Developed post-2015
SPLiT-seq	Combinatorial barcoding; no physical isolation required	Ultra-high throughput; low equipment requirement	Sparse gene detection; complex processing	Mapping neuron subtypes in mouse spinal cord development	Introduced ~2018
10x Visium/BGI Stereo-seq	Spatial transcriptomics; spatial gene expression in situ	High spatial resolution; large capture area	Requires tissue sectioning; limited to spatial resolution	Mapping spatial gene expression in human spinal cord	Emerging since ~2020

2.3. Technology Selection Strategies and Application Cases in Spinal Cord Research

In spinal cord research, the selection of an appropriate scRNA-seq technology must be carefully tailored to specific research objectives and the nature of the biological samples. For studies aiming to identify novel transcript isoforms or rare cell populations, SMART-seq is often preferred because of its high sensitivity and ability to capture full-length transcripts. In contrast, high-throughput platforms, such as 10x Genomics Chromium or Drop-seq, are more suitable for constructing large-scale single-cell atlases of the spinal cord.

In the context of spinal cord development, Delile et al. employed the 10x Genomics Chromium platform to perform scRNA-seq on the cervical and thoracic regions of the embryonic mouse spinal cord, covering multiple developmental stages from embryonic day 9.5 to 13.5. This study generated a high-resolution transcriptional atlas, revealing the spatial and temporal dynamics of neural progenitors and neuronal populations. Moreover, this study identified novel marker genes and regulatory sequences involved in neuronal differentiation, providing a valuable resource for elucidating the cellular diversity and molecular mechanisms underlying spinal cord development [17].

To construct a comprehensive cellular map of the human spinal cord, Andersen et al. applied the 10x Genomics Chromium system to systematically profile the human fetal spinal cord. Their study uncovered substantial cellular heterogeneity and regional specialization during the development process. By integrating data across multiple developmental time points, the authors tracked temporal changes in cell types and mapped disease-associated genes to specific cellular populations, offering novel insights into the cellular architecture of motor control and the pathogenesis of neurological disorders [34]. This study represents a critical step toward building a detailed single-cell atlas of human spinal cord development.

In studies of SCI, researchers have utilized single-nucleus RNA sequencing (snRNA-seq) to profile transcriptional changes in the injured spinal cord at high resolution. For example, Matson et al. used the 10x Genomics Chromium platform to generate a cellular map of the lumbar spinal cord following injury in a mouse model. Their findings revealed injury-induced alterations in distinct neuronal subpopulations and highlighted specific neuronal subsets with regenerative potential [35]. These results provide a theoretical framework for future therapeutic strategies targeting specific neuronal groups to promote spinal cord repair.

In summary, the 10x Genomics Chromium system has emerged as the predominant platform used in most spinal cord studies because of its balance of throughput and resolution. While it is crucial to maintain a balanced perspective and recognize the platform's limitations, it is equally important to acknowledge the significant advancements that 10x Genomics has brought to this field. The thoughtful selection and application of scRNA-seq technologies, aligned with the specific aims and sample characteristics of a study, are essential for advancing our understanding of spinal cord development, function, and regeneration.

3. Applications of Single-Cell Transcriptomics in Spinal Cord Research

Single-cell transcriptomics has rapidly evolved over the past decade, giving rise to a diverse set of technologies that have been instrumental in advancing spinal cord research. These platforms vary in resolution, throughput, and scalability, enabling applications ranging from transcript isoform discovery to the construction of comprehensive cellular atlas. In spinal cord studies, scRNA-seq technologies have been applied across multiple major domains, including the characterization of developmental processes, the elucidation of injury-induced molecular changes, the investigation of neurodegenerative disease mechanisms, and integration with spatial transcriptomics and multi-omics approaches for translational insights. Figure 1 illustrates the chronological development of key scRNA-seq platforms and their representative applications in spinal cord research.

3.1. Decoding Spinal Cord Developmental Biology

Single-cell transcriptomic technologies have emerged as pivotal tools for deciphering cellular diversity and lineage specification during spinal cord development. By generating cell type-specific gene expression maps, researchers can precisely track processes such as neural tube formation, fate specification, and circuit assembly. This section highlights key discoveries enabled by single-cell approaches, focusing on the temporal regulation of neurogenesis, mechanisms underlying cell type diversification, and the dissection of critical signaling pathways at single-cell resolution (Table 2).

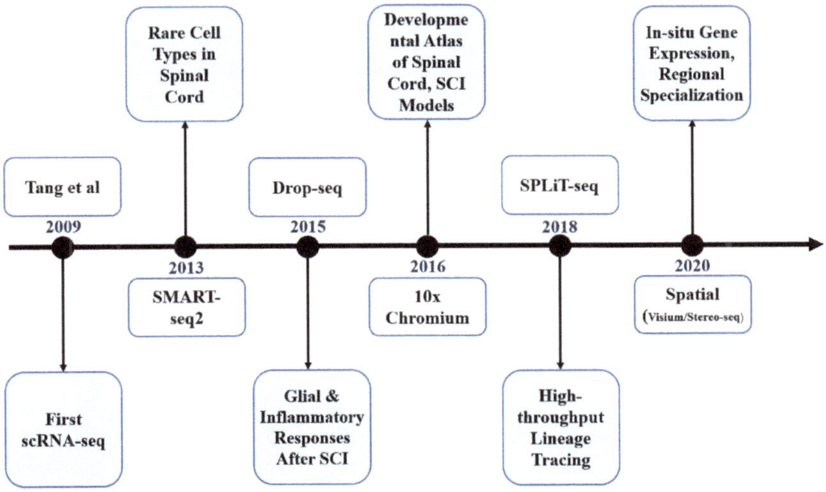

Figure 1. Timeline of scRNA-seq Platforms and Their Contributions to Spinal Cord Research.

Table 2. Representative Single-Cell Transcriptomic Studies of Spinal Cord Development.

Study	Species	Developmental Stage	Platform	Key Findings
Delile et al. [17]	Mouse	E9.5–E13.5	10x Genomics Chromium	Spatial/temporal dynamics of neural progenitors; novel markers in dorsoventral domains
Andersen et al. [34]	Human	~22 weeks gestation	10x Genomics Chromium	Glial heterogeneity, astrocyte regionalization, disease gene mapping to specific cell types
Sathyamurthy et al. [19]	Mouse	Adult	Drop-Seq	43 neuronal subtypes, region-specific distribution, spinal neuron molecular map
Blum et al. [4]	Mouse	Adult	10x Genomics Chromium	Motor neuron heterogeneity, transcriptional profiles linked to axonal targeting and function
Cao et al. [36]	Mouse	Organogenesis (multi-stage)	SPLiT-seq	Spinal progenitor transcriptional transitions, Hox genes, Shh pathway in developmental regulation
Zhang et al. [18]	Human	Adult	10x Genomics Chromium	21 neuronal subtypes, spatial distribution, human-mouse comparison, sex-specific transcription

Delile et al. employed the 10x Genomics platform to perform scRNA-seq on cervical and thoracic spinal cord regions from embryonic day (E) 9.5 to 13.5 in mice, generating a high-resolution developmental atlas that revealed the spatial and temporal dynamics of neural progenitor and neuronal populations [17]. The study identified novel marker genes expressed in distinct dorsoventral domains of progenitors and neurons, providing insights into the transcriptional programs that guide cell fate decisions during spinal cord development [17].

In human spinal cord development, Andersen et al. conducted scRNA-seq analysis on mid-gestation (~22 weeks) fetal samples and uncovered pronounced heterogeneity between and within major cell types [34]. For example, glial cells show spatially dependent transcriptional programs along the dorsoventral and rostrocaudal axes, and astrocytes display differential gene expression based on their localization in the white versus gray matter. Moreover, the study integrated multiple existing datasets covering the same developmental window to trace temporal changes in cellular composition and mapped disease-related genes to specific cell types, offering novel insights into the cellular basis of motor control and opportunities for stem cell-based disease modeling [34].

A comparative analysis of these studies revealed both similarities and differences in spinal cord development between mice and humans. While Delile et al. [17] identified distinct spatial and temporal dynamics of neural progenitor and neuronal populations in the mouse spinal cord, Andersen et al. [34] demonstrated significant heterogeneity within and between the major cell types in the human spinal cord. Notably, the identification of novel marker genes in mouse models by Delile et al. [17] aligns with the discovery of distinct transcriptional programs in human glial cells by Andersen et al. [34]. However, there are notable differences. For instance, the spatial dependency of transcriptional programs in human glial cells along the dorsoventral and rostrocaudal axes [34] contrasts with the more generalized spatial dynamics observed in mouse models [17]. These differences highlight the importance of considering species-specific variations when interpreting findings from mouse models in the context of human spinal cord development. A critical reflection on these variations suggests that while mouse models provide valuable insights into the fundamental mechanisms of spinal cord development, they may not fully capture the complexity and heterogeneity observed in human tissue. This underscores the need for further comparative studies to bridge the gap between mouse and human data, particularly in understanding neuronal subtypes and glial cell heterogeneity.

Significant progress has been made in resolving motor neuron diversity at the single-cell level. Sathyamurthy et al. utilized massively parallel snRNA-seq to profile spinal cord neurons in adult mice and identified 43 transcriptionally distinct neuronal subtypes with region-specific distributions [19]. Their work established a comprehensive molecular map of spinal neurons and linked transcriptional identities to the anatomical and functional domains. These findings have led to new hypotheses regarding the functional roles of specific neuronal subtypes in motor control and disease development. For example, the identification of distinct transcriptional programs in motor neurons suggests that targeted interventions can be developed to modulate the activity of specific neuronal populations, potentially offering new therapeutic directions for motor neuron diseases [19].

Building on these findings, Blum et al. focused specifically on adult spinal motor neurons and used scRNA-seq to investigate transcriptional heterogeneity within this population [4]. Despite similar morphological and positional features, individual motor neurons exhibit distinct transcriptional programs related to their electrophysiological properties, spatial organization, and axonal targeting [4]. This study provides a detailed molecular atlas of motor neurons, laying the groundwork for future investigations into motor system function and dysfunction. The detailed molecular atlas generated by Blum et al. [4] has opened new avenues for exploring the mechanisms underlying motor neuron diversity and has led to new hypotheses regarding the role of transcriptional heterogeneity in motor neuron function. For instance, this study suggests that specific transcriptional profiles may be associated with different motor neuron subtypes, which may have implications for understanding the pathophysiology of motor neuron diseases [4].

In a broader developmental context, Cao et al. performed large-scale scRNA-seq during mouse organogenesis and constructed a multi-organ cell atlas, identifying developmental trajectories across multiple germ layers [36]. Their analysis showed that spinal cord progenitors undergo clear transcriptional transitions over time, driven by key regulatory factors such as Hox genes and the Sonic Hedgehog (Shh) signaling pathway [36]. This study offers a valuable resource for studying neural development and the emergence of cellular heterogeneity in the spinal cord. The identification of key regulatory factors and developmental trajectories in the spinal cord has led to new hypotheses regarding the mechanisms driving cellular differentiation and tissue organization. For example, the role of Hox genes and the Shh signaling pathway in spinal cord development suggests potential targets for therapeutic interventions aiming to promote neural regeneration and tissue repair [36].

Advances in computational methods have further enhanced our ability to analyze the developmental dynamics. The reversed graph embedding (RGE) algorithm proposed by Qiu et al. enables the reconstruction of pseudotemporal trajectories from scRNA-seq data without time labels [37]. Implemented in Monocle 2, this method has been widely used to model nonlinear developmental progressions in complex systems, including the differentiation of spinal cord progenitors into diverse neuronal lineages. Computational tools, such as pseudotime analysis, provide a more detailed understanding of cellular transitions and developmental trajectories in the spinal cord [38]. Pseudotime analysis allows researchers to infer the temporal sequence of cellular states from static scRNA-seq data, offering insights into the dynamic processes underlying spinal cord development and regeneration [37]. This is particularly valuable for studying the differentiation of neural progenitors and the maturation of neuronal subtypes, which are critical for understanding the functional organization of the spinal cord [17].

Recently, Zhang et al. combined snRNA-seq with spatial transcriptomics to profile spinal cord neurons from nine human donors, identifying 21 neuronal subtypes and defining their spatial distribution [18]. Comparative analysis between human and mouse spinal cords revealed broad similarities in neuronal composition and species-specific transcriptional signatures. Notably, sex-specific differences in gene expression were also observed in motor neurons, including SCN10A and HCN1 [18]. Spatial transcriptomics has revolutionized our understanding of the spatial organization of cells within the spinal cord. Unlike traditional scRNA-seq, spatial transcriptomics provides a snapshot of gene expression without spatial context and allows for the simultaneous measurement of gene expression and cellular localization [39].

The studies selected for this review were chosen to represent the latest advancements in single-cell transcriptomics applied to spinal cord research. Collectively, these studies illustrate the power of single-cell and spatial transcriptomics in resolving the cellular diversity and complex regulatory networks underpinning spinal cord development. These findings provide foundational insights into neurodevelopmental biology and inform our understanding of the molecular basis of spinal disorders.

Despite these advancements, current single-cell approaches in spinal cord studies face several limitations. For instance, spatial resolution remains a challenge, as many single-cell techniques do not provide detailed spatial information regarding cell locations within tissues [14]. This limitation is particularly significant in the spinal cord, where spatial organization is crucial for understanding cellular interactions and tissue function. Sampling bias can also affect the representativeness of the data, particularly when dealing with rare cell types or small sample sizes [40]. Additionally, the technical noise inherent in single-cell sequencing can introduce variability and limit the accuracy of gene expression measurements [41]. These limitations highlight the need for continued technological improvements and methodological refinements to enhance the reliability and applicability of single-cell studies in spinal cord research. Future studies should address these challenges by integrating spatial transcriptomics with single-cell RNA sequencing and developing new computational tools to mitigate sampling biases and technical noise.

3.2. Investigating Spinal Cord Injury and Neurodegeneration

scRNA-seq has provided unprecedented resolution for dissecting the cellular heterogeneity and molecular mechanisms underlying SCI and neurodegenerative diseases. By enabling the high-throughput profiling of cell populations at various time points and pathological states, researchers can identify critical cellular subtypes, signaling pathways, and transcription factors, thereby advancing the understanding of disease mechanisms and uncovering potential therapeutic targets.

SCI is a severe CNS disorder characterized by disruption of spinal cord structure and function, leading to impairments in motor, sensory, and autonomic functions [42]. The application of single-cell transcriptomic technologies offers unique opportunities to explore cellular responses following SCI at single-cell resolution, allowing for the characterization of cellular composition, state transitions, and gene expression dynamics in injured spinal tissues [43]. Following SCI, the spinal cord undergoes complex cellular and molecular alterations, with glial cell responses being particularly well delineated by scRNA-seq studies.

Research has demonstrated that microglia and astrocytes exhibit pronounced dynamic changes after injury, forming multiple functional subtypes. Notably, microglia undergo a second wave of activation approximately 14 days post-injury, coinciding with secondary neuronal damage [44]. While other cell types tend to revert to their pre-injury states, microglia remain persistently activated, potentially altering the immune microenvironment in the long term [44]. For instance, certain microglial subtypes with high Cd68 expression may suppress regenerative capacities [44]. Zhang et al. (2025) conducted a cross-species study utilizing single-nucleus RNA sequencing and lineage tracing to comprehensively characterize neural cells in the spinal cord from development to injury [45]. This study revealed, for the first time, that astrocytes in the adult spinal cord possess significant multipotency and can transdifferentiate into mature oligodendrocytes following injury [45].

Oligodendrocytes and their progenitors are crucial for maintaining myelin integrity and promoting axonal regeneration [46]. scRNA-seq has enabled an in-depth analysis of the cellular heterogeneity within the oligodendrocyte lineage, elucidating their functional changes and intercellular interactions at various post-injury stages [47]. Notably, the Psap (prosaposin)/Gpr37l1 and Psap/Gpr37 ligand-receptor pairs have been implicated as key regulators within the oligodendrocyte lineage, and several transcription factors with potential regulatory roles have been predicted [47].

Moreover, snRNA-seq has revealed significant alterations in cells distal to the injury sites. In a mouse model, researchers analyzed lumbar spinal cord cells following thoracic SCI and observed dynamic responses across multiple cell types during the acute, subacute, and chronic phases [35]. Particularly, a rare neuronal population, including major spinocerebellar projection neurons, was found to express regeneration-associated genes post-injury [35]. These neurons demonstrated axonal preservation, extension, and remodeling, suggesting a potential role in functional recovery [35].

snRNA-seq has also shown that neurons exhibit distinct gene expression changes and structural and functional plasticity after SCI. Sarathwath et al. (2024) investigated regenerative processes in adult zebrafish six weeks post-severe SCI and discovered that adult neurogenesis and neural plasticity synergistically promote spinal repair [48]. They identified a population of injury-responsive neurons (iNeurons), which displayed significant plasticity and adopted neuroblast-like transcriptional features one week post-injury, thereby establishing zebrafish as a model for plasticity-driven neural repair [48].

Single-cell transcriptomic analyses have elucidated the complex immune responses following SCI. Gillespie et al. (2024), using humanized NSG-SGM3 mice, systematically analyzed human immune responses post-SCI via scRNA-seq [49]. The study identified diverse immune cell subpopulations and revealed the downregulation of T cell receptor signaling and antigen presentation, confirming the presence of spinal cord injury-induced immune deficiency syndrome (SCI-IDS) [49]. Concurrently, immune cells in the local lesion microenvironment are activated, with an upregulation of genes associated with proliferation and oxidative phosphorylation [49]. This model provides a valuable platform for translational studies on immune responses in SCI.

The disruption of the spinal vascular system is a critical pathological factor post-SCI, contributing to a deteriorating microenvironment, exacerbated inflammation, and nutrient deprivation at the injury site. Yao et al. (2022) conducted a comprehensive scRNA-seq analysis of vascular endothelial cells (ECs) in a rat SCI model and discovered that microglia and macrophages regulate specific EC subtypes via the SPP1 and IGF signaling pathways, promoting endogenous angiogenesis [50]. This study highlights the importance of immune–endothelial interactions in post-SCI vascular remodeling, offering new avenues for therapies aiming to enhance angiogenesis after SCI. Recent studies have also highlighted the role of inflammatory signaling pathways, such as IL-33/ST2, in modulating neuroinflammation and neuroprotection after SCI, offering potential therapeutic targets for intervention [51].

Spinal muscular atrophy (SMA) is a neurodegenerative disorder caused by the loss of SMN protein function [52]. While SMA primarily affects anterior horn motor neurons, non-neuronal cells also play significant roles in its pathogenesis [53]. The single-cell RNA sequencing (scRNA-seq) of spinal cord cells from severe SMA mouse models identified ten distinct cell types and their differentially expressed genes. CellChat analysis has revealed markedly reduced intercellular communication in SMA spinal cords [54]. Furthermore, a specific vascular fibroblast subpopulation is significantly diminished, potentially leading to vascular defects and widespread impairments in protein synthesis and energy metabolism [54].

During the secondary phase of SCI, the immune microenvironment at the injury site plays a critical role in spinal regeneration [55]. Previous studies have shown that macrophages/microglia exhibit both pro-inflammatory and anti-inflammatory functions during the subacute phase of SCI [56]. Zhang et al. (2023) integrated scRNA-seq and bulk RNA-seq data to delineate the immune infiltration landscape and identify potential therapeutic agents for SCI [57]. During the subacute phase, B2m, Itgb5, and Vav1 were identified as key molecular targets in macrophages/microglia, potentially promoting neural regeneration [57]. Additionally, low-dose decitabine modulates immune cell polarization and enhances the regenerative capacity of spinal tissues [57]. Identifying hub genes and targeted therapies for macrophage/microglial responses post-SCI is crucial for advancing neural regeneration.

In conclusion, single-cell transcriptomic technologies have been instrumental in unveiling cellular heterogeneity, disrupted cell-cell communication, and immune microenvironmental changes associated with SCI and neurodegenerative diseases. These tools have enabled researchers to dissect complex cellular responses and molecular mechanisms with unprecedented resolution. These studies have elucidated critical processes such as neuronal plasticity, glial trans-differentiation, immune modulation, and vascular remodeling, providing a robust theoretical foundation for the development of novel therapeutic strategies.

3.3. Spatial Transcriptomics and Multi-Omics Integration

scRNA-seq has significantly advanced our understanding of the cellular heterogeneity within the spinal cord. Proteomic insights, such as UBL3-mediated EV protein secretion in cancer cells [58], highlight the value of integrating transcriptomics with extracellular vesicle studies in spinal cord research. However, the lack of spatial context limits comprehensive insights into the localization and interactions of cells within the tissue microenvironment. The emergence of spatial transcriptomics addresses this limitation by enabling transcriptome-wide gene expression profiling while preserving the spatial organization of the tissues. This allows researchers to capture the spatial distribution of gene expression at a high resolution within tissue sections, facilitating a deeper understanding of spinal cord structure and function. Compared to conventional scRNA-seq, spatial transcriptomics can precisely map the spatial distribution patterns of distinct cell types within the spinal cord, retaining structural information and thereby offering a more holistic view of the spatiotemporal characteristics of neurons and glial cells.

Delile et al. (2019) utilized single-cell transcriptomic analysis to construct a spatiotemporal atlas of gene expression during mouse spinal cord development, identifying various neuronal subtypes and their spatial localization, thereby enhancing our understanding of spinal cord developmental mechanisms [17]. Russ et al. (2021) integrated six independent scRNA-seq datasets to generate a comprehensive cell-type atlas of the mouse spinal cord [59]. This study identified 84 distinct cell types and validated their spatial distribution using high-throughput in situ hybridization, revealing cell-type-specific distributions along the dorsoventral and rostrocaudal axes [59]. Additionally, they developed an open-source cell type classification tool, SeqSeek, to facilitate standardized cell type identification, providing a valuable resource for exploring spinal cord cellular diversity and spatial organization [59].

Zhang et al. (2024) applied the 10x Genomics Visium platform to generate spatial transcriptomes of lumbar spinal cord sections from six adult human donors (three males and three females, aged 47–59 years) [18]. They identified 17 cell types with distinct spatial distribution patterns and further integrated spatial transcriptomics with single-nucleus RNA sequencing to systematically classify neuronal subtypes and map their localization [18]. Notably, they uncovered sex-specific differences in gene expression within motor neurons, including the differential expression of SCN10A and HCN1, providing a molecular basis for sex-specific spinal cord functions [18].

Han et al. (2024) employed spatial transcriptomics and scRNA-seq to analyze the molecular characteristics of ischemic hemispheres in a mouse stroke model [60]. They found that the interaction between galectin-9 (LGALS9) and CD44 plays a crucial role in post-stroke neuroinflammation and repair [60]. Lgals9, delivered via extracellular vesicles, promoted functional recovery, highlighting its potential as a therapeutic target and revealing cell-type-specific responses in injured regions [60].

Recent studies have underscored the value of integrating mass spectrometry-based omics with spatial and transcriptomic approaches to enhance the molecular resolution of spinal cord research. For instance, matrix-assisted laser desorption/ionization (MALDI)-based mass spectrometry imaging (MSI) has provided spatially resolved metabolic profiles of the spinal cord, enabling the detection of molecular alterations following injury or disease progression [61]. In vascular and neuroinflammatory contexts, mass spectrometry proteomics has facilitated the identification of lipid and protein regulators involved in immune and endothelial cell signaling [62,63]. Complementing transcriptomic insights, these mass spectrometry-based approaches offer orthogonal validation of cell type-specific functional states and metabolic reprogramming processes. Notably, reviews and protocols emphasize the synergy between MSI and omics, promoting the high-content spatial mapping of nervous tissue [64]. Therefore, the convergence of MSI with single-cell and spatial transcriptomics is poised to enrich multi-omics analyses of spinal cord tissues, providing a deeper understanding of cell metabolism, signaling dynamics, and therapeutic targets [65,66].

Spatial multi-omics has demonstrated unique value in SCI research. By integrating immunohistochemistry with multiparametric analyses, spatial multi-omics can delineate microenvironmental changes during the secondary injury phase post-SCI [67]. Researchers have systematically reviewed advancements in spatial multi-omics for studying the post-injury spinal microenvironment, including alterations in the immune milieu, and have discussed potential future therapeutic strategies [67].

Spatial multi-omics technologies offer novel perspectives for understanding spinal cord disease. In amyotrophic lateral sclerosis (ALS) research, spatial transcriptomic analyses have shown that ALS risk genes are enriched in motor neurons and microglia [68]. This pioneering study introduced TF-seqFISH, an image-based single-cell transcription factor spatial decoding method, to investigate the spatial expression and regulation of transcription factors during human spinal cord development [68]. By integrating TF-seqFISH data with scRNA-seq, researchers have elucidated the spatial distribution of neural progenitors along the dorsoventral axis and identified the molecular and spatial characteristics governing neuronal generation, migration, and differentiation along the mediolateral axis [68]. These findings provide a valuable spatiotemporal transcriptomic resource for the developing human spinal cord and offer potential intervention strategies for SCI repair and ALS treatment.

These studies, despite their diverse methodologies, collectively illustrate how spatial transcriptomics can unravel the spatiotemporal features unique to both development and injury. For instance, developmental studies in mice (e.g., Delile et al. [17]) focus on progenitor domain patterning, whereas injury models (e.g., Zhang et al. [45]) explore dynamic cell fate changes, such as astrocyte-to-oligodendrocyte transitions. Moreover, human-focused datasets (Zhang et al. [18]) reveal conserved and species-specific molecular features that are not captured in rodent models.

Spatial transcriptomics complements single-cell and single-nucleus RNA-seq by preserving anatomical architecture and enabling the localization of transcriptomic signatures identified earlier. Together, these methods offer a layered understanding of spinal cord biology; scRNA-seq identifies cell types, whereas spatial tools anchor these identities in the tissue space. Recent advances in omics technologies have revolutionized the understanding of spinal cord biology. As illustrated in Figure 2, scRNA-seq and snRNA-seq have enabled the comprehensive characterization of cellular diversity, whereas spatial transcriptomics provides crucial spatial context by preserving tissue architecture. Proteomics and multi-omics approaches further expand this toolkit by integrating protein-level data and cross-omics correlations, offering unprecedented insights into the function and disease mechanisms of the spinal cord.

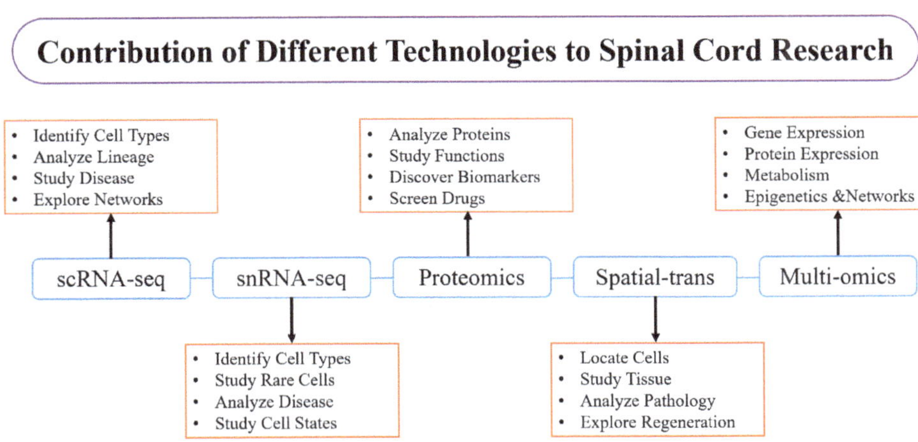

Figure 2. Contributions of representative technologies to spinal cord biology research.

Despite its promise, spatial transcriptomics faces key limitations, including low resolution, limited gene throughput, and spatial misregistration errors [69,70]. Innovations such as Slide-seq V2, Stereo-seq, and spatial deconvolution algorithms (e.g., Tangram and SIMO) aim to enhance accuracy and scalability [71,72]. These improvements are critical for mapping fine-grained structures, such as the laminar spinal cord architecture.

The integration of single-cell/nucleus transcriptomics with spatial transcriptomics offers complementary insights: while scRNA-seq provides high-resolution cell type definitions, spatial transcriptomics preserves the spatial context within the tissues [73]. Effective multi-omics data integration requires robust computational methods to handle the heterogeneity and high dimensionality of various data types. Gao et al. (2023) proposed a graph convolutional network-based framework (GCN-SC) for single-cell multi-omics integration, effectively fusing diverse modalities and enhancing the accuracy of cell type classification [74]. Wang et al. developed contrastive cycle adversarial autoencoders to align and integrate single-cell multi-omics data, maintaining structural integrity while minimizing cross-modality discrepancies, thereby improving integration performance [75].

Cao et al. (2022) introduced GLUE (graph-linked unified embedding), a modular framework for integrating unpaired single-cell multi-omics data and inferring regulatory interactions [76]. By explicitly modeling cross-omics regulatory interactions, GLUE bridges the gap between data types, outperforming existing tools in terms of accuracy, robustness, and scalability [76]. Recently, Yang et al. (2025) proposed SIMO, a probabilistic alignment-based computational method for multi-omics spatial integration, which is capable of efficiently addressing modality differences and uncovering multi-level cellular topologies and regulatory patterns [71]. GLUE integrates unpaired omics data by modeling the regulatory interactions. Although it has not yet been specifically applied to spinal cord datasets, it has the potential to infer gene regulation in injury models. SIMO, a spatial multi-omics integrator, has demonstrated efficacy in spatial heterogeneity detection, which could enhance the mapping of regenerating regions post-SCI. GCN-SC employs graph convolution to improve cell type classification, which is crucial in tissues such as the spinal cord, where rare cell types are of interest [71,74,76]. Applications in both simulated and real biological datasets have demonstrated its high accuracy and robustness in capturing multimodal spatial heterogeneity, providing a powerful computational tool for dissecting molecular spatial organization [71].

Deep learning approaches have also shown tremendous potential for the integration of spatial transcriptomics with other modalities. Researchers have systematically reviewed deep learning applications for integrating spatial transcriptomics with histological images, chromatin imaging, and scRNA-seq data [77]. These methods can enhance spatial transcriptomic data interpretation, while spatial data provide essential contextual information for single-cell analyses.

In summary, spatial transcriptomics and multi-omics integration offer unprecedented opportunities for spinal cord research, enabling the comprehensive molecular and functional characterization of spinal cells within their native spatial contexts. With continuous advancements in technology and computational methodologies, this field is poised to make significant breakthroughs in spinal cord development, disease, and injury research.

4. Current Challenges and Limitations

Single-cell transcriptomics in spinal cord research faces several key limitations that hinder both experimental and clinical advancements. These challenges can be broadly categorized into technical, computational, and biological aspects (Figure 3).

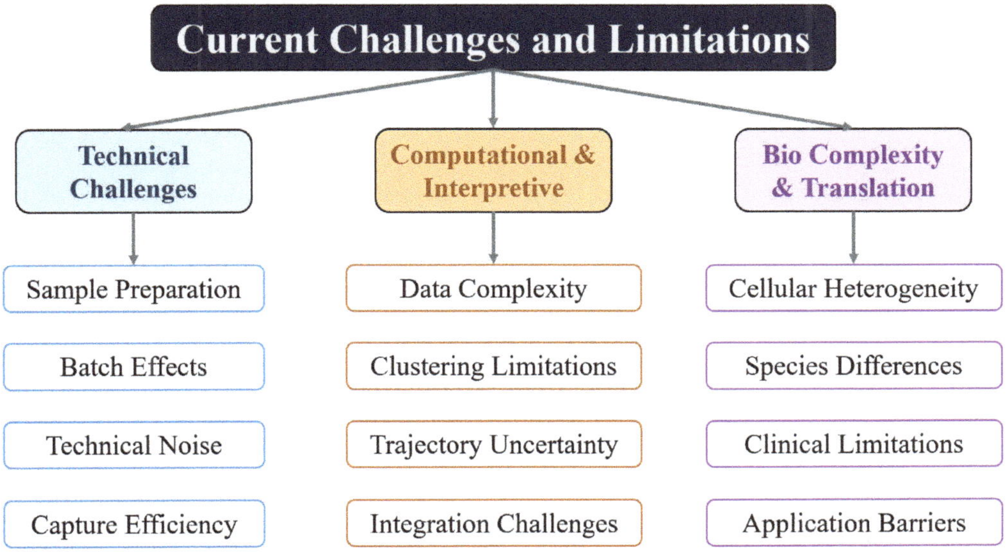

Figure 3. Current Challenges and Limitations in Applying Single-Cell Transcriptomics to Spinal Cord Research.

4.1. Technical Challenges

scRNA-seq has demonstrated tremendous potential in spinal cord research; however, it still faces numerous technical challenges and limitations in its application. This section systematically discusses these obstacles to provide insights into future studies. Key technical issues include sample preparation and cell dissociation, data sparsity and technical noise, difficulty in detecting rare cell types, a loss of spatial information, batch effects, and standardization concerns.

The preparation of spinal cord tissue samples presents unique challenges. The spinal cord is characterized by high structural complexity and cellular heterogeneity, comprising tightly packed neurons, glial cells, and vascular components [78]. Moreover, distal neuronal structures, such as axons and dendrites, are rich in RNA [79]. Conventional tissue dissociation methods used in scRNA-seq can disrupt these structures, leading to the loss of specific RNA species and compromising data integrity and completeness. Ament et al. reported that scRNA-seq often fails to capture transcripts localized in distal compartments of neurons, including dendrites, axons, growth cones, synapses, and endfeet—collectively referred to as the "dark transcriptome"—which play crucial roles in neural development and function [80]. In injured spinal cord tissue, additional challenges arise: many cell types do not survive the harsh dissociation protocols, while surviving cells often exhibit stress-induced transcriptional artifacts [81].

To address this, researchers have adopted snRNA-seq, which does not require intact cells and better preserves cellular diversity, particularly in tissues that are difficult to dissociate. For instance, Sathyamurthy et al. utilized snRNA-seq to construct a cellular atlas of the adult mouse spinal cord, identifying 43 neuronal populations, thus demonstrating the utility of this method in complex tissues [19].

The detection of rare cell types remains a critical limitation. The spinal cord contains rare but functionally important cell types, such as specific neuronal and glial subpopulations [35]. Due to their low abundance, conventional scRNA-seq often fails to efficiently capture these cells, limiting our understanding of their biological functions. Although high-throughput platforms such as 10x Genomics have improved detection sensitivity, enrichment strategies or specific markers are still required to reliably identify rare populations [35]. For example, Milich et al. investigated the origin and fate of macrophages after SCI and highlighted their critical role in injury pathology, although their heterogeneity and dynamic changes remain insufficiently characterized [82].

Data sparsity and technical noise present significant analytical challenges. scRNA-seq datasets are inherently sparse, with many genes exhibiting zero expression in individual cells. These zero inflations result from low mRNA content, inefficient reverse transcription, and amplification bias [83]. High technical noise and biological variability further complicate data interpretation. Standard RNA-seq analytical approaches may be biased by cell-to-cell variability. Kharchenko et al. modeled single-cell measurements as a mixture of success and failure in transcript detection and proposed a Bayesian framework to mitigate expression distortions and enhance the accuracy of differential expression analysis [84]. Additionally, Liu et al. noted that in small-scale organisms, single-cell transcriptomics is severely affected by high variability, which affects data reliability and interpretation [85]. The heterogeneity of spinal cord cell populations further exacerbates these issues. Studies on SCI and neurodegenerative diseases have indicated that different cell types exhibit distinct responses to SMN protein deficiency, necessitating highly sensitive techniques to discern subtle transcriptional differences [54].

The growing volume of scRNA-seq datasets enables researchers to systematically map cellular transcriptomes in diverse biological contexts. However, integrating multiple datasets remains challenging because of the biological and technical heterogeneity across platforms [86].

Another critical limitation is the loss of spatial context of the data. Traditional scRNA-seq dissociates cells from their native tissue architecture, thus losing spatial information essential for understanding cell-cell interactions and microenvironments within the spinal cord. To overcome this, spatial transcriptomics techniques, such as Visium and LCM-seq, have been developed to capture gene expression data while preserving tissue structure [14,87]. Despite advancements in spatial methods such as Slide-seq and MERFISH, challenges remain in terms of resolution, throughput, and cost, preventing them from fully replacing scRNA-seq [39,69]. As Marx (2021) noted, while spatial transcriptomics allows transcriptome profiling within the tissue context, further technological improvements are needed for broader applications [70].

Batch effects also threaten the reliability of scRNA-seq data. Technical variability between experimental batches can introduce significant batch effects that compromise data comparability and integration. Although several algorithms (e.g., MNN, Harmony, and Seurat) exist to correct batch effects, challenges persist, especially when integrating data across conditions, platforms, or species [88]. Furthermore, the lack of standardized workflows limits the comparability of datasets between studies. Luecken and Theis outlined the current best practices in scRNA-seq analysis, including preprocessing, normalization, and batch correction strategies [89].

Despite these challenges, ongoing methodological advancements and computational innovations are gradually overcoming these limitations, offering new opportunities for mechanistic insights and therapeutic development for spinal cord diseases.

4.2. Computational and Interpretive Bottlenecks

The application of scRNA-seq in spinal cord research has expanded significantly; however, it faces numerous computational and interpretive challenges. Compared to traditional bulk RNA sequencing, scRNA-seq data are inherently noisier and more complex, presenting substantial analytical difficulties [90]. Due to the limited starting material, scRNA-seq data face various computational challenges, including normalization, differential gene expression analysis, and dimensionality reduction [91]. These bottlenecks not only compromise the accuracy and reproducibility of the data but also become critical limiting factors for the broader application of single-cell transcriptomics in spinal cord studies.

A major issue in scRNA-seq data is their high sparsity and technical noise, which complicate data preprocessing and quality control [92]. For instance, the dropout phenomenon of lowly expressed genes and batch effects significantly impact the accuracy of downstream analyses [92,93]. Therefore, quality control (QC) is essential for identifying and removing low-quality cells to ensure the reliability and reproducibility of the results. Although various normalization and batch correction methods, such as Seurat and Harmony, have been developed, their efficacy remains limited in complex tissues, such as the spinal cord [88].

The high sparsity of scRNA-seq data is one of the principal barriers to analysis. The imputation of missing values is a crucial strategy for addressing this issue. Several imputation methods tailored for scRNA-seq data, including ScImpute, SAVER, and MAGIC, have been proposed [94–96]. These methods vary in their assumptions and scalability, which influence their performance. However, some researchers argue that dropout patterns might carry biologically meaningful information useful for cell type identification, offering new directions for computational method development [97].

In the context of spinal cord research, these computational and interpretive bottlenecks are particularly pronounced. A study on the single-cell transcriptomics of the developing human spinal cord revealed a significantly higher number of cells detected in the CS17 sample than in the others, potentially due to environmental RNA accumulation [98]. This phenomenon was closely linked to a 24 h delay in sample preparation, underscoring the critical impact of sample preservation conditions and processing timeliness on data quality [98]. Moreover, the complexity and high heterogeneity of spinal cord tissues impose greater demands for the interpretation of single-cell sequencing data.

Integrating scRNA-seq with other omics data, such as ATAC-seq and proteomics, can provide a comprehensive understanding of cellular function and regulatory mechanisms. However, differences in scale, noise, and missing data across these data types pose significant challenges for multi-omics integration. Although methods such as MOFA and Seurat v3 have attempted to address these issues, limitations persist in their practical applications [99,100].

The biological interpretation and validation of scRNA-seq results are crucial, yet challenging steps. The potential for false positives, technical artifacts, and dependency on extensive biological knowledge complicates the interpretation process. Furthermore, the lack of standardized validation protocols and reference databases limits the reproducibility and credibility of the findings.

In summary, the computational and interpretive bottlenecks in spinal cord scRNA-seq research include complex data preprocessing, difficulties in cell type identification, limitations in trajectory inference, challenges in multi-omics integration, computational resource constraints, and obstacles in result interpretation. Addressing these issues requires the development of more efficient, accurate, and interpretable analytical methods to advance the application of scRNA-seq in spinal cord research.

4.3. Biological Complexity and Clinical Translation

scRNA-seq has significantly advanced our understanding of the cellular heterogeneity of the spinal cord. However, the clinical translation of these findings is impeded by several factors, including the intricacies of biological complexity, interspecies differences, and barriers to transitioning from experimental research to clinical applications.

Cellular heterogeneity and dynamic changes within the spinal cord represent major biological challenges. The spinal cord consists of various cell types, including neurons, astrocytes, oligodendrocytes, microglia, and vascular-associated cells [101,102]. These cells exhibit substantial heterogeneity and undergo dynamic shifts in different physiological and pathological conditions. For example, in SCI models, microglia and astrocytes have been shown to display distinct subtypes and functional states at various stages post-injury, indicating the presence of complex regulatory mechanisms in response to injury and repair [44]. Moreover, Fan et al. (2023) applied single-cell molecular profiling to rhesus macaque SCI models and identified unique molecular heterogeneity, spatiotemporal cellular dynamics, and intricate intercellular interactions, further highlighting the dynamic complexity of spinal cord tissue during injury [103].

Interspecies differences and their impact on clinical translation are critical challenges. Although animal models, particularly mice, are widely employed in spinal cord studies, substantial discrepancies exist in cell composition and gene expression profiles between species. Zhang et al. (2024) reported that certain genes highly expressed in the mouse spinal cord are minimally expressed or entirely absent in humans and vice versa [18]. A comprehensive transcriptomic comparison of canonical neuronal markers between human and mouse spinal cords revealed notable divergences in genes such as CALCA and GRP, underscoring the need for caution regarding species-specific gene expression in clinical translation [18]. These interspecies variations may result in the failure of therapeutic strategies validated in animal models during human clinical trials. Additionally, sex differences further complicate translational efforts, as Zhang et al. (2024) demonstrated sex-specific gene expression patterns in the human spinal cord that could influence disease susceptibility and treatment efficacy [18].

Furthermore, differences in the timing of glial cell development between humans and mice add another layer of complexity. Zhang et al. (2021) found that human astrocytes and oligodendrocytes begin differentiating as early as the 8th gestational week, whereas in mice, GFAP expression—an astrocytic marker—appears only at embryonic day 16 (E16) [104]. Such species-specific developmental timelines complicate the translation of findings from animal models to human contexts, as emphasized by Kushnarev et al. [105].

Barriers to translating experimental research into clinical applications further impede progress. One major limitation is the difficulty in acquiring human spinal cord samples, which restricts large-scale single-cell studies in clinical settings. Although Zhang et al. (2024) successfully sequenced lumbar spinal cord samples from nine human donors, such studies are rare [18]. The acquisition of paired pre- and post-treatment clinical samples is particularly challenging, limiting the exploration of therapeutic mechanisms at the molecular level in humans.

Although scRNA-seq offers new avenues for precision medicine in spinal cord disorders, translation from bench to bedside remains a formidable task. Awuah et al. (2023) highlighted the potential of scRNA-seq to revolutionize clinical practice in neurological diseases by providing personalized insights and improving prognoses [106]. However, most current research remains descriptive, and substantial work is required to translate these data into diagnostic tools, prognostic biomarkers, or therapeutic targets.

Finally, technical and operational hurdles in clinical settings remain to be significant. There is no universally accepted standard for sample collection and processing, resulting in difficulties in comparing or integrating data across studies [107]. Additionally, the analysis and interpretation of scRNA-seq data require sophisticated bioinformatics expertise, which currently limits its widespread clinical implementation [107]. The high cost and technical complexity of scRNA-seq also pose barriers, although technological advancements have somewhat mitigated these issues. Further efforts are required to decrease costs and simplify the workflows for large-scale clinical research.

Although scRNA-seq excels at identifying rare cell types and transient states, offering a resolution that bulk RNA sequencing cannot match, it also suffers from significant limitations. The high cost and technical complexity of scRNA-seq render it inaccessible to many laboratories lacking specialized equipment or expertise [92]. Moreover, the process of dissociating spinal cord tissue into single cells can lead to cellular stress and changes in gene expression, introducing artifacts that may compromise data accuracy [108]. Additionally, sampling biases can result in the underrepresentation of rare cell types or specific developmental stages, while technical noise inherent to single-cell sequencing can obscure low-abundance transcripts and subtle gene expression changes.

The integration of spatial transcriptomics with scRNA-seq has begun to address some of these issues by providing a spatial context for gene expression data. This integration enhances our understanding of cellular interactions and tissue architecture. However, spatial resolution remains a significant challenge. Current techniques often fail to capture the intricate spatial organization of cells within the spinal cord, which is critical for understanding cellular functions and disease mechanisms [109]. Furthermore, the integration of multi-omics data, including proteomics and epigenomics, is essential for a comprehensive understanding of cellular regulation and function. This integration could potentially reveal new layers of complexity and provide a more comprehensive picture of spinal cord biology.

Despite the promise of scRNA-seq in identifying novel therapeutic targets, significant challenges remain in translating these findings into clinical applications. Bridging the gap between basic research and clinical practice remains a major hurdle. While scRNA-seq can identify distinct transcriptional programs in motor neurons, suggesting potential targets for modulating neuronal activity, the path from discovery to effective treatment is fraught with difficulties [19]. The development of advanced computational tools is crucial for handling the large-scale data generated by scRNA-seq and mitigating biases and noise. These tools must be refined to improve the reliability and applicability of single-cell studies in spinal cord research in the future. As the field continues to evolve, addressing these challenges is essential to fully realize the potential of scRNA-seq in advancing our understanding of spinal cord biology and disease.

In summary, despite these challenges, continued technological innovation and interdisciplinary collaboration are expected to enable scRNA-seq to drive transformative progress in spinal cord research, ultimately benefiting patients with spinal cord disorders.

5. Future Directions and Perspectives

5.1. Toward a Comprehensive Spinal Cord Cell Atlas

The construction of a comprehensive spinal cord cell atlas is one of the primary goals of current single-cell transcriptomic research. Although several studies have classified the cell types of the mouse spinal cord, these datasets have not yet been integrated into a unified reference framework. For instance, Russ et al. compiled existing single-cell transcriptomic data to construct a mouse spinal cord cell atlas, revealing the hierarchical organization and spatial distribution of diverse cell types [59].

In human spinal cord studies, Yadav et al. employed single-nucleus RNA sequencing combined with spatial transcriptomics to classify adult spinal cord cell types, enriching our understanding of human spinal cord cellular diversity [78]. Zhang et al. utilized single-nucleus RNA sequencing and spatial transcriptomics to create a detailed cell atlas of the adult human spinal cord, identifying 21 neuronal subtypes and elucidating their spatial distributions and sex-specific differences [18]. Sex-specific transcription in motor neurons, such as the differential expression of SCN10A and HCN1 [18], may underlie the sex-biased prevalence of disorders such as ALS and neuropathic pain. While initial findings come from human datasets, comparative analyses with mouse models suggest a partial conservation of these differences, although further investigation is required to determine their developmental regulation and functional relevance.

To achieve a fully comprehensive spinal cord cell atlas, future research must integrate single-cell data from diverse studies and establish standardized criteria for cell type classification. It is also essential to consider interindividual variability, including factors such as sex, age, and disease states, which may influence cellular composition. For example, the MOp census and atlas provide a robust framework for integrating cross-species cell-type data, offering a model that spinal cord research can adopt to standardize annotation and enhance cross-study compatibility [110].

Moreover, future spinal cord atlases should integrate multi-omics data that extend beyond transcriptomics alone. A recent example, Tabulae Paralytica, exemplifies this trend by incorporating four SCI datasets, including a single-nucleus transcriptomic atlas of 500,000 cells, multi-omics data combining transcriptomic and epigenomic measurements, and two spatial transcriptomic maps covering four spatial and temporal dimensions [81]. This platform facilitates the integration of four-dimensional, multimodal, and genome-scale datasets for biological and medical research [81].

With the growing availability of spinal cord single-cell datasets, establishing a unified reference framework and standardized cell-type identification methodologies has become increasingly critical. Additionally, future research should focus on the development of more advanced computational tools, such as SeqSeek, an open-source cell type classifier, to promote standardized cell type annotation [59].

These efforts will collectively contribute to a deeper understanding of the cellular composition and functional organization of the spinal cord, laying the groundwork for advancing research and therapies targeting neurological diseases.

5.2. *AI and Machine Learning in Single-Cell Analysis*

With the exponential growth of single-cell transcriptomic data, artificial intelligence (AI) and machine learning (ML) techniques have emerged as critical tools for managing and analyzing complex datasets, offering vast potential for discovery. Deep learning, in particular, has demonstrated strong promise in single-cell data analysis and is expected to unlock further opportunities in the future [111]. In spinal cord research, these computational methods have revolutionized research paradigms by optimizing data preprocessing, enhancing cell type identification and marker gene prediction, improving annotation accuracy, elucidating dynamic trajectories, and integrating multimodal data.

Recent advances in deep learning, especially graph neural networks (GNNs) and autoencoders, have shown great efficacy in preprocessing and denoising scRNA-seq data. For example, Wang et al. introduced the scGNN framework, which constructs graph-based representations of cellular relationships and applies multimodal autoencoders to mitigate data sparsity, improving both cell clustering and gene expression imputation, with broad applicability in single-cell neural studies [112]. Furthermore, Xi et al. systematically evaluated various autoencoder architectures for single-cell data imputation, and their optimized designs significantly enhanced downstream analysis accuracy, providing a solid foundation for high-quality data mining [113].

AI/ML methods have also excelled in cell-type identification and marker gene prediction. Cao et al. systematically assessed 13 supervised learning algorithms for phenotypic classification in scRNA-seq data and found that ElasticNet and XGBoost consistently outperformed the others across diverse dataset scales, demonstrating their potential for robust cell-type classification [114].

For cell type annotation, tools such as scDeepSort, ACTINN, and scBERT enable highly accurate predictions across platforms. Meanwhile, preprocessing and noise reduction are addressed by models such as scGNN and autoencoder-based frameworks. scDeepSort, a weighted GNN-based pretrained annotation tool, enables high-accuracy cell type labeling without relying on predefined marker genes or RNA-seq references, making it broadly applicable across tissues and platforms [115]. Similarly, Automated Cell Type Identification using Neural Networks (ACTINN) offers the rapid and accurate identification of cell types, validated across murine leukocytes, human PBMCs, and human T-cell subtypes, suggesting its utility as an adjunct in scRNA-seq pipelines [116]. Yang et al. developed scBERT, a BERT-inspired model for gene expression pattern recognition [117]. Separately, Wang and Du introduced WCSGNet, a GNN-based approach that focuses on cell-specific gene networks, enabling precise annotation in diverse cellular contexts [118].

With established cell atlases, the full potential of gene expression analysis can be leveraged to define and decode spinal cord cell types [119]. The ScnML model, specifically designed for spinal cord neuron subpopulation recognition, enhances the classification accuracy in high-dimensional feature spaces and addresses the challenges of computational efficiency and overfitting seen in conventional models [120].

Developmental trajectory inference and dynamic process modeling are crucial for uncovering cell-state transitions, fate decisions, and tissue developmental mechanisms. Fang et al. introduced Chronocell, a trajectory inference approach based on "process time," which improves the biological interpretability of developmental pathways and is particularly suited to complex dynamic reconstructions [121]. In spinal cord development, Shi et al. integrated scRNA-seq, spatial transcriptomics, and TF-seqFISH to delineate the migration and differentiation trajectories of neural progenitors along the dorsoventral and mediolateral axes, highlighting the finely tuned spatiotemporal regulation by transcription factors [68]. This study deepens the understanding of spinal cord developmental dynamics and proposes novel time-space-integrated modeling approaches [68].

Deep learning also shows great potential for multimodal integration and spatial transcriptomics. Biancalani et al. developed Tangram, a deep learning-based framework for the precise alignment of scRNA-seq data with various spatial transcriptomics technologies (e.g., MERFISH, Visium), enabling genome-wide spatial expression reconstruction at single-cell resolution [72]. Recently, researchers introduced scNET, which combines scRNA-seq data with protein–protein interaction networks, leveraging GNNs to enhance gene function representation and pathway identification, providing a novel tool for multi-omics integration [122].

In the context of SCI, researchers have pursued multidimensional explorations, from molecular characterization to clinical trajectory analysis. Liu et al. applied the ScnML model to analyze scRNA-seq data, achieving a high-precision prediction of neuronal subpopulations and identifying functional state-related marker genes, thus offering molecular insights into post-injury cellular dynamics [120]. Concurrently, Jaja et al. developed a classification system for recovery trajectories in patients with cervical SCI based on longitudinal clinical data, revealing dynamic heterogeneity in functional recovery and underscoring the prognostic significance of recovery patterns over initial injury severity [123]. These studies jointly establish a comprehensive framework linking single-cell molecular features with clinical outcomes, thereby advancing precision diagnostics and therapy for SCI.

The combined application of ML, statistical methods, and AI in scRNA-seq analysis significantly enhances cell type identification and an understanding of cellular heterogeneity and provides insights into biological processes [124]. These approaches are anticipated to improve disease stratification, foster innovative therapeutic strategies, and propel precision medicine [125].

Despite these advancements, AI/ML techniques still face challenges in terms of data quality control, model interpretability, and cross-platform data integration. Future research should focus on developing more robust algorithms, enhancing interpretability, and standardizing data processing protocols to facilitate the widespread application of AI/ML in single-cell spinal cord research.

In conclusion, AI and machine learning hold great promise for single-cell transcriptomic analysis, particularly in spinal cord studies. Through continuous algorithm refinement and multisource data integration, these technologies are poised to drive spinal cord disease diagnosis and treatment into a new era of precision medicine.

5.3. Personalized Medicine and Regenerative Therapies

While AI/ML tools significantly enhance the analysis of scRNA-seq data, their full potential is realized when applied in translational contexts, such as regenerative medicine, where single-cell insights can guide therapeutic innovations. With the rapid advancement of scRNA-seq technologies, spinal cord research has entered a new era characterized by personalized medicine and regenerative therapies. By enabling a detailed analysis of cellular heterogeneity and molecular signatures, scRNA-seq provides a robust foundation for precise diagnosis and individualized therapeutic strategies in cancer.

The application of scRNA-seq in personalized medicine lies in its capacity to delineate the transcriptional profiles of diverse cell types within spinal cord tissues with high resolution. In particular, following SCI, scRNA-seq facilitates the identification of specific cellular subpopulations and their dynamic responses. For instance, during the subacute phase of SCI, immune-related genes such as B2m, Itgb5, and Vav1 are upregulated in microglia and macrophages, suggesting their potential as key targets for modulating immune responses and promoting neural regeneration [57]. Furthermore, machine learning models, such as ScnML, integrated with scRNA-seq data, have demonstrated high efficiency in predicting spinal neural cell subtypes and identifying novel marker genes, thereby offering powerful tools for personalized interventions [120]. Moreover, scRNA-seq has revealed significant heterogeneity and dynamic changes among various cell types in the post-injury microenvironment, underscoring its pivotal role in elucidating the mechanisms of injury and guiding regenerative strategies. [126].

In regenerative medicine, scRNA-seq is a powerful tool for understanding cellular diversity and developing novel therapeutic approaches. In neural stem cell studies, scRNA-seq has shown that Nestin-positive neural stem cells are activated post-SCI, exhibiting regenerative potential, implying that these endogenous cells may contribute to neural repair [127]. Additionally, astrocytes have been found to transdifferentiate into oligodendrocytes following injury, participating in remyelination processes and offering new strategies for regeneration [104]. Notably, adult zebrafish exhibit an intrinsic capacity for spinal cord regeneration, with the neurogenesis of glutamatergic and GABAergic neurons helping to restore excitatory/inhibitory balance, making them an ideal model for plasticity-driven repair [48]. Furthermore, Li et al. utilized human pluripotent stem cell-derived spinal cord organoid (SSCO) models in conjunction with scRNA-seq to map the developmental trajectory of the spine and spinal cord, identifying HMMR-positive bipotent neuro-mesodermal progenitors, thereby offering novel tools for regenerative therapies and disease modeling [128].

Looking ahead, the integration of scRNA-seq with spatial transcriptomics, epigenomics, and proteomics promises a more comprehensive understanding of the molecular mechanisms underlying SCI, thereby accelerating the development of personalized medicine and regenerative treatment. For example, the integration of single-cell and spatial transcriptomics has enabled the construction of a cellular atlas of the adult human spinal cord, providing new insights into its complex architecture [78]. Moreover, the application of artificial intelligence and machine learning will further expedite the extraction of biologically meaningful information from large-scale single-cell datasets, driving the realization of precision medicine applications.

In summary, single-cell transcriptomics has brought about transformative changes in spinal cord research, deepening our understanding of spinal development and pathology, and laying a solid foundation for the advancement of personalized medicine and regenerative therapies. As technologies continue to evolve and interdisciplinary collaborations intensify, scRNA-seq will undoubtedly drive further progress in both basic research and clinical applications related to spinal cord injury.

6. Conclusions

In this review, we systematically summarize recent advancements in scRNA-seq in the field of spinal cord research. By integrating findings from recent studies, it is evident that scRNA-seq has led to groundbreaking progress in various areas, including spinal cord developmental biology [17,68], mechanisms of injury repair [126,129], and neurodegenerative disease research [54]. scRNA-seq has revealed cellular heterogeneity, developmental trajectories, and intercellular interactions within spinal cord tissues through high-throughput sequencing of transcripts at the single-cell level, offering unprecedented resolution for deciphering the complexity of spinal cord biology. Clinically, single-cell technologies are reshaping the paradigms of diagnosis and treatment of spinal cord disorders.

Despite these notable advantages, the application of single-cell transcriptomics in spinal cord research faces several challenges. These include difficulties in sample acquisition, the computational complexity of data analysis, and the necessity for a biological validation of the findings. Moreover, high technical costs and a lack of standardized protocols limit its broader implementation in clinical settings.

Looking forward, the integration of scRNA-seq with emerging spatial transcriptomics and multi-omics technologies is expected to facilitate the construction of a more comprehensive cellular atlas of the spinal cord, elucidating the spatiotemporal dynamics of the cellular processes. The adoption of artificial intelligence and machine learning approaches will further enhance the efficiency and precision of data analysis, thereby advancing personalized medicine and regenerative therapies. The convergence of single-cell transcriptomics with these cutting-edge technologies is poised to propel spinal cord research into a new era of discovery.

In conclusion, single-cell transcriptomics offers a powerful tool for spinal cord research, deepening our understanding of spinal cord development, injury, and disease. With continuous technological innovations and interdisciplinary collaborations, scRNA-seq is anticipated to play an increasingly vital role in both fundamental research and clinical applications, fostering sustained progress in neuroscience.

Author Contributions: Conceptualization, M.M.H.; methodology, M.A.M. and M.M.H.; writing—review and editing, M.A.M. and M.M.H.; supervision, M.M.H. All authors have read and agreed to the published version of the manuscript.

Funding: This study received no external funding.

Data Availability Statement: No new data were created or analyzed in this study. Data sharing is not applicable to this article.

Conflicts of Interest: The authors declare no conflict of interest.

Abbreviations

The following abbreviations are used in this manuscript:

scRNA-seq	Single-cell RNA sequencing
CNS	Central Nervous System
SCI	Spinal Cord Injury
SMA	Spinal Muscular Atrophy
QC	Quality Control
AI	Artificial Intelligence
ML	Machine Learning
PBMCs	Peripheral Blood Mononuclear Cells
SMN	Survival Motor Neuron
ALS	Amyotrophic Lateral Sclerosis
GNNs	Graph Neural Networks
snRNA-seq	Single-nucleus RNA sequencing
MSI	mass spectrometry imaging

References

1. Harrow-Mortelliti, M.; Reddy, V.; Jimsheleishvili, G. Physiology, Spinal Cord. [Updated 2023 Mar 17]. In *StatPearls [Internet]*; StatPearls Publishing: Treasure Island, FL, USA, 2025. Available online: https://www.ncbi.nlm.nih.gov/books/NBK544267/ (accessed on 28 April 2025).
2. Bahney, J.; von Bartheld, C.S. The Cellular Composition and Glia-Neuron Ratio in the Spinal Cord of a Human and a Nonhuman Primate: Comparison with Other Species and Brain Regions. *Anat. Rec.* **2018**, *301*, 697–710. [CrossRef] [PubMed]
3. Liau, E.S.; Jin, S.; Chen, Y.C.; Liu, W.S.; Calon, M.; Nedelec, S.; Nie, Q.; Chen, J.A. Single-cell transcriptomic analysis reveals diversity within mammalian spinal motor neurons. *Nat. Commun.* **2023**, *14*, 46. [CrossRef]
4. Blum, J.A.; Klemm, S.; Shadrach, J.L.; Guttenplan, K.A.; Nakayama, L.; Kathiria, A.; Hoang, P.T.; Gautier, O.; Kaltschmidt, J.A.; Greenleaf, W.J.; et al. Single-cell transcriptomic analysis of the adult mouse spinal cord reveals molecular diversity of autonomic and skeletal motor neurons. *Nat. Neurosci.* **2021**, *24*, 572–583. [CrossRef] [PubMed]
5. Wang, R.; Zhou, R.; Chen, Z.; Gao, S.; Zhou, F. The Glial Cells Respond to Spinal Cord Injury. *Front. Neurol.* **2022**, *13*, 844497. [CrossRef] [PubMed]
6. Jäkel, S.; Dimou, L. Glial Cells and Their Function in the Adult Brain: A Journey through the History of Their Ablation. *Front. Cell. Neurosci.* **2017**, *11*, 24. [CrossRef]
7. Gaudet, A.D.; Fonken, L.K. Glial Cells Shape Pathology and Repair After Spinal Cord Injury. *Neurother. J. Am. Soc. Exp. Neurother.* **2018**, *15*, 554–577. [CrossRef]
8. Wang, N.; Zheng, J.; Chen, Z.; Liu, Y.; Dura, B.; Kwak, M.; Xavier-Ferrucio, J.; Lu, Y.C.; Zhang, M.; Roden, C.; et al. Single-cell microRNA-mRNA co-sequencing reveals non-genetic heterogeneity and mechanisms of microRNA regulation. *Nat. Commun.* **2019**, *10*, 95. [CrossRef]
9. Rosenberg, A.B.; Roco, C.M.; Muscat, R.A.; Kuchina, A.; Sample, P.; Yao, Z.; Graybuck, L.T.; Peeler, D.J.; Mukherjee, S.; Chen, W.; et al. Single-cell profiling of the developing mouse brain and spinal cord with split-pool barcoding. *Science* **2018**, *360*, 176–182. [CrossRef]
10. Chuang, H.C.; Li, R.; Huang, H.; Liu, S.W.; Wan, C.; Chaudhuri, S.; Yue, L.; Wong, T.; Dominical, V.; Yen, R.; et al. Single-cell sequencing of full-length transcripts and T-cell receptors with automated high-throughput Smart-seq3. *BMC Genom.* **2024**, *25*, 1127. [CrossRef]
11. Tang, F.; Barbacioru, C.; Wang, Y.; Nordman, E.; Lee, C.; Xu, N.; Wang, X.; Bodeau, J.; Tuch, B.B.; Siddiqui, A.; et al. mRNA-Seq whole-transcriptome analysis of a single cell. *Nat. Methods* **2009**, *6*, 377–382. [CrossRef]
12. Shen, X.; Zhao, Y.; Wang, Z.; Shi, Q. Recent advances in high-throughput single-cell transcriptomics and spatial transcriptomics. *Lab Chip* **2022**, *22*, 4774–4791. [CrossRef] [PubMed]
13. Wirz, J.; Fessler, L.I.; Gehring, W.J. Localization of the Antennapedia protein in Drosophila embryos and imaginal discs. *EMBO J.* **1986**, *5*, 3327–3334. [CrossRef] [PubMed]
14. Ståhl, P.L.; Salmén, F.; Vickovic, S.; Lundmark, A.; Navarro, J.F.; Magnusson, J.; Giacomello, S.; Asp, M.; Westholm, J.O.; Huss, M.; et al. Visualization and analysis of gene expression in tissue sections by spatial transcriptomics. *Science* **2016**, *353*, 78–82. [CrossRef]
15. Molla Desta, G.; Birhanu, A.G. Advancements in single-cell RNA sequencing and spatial transcriptomics: Transforming biomedical research. *Acta Biochim. Pol.* **2025**, *72*, 13922. [CrossRef]
16. Aldridge, S.; Teichmann, S.A. Single cell transcriptomics comes of age. *Nat. Commun.* **2020**, *11*, 4307. [CrossRef] [PubMed]

17. Delile, J.; Rayon, T.; Melchionda, M.; Edwards, A.; Briscoe, J.; Sagner, A. Single cell transcriptomics reveals spatial and temporal dynamics of gene expression in the developing mouse spinal cord. *Development* **2019**, *146*, dev173807. [CrossRef]
18. Zhang, D.; Chen, Y.; Wei, Y.; Chen, H.; Wu, Y.; Wu, L.; Li, J.; Ren, Q.; Miao, C.; Zhu, T.; et al. Spatial transcriptomics and single-nucleus RNA sequencing reveal a transcriptomic atlas of adult human spinal cord. *eLife* **2024**, *12*, RP92046. [CrossRef]
19. Sathyamurthy, A.; Johnson, K.R.; Matson, K.J.E.; Dobrott, C.I.; Li, L.; Ryba, A.R.; Bergman, T.B.; Kelly, M.C.; Kelley, M.W.; Levine, A.J. Massively Parallel Single Nucleus Transcriptional Profiling Defines Spinal Cord Neurons and Their Activity during Behavior. *Cell Rep.* **2018**, *22*, 2216–2225. [CrossRef]
20. Jovic, D.; Liang, X.; Zeng, H.; Lin, L.; Xu, F.; Luo, Y. Single-cell RNA sequencing technologies and applications: A brief overview. *Clin. Transl. Med.* **2022**, *12*, e694. [CrossRef]
21. Ramsköld, D.; Luo, S.; Wang, Y.C.; Li, R.; Deng, Q.; Faridani, O.R.; Daniels, G.A.; Khrebtukova, I.; Loring, J.F.; Laurent, L.C.; et al. Full-length mRNA-Seq from single-cell levels of RNA and individual circulating tumor cells. *Nat. Biotechnol.* **2012**, *30*, 777–782. [CrossRef]
22. Wu, X.; Yang, B.; Udo-Inyang, I.; Ji, S.; Ozog, D.; Zhou, L.; Mi, Q.S. Research Techniques Made Simple: Single-Cell RNA Sequencing and its Applications in Dermatology. *J. Investig. Dermatol.* **2018**, *138*, 1004–1009. [CrossRef] [PubMed]
23. Macosko, E.Z.; Basu, A.; Satija, R.; Nemesh, J.; Shekhar, K.; Goldman, M.; Tirosh, I.; Bialas, A.R.; Kamitaki, N.; Martersteck, E.M.; et al. Highly Parallel Genome-wide Expression Profiling of Individual Cells Using Nanoliter Droplets. *Cell* **2015**, *161*, 1202–1214. [CrossRef] [PubMed]
24. Li, X.; Wang, C.Y. From bulk, single-cell to spatial RNA sequencing. *Int. J. Oral Sci.* **2021**, *13*, 36. [CrossRef]
25. Zheng, G.X.; Terry, J.M.; Belgrader, P.; Ryvkin, P.; Bent, Z.W.; Wilson, R.; Ziraldo, S.B.; Wheeler, T.D.; McDermott, G.P.; Zhu, J.; et al. Massively parallel digital transcriptional profiling of single cells. *Nat. Commun.* **2017**, *8*, 14049. [CrossRef] [PubMed]
26. Anaparthy, N.; Ho, Y.J.; Martelotto, L.; Hammell, M.; Hicks, J. Single-Cell Applications of Next-Generation Sequencing. *Cold Spring Harb. Perspect. Med.* **2019**, *9*, a026898. [CrossRef]
27. Wang, X.; He, Y.; Zhang, Q.; Ren, X.; Zhang, Z. Direct Comparative Analyses of 10X Genomics Chromium and Smart-seq2. *Genom. Proteom. Bioinform.* **2021**, *19*, 253–266. [CrossRef]
28. Salcher, S.; Heidegger, I.; Untergasser, G.; Fotakis, G.; Scheiber, A.; Martowicz, A.; Noureen, A.; Krogsdam, A.; Schatz, C.; Schäfer, G.; et al. Comparative analysis of 10X Chromium vs. BD Rhapsody whole transcriptome single-cell sequencing technologies in complex human tissues. *Heliyon* **2024**, *10*, e28358. [CrossRef]
29. Picelli, S.; Björklund, Å.K.; Faridani, O.R.; Sagasser, S.; Winberg, G.; Sandberg, R. Smart-seq2 for sensitive full-length transcriptome profiling in single cells. *Nat. Methods* **2013**, *10*, 1096–1098. [CrossRef]
30. Hagemann-Jensen, M.; Ziegenhain, C.; Chen, P.; Ramsköld, D.; Hendriks, G.J.; Larsson, A.J.M.; Faridani, O.R.; Sandberg, R. Single-cell RNA counting at allele and isoform resolution using Smart-seq3. *Nat. Biotechnol.* **2020**, *38*, 708–714. [CrossRef]
31. Klein, A.M.; Mazutis, L.; Akartuna, I.; Tallapragada, N.; Veres, A.; Li, V.; Peshkin, L.; Weitz, D.A.; Kirschner, M.W. Droplet barcoding for single-cell transcriptomics applied to embryonic stem cells. *Cell* **2015**, *161*, 1187–1201. [CrossRef]
32. Ziegenhain, C.; Vieth, B.; Parekh, S.; Reinius, B.; Guillaumet-Adkins, A.; Smets, M.; Leonhardt, H.; Heyn, H.; Hellmann, I.; Enard, W. Comparative Analysis of Single-Cell RNA Sequencing Methods. *Mol. Cell* **2017**, *65*, 631–643.e4. [CrossRef] [PubMed]
33. Chen, A.; Liao, S.; Cheng, M.; Ma, K.; Wu, L.; Lai, Y.; Qiu, X.; Yang, J.; Xu, J.; Hao, S.; et al. Spatiotemporal transcriptomic atlas of mouse organogenesis using DNA nanoball-patterned arrays. *Cell* **2022**, *185*, 1777–1792.e21. [CrossRef] [PubMed]
34. Andersen, J.; Thom, N.; Shadrach, J.L.; Chen, X.; Onesto, M.M.; Amin, N.D.; Yoon, S.J.; Li, L.; Greenleaf, W.J.; Müller, F.; et al. Single-cell transcriptomic landscape of the developing human spinal cord. *Nat. Neurosci.* **2023**, *26*, 902–914. [CrossRef] [PubMed]
35. Matson, K.J.E.; Russ, D.E.; Kathe, C.; Hua, I.; Maric, D.; Ding, Y.; Krynitsky, J.; Pursley, R.; Sathyamurthy, A.; Squair, J.W.; et al. Single cell atlas of spinal cord injury in mice reveals a pro-regenerative signature in spinocerebellar neurons. *Nat. Commun.* **2022**, *13*, 5628. [CrossRef]
36. Cao, J.; Spielmann, M.; Qiu, X.; Huang, X.; Ibrahim, D.M.; Hill, A.J.; Zhang, F.; Mundlos, S.; Christiansen, L.; Steemers, F.J.; et al. The single-cell transcriptional landscape of mammalian organogenesis. *Nature* **2019**, *566*, 496–502. [CrossRef]
37. Qiu, X.; Mao, Q.; Tang, Y.; Wang, L.; Chawla, R.; Pliner, H.A.; Trapnell, C. Reversed graph embedding resolves complex single-cell trajectories. *Nat. Methods* **2017**, *14*, 979–982. [CrossRef]
38. Trapnell, C.; Cacchiarelli, D.; Grimsby, J.; Pokharel, P.; Li, S.; Morse, M.; Lennon, N.J.; Livak, K.J.; Mikkelsen, T.S.; Rinn, J.L. The dynamics and regulators of cell fate decisions are revealed by pseudotemporal ordering of single cells. *Nat. Biotechnol.* **2014**, *32*, 381–386. [CrossRef]
39. Rodriques, S.G.; Stickels, R.R.; Goeva, A.; Martin, C.A.; Murray, E.; Vanderburg, C.R.; Welch, J.; Chen, L.M.; Chen, F.; Macosko, E.Z. Slide-seq: A scalable technology for measuring genome-wide expression at high spatial resolution. *Science* **2019**, *363*, 1463–1467. [CrossRef]
40. Svensson, V.; Vento-Tormo, R.; Teichmann, S.A. Exponential scaling of single-cell RNA-seq in the past decade. *Nat. Protoc.* **2018**, *13*, 599–604. [CrossRef]

41. Hwang, B.; Lee, J.H.; Bang, D. Single-cell RNA sequencing technologies and bioinformatics pipelines. *Exp. Mol. Med.* **2018**, *50*, 1–14. [CrossRef]
42. Ahuja, C.S.; Wilson, J.R.; Nori, S.; Kotter, M.R.N.; Druschel, C.; Curt, A.; Fehlings, M.G. Traumatic spinal cord injury. Nature reviews. *Dis. Primers* **2017**, *3*, 17018. [CrossRef] [PubMed]
43. Wahane, S.; Zhou, X.; Zhou, X.; Guo, L.; Friedl, M.S.; Kluge, M.; Ramakrishnan, A.; Shen, L.; Friedel, C.C.; Zhang, B.; et al. Diversified transcriptional responses of myeloid and glial cells in spinal cord injury shaped by HDAC3 activity. *Sci. Adv.* **2021**, *7*, eabd8811. [CrossRef] [PubMed]
44. Li, C.; Wu, Z.; Zhou, L.; Shao, J.; Hu, X.; Xu, W.; Ren, Y.; Zhu, X.; Ge, W.; Zhang, K.; et al. Temporal and spatial cellular and molecular pathological alterations with single-cell resolution in the adult spinal cord after injury. *Signal Transduct. Target. Ther.* **2022**, *7*, 65. [CrossRef] [PubMed]
45. Zhang, Q.; Wu, X.; Fan, Y.; Zhang, H.; Yin, M.; Xue, X.; Yin, Y.; Jin, C.; Quan, R.; Jiang, P.; et al. Characterizing progenitor cells in developing and injured spinal cord: Insights from single-nucleus transcriptomics and lineage tracing. *Proc. Natl. Acad. Sci. USA* **2025**, *122*, e2413140122. [CrossRef]
46. Hildebrand, C.; Remahl, S.; Persson, H.; Bjartmar, C. Myelinated nerve fibres in the CNS. *Prog. Neurobiol.* **1993**, *40*, 319–384. [CrossRef]
47. Wu, Z.; Li, G.; Wang, S.; Zhang, N.; Li, X.; Zhang, F.; Niu, J.; Wang, N.; Zu, J.; Wang, Y. Single-cell analysis of spinal cord injury reveals functional heterogeneity of oligodendrocyte lineage cells. *Gene* **2023**, *886*, 147713. [CrossRef]
48. Saraswathy, V.M.; Zhou, L.; Mokalled, M.H. Single-cell analysis of innate spinal cord regeneration identifies intersecting modes of neuronal repair. *Nat. Commun.* **2024**, *15*, 6808. [CrossRef]
49. Gillespie, E.R.; Grice, L.F.; Courtney, I.G.; Lao, H.W.; Jung, W.; Ramkomuth, S.; Xie, J.; Brown, D.A.; Walsham, J.; Radford, K.J. et al. Single-cell RNA sequencing reveals peripheral blood leukocyte responses to spinal cord injury in mice with humanised immune systems. *J. Neuroinflamm.* **2024**, *21*, 63. [CrossRef]
50. Yao, C.; Cao, Y.; Wang, D.; Lv, Y.; Liu, Y.; Gu, X.; Wang, Y.; Wang, X.; Yu, B. Single-cell sequencing reveals microglia induced angiogenesis by specific subsets of endothelial cells following spinal cord injury. *FASEB J. Off. Publ. Fed. Am. Soc. Exp. Biol.* **2022**, *36*, e22393. [CrossRef]
51. Chen, B.; Hasan, M.M.; Zhang, H.; Zhai, Q.; Waliullah, A.S.M.; Ping, Y.; Zhang, C.; Oyama, S.; Mimi, M.A.; Tomochika, Y.; et al. UBL3 Interacts with Alpha-Synuclein in Cells and the Interaction Is Downregulated by the EGFR Pathway Inhibitor Osimertinib. *Biomedicines* **2023**, *11*, 1685. [CrossRef]
52. Lefebvre, S.; Bürglen, L.; Reboullet, S.; Clermont, O.; Burlet, P.; Viollet, L.; Benichou, B.; Cruaud, C.; Millasseau, P.; Zeviani, M. Identification and characterization of a spinal muscular atrophy-determining gene. *Cell* **1995**, *80*, 155–165. [CrossRef] [PubMed]
53. Lobsiger, C.S.; Cleveland, D.W. Glial cells as intrinsic components of non-cell-autonomous neurodegenerative disease. *Nat. Neurosci.* **2007**, *10*, 1355–1360. [CrossRef]
54. Sun, J.; Qiu, J.; Yang, Q.; Ju, Q.; Qu, R.; Wang, X.; Wu, L.; Xing, L. Single-cell RNA sequencing reveals dysregulation of spinal cord cell types in a severe spinal muscular atrophy mouse model. *PLoS Genet.* **2022**, *18*, e1010392. [CrossRef] [PubMed]
55. Feng, Y.; Peng, Y.; Jie, J.; Yang, Y.; Yang, P. The immune microenvironment and tissue engineering strategies for spinal cord regeneration. *Front. Cell. Neurosci.* **2022**, *16*, 969002. [CrossRef] [PubMed]
56. Fu, S.P.; Chen, S.Y.; Pang, Q.M.; Zhang, M.; Wu, X.C.; Wan, X.; Wan, W.H.; Ao, J.; Zhang, T. Advances in the research of the role of macrophage/microglia polarization-mediated inflammatory response in spinal cord injury. *Front. Immunol.* **2022**, *13*, 1014013. [CrossRef]
57. Zhang, Q.; Yu, B.; Zhang, Y.; Tian, Y.; Yang, S.; Chen, Y.; Wu, H. Combination of single-cell and bulk RNA seq reveals the immune infiltration landscape and targeted therapeutic drugs in spinal cord injury. *Front. Immunol.* **2023**, *14*, 1068359. [CrossRef]
58. Mimi, M.A.; Hasan, M.M.; Takanashi, Y.; Waliullah, A.S.M.; Mamun, M.A.; Chi, Z.; Kahyo, T.; Aramaki, S.; Takatsuka, D.; Koizumi, K.; et al. UBL3 overexpression enhances EV-mediated Achilles protein secretion in conditioned media of MDA-MB-231 cells. *Biochem. Biophys. Res. Commun.* **2024**, *738*, 150559. [CrossRef]
59. Russ, D.E.; Cross, R.B.P.; Li, L.; Koch, S.C.; Matson, K.J.E.; Yadav, A.; Alkaslasi, M.R.; Lee, D.I.; Le Pichon, C.E.; Menon, V.; et al. A harmonized atlas of mouse spinal cord cell types and their spatial organization. *Nat. Commun.* **2021**, *12*, 5722. [CrossRef]
60. Han, B.; Zhou, S.; Zhang, Y.; Chen, S.; Xi, W.; Liu, C.; Zhou, X.; Yuan, M.; Yu, X.; Li, L.; et al. Integrating spatial and single-cell transcriptomics to characterize the molecular and cellular architecture of the ischemic mouse brain. *Sci. Transl. Med.* **2024**, *16*, eadg1323. [CrossRef]
61. Hasan, M.M.; Eto, F.; Mamun, M.A.; Sato, S.; Islam, A.; Waliullah, A.S.M.; Chi, D.H.; Takahashi, Y.; Kahyo, T.; Naito, Y.; et al. Desorption ionization using through-hole alumina membrane offers higher reproducibility than 2,5-dihydroxybenzoic acid, a widely used matrix in Fourier transform ion cyclotron resonance mass spectrometry imaging analysis. *Rapid Commun. Mass Spectrom. RCM* **2021**, *35*, e9076. [CrossRef]

62. Chi, D.H.; Kahyo, T.; Islam, A.; Hasan, M.M.; Waliullah, A.S.M.; Mamun, M.A.; Nakajima, M.; Ikoma, T.; Akita, K.; Maekawa, Y.; et al. NAD$^+$ Levels Are Augmented in Aortic Tissue of ApoE$^{-/-}$ Mice by Dietary Omega-3 Fatty Acids. *Arterioscler. Thromb. Vasc. Biol.* **2022**, *42*, 395–406. [CrossRef] [PubMed]
63. Mamun, M.A.; Islam, A.; Hasan, M.M.; Waliullah, A.S.M.; Tamannaa, Z.; Huu Chi, D.; Sato, T.; Kahyo, T.; Kikushima, K.; Takahashi, Y.; et al. The human vermilion surface contains a rich amount of cholesterol sulfate than the skin. *J. Dermatol. Sci.* **2021**, *103*, 143–150. [CrossRef] [PubMed]
64. Hasan, M.; Mimi, A.; Kikushima, K.; Kahyo, T.; Setou, M. Mass spectrometry imaging for glycosphingolipids. In *Glycoscience Protocols (GlycoPODv2)*; Japan Consortium for Glycobiology and Glycotechnology: Saitama, Japan, 2021. Available online: https://www.ncbi.nlm.nih.gov/books/NBK593865/ (accessed on 28 April 2025).
65. Khan, Y.S.; Lui, F. Neuroanatomy, Spinal Cord. In *StatPearls [Internet]*; StatPearls Publishing: Treasure Island, FL, USA, 2025. Available online: https://www.ncbi.nlm.nih.gov/books/NBK559056/ (accessed on 28 April 2025).
66. Nabi, M.; Mamun, A.; Islam, A.; Hasan, M.; Waliullah, A.S.M.; Tamannaa, Z.; Sato, T.; Kahyo, T.; Setou, M. Mass spectrometry in the lipid study of cancer. *Expert Rev. Proteom.* **2021**, *18*, 201–219. [CrossRef]
67. Peng, R.; Zhang, L.; Xie, Y.; Guo, S.; Cao, X.; Yang, M. Spatial multi-omics analysis of the microenvironment in traumatic spinal cord injury: A narrative review. *Front. Immunol.* **2024**, *15*, 1432841. [CrossRef]
68. Shi, Y.; Huang, L.; Dong, H.; Yang, M.; Ding, W.; Zhou, X.; Lu, T.; Liu, Z.; Zhou, X.; Wang, M.; et al. Decoding the spatiotemporal regulation of transcription factors during human spinal cord development. *Cell Res.* **2024**, *34*, 193–213. [CrossRef]
69. Chen, K.H.; Boettiger, A.N.; Moffitt, J.R.; Wang, S.; Zhuang, X. RNA imaging. Spatially resolved, highly multiplexed RNA profiling in single cells. *Science* **2015**, *348*, aaa6090. [CrossRef] [PubMed]
70. Marx, V. Method of the Year: Spatially resolved transcriptomics. *Nat. Methods* **2021**, *18*, 9–14. [CrossRef]
71. Yang, P.; Jin, K.; Yao, Y.; Jin, L.; Shao, X.; Li, C.; Lu, X.; Fan, X. Spatial integration of multi-omics single-cell data with SIMO. *Nat. Commun.* **2025**, *16*, 1265. [CrossRef]
72. Biancalani, T.; Scalia, G.; Buffoni, L.; Avasthi, R.; Lu, Z.; Sanger, A.; Tokcan, N.; Vanderburg, C.R.; Segerstolpe, Å.; Zhang, M.; et al. Deep learning and alignment of spatially resolved single-cell transcriptomes with Tangram. *Nat. Methods* **2021**, *18*, 1352–1362. [CrossRef]
73. Miao, Z.; Humphreys, B.D.; McMahon, A.P.; Kim, J. Multi-omics integration in the age of million single-cell data. *Nat. Rev. Nephrol.* **2021**, *17*, 710–724. [CrossRef]
74. Gao, H.; Zhang, B.; Liu, L.; Li, S.; Gao, X.; Yu, B. A universal framework for single-cell multi-omics data integration with graph convolutional networks. *Brief. Bioinform.* **2023**, *24*, bbad081. [CrossRef] [PubMed]
75. Wang, X.; Hu, Z.; Yu, T.; Wang, Y.; Wang, R.; Wei, Y.; Shu, J.; Ma, J.; Li, Y. Con-AAE: Contrastive cycle adversarial autoencoders for single-cell multi-omics alignment and integration. *Bioinformatics* **2023**, *39*, btad162. [CrossRef] [PubMed]
76. Cao, Z.J.; Gao, G. Multi-omics single-cell data integration and regulatory inference with graph-linked embedding. *Nat. Biotechnol.* **2022**, *40*, 1458–1466. [CrossRef] [PubMed]
77. Luo, J.; Fu, J.; Lu, Z.; Tu, J. Deep learning in integrating spatial transcriptomics with other modalities. *Brief. Bioinform.* **2024**, *26*, bbae719. [CrossRef]
78. Yadav, A.; Matson, K.J.E.; Li, L.; Hua, I.; Petrescu, J.; Kang, K.; Alkaslasi, M.R.; Lee, D.I.; Hasan, S.; Galuta, A.; et al. A cellular taxonomy of the adult human spinal cord. *Neuron* **2023**, *111*, 328–344.e7. [CrossRef]
79. Taylor, R.; Nikolaou, N. RNA in axons, dendrites, synapses and beyond. *Front. Mol. Neurosci.* **2024**, *17*, 1397378. [CrossRef]
80. Ament, S.A.; Poulopoulos, A. The brain's dark transcriptome: Sequencing RNA in distal compartments of neurons and glia. *Curr. Opin. Neurobiol.* **2023**, *81*, 102725. [CrossRef]
81. Skinnider, M.A.; Gautier, M.; Teo, A.Y.Y.; Kathe, C.; Hutson, T.H.; Laskaratos, A.; de Coucy, A.; Regazzi, N.; Aureli, V.; James, N.D.; et al. Single-cell and spatial atlases of spinal cord injury in the Tabulae Paralytica. *Nature* **2024**, *631*, 150–163. [CrossRef]
82. Milich, L.M.; Ryan, C.B.; Lee, J.K. The origin, fate, and contribution of macrophages to spinal cord injury pathology. *Acta Neuropathol.* **2019**, *137*, 785–797. [CrossRef]
83. Jiang, R.; Sun, T.; Song, D.; Li, J.J. Statistics or biology: The zero-inflation controversy about scRNA-seq data. *Genome Biol.* **2022**, *23*, 31. [CrossRef]
84. Kharchenko, P.V.; Silberstein, L.; Scadden, D.T. Bayesian approach to single-cell differential expression analysis. *Nat. Methods* **2014**, *11*, 740–742. [CrossRef] [PubMed]
85. Liu, Z.; Hu, S.K.; Campbell, V.; Tatters, A.O.; Heidelberg, K.B.; Caron, D.A. Single-cell transcriptomics of small microbial eukaryotes: Limitations and potential. *ISME J.* **2017**, *11*, 1282–1285. [CrossRef]
86. Korsunsky, I.; Millard, N.; Fan, J.; Slowikowski, K.; Zhang, F.; Wei, K.; Baglaenko, Y.; Brenner, M.; Loh, P.R.; Raychaudhuri, S. Fast, sensitive and accurate integration of single-cell data with Harmony. *Nat. Methods* **2019**, *16*, 1289–1296. [CrossRef] [PubMed]
87. Nichterwitz, S.; Chen, G.; Aguila Benitez, J.; Yilmaz, M.; Storvall, H.; Cao, M.; Sandberg, R.; Deng, Q.; Hedlund, E. Laser capture microscopy coupled with Smart-seq2 for precise spatial transcriptomic profiling. *Nat. Commun.* **2016**, *7*, 12139. [CrossRef]

88. Tran, H.T.N.; Ang, K.S.; Chevrier, M.; Zhang, X.; Lee, N.Y.S.; Goh, M.; Chen, J. A benchmark of batch-effect correction methods for single-cell RNA sequencing data. *Genome Biol.* **2020**, *21*, 12. [CrossRef]
89. Luecken, M.D.; Theis, F.J. Current best practices in single-cell RNA-seq analysis: A tutorial. *Mol. Syst. Biol.* **2019**, *15*, e8746. [CrossRef]
90. Chen, G.; Ning, B.; Shi, T. Single-Cell RNA-Seq Technologies and Related Computational Data Analysis. *Front. Genet.* **2019**, *10*, 317. [CrossRef]
91. Adil, A.; Kumar, V.; Jan, A.T.; Asger, M. Single-Cell Transcriptomics: Current Methods and Challenges in Data Acquisition and Analysis. *Front. Neurosci.* **2021**, *15*, 591122. [CrossRef] [PubMed]
92. Haque, A.; Engel, J.; Teichmann, S.A.; Lönnberg, T. A practical guide to single-cell RNA-sequencing for biomedical research and clinical applications. *Genome Med.* **2017**, *9*, 75. [CrossRef]
93. Haghverdi, L.; Lun, A.T.L.; Morgan, M.D.; Marioni, J.C. Batch effects in single-cell RNA-sequencing data are corrected by matching mutual nearest neighbors. *Nat. Biotechnol.* **2018**, *36*, 421–427. [CrossRef]
94. Li, W.V.; Li, J.J. An accurate and robust imputation method scImpute for single-cell RNA-seq data. *Nat. Commun.* **2018**, *9*, 997. [CrossRef]
95. Huang, M.; Wang, J.; Torre, E.; Dueck, H.; Shaffer, S.; Bonasio, R.; Murray, J.I.; Raj, A.; Li, M.; Zhang, N.R. SAVER: Gene expression recovery for single-cell RNA sequencing. *Nat. Methods* **2018**, *15*, 539–542. [CrossRef] [PubMed]
96. van Dijk, D.; Sharma, R.; Nainys, J.; Yim, K.; Kathail, P.; Carr, A.J.; Burdziak, C.; Moon, K.R.; Chaffer, C.L.; Pattabiraman, D.; et al. Recovering Gene Interactions from Single-Cell Data Using Data Diffusion. *Cell* **2018**, *174*, 716–729.e27. [CrossRef] [PubMed]
97. Qiu, P. Embracing the dropouts in single-cell RNA-seq analysis. *Nat. Commun.* **2020**, *11*, 1169. [CrossRef]
98. Rayon, T.; Maizels, R.J.; Barrington, C.; Briscoe, J. Single-cell transcriptome profiling of the human developing spinal cord reveals a conserved genetic programme with human-specific features. *Development* **2021**, *148*, dev199711. [CrossRef] [PubMed]
99. Argelaguet, R.; Velten, B.; Arnol, D.; Dietrich, S.; Zenz, T.; Marioni, J.C.; Buettner, F.; Huber, W.; Stegle, O. Multi-Omics Factor Analysis-a framework for unsupervised integration of multi-omics data sets. *Mol. Syst. Biol.* **2018**, *14*, e8124. [CrossRef] [PubMed] [PubMed Central]
100. Stuart, T.; Butler, A.; Hoffman, P.; Hafemeister, C.; Papalexi, E.; Mauck, W.M.; 3rd Hao, Y.; Stoeckius, M.; Smibert, P.; Satija, R. Comprehensive Integration of Single-Cell Data. *Cell* **2019**, *177*, 1888–1902.e21. [CrossRef]
101. Sofroniew, M.V.; Vinters, H.V. Astrocytes: Biology and pathology. *Acta Neuropathol.* **2010**, *119*, 7–35. [CrossRef]
102. Peirs, C.; Seal, R.P. Neural circuits for pain: Recent advances and current views. *Science* **2016**, *354*, 578–584. [CrossRef]
103. Fan, Y.; Wu, X.; Han, S.; Zhang, Q.; Sun, Z.; Chen, B.; Xue, X.; Zhang, H.; Chen, Z.; Yin, M.; et al. Single-cell analysis reveals region-heterogeneous responses in rhesus monkey spinal cord with complete injury. *Nat. Commun.* **2023**, *14*, 4796. [CrossRef]
104. Zhang, Q.; Wu, X.; Fan, Y.; Jiang, P.; Zhao, Y.; Yang, Y.; Han, S.; Xu, B.; Chen, B.; Han, J.; et al. Single-cell analysis reveals dynamic changes of neural cells in developing human spinal cord. *EMBO Rep.* **2021**, *22*, e52728. [CrossRef] [PubMed]
105. Kushnarev, M.; Pirvulescu, I.P.; Candido, K.D.; Knezevic, N.N. Neuropathic pain: Preclinical and early clinical progress with voltage-gated sodium channel blockers. *Expert Opin. Investig. Drugs* **2020**, *29*, 259–271. [CrossRef] [PubMed]
106. Awuah, W.A.; Ahluwalia, A.; Ghosh, S.; Roy, S.; Tan, J.K.; Adebusoye, F.T.; Ferreira, T.; Bharadwaj, H.R.; Shet, V.; Kundu, M.; et al. The molecular landscape of neurological disorders: Insights from single-cell RNA sequencing in neurology and neurosurgery. *Eur. J. Med. Res.* **2023**, *28*, 529. [CrossRef] [PubMed]
107. Lähnemann, D.; Köster, J.; Szczurek, E.; McCarthy, D.J.; Hicks, S.C.; Robinson, M.D.; Vallejos, C.A.; Campbell, K.R.; Beerenwinkel, N.; Mahfouz, A.; et al. Eleven grand challenges in single-cell data science. *Genome Biol.* **2020**, *21*, 31. [CrossRef] [PubMed]
108. Wu, A.R.; Neff, N.F.; Kalisky, T.; Dalerba, P.; Treutlein, B.; Rothenberg, M.E.; Mburu, F.M.; Mantalas, G.L.; Sim, S.; Clarke, M.F.; et al. Quantitative assessment of single-cell RNA-sequencing methods. *Nat. Methods* **2014**, *11*, 41–46. [CrossRef]
109. Moses, L.; Pachter, L. Museum of spatial transcriptomics. *Nat. Methods* **2022**, *19*, 534–546. [CrossRef]
110. BRAIN Initiative Cell Census Network (BICCN). A multimodal cell census and atlas of the mammalian primary motor cortex. *Nature* **2021**, *598*, 86–102. [CrossRef]
111. Ma, Q.; Xu, D. Deep learning shapes single-cell data analysis. *Nat. Rev. Mol. Cell Biol.* **2022**, *23*, 303–304. [CrossRef]
112. Wang, J.; Ma, A.; Chang, Y.; Gong, J.; Jiang, Y.; Qi, R.; Wang, C.; Fu, H.; Ma, Q.; Xu, D. Author Correction: scGNN is a novel graph neural network framework for single-cell RNA-Seq analyses. *Nat. Commun.* **2022**, *13*, 2554. [CrossRef]
113. Xi, N.M.; Li, J.J. Exploring the optimization of autoencoder design for imputing single-cell RNA sequencing data. *Comput. Struct. Biotechnol. J.* **2023**, *21*, 4079–4095. [CrossRef]
114. Cao, X.; Xing, L.; Majd, E.; He, H.; Gu, J.; Zhang, X. A Systematic Evaluation of Supervised Machine Learning Algorithms for Cell Phenotype Classification Using Single-Cell RNA Sequencing Data. *Front. Genet.* **2022**, *13*, 836798. [CrossRef] [PubMed]
115. Shao, X.; Yang, H.; Zhuang, X.; Liao, J.; Yang, P.; Cheng, J.; Lu, X.; Chen, H.; Fan, X. scDeepSort: A pre-trained cell-type annotation method for single-cell transcriptomics using deep learning with a weighted graph neural network. *Nucleic Acids Res.* **2021**, *49*, e122. [CrossRef] [PubMed]

116. Ma, F.; Pellegrini, M. ACTINN: Automated identification of cell types in single cell RNA sequencing. *Bioinformatics* **2020**, *36*, 533–538. [CrossRef] [PubMed]
117. Yang, F.; Wang, W.; Wang, F.; Fang, Y.; Tang, D.; Huang, J.; Lu, H.; Yao, J. scBERT as a large-scale pretrained deep language model for cell type annotation of single-cell RNA-seq data. *Nat. Mach. Intell.* **2022**, *4*, 852–866. [CrossRef]
118. Wang, Y.R.; Du, P.F. WCSGNet: A graph neural network approach using weighted cell-specific networks for cell-type annotation in scRNA-seq. *Front. Genet.* **2025**, *16*, 1553352. [CrossRef]
119. Dobrott, C.I.; Sathyamurthy, A.; Levine, A.J. Decoding Cell Type Diversity Within the Spinal Cord. *Curr. Opin. Physiol.* **2019**, *8*, 1–6. [CrossRef]
120. Liu, L.; Huang, Y.; Zheng, Y.; Liao, Y.; Ma, S.; Wang, Q. ScnML models single-cell transcriptome to predict spinal cord neuronal cell status. *Front. Genet.* **2024**, *15*, 1413484. [CrossRef]
121. Fang, M.; Gorin, G.; Pachter, L. Trajectory inference from single-cell genomics data with a process time model. *PLoS Comput. Biol.* **2025**, *21*, e1012752. [CrossRef] [PubMed]
122. A graph neural network that combines scRNA-seq and protein-protein interaction data. *Nat. Methods* **2025**, *22*, 660–661. [CrossRef]
123. Jaja, B.N.R.; Badhiwala, J.; Guest, J.; Harrop, J.; Shaffrey, C.; Boakye, M.; Kurpad, S.; Grossman, R.; Toups, E.; Geisler, F.; et al. Trajectory-Based Classification of Recovery in Sensorimotor Complete Traumatic Cervical Spinal Cord Injury. *Neurology* **2021**, *96*, e2736–e2748. [CrossRef]
124. Patel, M.; Magre, N.; Motwani, H.; Brown, N.B. Advances in machine learning, statistical methods, and AI for single-cell RNA annotation using raw count matrices in scRNA-seq data. *arXiv* **2024**, arXiv:2406.05258. [CrossRef]
125. Shen, X.; Li, X. Deep-learning methods for unveiling large-scale single-cell transcriptomes. *Cancer Biol. Med.* **2024**, *20*, 972–980. [CrossRef] [PubMed]
126. Cao, Y.; Zhu, S.; Yu, B.; Yao, C. Single-cell RNA sequencing for traumatic spinal cord injury. *FASEB J. Off. Publ. Fed. Am. Soc. Exp. Biol.* **2022**, *36*, e22656. [CrossRef] [PubMed]
127. Shu, M.; Xue, X.; Nie, H.; Wu, X.; Sun, M.; Qiao, L.; Li, X.; Xu, B.; Xiao, Z.; Zhao, Y.; et al. Single-cell RNA sequencing reveals Nestin+ active neural stem cells outside the central canal after spinal cord injury. *Sci. China Life Sci.* **2022**, *65*, 295–308. [CrossRef]
128. Li, D.; Ma, Y.; Huang, W.; Li, X.; Liu, H.; Xiong, C.; Zhao, Q.; Wang, B.; Lai, X.; Huang, S.; et al. Modeling human spine-spinal cord organogenesis by hPSC-derived neuromesodermal progenitors. *bioRxiv* **2023**. [CrossRef]
129. Chelyshev, Y.A.; Ermolin, I.L. RNA Sequencing and Spatial Transcriptomics in Traumatic Spinal Cord Injury (Review). *Sovrem. Tekhnologii Meditsine* **2023**, *15*, 75–86. [CrossRef]

BioChem

MDPI

Article

HCC in MASLD and ALD: Biochemical Pathways, Epidemiology, Diagnosis, and Treatment

Sheel Patel [1,*], Fares Kasem [1], Dylan Flaherty [2] and Ashutosh Barve [2]

[1] Department of Medicine, University of Louisville, Louisville, KY 40202, USA; fares.kasem@louisville.edu
[2] Department of Medicine, Division of Gastroenterology, Hepatology, and Nutrition, University of Louisville, Louisville, KY 40202, USA; dylan.flaherty@louisville.edu (D.F.); ashutosh.barve@louisville.edu (A.B.)
* Correspondence: sppate07@louisville.edu

Abstract

Hepatocellular carcinoma (HCC) is a leading cause of cancer-related mortality globally, with metabolic-dysfunction-associated steatohepatitis (MASH) and alcohol-related liver disease (ALD) emerging as major etiologies. This review explores the epidemiological trends, pathogenesis, and clinical management of HCC arising from MASH and ALD, highlighting both the shared and distinct mechanisms. MASH-HCC is driven by metabolic dysregulation, including obesity, insulin resistance, and lipotoxicity, with genetic polymorphisms such as PNPLA3 and TM6SF2 playing critical roles in disease progression. ALD-HCC, in contrast, is propelled by the toxic byproducts of ethanol metabolism, including acetaldehyde and reactive oxygen species, which induce chronic inflammation, and fibrosis. Both conditions also involve immune dysregulation, gut dysbiosis, and increased intestinal permeability, contributing to hepatic carcinogenesis. The review emphasizes that, while there is consensus regarding the screening of HCC in cirrhosis patients, there is lack of consensus on screening strategies for non-cirrhotic MASH patients who are also at risk for HCC. This underscores the importance of the early detection of cirrhosis using advanced diagnostic tools such as transient elastography and fibrosis scores. Current therapeutic approaches, ranging from surgical resection, liver transplantation, and locoregional therapies to systemic therapies like immune checkpoint inhibitors, are discussed, with an emphasis on the need for personalized treatment strategies. Finally, the review highlights future research priorities, including the development of novel biomarkers, exploration of the gut–liver axis, and deeper investigation of the interplay between genetic predisposition and environmental factors. By synthesizing these insights, the review aims to inform multidisciplinary approaches to reduce the global burden of MASH- and ALD-related HCC and improve patient outcomes.

Keywords: hepatocellular carcinoma; HCC; liver cancer; MASLD; MASH; alcohol-related liver disease; ALD

1. Introduction

Primary liver carcinomas represent the sixth most common cancer worldwide and the third leading cause of cancer-related death worldwide, with an incidence of around 1 million cases annually [1,2]. Hepatocellular carcinoma accounts for approximately 90% of primary liver cancers, usually in the setting of chronic liver disease caused by hepatitis B virus (HBV), hepatitis C virus (HCV), alcohol-associated liver disease (ALD), and MASLD (metabolic-dysfunction-associated steatotic liver disease) [3,4]. MASLD and ALD are now classified under the overarching term Steatotic Liver Disease (SLD), which encompasses all the different etiologies of hepatic steatosis and is further classified into multiple subtypes [5,6]. Apart from MASLD and ALD, a new entity called MetALD has been defined to describe patients with features of MASLD who also have a significant intake of alcohol (140–350 g per week for women and 210–420 g per week for men) [5,6]. Additional sub-classes of SLD are 'cryptogenic SLD' and 'specific etiology SLD', which is further subdivided into 'drug-induced liver injury', monogenic diseases (e.g., lysosomal acid lipase deficiency, Wilson's disease, etc.), and 'miscellaneous diseases' (e.g., celiac disease, malnutrition, chronic hepatitis C, etc.) [5,6]. In the present review, we will focus on MASLD and ALD. MASLD is defined as at least 5% hepatic steatosis with one or more conditions: T2DM, obesity, or metabolic dysregulation [7,8].

MASLD is defined as at least 5% hepatic steatosis with one or more conditions: T2DM, obesity, or metabolic dysregulation [7,8]. It is a spectrum of liver diseases, ranging from simple steatosis to its more aggressive form, metabolic-dysfunction-associated steatohepatitis (MASH), which is defined as the presence of inflammation and cellular injury (ballooning), with or without fibrosis. MASH can progress to cirrhosis and, ultimately, HCC, even in patients without advanced fibrosis. Alarmingly, MASH-HCC has become the fastest-growing etiology of HCC in the Western world, reflecting the global obesity epidemic and increased prevalence of metabolic disorders.

In contrast, ALD results from the hepatotoxic effects of chronic alcohol consumption, which triggers oxidative stress, immune dysregulation, and fibrogenic pathways. ALD-HCC remains a significant global issue, with alcohol use ranking among the leading modifiable risk factors for liver cancer. Although historically more prevalent in men, rising alcohol consumption among women has led to an increasing incidence of ALD-HCC in female populations, further exacerbating its public health impact.

While MASLD and ALD arise from distinct etiologies, they share critical pathogenic mechanisms, including oxidative stress, mitochondrial dysfunction, and chronic inflammation, that drive liver injury and promote carcinogenesis. However, differences in genetic predispositions, clinical presentations, and the impact of lifestyle factors demand individualized approaches to their management. This review explores the epidemiology, pathogenesis, and management of HCC in MASLD and ALD, emphasizing their shared and distinct features. By synthesizing the current evidence, we aim to provide a comprehensive understanding of these conditions, identify the gaps in our knowledge, and highlight potential avenues for further research.

2. Epidemiology

2.1. Overview of Epidemiological Trends of MASLD, MASH, and ALD HCC

It is estimated that more than one million individuals worldwide will be affected by liver cancer annually by 2025 [9]. Hepatocellular carcinoma is the most common form of liver cancer and accounts for around 90% of cases [10]. Although hepatocellular carcinoma (HCC) generally has a poor prognosis, with a 5-year survival rate below 20%, outcomes vary significantly based on the tumor stage. Patients with early-stage HCC can achieve a median survival of over 10 years, whereas those with advanced disease typically have a survival of only 1 to 3 years [11]. Studies comparing the overall survival of HCC patients caused by MASH have found no significant differences between the two groups [12].

MASLD has become the most prevalent chronic liver disease in history, affecting 1.66 billion individuals globally [13,14]. It is estimated that around 10% of all HCC cases are secondary to MASLD [15]. Data from the 2019 Global Burden of Disease (GBD) study estimated that the global prevalence of MASLD was 30.05% [16]. Western countries, with a high intake of high-fructose and processed foods, are experiencing rising rates of

MASLD [17]. It is believed that approximately 20% of patients diagnosed with MASLD will progress to MASH [13], which has a global prevalence of 5.27% with significant geographic variations [16]. For example, Latin America has the highest prevalence of MASH at 7.11%, while Western Europe has the lowest prevalence at 4.02% [16]. Between 2010 and 2019, MASH was not only the fastest-growing etiology of liver cancer cases worldwide but also the leading cause of liver cancer deaths [18]. By 2030, it is projected that the incidence of MASH-HCC will increase by 130% in the United States [19]. This rise reflects a true epidemiologic trend rather than improved detection or changes in the diagnostic criteria [18].

Even as the MASLD prevalence evolves, ALD remains a highly significant global health problem, as it has been for millennia [20]. It was found that 25% of global deaths due to cirrhosis were associated with alcohol use. Alcohol use has been steadily rising globally, with alcohol consumption per capita in 2005 at 5.5 L, which is expected to increase to 7.6 L by 2030 [21,22]. A study by the USA Scientific Registry of Transplant Recipients from 2002 to 2016 found that 10.3% of the patients listed for a transplant had HCC secondary to ALD [23]. After MASH, alcohol use is the second-fastest-growing cause of liver cancer death [24]. ALD accounts for around 30% of HCC cases [25]. In 2019, alcohol was associated with an estimated 19% of deaths from liver cancer globally, with considerable regional variation [26,27]. Eastern Europe and parts of Asia have higher ALD-related HCC due to the elevated per capita alcohol consumption [24]. For example, the highest percentage of HCC mortality that was associated with alcohol was in Europe (35%), and the lowest percentage was in the Eastern Mediterranean region (10%) [24].

2.2. Sex Distribution in MASLD, ALD, and HCC

A meta-analysis of 22 studies found no difference in the prevalence of MASH between men and women who had biopsy-proven MASLD [28]. While it is believed that men generally have a higher prevalence of MASLD compared to women [29], studies have shown that post-menopausal women have a higher prevalence of MASLD compared to men [13,30]. This change in prevalence after menopause is hypothesized to be due to a protective effect that estrogen has during the pre-menopausal years. Regarding ALD, men have a higher rate of developing alcohol-related cirrhosis compared to women [31]. However, recent studies have shown a higher rate of increase in the prevalence of alcohol-related cirrhosis in women compared to men. In a retrospective study of privately insured persons in the US, between 2009 and 2015, there was a 50% increase in alcohol-related cirrhosis prevalence in women compared to a 30% increase in men [32]. This is postulated to be due to the increasing rate of alcohol use in women compared to men. A meta-analysis of six nationwide surveys in the US found that, between 2000 and 2016, the prevalence of alcohol use in women increased by approximately 0.6% per year, but no significant increase was found in men [33]. HCC has been well-known to disproportionately affect men more than women, with studies finding that men are two to four times more likely to develop liver cancer [34].

2.3. Risk Factors in MASH

Understanding the risk factors for the progression of MASH to cirrhosis is crucial, as they help identify individuals at a higher risk for developing advanced liver disease and guide targeted prevention and management strategies. The high prevalence of MASLD and its progression to MASH is strongly linked to chronic health conditions associated with metabolic syndrome (MetS): T2DM, HTN, HLD, heart disease, and obesity [35]. A meta-analysis of nine observational studies consisting of more than 1.5 million patients found that obese patients had a two-fold increase in HCC-associated mortality [4,36]. MASLD is characterized by an increase in hepatic lipid and triglyceride accumulation. Without early intervention, steatosis can progress to hepatocyte injury and, eventually, fibrosis, leading to liver cirrhosis and even higher chances of progression to HCC. The risk for developing HCC is increased 2.6-fold when obesity occurs with diabetes, hypertension, and hyperlipidemia [37]. Studies have found there to be a gender difference as well. A meta-analysis of 54 studies consisting of more than sixty-two thousand patients found that women had a 19% lower risk of MASLD but a 37% higher risk of advanced fibrosis than men [28]. In the US, around 30% of the adult population has MASLD, and 5% have progressed to MASH [38]. This percentage is vastly impacted by race in the US, with Hispanic individuals having the highest rates of MASLD, followed by White and Black individuals (21%, 12.5%, and 11.6%, respectively), which highlights the consideration of socioeconomic factors in patients [39]. Furthermore, genetic risk factors such as PNPLA3 variants are disproportionately prevalent in Hispanic populations, contributing to a higher MASLD-HCC incidence. Given the complex interplay of genetics, metabolic, and environmental factors in the progression of MASLD to MASH and cirrhosis, a multidisciplinary approach to early identification, risk stratification, and intervention is essential in order to mitigate the growing global burden of liver-related morbidity and mortality.

2.4. Risk Factors in ALD

Alcohol itself has been associated with an increased risk of cancer, and studies have found specific mechanisms that lead to the pathogenesis of HCC with alcohol use and ALD, discussed elsewhere in this article. Hence, the alarming increase in alcohol intake noted above increases the chances of ALD cirrhosis and progression to HCC. Females are more susceptible to ALD with lower volumes of alcohol, while males have a higher incidence of ALD-HCC [40]. Smoking and concomitant liver disease have also been shown to increase HCC with alcohol use. While the exact mechanism is unknown, it is believed that cigarette smoking and alcohol use have a synergetic carcinogenic effect [41]. ALD with a history of hepatitis B and C has also been shown to increase the risk of HCC [42]. The rising global alcohol consumption and its complex interplay with other risk factors underscore a critical need for increased awareness and targeted interventions to address these risk factors.

3. Pathogenesis

3.1. MASH-HCC

MASH is a multifactorial disease that involves a delicate relationship between numerous pathological mechanisms contributing to injury and fibrosis. Metabolic dysregulation (insulin resistance and lipotoxicity), oxidative stress, and increased pro-inflammatory cells are believed to be among the most vital contributors to chronic inflammation, which leads to hepatocyte damage and the potential progression to HCC [43–46].

3.2. Metabolic Dysregulation

A population-based case–control study using data from the Medicare database found that metabolic syndrome is associated with a 2.13-fold increase in HCC [47]. Interestingly, hypertension and hypercholesteremia did not show a correlation, while hyperglycemia and obesity increased HCC incidence, illustrating the key role glucose plays in hepatocarcinogenesis [48,49]. In T2DM patients, there is a 2-fold increase in the risk of HCC [50,51]. Physiologically, insulin promotes lipid synthesis and storage and reduces the formation of free fatty acids by inhibiting or suppressing lipolysis. In a state of insulin resistance, as in patients with diabetes, hyperinsulinemia and the subsequent hyperglycemia promote lipogenesis in hepatocytes.

Hepatic lipogenesis is modulated by two proteins—ChRBP (carbohydrate-response element-binding protein) and SREBP-1c (sterol regulatory element-binding protein 1c), both of which lead to overall lipid accumulation in hepatocytes [52]. To counteract the lipid-overloaded liver, adaptive changes in free fatty acid (FFA) metabolism must occur. It has been postulated that the total amount of triglycerides stored in hepatocytes is not the major determinant of lipotoxicity, and that specific lipid classes are agents of hepatocyte injury. Examples of such lipids are cholesterol, lysophosphatidylcholine, ceramides, and free fatty acids such as palmitic acid, which have been linked to the progression of MASH [53]. The increased lipids and FFAs in hepatocytes are believed to induce mitochondrial damage by producing mitochondrial reactive oxygen species (ROS). This can lead to protein and lipid peroxidation and impede β-oxidation, causing mitochondrial damage and, ultimately, resulting in cell death [54].

Previous studies have shown that saturated fatty acids induce liver damage by triggering pro-inflammatory macrophage activation through various signaling cascades, such as IKK/NF- kB and JAK/STAT3 [55]. While the specific biochemical pathways are beyond the scope of this article, pro-inflammatory cytokines such as TNF-α (tumor necrosis factor-alpha) and IL-6 (interleukin 6) are eventually produced, inducing a state of chronic inflammation in MASH. Furthermore, TNF-α has been shown to reprogram metabolic pathways by deactivating insulin signaling, thus enhancing lipolysis in adipocytes and delaying glucose clearance [43,44]. Mice studies have shown that the knockout of TNF-α protected them from insulin resistance and reduced serum fatty acid levels and oxidative stress [56–60]. Interestingly, studies by Nakagawa et al. [58] in 2014 showed that mice with TNF1 receptor knockout decreased the HCC development in obese mice. This complex interplay between biochemical pathways leads to chronic, low-grade inflammation, contributing to hepatocyte fibrosis and the eventual progression to HCC from MASH [61,62].

3.3. *Oxidative Stress in Hepatocytes*

Patients with MASH have higher levels of oxidative DNA damage than patients with other liver diseases [46,63]. As stated above, increased amounts of lipid and FFA lead to the production of mitochondrial reactive oxygen species due to mitochondrial damage. Increased lipid peroxidation and oxidative stress can further diminish mitochondrial function and affect the respiratory chain activity [64]. When lipotoxicity or oxidative damage occurs, an adaptive response is activated called the unfolded protein response (UPR), which is regulated by the endoplasmic reticulum (ER) [65]. It serves to bring homeostasis to the ER. However, during extended periods of ER stress, such as when hepatocytes are lipid-overloaded, the UPR is unable to restore homeostasis, and apoptosis occurs through a variety of proteins such as JNK (c-Jun N-terminal kinase), and CHOP (C/EBP homologous protein), as well as Bcl2 (B-cell lymphoma-2), PUMA (p53 up-regulated modulator of apoptosis), and DP5 (death protein 5) [66–68]. While ROS is usually beneficial to macrophages and is the first line against pathogens, DNA damage can occur as a bystander effect and likely plays a role in the progression of MASH to HCC and the associated proliferative response [69,70].

3.4. *Gut Microbiome*

The microbiome is a key modulator of metabolism and plays an essential role in immune system modulation, health maintenance, tolerance development, and the prevention of colonization by pathogens [71]. Studies have shown that the dysregulation of the normal gut microbiome, or dysbiosis, can contribute to chronic liver conditions such as MASH, ALD, and cirrhosis [72–74]. In MASH, the reduced butyrate-producing bacteria and elevated endotoxin-producing species create a pro-inflammatory state that promotes carcinogenesis. In ALD, alcohol disrupts the gut barrier integrity, allowing LPS to translocate to the liver and activate Kupffer cells. The gut microbiota affects the production of bacteria-derived metabolites like bile acid [75]. Recent studies have shown that regulating bile acid metabolism affects hepatic fibrogenesis and liver injury, particularly when there is an increased abundance of bacterial strains such as Lactobacilli and Bacteroides [76]. While the exact mechanism leading to liver injury is unknown, it is hypothesized that the tight junction within the intestinal epithelial barrier may play a role. Evidence suggests that dysfunctional gut permeability is present in patients with MASLD or MASH compared with healthy individuals, and the chronic inflammatory state may also be attributed to the epithelial barrier [77–80]. In mice studies, it was found that a compromised gut barrier led to the translocation of bacteria that could reach the liver through the portal vein and exacerbate liver inflammation and fibrosis [81]. While the exact mechanism that the gut microbiome plays in hepatic fibrosis is still under research, studies have shown a potential connection to fibrosis-related carcinogenesis in MASH-HCC [82,83].

3.5. *Genetic Factors*

Genetic predisposition significantly impacts the progression of MASH towards HCC. Several genetic polymorphisms have been associated with the progression of MASH fibrosis. The best known of these polymorphisms is PNPLA3 on chromosome 22 [84]. This variant leads to the impaired triglyceride mobilization of hepatic lipid droplets with increased hepatic lipid accumulation [84], eventually leading to increased oxidative stress. A European study found that PNPLA3 leads to a three-fold increase in HCC [71]. Furthermore, this increase is independent of other risk factors such as BMI, diabetes, and advanced fibrosis [85]. Another polymorphism is TM6SF2, located on chromosome 19. TM6SF2 causes the impaired secretion of very-low-density lipoproteins (VLDLs), thereby increasing hepatic lipid retention, leading to liver fibrosis and HCC [86,87]. On the other hand, variants in the HSD17B13 gene (17-beta hydroxysteroid dehydrogenase 13) located in chromosome 4 attenuate the risk of steatohepatitis and the progression to HCC [88,89]. Genomic mapping has been a valuable tool in identifying genetic variations linked to MASH-HCC. Research has found that the most frequent HCC drivers in MASH patients were mutations in TERT (telomerase reverse-transcriptase) promoter genes, CTNNB1 (catenin beta 1), TP53 single-nucleotide variants, and a TGF family activin receptor, ACVR2A, mutation [90]. While these polymorphisms and mutations are just a few associated with MASH and HCC, they highlight the complex pathogenesis involved in the progression to HCC. Incorporating genetic markers such as PNPLA3, TM6SF2, and HSD17B13 into polygenic risk models holds promise for the early stratification of HCC risk in MASLD and ALD patients. Future clinical tools may leverage these variants to personalize screening strategies and guide surveillance decisions.

3.6. *Immune Response*

Lobular inflammatory responses are believed to play an important role in hepatic fibrosis, cirrhosis, and HCC [91]. Both the innate and adaptive immune responses play a role in promoting hepatic inflammation in MASH. Kupffer cells, leukocytes, and natural killer cells all promote a pro-inflammatory state by releasing eicosanoids, nitric oxide, reactive oxygen species, and cytokines (e.g., TNF, IL-6, and chemokines) [53,92]. The continued liver injury sustained in MASH is believed to release damage-associated molecular patterns (DAMPs), which, in turn, causes the assembly of inflammatory corpuscles, activation of the inflammatory response, and further amplification of liver injury [93,94].

4. Pathogenesis

4.1. ALD-HCC

Alcohol promotes liver carcinogenesis through the formation of acetaldehyde and ROS, changes to the immune system, induction of chronic inflammation, and alterations to gene expression [95]. Chronic alcohol intake alters the architecture and compromises the functional capacity of the liver by triggering steatosis, steatohepatitis, and cirrhosis [95].

4.2. Alcohol Metabolism

Ethanol is metabolized into acetaldehyde by alcohol dehydrogenase (ADH) in the cytosol of hepatocytes. Acetaldehyde then enters the mitochondria, where it is oxidized to acetate by mitochondrial aldehyde dehydrogenase (ALDH) [91]. Acetaldehyde is a reactive and mutagenic compound that can promote DNA repair failure, lipid peroxidation, and mitochondrial damage by causing inflammation, extracellular matrix remolding, and promoting fibrogenesis, eventually leading to carcinogenesis [96,97]. While most alcohol is metabolized by ADH, CYP2E1 also metabolizes it, especially with excessive alcohol consumption [98]. The catalytic cycle of CYP2E1 produces ROS, contributing further to hepatic steatosis and DNA damage. Chronic alcohol exposure increases the levels of CYPE21, potentiating its damaging effects [99]. The enhanced alcohol metabolism in individuals with alcohol use disorder is due to this increase in CYP2E1 levels.

4.3. Oxidative Stress

Similar to the mechanism of cellular damage in MASLD, ROS generated as byproducts of alcohol metabolism in ALD cause cellular damage at the DNA level, impacting functions such as transcription and replication that can lead to ALD and hepatocarcinogenesis. Further, ROS induces the activation of cytokines and immune cells, leading to the upregulation of angiogenesis and metastatic processes [100]. In addition, ROS can lead to hepatic fibrosis via the activation of hepatic stellate cells through TGF-beta production. The ROS byproducts of CYP2E1 alcohol metabolism include H_2O_2, hydroxyl (OH^-), and carbon-centered OH^- [101]. In a healthy liver, antioxidants typically neutralize these ROS. However, chronic exposure to oxidative stress depletes the antioxidant system [99]. Downstream effects of oxidative stress include lipid peroxidation, which creates byproducts such as MDA (Malonaldehyde) and 4-HNE (trans-4-hydrox-2-nonenal) that have been known to cause mutations in p53 associated with HCC [102]. Further, the upregulation of VEGF (vascular endothelial growth factor) and MCP-1 (monocyte chemotactic protein-1) are also associated with tumor angiogenesis and metastasis [98]. Chronic alcohol consumption leads to structural and functional abnormalities in the hepatic mitochondria: it increases their size, impairs the integrity of the mitochondrial DNA (mtDNA), reduces hepatic ATP levels, and impairs the biosynthesis of mitochondrial proteins [99]. As noted above, chronic alcohol metabolism increases the levels of mitochondrial CYP2E1, which, in parallel with the associated mitochondrial dysfunction, further intensifies the generation and accumulation of ROS in hepatic mitochondria [98].

4.4. Role of the Immune System and Gut Microbiome

Like MASH, there is a complex interplay between the gut microbiota and the immune system with ALD. Alcohol consumption can disrupt the intestinal barrier, predisposing one to gut dysbiosis and leading to an influx of endotoxins and inflammatory cytokines in the liver [103–105]. With chronic alcohol use, there is an increased translocation of bacteria-derived lipopolysaccharide (LPS) from the gut to the liver, leading to the activation of the innate immune system by Kupffer cells [106]. Downstream molecular mechanisms such as the IKK/NF-Kb pathway lead to the release of pro-inflammatory cytokines such as TNF-alpha and IL-1β, contributing to hepatocyte injury [107]. Kupffer cells also impact anti-inflammatory cytokines such as IL-10, which have downstream effects on the signaling protein STAT3. Under normal conditions, STAT3 has a hepatoprotective effect but it is linked to an oncogenic effect in HCC [108]. NK cells are naturally part of the innate immune system and play a protective role against infection, but they also have a role in cancer surveillance [109]. Studies have shown that not only can alcohol consumption impair NK cell function, leading to hepatocyte injury, but their quantity is also reduced in HCC [103,104]. Regarding the gut microbiota, studies have found a difference in the gut microflora in patients with HCC compared to health controls. Among patients with early-stage HCC, the levels of butyrate-producing bacterial families (protective to the intestinal barrier) decreased, whereas Klebsiella and Haemophilus species, which produce LPS, increased [110,111]. As mentioned previously, LPS activates pro-inflammatory cytokines, leading to hepatocyte injury [112]. While further research is needed to elucidate the impact of the gut microbiota on HCC, current studies have shown that the intestinal microbiome plays a significant role in ALD and ALD-associated carcinogenesis to HCC.

4.5. Genetic Variations

Growing evidence shows that genetic factors play a role in developing HCC in ALD. A genome-wide association study was performed on individuals of European descent, which found that the rs738409 variant of the PNPLA3 gene increased the risk of developing cirrhosis after alcohol consumption [113]. This gene variant has also been linked to MASH-HCC, as described previously. In MASH, PNPLA3 impacts intrahepatic triglyceride accumulation, but its role in ALD remains unclear. Further, a meta-analysis of five studies has found that the GG phenotype, compared to the CC phenotype, has a higher prevalence among patients with alcohol-associated cirrhosis with HCC [114]. Other genome-wide association studies have found that MBOAT7 (membrane-bound O-acyltransferase domain-containing protein 7) and TM6SF2 are genetic factors that can increase the risk of alcohol-associated cirrhosis [115]. However, FAF2 (Fas-Associated Factor 2), which plays a role in lipid droplet regulation, is protective against alcohol-associated cirrhosis [115]. Researchers have also found the rs708113 [T] minor allele of the WNT3A-WNT9A gene on chromosome 1 is associated with a lower risk of developing alcohol-associated HCC [116]. The human telomerase reverse-transcriptase (TERT) gene on chromosome 5 is crucial in the proliferation of multiple cancers. It encodes a ribonuclease telomerase enzyme that prevents the shortening of telomeres. The rs2736098*A allele and rs2736100*T allele are believed to be associated with telomere shortening and an increased risk of hepatocellular carcinoma [117]. Genetic polymorphisms in alcohol dehydrogenase (ADH), specifically ADH1C*1, have also been linked to carcinogenesis due to its ability to generate ROS through acetaldehyde [118]. Studies have shown the vital interplay between genetics and the development of ALD-HCC, highlighting the need for personalized approaches to risk assessment and management in affected individuals.

5. Clinical Management

5.1. Surveillance

The preferred imaging modality for HCC surveillance remains an abdominal ultrasound [119,120]. Studies have shown that an abdominal ultrasound with alpha-fetoprotein (AFP) levels provides a better diagnostic odds ratio than an ultrasound alone [119]. Hence, the AASLD guidance recommends both an ultrasound and AFP should be used for HCC surveillance. Surveillance should occur at 6-month intervals indefinitely in patients with liver cirrhosis of any etiology, including MASH and ALD [119].

While MRI is the most accurate modality for detecting and characterizing HCC, with a sensitivity ranging from 47% to 95% for lesions smaller than 2 cm, it is typically used as a diagnostic test rather than a primary screening tool due to its cost and limited availability [121]. Abbreviated MRI (AMRI) has emerged as a potential alternative for HCC surveillance. AMRI involves a reduced number of sequences, which shortens the acquisition time and improves cost-effectiveness while maintaining a high sensitivity and specificity for early-stage HCC detection [122]. Currently, the AASLD recommends against the use of MRI for routine HCC screening due to radiology service capacity, patient acceptance, and cost-effectiveness [119].

Despite the widespread global prevalence of MASLD and ALD, current guidelines do not recommend screening in non-cirrhotic MASH due to the lower absolute risk, lack of validated risk models, and cost-effectiveness concerns [123]. In contrast, ALD-HCC typically arises in cirrhotic patients, who are already included in standard surveillance protocols. Additionally, ultrasound performs poorly in obese individuals, a common feature in MASLD, further limiting effective screening. Emerging tools, such as FIB-4, elastography, and genetic risk scores, are promising but not yet endorsed by guidelines [124, 125]. Hence, there is a need to develop biomarkers or risk stratification models which will help us identify MASLD and ALD patients with a high-enough risk of HCC to warrant surveillance.

5.2. Diagnosis

Currently, there is no difference in the diagnosis of MASH/ALD-associated HCC compared to other etiologies. The diagnosis of HCC is primarily based on imaging without the need for pathologic confirmation. However, imaging can be used for diagnosis only in patients with cirrhosis (including ALD-cirrhosis and MASH-cirrhosis), chronic HBV, or a history of HCC. The preferred imaging modality is contrast-enhanced multiphase CT or MRI of the abdomen. With both modalities, a finding of arterial phase hyperenhancement (APHE) and washout on the portal venous or delayed phases is considered a radiological hallmark of HCC with a high predictive value and sensitivity for lesions greater than 1 cm in size in patients with chronic hepatitis B, cirrhosis, or a history of HCC [126,127]. Evaluating non-invasive imagining is based on the Liver Imaging Reporting and Data System (LI-RADS) criteria ranging from LR-1 (benign) to LR-5 (HCC) [128,129]. Figure 1 outlines the LI-RADS criteria in further detail.

A histological diagnosis is required if cirrhosis, HBV, or a history of HCC is not present, as the LI-RADS criteria do not apply to this population. One caveat to this general rule is that the LI-RADS criteria also do not apply to patients with cirrhosis from vascular etiologies (e.g., cardiac cirrhosis), and the diagnosis in these patients should also be made with histopathology. Diagnosis should be based on the International Consensus Group for Hepatocellular Neoplasia using histological and immunohistochemical analyses [130]. The exact histological findings are beyond the scope of this article. A negative liver biopsy does not exclude the diagnosis of HCC in inconclusive cases; a second biopsy is recommended, especially in cases of high clinical suspicion. Previously, AFP (alpha-fetoprotein) levels greater than 400 ng/mL were regarded as a diagnostic criterion, but, due to its inadequate accuracy, current guidelines recommend against using AFP for diagnosis [131,132].

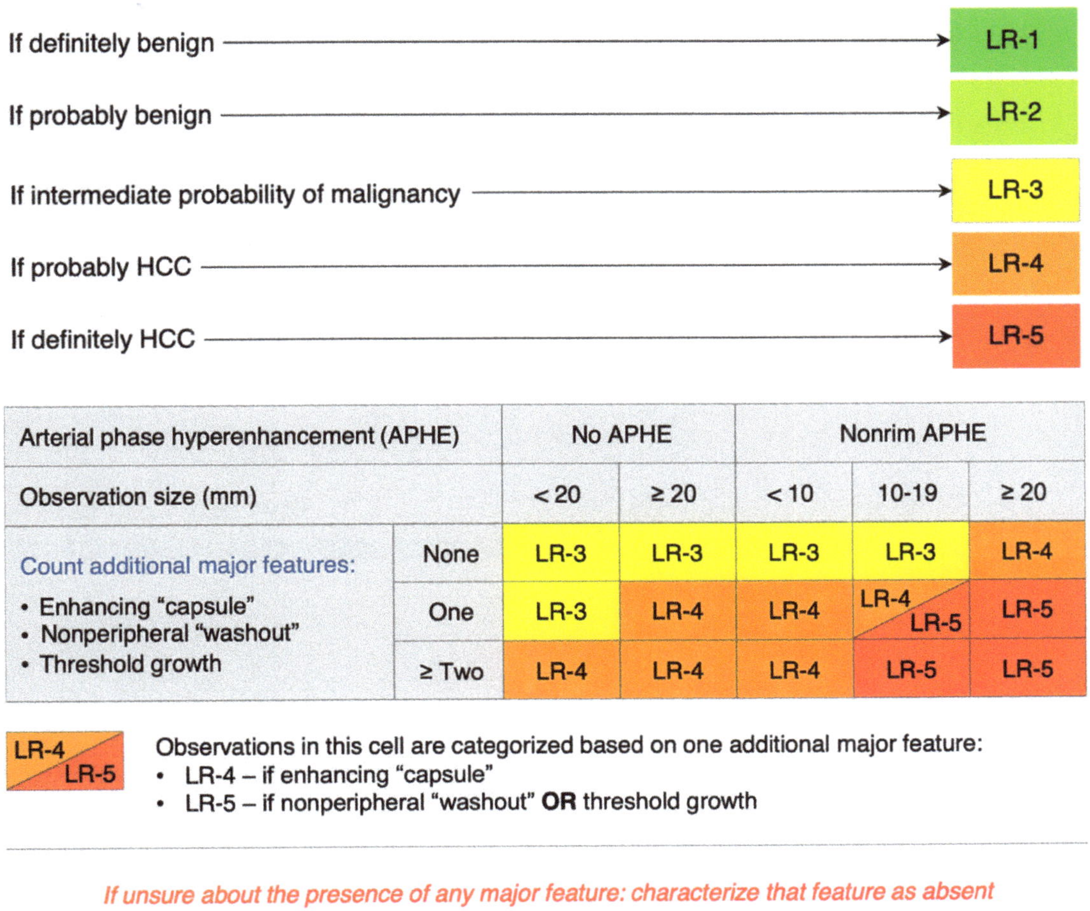

Figure 1. CT/MRI LI-RADS diagnostic algorithm [121,128].

5.3. Staging

All patients with HCC should undergo cancer staging with contrast-enhanced multiphase CT or MRI of the abdomen to assess the extent of malignancy. The TNM staging system is not typically used for HCC. The most universally accepted staging system for HCC is the Barcelona Clinic Liver Cancer (BCLC) staging system (Figure 2) [133]. The BCLC classifies tumors from stage 0 to stage D, with D being the terminal stage based on end-stage liver dysfunction, irrespective of tumor burden. Patients beyond Stage A should undergo noncontract CT of the chest to evaluate for metastatic disease. At the initial evaluation, the tumor staging, burden, degree of liver dysfunction, and Eastern Cooperative Oncology Group (ECOG) performance status (PS) should be documented for baseline measurements as this helps delineate appropriate treatment regimens for patients [119].

5.4. Treatment

The treatment of HCC can be broadly divided into surgical, locoregional, and systemic therapies. A multidisciplinary approach to care is critical. A multidisciplinary team should be established not only to opine on the initial diagnosis and decide on the initial treatment plan but also to follow the patient over time to adjust the management plan as the tumor burden and the patient's condition change.

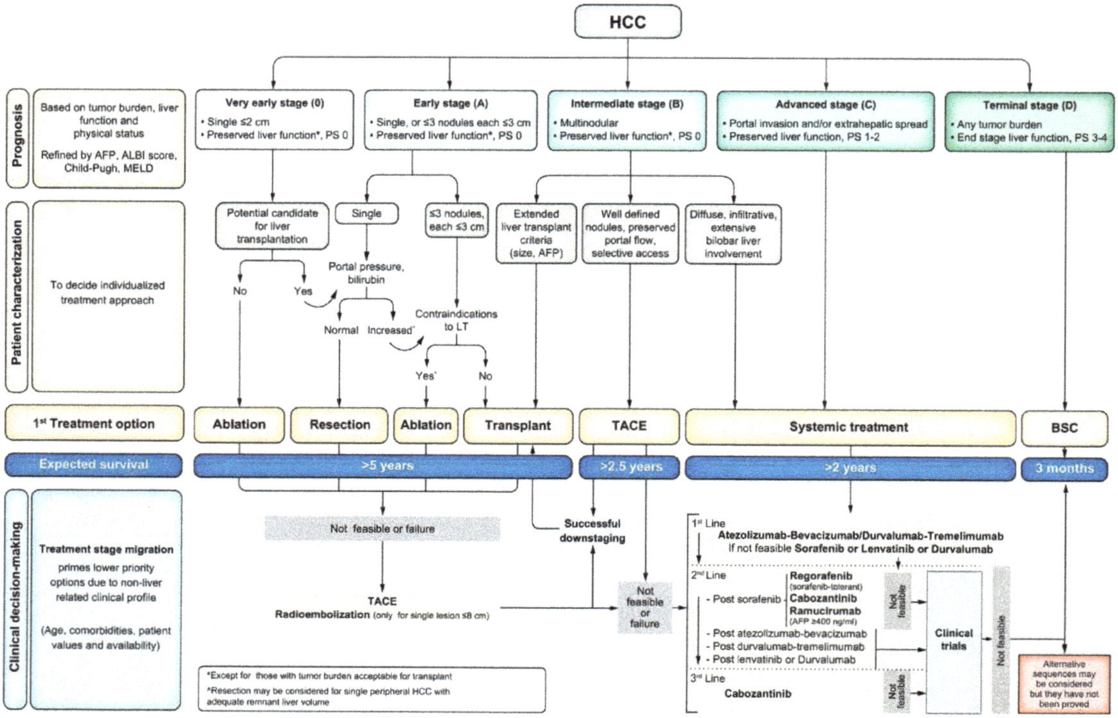

Figure 2. BCLC staging [133].

5.5. Surgical Resection

In patients with HCC without cirrhosis, the treatment of choice is surgical resection for localized tumors [119]. In those with cirrhosis, surgical resection should be considered the treatment of choice for patients with a limited tumor burden and an amenable anatomic location in well-compensated cirrhosis without clinically significant portal hypertension and an adequate future liver remnant [134]. Surgical resection can be accomplished with minimally invasive surgery with laparoscopic or robotic approaches. The 5-year survival rate in these patients exceeds 70% [119]. The Child–Turcotte–Pugh (CTP) score has been used to assess patients for surgical candidacy. Typically, patients in CTP Class A can be considered for surgery since they usually have an appropriate liver reserve and are amenable to surgical resection. The recurrence of HCC after resection remains a significant obstacle, with a recurrence rate of as high as 70% at five years [119]. The risk of recurrence is the highest in the first year post-resection, which fortifies the importance of continued surveillance with cross-sectional imaging in addition to AFP [119].

5.6. Liver Transplantation

In patients with early-stage HCC who do not qualify for resection due to underlying liver disease, liver transplantation (LT) should be considered. Doyle et al., in 2012, reported a 69.1% 10-year survival rate in HCC patients who underwent LT [135]. Importantly, LT has a lower recurrence rate compared to patients undergoing liver resection or ablation. A meta-analysis of 18,421 patients found that the risk of recurrence after liver resection is threefold that of LT [136]. The same analysis found that the mortality rate after liver resection is about 50% higher than LT [135]. To optimize the outcomes after LT, the Milan Criteria (a single tumor with a diameter ≤ 5 cm, or ≤3 tumors with none exceeding 3 cm in diameter) have been the gold standard for patient selection [119]. Patients who are beyond the Milan criteria can be "downstaged", and then considered for liver transplantation. In these patients who do not initially qualify for an LT due tumor burden, and in those who do qualify but must wait for allograft availability, bridge therapy to downstage and control the tumor size is performed using locoregional therapies (LRTs) as noted below. Understanding the etiologic background of HCC has important clinical implications. In MASH-HCC, metabolic comorbidities such as obesity and cardiovascular disease can complicate surgical or transplant eligibility. In contrast, ALD-HCC patients often face challenges related to alcohol relapse risk, malnutrition, and psychosocial evaluation. Tailoring surveillance, therapeutic selection, and perioperative planning to etiology can improve outcomes. Socioeconomic and clinical barriers influence transplant eligibility. ALD patients may face stigma and sobriety-related delays, while MASH patients often have comorbidities (e.g., obesity and cardiovascular disease) that increase the surgical risk. Furthermore, race, insurance status, and geographic location impact both transplant access and post-transplant outcomes [137].

5.7. Surgical Risk

The management of patients with MASLD-HCC and ALD-HCC may differ in surgical risk evaluation. Studies have found that patients with MASLD-HCC have a higher rate of post-surgical complications compared to patients with HCC from other etiologies [138,139]. It is estimated that the 90-day mortality in this subgroup is 11%, with a major complication rate of 31% after surgical resection [140]. Chronic conditions such as T2DM, HTN, and HLD are increased in this subgroup, and studies have shown these conditions have a negative post-operative effect [141]. Interestingly, it was found that the degree of steatosis also affected post-operative outcomes—patients with a higher degree of steatosis were found to have increased morbidity and mortality [140]. Regarding the surgical risk in ALD-HCC, studies have found that patients with ALD have more adverse effects and a higher in-hospital mortality that is estimated to be 2.6-fold higher compared to patients without ALD [142]. One potential reason for this is the degree of malnutrition in this cohort, which is known to increase the post-operative risk significantly [143,144]. While no direct comparison of surgical risk between MASH and ALD-HCC has been made, peri-operative risk management is imperative, and further research is needed in order to identify areas for improvement.

5.8. Locoregional Therapy

Locoregional therapies are imaging-guided, tumor-specific procedures used in about half of the patients with HCC [145]. The mainstay modalities in these scenarios include radiofrequency ablation (RFA), where thermal damage is delivered to the tumors using electromagnetic energy microwave ablation (MWA), which also relies on thermal damage but differs in its energy delivery methods [146–148]; TACE (transarterial chemoembolization), which embolizes the distal branches of the hepatic artery supplying the tumor using embolic particles like gelatin sponges, or synthetic polymer beads, and chemotherapeutic agents like doxorubicin, cisplatin, or irinotecan [146,149]; TARE (transarterial radioembolization), which delivers a high dose of targeted radiation to the tumors utilizing Y90 (yttrium-90)-impregnated microspheres [146,150]; and SBRT (stereotactic body radiotherapy), which delivers high doses of radiation with a high precision to a targeted tumor in a small number of fractions, typically between 1 to 5 sessions [151].

5.9. Systemic Therapy

Systemic therapy is reserved for patients with unresectable HCC who are not candidates for locoregional therapy. Systemic treatment for MASH-HCC and ALD-HCC is the same. Medications used in systemic treatment can be divided into the following categories: immune checkpoint inhibitors (ICIs), multi-target tyrosine kinase inhibitors (MTKIs), and monoclonal antiangiogenic antibodies (anti-VEGF antibodies). ICIs include PD1 inhibitors (programmed death 1) like pembrolizumab and nivolumab, inhibitors of the ligand of PD1 (PD-L1) like durvalumab and atezolizumab, or CTLA4 inhibitors (cytotoxic T lymphocyte-associated protein 4) like tremelimumab and ipilimumab [119]. Examples of MTKIs are sorafenib, lenvatinib, cabozantinib, and regorafenib. The monoclonal antiangiogenic antibodies comprise ramucirumab and bevacizumab, and they can be used in combination with ICIs or as a monotherapy [119]. Intravenous ICI treatment regimens are recommended as a first-line therapy since they have a higher efficacy and are better tolerated than MTKIs. However, they are associated with immune-related adverse effects. Since ICIs enhance T-cell responses, they are not recommended in post-transplant patients due to a high risk of graft loss, as multiple cases of severe rejection after LT have been reported [152]. MTKIs are oral agents but have a poorer tolerance profile than the ICIs. Prominent MTKI adverse side effects include hand–foot skin reaction, diarrhea, hypertension, proteinuria, gastrointestinal side effects, the inhibition of wound healing, hemorrhage, and thromboembolism [119]. Notably, the immunotherapy response may differ by etiology [153]. MASH-HCC appears less responsive to ICIs due to an immunosuppressive microenvironment dominated by dysfunctional T cells. In contrast, ALD-HCC may exhibit a more inflamed immune profile that enhances ICI efficacy, although direct comparative trials remain limited. While the specific details of each of these therapies are beyond the scope of this article, the correct systemic treatment should be chosen in a multidisciplinary manner as each of these medications has specific indications and side effect profiles. Clinical trials often underrepresent patients with significant metabolic comorbidities or advanced liver dysfunction, which are common in MASH-HCC. This limits the generalizability and underscores the need for etiology-specific trial stratification and broader inclusion criteria. Therapy needs to be tailored to each patient (Table 1).

Table 1. Comparison between MASH-HCC and ALD-HCC.

Category	MASH-HCC	ALD-HCC	Shared Features
Epidemiology	• 10% of global HCC cases are secondary to MASLD [15]. • Projected to become the leading cause of HCC by 2030 [18]. • Higher prevalence in Hispanic populations in the U.S. (21%) compared to White (12.5%) and Black individuals (11.6%) [38].	• Accounts for 30% of global HCC cases [24]. • Alcohol use is the second-fastest-growing cause of liver cancer deaths globally [23]. • Alcohol-related cirrhosis and HCC are more prevalent in men, though the gap is narrowing with rising alcohol consumption in women [13,28,29].	• Both are influenced by lifestyle factors (diet, and alcohol use). • Geographic variability. • Socioeconomic disparities in prevalence and access to treatment.

Table 1. *Cont.*

Category	MASH-HCC	ALD-HCC	Shared Features
Pathogenesis	• Driven by insulin resistance, lipotoxicity, hyperglycemia, and adipose tissue dysfunction. • Accumulation of free fatty acids (FFAs) leads to hepatic lipotoxicity and mitochondrial dysfunction. • Chronic inflammation results in fibrogenesis, eventually leading to cirrhosis and HCC. • Key pathways include oxidative stress, ER stress, and activation of pro-inflammatory cytokines like TNF-α.	• Ethanol metabolism leads to acetaldehyde production, a toxic and mutagenic compound. • Acetaldehyde induces DNA damage, lipid peroxidation, and mitochondrial dysfunction. • Chronic oxidative stress triggers immune dysregulation, contributing to fibrosis and HCC. • Key pathways include ROS generation, immune cell activation, and cytokine release.	• Both share mechanisms involving oxidative stress, chronic inflammation, and mitochondrial dysfunction. • Both are associated with gut dysbiosis, leading to increased intestinal permeability and translocation of endotoxins (e.g., LPS), which exacerbate liver injury.
Genetics	• Common genetic variants include PNPLA3 (associated with increased hepatic triglyceride accumulation) and TM6SF2 (impairs VLDL secretion) [83,85,86]. • Variants in HSD17B13 are protective against progression to HCC in MASH [87,88]. • Genetic predisposition modulates lipid metabolism and the inflammatory response, influencing disease severity.	• ADH gene variants, especially ADH1C*1, are linked to increased acetaldehyde production, exacerbating liver damage [90]. • TERT gene mutations, particularly in the promoter region, are linked to telomere shortening and increased cancer risk [116].	• Genetic factors such as PNPLA3 are common to both conditions, affecting lipid metabolism and fibrosis progression. • Both conditions involve mutations that alter metabolic pathways, leading to hepatic fat accumulation and fibrosis.
Management	• Focus on metabolic control through lifestyle changes, anti-diabetic medications, and lipid-lowering agents. • Treat underlying MASH with lifestyle modifications and pharmacotherapy. • Bariatric surgery, endoscopic bariatric procedures, and pharmacotherapy may be considered.	• Alcohol cessation is the cornerstone of treatment; nutritional support is critical due to the high prevalence of malnutrition in ALD patients [25,154].	• Management of the cancer in both involves identical treatment modalities. • Liver transplantation is an important curative option in both diseases [119].

Table 1. *Cont.*

Category	MASH-HCC	ALD-HCC	Shared Features
Prognosis	• Progression to cirrhosis and HCC is slow but can be accelerated in the presence of diabetes, hypertension, and obesity [8,155].	• Prognosis is significantly worsened if alcohol use continues [156]. • Patients with ongoing alcohol use disorder have higher postoperative mortality and recurrence rates following resection [137].	• Both conditions show high recurrence rates after treatment, with the exception of liver transplantation [131,132]. • Early-stage detection and effective treatment of underlying liver disease can significantly improve prognosis [157,158]. • Studies comparing overall survival have found no significant differences between the two groups [12].

6. Conclusions

The global rise in hepatocellular carcinoma associated with MASH and ALD represents a critical challenge that demands urgent, multifaceted intervention. While MASH-HCC is projected to become the predominant etiology of liver cancer in the Western world by 2030, ALD-HCC continues to account for a significant proportion of cases worldwide, driven by high and rising alcohol consumption. These trends emphasize the need for early detection and treatment strategies tailored to the unique characteristics of these conditions.

MASH-HCC highlights the pivotal role of metabolic dysregulation, with obesity, type 2 diabetes, and other components of metabolic syndrome serving as major drivers of hepatic inflammation and carcinogenesis. Genetic polymorphisms, including PNPLA3, TM6SF2, and HSD17B13, offer opportunities for personalized risk stratification and targeted therapies. In contrast, the pathogenesis of ALD-HCC is driven by alcohol metabolism and its toxic byproducts, such as acetaldehyde and reactive oxygen species (ROS), which lead to direct hepatocyte injury, immune dysregulation, and fibrogenesis. Furthermore, genetic insights, such as variations in ADH, TERT, and PNPLA3, should guide individualized care plans for those at heightened risk.

Both MASH-HCC and ALD-HCC share common pathways, such as oxidative stress, mitochondrial dysfunction, and the role of the gut–liver axis. Dysbiosis and increased intestinal permeability exacerbate hepatic inflammation and fibrosis, making the microbiome a promising area for therapeutic exploration. Research into microbiome modulation, such as probiotics, prebiotics, and fecal microbiota transplantation, could offer novel strategies to mitigate disease progression in both etiologies. The mitigation of MASLD and ALD will undoubtedly have a corresponding dampening effect on the disease burden of HCC related to these conditions.

Therapeutic advancements are vital to improving outcomes in HCC. Early-stage disease management should focus on curative options, such as surgical resection, liver transplantation, and ablation. However, for patients with advanced disease, locoregional therapies (e.g., TACE, TARE, and ablation) and systemic treatments (e.g., immune checkpoint inhibitors, tyrosine kinase inhibitors, and antiangiogenic agents) provide valuable pathways for disease control.

Future research should prioritize validated screening algorithms for non-cirrhotic MASLD, the integration of genetic and microbiome-based risk stratification tools, and etiology-specific therapeutic trials. The clinical trial design must also account for patient heterogeneity in liver function and metabolic status to improve external validity. Equally important is the need to address healthcare disparities that influence the outcomes in HCC, particularly in underserved populations disproportionately affected by MASH and ALD. By embracing a multidisciplinary and personalized approach to prevention, diagnosis, and treatment, we can reduce the burden of these diseases, improve survival rates, and enhance the quality of life for patients worldwide.

Author Contributions: Conceptualization, A.B.; methodology, A.B., S.P., F.K. and D.F.; investigation, A.B., S.P., F.K. and D.F.; resources, A.B., S.P., F.K. and D.F.; data curation, A.B., S.P., F.K. and D.F.; writing—original draft presentation, A.B., S.P., F.K. and D.F.; writing—review and editing, A.B., S.P., F.K. and D.F.; visualization, A.B., S.P., F.K. and D.F.; supervision, A.B.; project administration, A.B. All authors have read and agreed to the published version of the manuscript.

Funding: This research received no external funding.

Data Availability Statement: No new data were created in this article.

Conflicts of Interest: The authors declare no conflicts of interest.

References

1. Sung, H.; Ferlay, J.; Siegel, R.L.; Laversanne, M.; Soerjomataram, I.; Jemal, A.; Bray, F. Global Cancer Statistics 2020: GLOBOCAN Estimates of Incidence and Mortality Worldwide for 36 Cancers in 185 Countries. *CA Cancer J. Clin.* **2021**, *71*, 209–249. [CrossRef] [PubMed]
2. Ferlay, J.; Colombet, M.; Soerjomataram, I.; Mathers, C.; Parkin, D.M.; Piñeros, M.; Znaor, A.; Bray, F. Estimating the global cancer incidence and mortality in 2018: GLOBOCAN sources and methods. *Int. J. Cancer* **2019**, *144*, 1941–1953. [CrossRef] [PubMed]
3. Llovet, J.M.; Kelley, R.K.; Villanueva, A.; Singal, A.G.; Pikarsky, E.; Roayaie, S.; Lencioni, R.; Koike, K.; Zucman-Rossi, J.; Finn, R.S. Hepatocellular carcinoma. *Nat. Rev. Dis. Primers* **2021**, *7*, 6. [CrossRef]
4. Anstee, Q.M.; Reeves, H.L.; Kotsiliti, E.; Govaere, O.; Heikenwalder, M. From NASH to HCC: Current concepts and future challenges. *Nat. Rev. Gastroenterol. Hepatol.* **2019**, *16*, 411–428. [CrossRef]
5. Younossi, Z.M.; Rinella, M.E.; Sanyal, A.J.; Harrison, S.A.; Brunt, E.M.; Goodman, Z.; Cohen, D.E.; Loomba, R. From NAFLD to MAFLD: Implications of a Premature Change in Terminology. *Hepatology* **2021**, *73*, 1194–1198. [CrossRef]
6. Rinella, M.E.; Lazarus, J.V.; Ratziu, V.; Francque, S.M.; Sanyal, A.J.; Kanwal, F.; Romero, D.; Abdelmalek, M.F.; Anstee, Q.M.; Arab, J.P.; et al. A multisociety Delphi consensus statement on new fatty liver disease nomenclature. *Hepatology* **2023**, *78*, 1966–1986. [CrossRef] [PubMed]
7. Boccatonda, A.; Andreetto, L.; D'Ardes, D.; Cocco, G.; Rossi, I.; Vicari, S.; Schiavone, C.; Cipollone, F.; Guagnano, M.T. From NAFLD to MAFLD: Definition, Pathophysiological Basis and Cardiovascular Implications. *Biomedicines* **2023**, *11*, 883. [CrossRef]
8. Llovet, J.M.; Willoughby, C.E.; Singal, A.G.; Greten, T.F.; Heikenwälder, M.; El-Serag, H.B.; Finn, R.S.; Friedman, S.L. Nonalcoholic steatohepatitis-related hepatocellular carcinoma: Pathogenesis and treatment. *Nat. Rev. Gastroenterol. Hepatol.* **2023**, *20*, 487–503. [CrossRef]
9. Cancer Today. Available online: https://gco.iarc.who.int/today/ (accessed on 25 September 2024).
10. Ioannou, G.N.; Splan, M.F.; Weiss, N.S.; McDonald, G.B.; Beretta, L.; Lee, S.P. Incidence and Predictors of Hepatocellular Carcinoma in Patients with Cirrhosis. *Clin. Gastroenterol. Hepatol.* **2007**, *5*, 938–945.e4. [CrossRef]
11. Kanneganti, M.; Singal, A.G. Diagnosis and management of indeterminate liver nodules in patients with cirrhosis. *Clin. Liver Dis.* **2023**, *22*, 181. [CrossRef]
12. Holzner, M.L.; Florman, S.; Schwartz, M.E.; Tabrizian, P. Outcomes of liver transplantation for nonalcoholic steatohepatitis-associated hepatocellular carcinoma. *HPB* **2022**, *24*, 470–477. [CrossRef]
13. Younossi, Z.M.; Koenig, A.B.; Abdelatif, D.; Fazel, Y.; Henry, L.; Wymer, M. Global epidemiology of nonalcoholic fatty liver disease-Meta-analytic assessment of prevalence, incidence, and outcomes. *Hepatology* **2016**, *64*, 73–84. [CrossRef]
14. Lazarus, J.V.; Colombo, M.; Cortez-Pinto, H.; Huang, T.T.-K.; Miller, V.; Ninburg, M.; Schattenberg, J.M.; Seim, L.; Wong, V.W.S.; Zelber-Sagi, S. NAFLD—sounding the alarm on a silent epidemic. *Nat. Rev. Gastroenterol. Hepatol.* **2020**, *17*, 377–379. [CrossRef]
15. Huang, D.Q.; El-Serag, H.B.; Loomba, R. Global epidemiology of NAFLD-related HCC: Trends, predictions, risk factors and prevention. *Nat. Rev. Gastroenterol. Hepatol.* **2021**, *18*, 223–238. [CrossRef] [PubMed]
16. Younossi, Z.M.; Golabi, P.; Paik, J.M.; Henry, A.; Van Dongen, C.; Henry, L. The global epidemiology of nonalcoholic fatty liver disease (NAFLD) and nonalcoholic steatohepatitis (NASH): A systematic review. *Hepatology* **2023**, *77*, 1335. [CrossRef]

17. Yu, S.; Li, C.; Ji, G.; Zhang, L. The Contribution of Dietary Fructose to Non-alcoholic Fatty Liver Disease. *Front. Pharmacol.* **2021**, *12*, 783393. [CrossRef]
18. Huang, D.Q.; Singal, A.G.; Kono, Y.; Tan, D.J.H.; El-Serag, H.B.; Loomba, R. Changing global epidemiology of liver cancer from 2010 to 2019: NASH is the fastest growing cause of liver cancer. *Cell Metab.* **2022**, *34*, 969–977.e2. [CrossRef]
19. Estes, C.; Anstee, Q.M.; Arias-Loste, M.T.; Bantel, H.; Bellentani, S.; Caballeria, J.; Colombo, M.; Craxi, A.; Crespo, J.; Day, C.P.; et al. Modeling NAFLD disease burden in China, France, Germany, Italy, Japan, Spain, United Kingdom, and United States for the period 2016–2030. *J. Hepatol.* **2018**, *69*, 896–904. [CrossRef]
20. Seitz, H.K.; Neuman, M.G. The History of Alcoholic Liver Disease: From an Unrecognized Disease to One of the Most Frequent Diseases in Hepatology. *J. Clin. Med.* **2021**, *10*, 858. [CrossRef]
21. Global Status Report on Alcohol and Health. 2018. Available online: https://www.who.int/publications/i/item/9789241565639 (accessed on 12 March 2025).
22. Manthey, J.; Shield, K.D.; Rylett, M.; Hasan, O.S.M.; Probst, C.; Rehm, J. Global alcohol exposure between 1990 and 2017 and forecasts until 2030: A modelling study. *Lancet* **2019**, *393*, 2493–2502. [CrossRef]
23. Younossi, Z.; Stepanova, M.; Ong, J.P.; Jacobson, I.M.; Bugianesi, E.; Duseja, A.; Eguchi, Y.; Wong, V.W.; Negro, F.; Yilmaz, Y.; et al. Nonalcoholic Steatohepatitis Is the Fastest Growing Cause of Hepatocellular Carcinoma in Liver Transplant Candidates. *Clin Gastroenterol. Hepatol.* **2019**, *17*, 748–755.e3. [CrossRef]
24. Huang, D.Q.; Mathurin, P.; Cortez-Pinto, H.; Loomba, R. Global epidemiology of alcohol-associated cirrhosis and HCC: Trends, projections and risk factors. *Nat. Rev. Gastroenterol. Hepatol.* **2023**, *20*, 37–49. [CrossRef]
25. Ganne-Carrié, N.; Nahon, P. Hepatocellular carcinoma in the setting of alcohol-related liver disease. *J. Hepatol.* **2019**, *70*, 284–293. [CrossRef]
26. GBD 2019 Diseases and Injuries Collaborators. Global burden of 369 diseases and injuries in 204 countries and territories, 1990-2019: A systematic analysis for the Global Burden of Disease Study 2019. *Lancet* **2020**, *396*, 1204–1222. [CrossRef]
27. Castro Narro, G.E.; Díaz, L.A.; Ortega, E.K.; Bautista Garín, M.F.; Reyes, E.C.; Martinez Delfin, P.S.; Arab, J.P.; Bataller, R. Alcohol-related liver disease: A global perspective. *Ann. Hepatol.* **2024**, *29*, 101499. [CrossRef]
28. Balakrishnan, M.; Patel, P.; Dunn-Valadez, S.; Dao, C.; Khan, V.; Ali, H.; El-Serag, L.; Hernaez, R.; Sisson, A.; Thrift, A.P.; et al. Women Have a Lower Risk of Nonalcoholic Fatty Liver Disease but a Higher Risk of Progression vs Men: A Systematic Review and Meta-analysis. *Clin. Gastroenterol. Hepatol.* **2021**, *19*, 61–71.e15. [CrossRef]
29. Burra, P.; Bizzaro, D.; Gonta, A.; Shalaby, S.; Gambato, M.; Morelli, M.C.; Trapani, S.; Floreani, A.; Marra, F.; Brunetto, M.R.; et al. Clinical impact of sexual dimorphism in non-alcoholic fatty liver disease (NAFLD) and non-alcoholic steatohepatitis (NASH). *Liver Int.* **2021**, *41*, 1713–1733. [CrossRef]
30. Eng, P.C.; Forlano, R.; Tan, T.; Manousou, P.; Dhillo, W.S.; Izzi-Engbeaya, C. Non-alcoholic fatty liver disease in women—Current knowledge and emerging concepts. *JHEP Rep.* **2023**, *5*, 100835. [CrossRef]
31. Lonardo, A.; Nascimbeni, F.; Ballestri, S.; Fairweather, D.; Win, S.; Than, T.A.; Abdelmalek, M.F.; Suzuki, A. Sex Differences in Nonalcoholic Fatty Liver Disease: State of the Art and Identification of Research Gaps. *Hepatology* **2019**, *70*, 1457. [CrossRef]
32. Carlini, L.E.; Fernandez, A.C.; Mellinger, J.L. Sex and gender in alcohol use disorder and alcohol-associated liver disease in the United States: A narrative review. *Hepatology* **2024**. Available online: https://journals.lww.com/hep/fulltext/9900/sex_and_gender_in_alcohol_use_disorder_and.864.aspx (accessed on 12 March 2025). [CrossRef]
33. Grucza, R.A.; Sher, K.J.; Kerr, W.C.; Krauss, M.J.; Lui, C.K.; McDowell, Y.E.; Hartz, S.; Virdi, G.; Bierut, L.J. Trends in Adult Alcohol Use and Binge Drinking in the Early 21st Century United States: A Meta-Analysis of Six National Survey Series. *Alcohol. Clin. Exp. Res.* **2018**, *42*, 1939–1950. [CrossRef]
34. Wu, E.M.; Wong, L.L.; Hernandez, B.Y.; Ji, J.-F.; Jia, W.; Kwee, S.A.; Kalathil, S. Gender differences in hepatocellular cancer: Disparities in nonalcoholic fatty liver disease/steatohepatitis and liver transplantation. *Hepatoma Res.* **2018**, *4*, 66. [CrossRef]
35. Paschos, P.; Paletas, K. Non alcoholic fatty liver disease and metabolic syndrome. *Hippokratia* **2009**, *13*, 9–19.
36. Gupta, A.; Das, A.; Majumder, K.; Arora, N.; Mayo, H.G.; Singh, P.P.; Beg, M.S.; Singh, S. Obesity is Independently Associated with Increased Risk of Hepatocellular Cancer–related Mortality: A Systematic Review and Meta-Analysis. *Am. J. Clin. Oncol.* **2018**, *41*, 874. [CrossRef]
37. Kanwal, F.; Kramer, J.R.; Mapakshi, S.; Natarajan, Y.; Chayanupatkul, M.; Richardson, P.A.; Li, L.; Desiderio, R.; Thrift, A.P.; Asch, S.M.; et al. Risk of Hepatocellular Cancer in Patients with Non-Alcoholic Fatty Liver Disease. *Gastroenterology* **2018**, *155*, 1828–1837.e2. [CrossRef]
38. Cotter, T.G.; Rinella, M. Nonalcoholic Fatty Liver Disease 2020: The State of the Disease. *Gastroenterology* **2020**, *158*, 1851–1864. [CrossRef]
39. Schneider, A.L.C.; Lazo, M.; Selvin, E.; Clark, J.M. Racial Differences in Nonalcoholic Fatty Liver Disease in the U.S. Population. *Obesity* **2014**, *22*, 292–299. [CrossRef]
40. Becker, U.; Deis, A.; Sørensen, T.I.; Grønbaek, M.; Borch-Johnsen, K.; Müller, C.F.; Schnohr, P.; Jensen, G. Prediction of risk of liver disease by alcohol intake, sex, and age: A prospective population study. *Hepatology* **1996**, *23*, 1025–1029. [CrossRef]

41. Marti-Aguado, D.; Clemente-Sanchez, A.; Bataller, R. Cigarette smoking and liver diseases. *J. Hepatol.* **2022**, *77*, 191–205. [CrossRef]
42. Donato, F.; Tagger, A.; Gelatti, U.; Parrinello, G.; Boffetta, P.; Albertini, A.; Decarli, A.; Trevisi, P.; Ribero, M.L.; Martelli, C.; et al. Alcohol and Hepatocellular Carcinoma: The Effect of Lifetime Intake and Hepatitis Virus Infections in Men and Women. *Am. J. Epidemiol.* **2002**, *155*, 323–331. [CrossRef]
43. Zhang, H.H.; Halbleib, M.; Ahmad, F.; Manganiello, V.C.; Greenberg, A.S. Tumor necrosis factor-alpha stimulates lipolysis in differentiated human adipocytes through activation of extracellular signal-related kinase and elevation of intracellular cAMP. *Diabetes* **2002**, *51*, 2929–2935. [CrossRef]
44. del Aguila, L.F.; Claffey, K.P.; Kirwan, J.P. TNF-alpha impairs insulin signaling and insulin stimulation of glucose uptake in C2C12 muscle cells. *Am. J. Physiol.* **1999**, *276*, E849–E855. [CrossRef]
45. Lauby-Secretan, B.; Scoccianti, C.; Loomis, D.; Grosse, Y.; Bianchini, F.; Straif, K. Body Fatness and Cancer—Viewpoint of the IARC Working Group. *N. Engl. J. Med.* **2016**, *375*, 794–798. [CrossRef] [PubMed]
46. Tanaka, S.; Miyanishi, K.; Kobune, M.; Kawano, Y.; Hoki, T.; Kubo, T.; Hayashi, T.; Sato, T.; Sato, Y.; Takimoto, R.; et al. Increased hepatic oxidative DNA damage in patients with nonalcoholic steatohepatitis who develop hepatocellular carcinoma. *J. Gastroenterol.* **2013**, *48*, 1249–1258. [CrossRef]
47. Welzel, T.M.; Graubard, B.I.; Zeuzem, S.; El-Serag, H.B.; Davila, J.A.; McGlynn, K.A. Metabolic syndrome increases the risk of primary liver cancer in the United States: A population-based case-control study. *Hepatology* **2011**, *54*, 463–471. [CrossRef] [PubMed]
48. Turati, F.; Talamini, R.; Pelucchi, C.; Polesel, J.; Franceschi, S.; Crispo, A.; Izzo, F.; La Vecchia, C.; Boffetta, P.; Montella, M. Metabolic syndrome and hepatocellular carcinoma risk. *Br. J. Cancer* **2013**, *108*, 222–228. [CrossRef] [PubMed]
49. Inoue, M.; Kurahashi, N.; Iwasaki, M.; Tanaka, Y.; Mizokami, M.; Noda, M.; Tsugane, S.; Japan Public Health Center-Based Prospective Study Group. Metabolic factors and subsequent risk of hepatocellular carcinoma by hepatitis virus infection status: A large-scale population-based cohort study of Japanese men and women (JPHC Study Cohort II). *Cancer Causes Control* **2009**, *20*, 741–750. [CrossRef] [PubMed]
50. Ajmera, V.; Cepin, S.; Tesfai, K.; Hofflich, H.; Cadman, K.; Lopez, S.; Madamba, E.; Bettencourt, R.; Richards, L.; Behling, C.; et al. A prospective study on the prevalence of NAFLD, advanced fibrosis, cirrhosis and hepatocellular carcinoma in people with type 2 diabetes. *J. Hepatol.* **2023**, *78*, 471–478. [CrossRef]
51. Siegel, A.B.; Zhu, A.X. Metabolic syndrome and hepatocellular carcinoma: Two growing epidemics with a potential link. *Cancer* **2009**, *115*, 5651–5661. [CrossRef]
52. Smith, G.I.; Shankaran, M.; Yoshino, M.; Schweitzer, G.G.; Chondronikola, M.; Beals, J.W.; Okunade, A.L.; Patterson, B.W.; Nyangau, E.; Field, T.; et al. Insulin resistance drives hepatic de novo lipogenesis in nonalcoholic fatty liver disease. *J. Clin. Investig.* **2020**, *130*, 1453–1460. [CrossRef]
53. Marra, F.; Svegliati-Baroni, G. Lipotoxicity and the gut-liver axis in NASH pathogenesis. *J. Hepatol.* **2018**, *68*, 280–295. [CrossRef]
54. Ipsen, D.H.; Lykkesfeldt, J.; Tveden-Nyborg, P. Molecular mechanisms of hepatic lipid accumulation in non-alcoholic fatty liver disease. *Cell. Mol. Life Sci.* **2018**, *75*, 3313–3327. [CrossRef]
55. Kawai, T.; Autieri, M.V.; Scalia, R. Adipose tissue inflammation and metabolic dysfunction in obesity. *Am. J. Physiol. Cell Physiol.* **2021**, *320*, C375–C391. [CrossRef]
56. Hotamisligil, G.S.; Arner, P.; Caro, J.F.; Atkinson, R.L.; Spiegelman, B.M. Increased adipose tissue expression of tumor necrosis factor-alpha in human obesity and insulin resistance. *J. Clin. Investig.* **1995**, *95*, 2409–2415. [CrossRef] [PubMed]
57. Barbuio, R.; Milanski, M.; Bertolo, M.B.; Saad, M.J.; Velloso, L.A. Infliximab reverses steatosis and improves insulin signal transduction in liver of rats fed a high-fat diet. *J. Endocrinol.* **2007**, *194*, 539–550. [CrossRef] [PubMed]
58. Nakagawa, H.; Umemura, A.; Taniguchi, K.; Font-Burgada, J.; Dhar, D.; Ogata, H.; Zhong, Z.; Valasek, M.A.; Seki, E.; Hidalgo, J.; et al. ER stress cooperates with hypernutrition to trigger TNF-dependent spontaneous HCC development. *Cancer Cell* **2014**, *26*, 331–343. [CrossRef] [PubMed]
59. Uysal, K.T.; Wiesbrock, S.M.; Marino, M.W.; Hotamisligil, G.S. Protection from obesity-induced insulin resistance in mice lacking TNF-alpha function. *Nature* **1997**, *389*, 610–614. [CrossRef]
60. Palisso, G.; Gambardella, A.; Tagliamonte, M.R.; Saccomanno, F.; Salvatore, T.; Gualdiero, P.; D'Onofrio, M.V.; Howard, B.V. Available online: https://academic.oup.com/jcem/article-abstract/81/12/4244/2650453?redirectedFrom=fulltext&login=false (accessed on 12 March 2025).
61. Koo, S.-Y.; Park, E.-J.; Lee, C.-W. Immunological distinctions between nonalcoholic steatohepatitis and hepatocellular carcinoma. *Exp. Mol. Med.* **2020**, *52*, 1209–1219. [CrossRef]
62. Torre, P.; Motta, B.M.; Sciorio, R.; Masarone, M.; Persico, M. Inflammation and Fibrogenesis in MAFLD: Role of the Hepatic Immune System. *Front. Med.* **2021**, *8*, 781567. [CrossRef]
63. Seki, S.; Kitada, T.; Yamada, T.; Sakaguchi, H.; Nakatani, K.; Wakasa, K. In situ detection of lipid peroxidation and oxidative DNA damage in non-alcoholic fatty liver diseases. *J. Hepatol.* **2002**, *37*, 56–62. [CrossRef]

64. Begriche, K.; Massart, J.; Robin, M.-A.; Bonnet, F.; Fromenty, B. Mitochondrial adaptations and dysfunctions in nonalcoholic fatty liver disease. *Hepatology* **2013**, *58*, 1497–1507. [CrossRef]
65. Fu, S.; Watkins, S.M.; Hotamisligil, G.S. The role of endoplasmic reticulum in hepatic lipid homeostasis and stress signaling. *Cell Metab.* **2012**, *15*, 623–634. [CrossRef]
66. Zhang, Z.; Zhang, L.; Zhou, L.; Lei, Y.; Zhang, Y.; Huang, C. Redox signaling and unfolded protein response coordinate cell fate decisions under ER stress. *Redox Biol.* **2019**, *25*, 101047. [CrossRef]
67. Henkel, A.; Green, R.M. The unfolded protein response in fatty liver disease. *Semin. Liver Dis.* **2013**, *33*, 321–329. [CrossRef]
68. Flessa, C.-M.; Kyrou, I.; Nasiri-Ansari, N.; Kaltsas, G.; Papavassiliou, A.G.; Kassi, E.; Randeva, H.S. Endoplasmic Reticulum Stress and Autophagy in the Pathogenesis of Non-alcoholic Fatty Liver Disease (NAFLD): Current Evidence and Perspectives. *Curr. Obes. Rep.* **2021**, *10*, 134–161. [CrossRef] [PubMed]
69. Yuan, D.; Huang, S.; Berger, E.; Liu, L.; Gross, N.; Heinzmann, F.; Ringelhan, M.; Connor, T.O.; Stadler, M.; Meister, M.; et al. Kupffer Cell-Derived Tnf Triggers Cholangiocellular Tumorigenesis through JNK due to Chronic Mitochondrial Dysfunction and ROS. *Cancer Cell* **2017**, *31*, 771–789.e6. [CrossRef] [PubMed]
70. Canli, Ö.; Nicolas, A.M.; Gupta, J.; Finkelmeier, F.; Goncharova, O.; Pesic, M.; Neumann, T.; Horst, D.; Löwer, M.; Sahin, U.; et al. Myeloid Cell-Derived Reactive Oxygen Species Induce Epithelial Mutagenesis. *Cancer Cell* **2017**, *32*, 869–883.e5. [CrossRef] [PubMed]
71. Leyh, C.; Coombes, J.D.; Schmidt, H.H.; Canbay, A.; Manka, P.P.; Best, J. MASLD-Related HCC—Update on Pathogenesis and Current Treatment Options. *J. Pers. Med.* **2024**, *14*, 370. [CrossRef] [PubMed]
72. Brenner, D.A.; Paik, Y.-H.; Schnabl, B. Role of Gut Microbiota in Liver Disease. *J. Clin. Gastroenterol.* **2015**, *49* (Suppl. S1), S25–S27. [CrossRef]
73. Etienne-Mesmin, L.; Chassaing, B.; Gewirtz, A.T. Tryptophan: A gut microbiota-derived metabolites regulating inflammation. *World J. Gastrointest. Pharmacol. Ther.* **2017**, *8*, 7–9. [CrossRef]
74. Konturek, P.C.; Harsch, I.A.; Konturek, K.; Schink, M.; Konturek, T.; Neurath, M.F.; Zopf, Y. Gut–Liver Axis: How Do Gut Bacteria Influence the Liver? *Med. Sci.* **2018**, *6*, 79. [CrossRef]
75. Chavez-Talavera, O. 2017. Available online: https://www.gastrojournal.org/article/S0016-5085(17)30157-9/fulltext?referrer=https://pubmed.ncbi.nlm.nih.gov/ (accessed on 12 March 2025).
76. Sydor, S.; Best, J.; Messerschmidt, I.; Manka, P.; Vilchez-Vargas, R.; Brodesser, S.; Lucas, C.; Wegehaupt, A.; Wenning, C.; Aßmuth, S.; et al. Altered Microbiota Diversity and Bile Acid Signaling in Cirrhotic and Noncirrhotic NASH-HCC. *Clin. Transl. Gastroenterol.* **2020**, *11*, e00131. [CrossRef] [PubMed]
77. Turner, J.R. Intestinal mucosal barrier function in health and disease. *Nat. Rev. Immunol.* **2009**, *9*, 799–809. [CrossRef] [PubMed]
78. Miele, L.; Valenza, V.; La Torre, G.; Montalto, M.; Cammarota, G.; Ricci, R.; Mascianà, R.; Forgione, A.; Gabrieli, M.L.; Perotti, G.; et al. Increased intestinal permeability and tight junction alterations in nonalcoholic fatty liver disease. *Hepatology* **2009**, *49*, 1877–1887. [CrossRef] [PubMed]
79. Volynets, V.; Küper, M.A.; Strahl, S.; Maier, I.B.; Spruss, A.; Wagnerberger, S.; Königsrainer, A.; Bischoff, S.C.; Bergheim, I. Nutrition, intestinal permeability, and blood ethanol levels are altered in patients with nonalcoholic fatty liver disease (NAFLD). *Dig. Dis. Sci.* **2012**, *57*, 1932–1941. [CrossRef]
80. Luther, J.; Garber, J.J.; Khalili, H.; Dave, M.; Bale, S.S.; Jindal, R.; Motola, D.L.; Luther, S.; Bohr, S.; Jeoung, S.W.; et al. Hepatic Injury in Nonalcoholic Steatohepatitis Contributes to Altered Intestinal Permeability. *Cell. Mol. Gastroenterol. Hepatol.* **2015**, *1*, 222–232. [CrossRef]
81. Kolodziejczyk, A.A.; Zheng, D.; Shibolet, O.; Elinav, E. The role of the microbiome in NAFLD and NASH. *EMBO Mol. Med.* **2019**, *11*, e9302. [CrossRef]
82. Bechmann, L.P.; Kocabayoglu, P.; Sowa, J.-P.; Sydor, S.; Best, J.; Schlattjan, M.; Beilfuss, A.; Schmitt, J.; Hannivoort, R.A.; Kilicarslan, A.; et al. Free fatty acids repress small heterodimer partner (SHP) activation and adiponectin counteracts bile acid-induced liver injury in superobese patients with nonalcoholic steatohepatitis. *Hepatology* **2013**, *57*, 1394–1406. [CrossRef]
83. Caussy, C.; Tripathi, A.; Humphrey, G.; Bassirian, S.; Singh, S.; Faulkner, C.; Bettencourt, R.; Rizo, E.; Richards, L.; Xu, Z.Z.; et al. A gut microbiome signature for cirrhosis due to nonalcoholic fatty liver disease. *Nat. Commun.* **2019**, *10*, 1406. [CrossRef]
84. Liu, Y.-L.; Patman, G.L.; Leathart, J.B.S.; Piguet, A.-C.; Burt, A.D.; Dufour, J.-F.; Day, C.P.; Daly, A.K.; Reeves, H.L.; Anstee, Q.M. Carriage of the PNPLA3 rs738409 C >G polymorphism confers an increased risk of non-alcoholic fatty liver disease associated hepatocellular carcinoma. *J. Hepatol.* **2014**, *61*, 75–81. [CrossRef]
85. Singh, S.; Anshita, D.; Ravichandiran, V. MCP-1: Function, regulation, and involvement in disease. *Int. Immunopharmacol.* **2021**, *101 Pt B*, 107598. [CrossRef]
86. Stickel, F.; Buch, S.; Nischalke, H.D.; Weiss, K.H.; Gotthardt, D.; Fischer, J.; Rosendahl, J.; Marot, A.; Elamly, M.; Casper, M.; et al. Genetic variants in PNPLA3 and TM6SF2 predispose to the development of hepatocellular carcinoma in individuals with alcohol-related cirrhosis. *Am. J. Gastroenterol.* **2018**, *113*, 1475–1483. [CrossRef] [PubMed]

87. Yang, J.; Trépo, E.; Nahon, P.; Cao, Q.; Moreno, C.; Letouzé, E.; Imbeaud, S.; Gustot, T.; Deviere, J.; Debette, S.; et al. PNPLA3 and TM6SF2 variants as risk factors of hepatocellular carcinoma across various etiologies and severity of underlying liver diseases. *Int. J. Cancer* **2019**, *144*, 533–544. [CrossRef]
88. Gellert-Kristensen, H.; Nordestgaard, B.G.; Tybjaerg-Hansen, A.; Stender, S. High Risk of Fatty Liver Disease Amplifies the Alanine Transaminase-Lowering Effect of a HSD17B13 Variant. *Hepatology* **2020**, *71*, 56–66. [CrossRef] [PubMed]
89. Abul-Husn, N.S.; Cheng, X.; Li, A.H.; Xin, Y.; Schurmann, C.; Stevis, P.; Liu, Y.; Kozlitina, J.; Stender, S.; Wood, G.C.; et al. A Protein-Truncating HSD17B13 Variant and Protection from Chronic Liver Disease. *N. Engl. J. Med.* **2018**, *378*, 1096–1106. [CrossRef] [PubMed]
90. Pinyol, R.; Torrecilla, S.; Wang, H.; Montironi, C.; Piqué-Gili, M.; Torres-Martin, M.; Wei-Qiang, L.; Willoughby, C.E.; Ramadori, P.; Andreu-Oller, C.; et al. Molecular characterisation of hepatocellular carcinoma in patients with non-alcoholic steatohepatitis. *J. Hepatol.* **2021**, *75*, 865–878. [CrossRef]
91. Niemelä, O.; Parkkila, S.; Pasanen, M.; Iimuro, Y.; Bradford, B.; Thurman, R.G. Early alcoholic liver injury: Formation of protein adducts with acetaldehyde and lipid peroxidation products, and expression of CYP2E1 and CYP3A. *Alcohol. Clin. Exp. Res.* **1998**, *22*, 2118–2124. [CrossRef]
92. Friedman, S.L.; Neuschwander-Tetri, B.A.; Rinella, M.; Sanyal, A.J. Mechanisms of NAFLD development and therapeutic strategies. *Nat. Med.* **2018**, *24*, 908–922. [CrossRef]
93. Mihm, S. Danger-Associated Molecular Patterns (DAMPs): Molecular Triggers for Sterile Inflammation in the Liver. *Int. J. Mol. Sci.* **2018**, *19*, 3104. [CrossRef]
94. Zhang, W.-J.; Chen, S.-J.; Zhou, S.-C.; Wu, S.-Z.; Wang, H. Inflammasomes and Fibrosis. *Front. Immunol.* **2021**, *12*, 643149. [CrossRef]
95. Seitz, H.K.; Bataller, R.; Cortez-Pinto, H.; Gao, B.; Gual, A.; Lackner, C.; Mathurin, P.; Mueller, S.; Szabo, G.; Tsukamoto, H. Alcoholic liver disease. *Nat. Rev. Dis. Primers* **2018**, *4*, 16. [CrossRef]
96. Cederbaum, A.I. CYP2E1—Biochemical and toxicological aspects and role in alcohol-induced liver injury. *Mt. Sinai J. Med. N. Y.* **2006**, *73*, 657–672.
97. Mello, T.; Ceni, E.; Surrenti, C.; Galli, A. Alcohol induced hepatic fibrosis: Role of acetaldehyde. *Mol. Asp. Med.* **2008**, *29*, 17–21. [CrossRef]
98. Lu, Y.; Cederbaum, A.I. CYP2E1 and oxidative liver injury by alcohol. *Free Radic. Biol. Med.* **2008**, *44*, 723–738. [CrossRef]
99. Tan, H.K.; Yates, E.; Lilly, K.; Dhanda, A.D. Oxidative stress in alcohol-related liver disease. *World J. Hepatol.* **2020**, *12*, 332–349. [CrossRef]
100. Hagymási, K.; Blázovics, A.; Lengyel, G.; Kocsis, I.; Fehér, J. Oxidative damage in alcoholic liver disease. *Eur. J. Gastroenterol. Hepatol.* **2001**, *13*, 49–53. [CrossRef]
101. Hu, W.; Feng, Z.; Eveleigh, J.; Iyer, G.; Pan, J.; Amin, S.; Chung, F.-L.; Tang, M.-S. The major lipid peroxidation product, trans-4-hydroxy-2-nonenal, preferentially forms DNA adducts at codon 249 of human p53 gene, a unique mutational hotspot in hepatocellular carcinoma. *Carcinogenesis* **2002**, *23*, 1781–1789. [CrossRef]
102. Wang, F.; Yang, J.-L.; Yu, K.; Xu, M.; Xu, Y.; Chen, L.; Lu, Y.; Fang, H.; Wang, X.; Hu, Z.; et al. Activation of the NF-κB pathway as a mechanism of alcohol enhanced progression and metastasis of human hepatocellular carcinoma. *Mol. Cancer* **2015**, *14*, 10. [CrossRef]
103. Bajaj, J.S. Alcohol, liver disease and the gut microbiota. *Nat. Rev. Gastroenterol. Hepatol.* **2019**, *16*, 235–246. [CrossRef]
104. Dubinkina, V.B.; Tyakht, A.V.; Odintsova, V.Y.; Yarygin, K.S.; Kovarsky, B.A.; Pavlenko, A.V.; Ischenko, D.S.; Popenko, A.S.; Alexeev, D.G.; Taraskina, A.Y.; et al. Links of gut microbiota composition with alcohol dependence syndrome and alcoholic liver disease. *Microbiome* **2017**, *5*, 141. [CrossRef] [PubMed]
105. Leclercq, S.; Matamoros, S.; Cani, P.D.; Neyrinck, A.M.; Jamar, F.; Stärkel, P.; Windey, K.; Tremaroli, V.; Bäckhed, F.; Verbeke, K.; et al. Intestinal permeability, gut-bacterial dysbiosis, and behavioral markers of alcohol-dependence severity. *Proc. Natl. Acad. Sci. USA* **2014**, *111*, E4485–E4493. [CrossRef] [PubMed]
106. Sakamoto, T.; Hara, M.; Higaki, Y.; Ichiba, M.; Horita, M.; Mizuta, T.; Eguchi, Y.; Yasutake, T.; Ozaki, I.; Yamamoto, K.; et al. Influence of alcohol consumption and gene polymorphisms of ADH2 and ALDH2 on hepatocellular carcinoma in a Japanese population. *Int. J. Cancer* **2006**, *118*, 1501–1507. [CrossRef]
107. Park, E.J.; Lee, J.H.; Yu, G.-Y.; He, G.; Ali, S.R.; Holzer, R.G.; Osterreicher, C.H.; Takahashi, H.; Karin, M. Dietary and genetic obesity promote liver inflammation and tumorigenesis by enhancing IL-6 and TNF expression. *Cell* **2010**, *140*, 197–208. [CrossRef]
108. Lee, C.; Cheung, S.T. STAT3: An Emerging Therapeutic Target for Hepatocellular Carcinoma. *Cancers* **2019**, *11*, 1646. [CrossRef]
109. Tian, Z.; Chen, Y.; Gao, B. Natural killer cells in liver disease. *Hepatology* **2013**, *57*, 1654–1662. [CrossRef]
110. Dapito, D.H.; Mencin, A.; Gwak, G.-Y.; Pradere, J.-P.; Jang, M.-K.; Mederacke, I.; Caviglia, J.M.; Khiabanian, H.; Adeyemi, A.; Bataller, R.; et al. Promotion of hepatocellular carcinoma by the intestinal microbiota and TLR4. *Cancer Cell* **2012**, *21*, 504–516. [CrossRef]
111. Schwabe, R.F.; Greten, T.F. Gut microbiome in HCC—Mechanisms, diagnosis and therapy. *J. Hepatol.* **2020**, *72*, 230–238. [CrossRef]

112. Ren, Z.; Li, A.; Jiang, J.; Zhou, L.; Yu, Z.; Lu, H.; Xie, H.; Chen, X.; Shao, L.; Zhang, R.; et al. Gut microbiome analysis as a tool towards targeted non-invasive biomarkers for early hepatocellular carcinoma. *Gut* **2019**, *68*, 1014–1023. [CrossRef]
113. Buch, S.; Stickel, F.; Trépo, E.; Way, M.; Herrmann, A.; Nischalke, H.D.; Brosch, M.; Rosendahl, J.; Berg, T.; Ridinger, M.; et al. A genome-wide association study confirms PNPLA3 and identifies TM6SF2 and MBOAT7 as risk loci for alcohol-related cirrhosis. *Nat. Genet.* **2015**, *47*, 1443–1448. [CrossRef]
114. Salameh, H.; Raff, E.; Erwin, A.; Seth, D.; Nischalke, H.D.; Falleti, E.; Burza, M.A.; Leathert, J.; Romeo, S.; Molinaro, A.; et al. PNPLA3 Gene Polymorphism Is Associated with Predisposition to and Severity of Alcoholic Liver Disease. *Am. J. Gastroenterol.* **2015**, *110*, 846–856. [CrossRef] [PubMed]
115. Schwantes-An, T.-H.; Darlay, R.; Mathurin, P.; Masson, S.; Liangpunsakul, S.; Mueller, S.; Aithal, G.P.; Eyer, F.; Gleeson, D.; Thompson, A.; et al. Genome-wide Association Study and Meta-analysis on Alcohol-Associated Liver Cirrhosis Identifies Genetic Risk Factors. *Hepatology* **2021**, *73*, 1920–1931. [CrossRef]
116. Trépo, E.; Caruso, S.; Yang, J.; Imbeaud, S.; Couchy, G.; Bayard, Q.; Letouzé, E.; Ganne-Carrié, N.; Moreno, C.; Oussalah, A.; et al. Common genetic variation in alcohol-related hepatocellular carcinoma: A case-control genome-wide association study. *Lancet Oncol.* **2022**, *23*, 161–171. [CrossRef]
117. Seif Eldin, W.R.; Saad, E.A.; Monier, A.; Elshazli, R.M. Association of TERT (rs2736098 and rs2736100) genetic variants with elevated risk of hepatocellular carcinoma: A retrospective case–control study. *Sci. Rep.* **2023**, *13*, 18382. [CrossRef]
118. Homann, N.; Stickel, F.; König, I.R.; Jacobs, A.; Junghanns, K.; Benesova, M.; Schuppan, D.; Himsel, S.; Zuber-Jerger, I.; Hellerbrand, C.; et al. Alcohol dehydrogenase 1C*1 allele is a genetic marker for alcohol-associated cancer in heavy drinkers. *Int. J. Cancer* **2006**, *118*, 1998–2002. [CrossRef]
119. Singal, A.G.; Llovet, J.M.; Yarchoan, M.; Mehta, N.; Heimbach, J.K.; Dawson, L.A.; Jou, J.H.; Kulik, L.M.; Agopian, V.G.; Marrero, J.A.; et al. AASLD Practice Guidance on prevention, diagnosis, and treatment of hepatocellular carcinoma. *Hepatology* **2023**, *78*, 1922–1965. [CrossRef]
120. Galle, P.R.; Forner, A.; Llovet, J.M.; Mazzaferro, V.; Piscaglia, F.; Raoul, J.-L.; Schirmacher, P.; Vilgrain, V. EASL Clinical Practice Guidelines: Management of hepatocellular carcinoma. *J. Hepatol.* **2018**, *69*, 182–236. [CrossRef]
121. ACR Appropriateness Criteria® Chronic Liver Disease—ClinicalKey. Available online: https://www.clinicalkey.com/#/content/playContent/1-s2.0-S1546144020301113?returnurl=https://linkinghub.elsevier.com/retrieve/pii/S1546144020301113?showall=true&referrer=https://pubmed.ncbi.nlm.nih.gov/ (accessed on 3 December 2024).
122. Gupta, P.; Soundararajan, R.; Patel, A.; Kumar-M, P.; Sharma, V.; Kalra, N. Abbreviated MRI for hepatocellular carcinoma screening: A systematic review and meta-analysis. *J. Hepatol.* **2021**, *75*, 108–119. [CrossRef]
123. Singal, A.G.; El-Serag, H.B. Rational HCC screening approaches for patients with NAFLD. *J. Hepatol.* **2022**, *76*, 195–201. [CrossRef]
124. Kjaergaard, M.; Lindvig, K.P.; Thorhauge, K.H.; Andersen, P.; Hansen, J.K.; Kastrup, N.; Jensen, J.M.; Hansen, C.D.; Johansen, S.; Israelsen, M.; et al. Using the ELF test, FIB-4 and NAFLD fibrosis score to screen the population for liver disease. *J. Hepatol.* **2023**, *79*, 277–286. [CrossRef] [PubMed]
125. Shrestha, R.; Kc, S.; Thapa, P.; Pokharel, A.; Karki, N.; Jaishi, B. Estimation of Liver Fat by FibroScan in Patients with Nonalcoholic Fatty Liver Disease. *Cureus* **2021**, *13*, e16414. [CrossRef] [PubMed]
126. Bolondi, L.; Gaiani, S.; Celli, N.; Golfieri, R.; Grigioni, W.F.; Leoni, S.; Venturi, A.M.; Piscaglia, F. Characterization of small nodules in cirrhosis by assessment of vascularity: The problem of hypovascular hepatocellular carcinoma. *Hepatology* **2005**, *42*, 27–34. [CrossRef]
127. Matsui, O.; Kobayashi, S.; Sanada, J.; Kouda, W.; Ryu, Y.; Kozaka, K.; Kitao, A.; Nakamura, K.; Gabata, T. Hepatocelluar nodules in liver cirrhosis: Hemodynamic evaluation (angiography-assisted CT) with special reference to multi-step hepatocarcinogenesis. *Abdom. Imaging* **2011**, *36*, 264–272. [CrossRef] [PubMed]
128. Mitchell, D.G.; Bruix, J.; Sherman, M.; Sirlin, C.B. LI-RADS (Liver Imaging Reporting and Data System): Summary, discussion, and consensus of the LI-RADS Management Working Group and future directions. *Hepatology* **2015**, *61*, 1056–1065. [CrossRef]
129. Tang, A.; Bashir, M.R.; Corwin, M.T.; Cruite, I.; Dietrich, C.F.; Do, R.K.G.; Ehman, E.C.; Fowler, K.J.; Hussain, H.K.; Jha, R.C.; et al. Evidence Supporting LI-RADS Major Features for CT- and MR Imaging-based Diagnosis of Hepatocellular Carcinoma: A Systematic Review. *Radiology* **2018**, *286*, 29–48. [CrossRef]
130. International Consensus Group for Hepatocellular Neoplasia the International Consensus Group for Hepatocellular Neoplasia. Pathologic diagnosis of early hepatocellular carcinoma: A report of the international consensus group for hepatocellular neoplasia. *Hepatology* **2009**, *49*, 658–664. [CrossRef] [PubMed]
131. European Association for The Study of The Liver; European Organisation for Research and Treatment of Cancer. EASL-EORTC clinical practice guidelines: Management of hepatocellular carcinoma. *J. Hepatol.* **2012**, *56*, 908–943. [CrossRef]
132. Heimbach, J.K.; Kulik, L.M.; Finn, R.S.; Sirlin, C.B.; Abecassis, M.M.; Roberts, L.R.; Zhu, A.X.; Murad, M.H.; Marrero, J.A. AASLD guidelines for the treatment of hepatocellular carcinoma. *Hepatology* **2018**, *67*, 358–380. [CrossRef]

133. Reig, M.; Forner, A.; Rimola, J.; Ferrer-Fàbrega, J.; Burrel, M.; Garcia-Criado, Á.; Kelley, R.K.; Galle, P.R.; Mazzaferro, V.; Salem, R.; et al. BCLC strategy for prognosis prediction and treatment recommendation: The 2022 update. *J. Hepatol.* **2022**, *76*, 681–693. [CrossRef]
134. Guglielmi, A.; Ruzzenente, A.; Conci, S.; Valdegamberi, A.; Iacono, C. How much remnant is enough in liver resection? *Dig. Surg.* **2012**, *29*, 6–17. [CrossRef]
135. Doyle, M.B.M.; Vachharajani, N.; Maynard, E.; Shenoy, S.; Anderson, C.; Wellen, J.R.; Lowell, J.A.; Chapman, W.C. Liver transplantation for hepatocellular carcinoma: Long-term results suggest excellent outcomes. *J. Am. Coll. Surg.* **2012**, *215*, 19–28; discussion 28–30. [CrossRef]
136. Koh, J.H.; Tan, D.J.H.; Ong, Y.; Lim, W.H.; Ng, C.H.; Tay, P.W.L.; Yong, J.N.; Muthiah, M.D.; Tan, E.X.; Pang, N.Q.; et al. Liver resection versus liver transplantation for hepatocellular carcinoma within Milan criteria: A meta-analysis of 18,421 patients. *Hepatobiliary Surg. Nutr.* **2022**, *11*, 78–93. [CrossRef] [PubMed]
137. Huang, D.C.; Yu, R.L.; Alqahtani, S.; Tamim, H.; Saberi, B.; Bonder, A. Racial, ethnic, and socioeconomic disparities impact post-liver transplant survival in patients with hepatocellular carcinoma. *Ann. Hepatol.* **2023**, *28*, 101127. [CrossRef] [PubMed]
138. Xin Koh, Y.; Jin Tan, H.; Xin Liew, Y.; Syn, N.; Yao Teo, J.; Yee Lee, S.; Goh, B.K.P.; Goh, G.B.B.; Yip Chan, C. Liver Resection for Nonalcoholic Fatty Liver Disease-Associated Hepatocellular Carcinoma. *J. Am. Coll. Surg.* **2019**, *229*, 467–478e1. [CrossRef] [PubMed]
139. Wakai, T.; Shirai, Y.; Sakata, J.; Korita, P.V.; Ajioka, Y.; Hatakeyama, K. Surgical outcomes for hepatocellular carcinoma in nonalcoholic fatty liver disease. *J. Gastrointest. Surg.* **2011**, *15*, 1450–1458. [CrossRef]
140. Cauchy, F.; Zalinski, S.; Dokmak, S.; Fuks, D.; Farges, O.; Castera, L.; Paradis, V.; Belghiti, J. Surgical treatment of hepatocellular carcinoma associated with the metabolic syndrome. *Br. J. Surg.* **2013**, *100*, 113–121. [CrossRef]
141. Balzan, S.; Nagarajan, G.; Farges, O.; Galleano, C.Z.; Dokmak, S.; Paugam, C.; Belghiti, J. Safety of liver resections in obese and overweight patients. *World J. Surg.* **2010**, *34*, 2960–2968. [CrossRef]
142. Wu, H.-Y.; Chang, C.-C.; Yeh, C.-C.; Chen, M.-Y.; Cherng, Y.-G.; Chen, T.-L.; Liao, C.-C. Adverse outcomes after non-hepatic surgeries in patients with alcoholic liver diseases: A propensity-score matched study. *BMC Gastroenterol.* **2022**, *22*, 475. [CrossRef]
143. McClain, C.J.; Barve, S.S.; Barve, A.; Marsano, L. Alcoholic liver disease and malnutrition. *Alcohol. Clin. Exp. Res.* **2011**, *35*, 815–820. [CrossRef]
144. Ford, K.L.; Prado, C.M.; Weimann, A.; Schuetz, P.; Lobo, D.N. Unresolved issues in perioperative nutrition: A narrative review. *Clin. Nutr.* **2022**, *41*, 1578–1590. [CrossRef]
145. Llovet, J.M.; De Baere, T.; Kulik, L.; Haber, P.K.; Greten, T.F.; Meyer, T.; Lencioni, R. Locoregional therapies in the era of molecular and immune treatments for hepatocellular carcinoma. *Nat. Rev. Gastroenterol. Hepatol.* **2021**, *18*, 293–313. [CrossRef]
146. Mazzaferro, V.; Regalia, E.; Doci, R.; Andreola, S.; Pulvirenti, A.; Bozzetti, F.; Montalto, F.; Ammatuna, M.; Morabito, A.; Gennari, L. Liver transplantation for the treatment of small hepatocellular carcinomas in patients with cirrhosis. *N. Engl. J. Med.* **1996**, *334*, 693–699. [CrossRef]
147. Lencioni, R.; Crocetti, L. Radiofrequency ablation of liver cancer. *Tech. Vasc. Interv. Radiol.* **2007**, *10*, 38–46. [CrossRef]
148. Gao, Q.; Anwar, I.J.; Abraham, N.; Barbas, A.S. Liver Transplantation for Hepatocellular Carcinoma after Downstaging or Bridging Therapy with Immune Checkpoint Inhibitors. *Cancers* **2021**, *13*, 6307. [CrossRef] [PubMed]
149. Glassberg, M.B.; Ghosh, S.; Clymer, J.W.; Qadeer, R.A.; Ferko, N.C.; Sadeghirad, B.; Wright, G.W.; Amaral, J.F. Microwave ablation compared with radiofrequency ablation for treatment of hepatocellular carcinoma and liver metastases: A systematic review and meta-analysis. *OncoTargets Ther.* **2019**, *12*, 6407–6438. [CrossRef]
150. Imai, N.; Ishigami, M.; Ishizu, Y.; Kuzuya, T.; Honda, T.; Hayashi, K.; Hirooka, Y.; Goto, H. Transarterial chemoembolization for hepatocellular carcinoma: A review of techniques. *World J. Hepatol.* **2014**, *6*, 844–850. [CrossRef]
151. Kallini, J.R.; Gabr, A.; Salem, R.; Lewandowski, R.J. Transarterial Radioembolization with Yttrium-90 for the Treatment of Hepatocellular Carcinoma. *Adv. Ther.* **2016**, *33*, 699–714. [CrossRef]
152. Qin, S.; Chen, M.; Cheng, A.-L.; Kaseb, A.O.; Kudo, M.; Lee, H.C.; Yopp, A.C.; Zhou, J.; Wang, L.; Wen, X.; et al. Atezolizumab plus bevacizumab versus active surveillance in patients with resected or ablated high-risk hepatocellular carcinoma (IMbrave050): A randomised, open-label, multicentre, phase 3 trial. *Lancet* **2023**, *402*, 1835–1847. [CrossRef]
153. Ganne-Carrié, N.; Nahon, P. Differences between hepatocellular carcinoma caused by alcohol and other aetiologies. *J. Hepatol.* **2025**, *82*, 909–917. [CrossRef]
154. Why Is Nutrition So Important in Alcohol Associated Liver Disease? | AASLD. Available online: https://www.aasld.org/liver-fellow-network/core-series/why-series/why-nutrition-so-important-alcohol-associated-liver (accessed on 14 August 2022).
155. Kim, H.; Lee, D.S.; An, T.H.; Park, H.-J.; Kim, W.K.; Bae, K.-H.; Oh, K.-J. Metabolic Spectrum of Liver Failure in Type 2 Diabetes and Obesity: From NAFLD to NASH to HCC. *Int. J. Mol. Sci.* **2021**, *22*, 4495. [CrossRef]
156. Kraglund, F.; Birn-Rydder, R.; Sørensen, M.D.; Abazi, R.; Villadsen, G.E.; Jepsen, P. Alcohol use and hepatocellular carcinoma risk in patients with alcohol-related cirrhosis. *Scand. J. Gastroenterol.* **2023**, *58*, 1321–1327. [CrossRef]

157. Wattacheril, J.J.; Abdelmalek, M.F.; Lim, J.K.; Sanyal, A.J. AGA Clinical Practice Update on the Role of Noninvasive Biomarkers in the Evaluation and Management of Nonalcoholic Fatty Liver Disease: Expert Review. *Gastroenterology* **2023**, *165*, 1080–1088. [CrossRef]
158. Jophlin, L.L.; Singal, A.K.; Bataller, R.; Wong, R.J.; Sauer, B.G.; Terrault, N.A.; Shah, V.H. ACG Clinical Guideline: Alcohol-Associated Liver Disease. *Am. J. Gastroenterol.* **2024**, *119*, 30–54. [CrossRef] [PubMed]

BioChem

MDPI

Review

The Occult Cascade That Leads to CTEPH

Charli Fox [1,2] and Lavannya M. Pandit [2,3,*]

[1] Department of Pediatrics, Baylor College of Medicine, Houston, TX 77030, USA
[2] Michael E. DeBakey Veterans Affairs Medical Center, Houston, TX 77030, USA
[3] Pulmonary/Critical Care/Sleep Medicine Section, Baylor College of Medicine, Houston, TX 77030, USA
* Correspondence: Lpandit@bcm.edu

Abstract

Chronic thromboembolic pulmonary hypertension (CTEPH) is a rare, progressive form of pre-capillary pulmonary hypertension characterized by persistent, organized thromboemboli in the pulmonary vasculature, leading to vascular remodeling, elevated pulmonary artery pressures, right heart failure, and significant morbidity and mortality if untreated. Despite advances, CTEPH remains underdiagnosed due to nonspecific symptoms and overlapping features with other forms of pulmonary hypertension. **Basic Methodology:** This review synthesizes data from large international registries, epidemiologic studies, translational research, and multicenter clinical trials. Key methodologies include analysis of registry data to assess incidence and risk factors, histopathological examination of lung specimens, and molecular studies investigating endothelial dysfunction and inflammatory pathways. Diagnostic modalities and treatment outcomes are evaluated through observational studies and randomized controlled trials. **Recent Advances and Affected Population:** Research has elucidated that CTEPH arises from incomplete resolution of pulmonary emboli, with subsequent fibrotic transformation mediated by dysregulated TGF-β/TGFBI signaling, endothelial dysfunction, and chronic inflammation. Affected populations are typically older adults, often with prior venous thromboembolism, splenectomy, or prothrombotic conditions, though up to 25% have no history of acute PE. The disease burden is substantial, with delayed diagnosis contributing to worse outcomes and higher societal costs. Microvascular arteriopathy and PAH-like lesions in non-occluded vessels further complicate the clinical picture. **Conclusions**: CTEPH is now recognized as a treatable disease, with multimodal therapies—surgical endarterectomy, balloon pulmonary angioplasty, and targeted pharmacotherapy—significantly improving survival and quality of life. Ongoing research into molecular mechanisms and biomarker-driven diagnostics promises earlier identification and more personalized management. Multidisciplinary care and continued translational investigation are essential to further reduce mortality and optimize outcomes for this complex patient population.

Keywords: pulmonary hypertension; pulmonary emboli; microthrombi

1. Introduction

Pulmonary hypertension (PH) is diagnosed in patients who have a mean pulmonary arterial pressure (mPAP) > 20 mmHg at rest, according to the proceedings of the sixth World Symposium on Pulmonary Hypertension, and affects approximately one percent of the population globally.

Pre-capillary PH is differentiated from post-capillary PH based on a PAWP (pulmonary artery wedge pressure) $\leq$ 15 mmHg and PVR (pulmonary vascular resistance)

> 3 Woods Units. Pulmonary hypertension (PH) is classified into five clinical groups based on etiology and pathophysiology, as updated by the 2022 ESC/ERS guidelines. Group 1 [1]), pulmonary arterial hypertension (PAH), includes idiopathic, heritable, drug-induced, and connective tissue disease-associated forms, with a prevalence of 1–15 cases per 100,000 [1,2]. Group 2 PH occurs due to left heart disease and is the most common, accounting for up to 68% of PH cases. Group 3 is associated with lung diseases or hypoxemia (e.g., COPD, interstitial lung disease), comprising about 47% of cases. Group 4, chronic thromboembolic pulmonary hypertension (CTEPH), results in part from recurrent and unresolved pulmonary emboli leading to chronic vascular remodeling and occurs in 0.5–3% of PE survivors, but is often underdiagnosed due to nonspecific symptoms and a lack of utilization of standard diagnostics. Group 5 includes multifactorial or unclear causes such as hematologic or systemic disorders. The ESC/ERS 2022 update lowered the mean pulmonary arterial pressure threshold to ≥20 mm Hg and revised pulmonary vascular resistance criteria, emphasizing earlier diagnosis and nuanced risk stratification. Underdiagnosis is particularly problematic in CTEPH, where unlike other forms of PH, curative surgery is possible in these patients, underscoring the need for heightened clinical suspicion and improved screening for patients with CTEPH. CTEPH is a rare form of pre-capillary PH characterized by progressive pulmonary arteriopathy similar to other forms of pre-capillary PH, but uniquely characterized by pulmonary arteriolar obstruction from organized thrombi and a progressive vascular remodeling process and distal small vessel arteriopathy. CTEPH is driven by the combined mechanisms of inflammation, abnormal fibrinolysis, and endothelial dysfunction [3]. Without treatment, CTEPH, similar to other forms of pulmonary hypertension (PH), can lead to right heart failure and death. Recent translational and epidemiologic advances have reshaped our understanding of its mechanisms, risk factors, and therapeutic outcomes.

2. Epidemiology of CTEPH

CTEPH affects between three and thirty individuals per million globally, with 0.4–14.7% of acute pulmonary embolism (PE) survivors developing the disease. This broad diagnostic incidence can be attributed to several factors in the challenges in diagnostic accuracy. More than half of patients with CTEPH have no history of clinically apparent acute PE, and there is overlap with other conditions that resemble CTEPH symptoms including dyspnea and fatigue, which lead to delayed diagnosis of CTEPH. An additional challenge in the accurate description of CTEPH prevalence stems from the variability of the cohorts from which CTEPH incidence has been reported, since the included patient populations vary in the diagnostic methods used to define CTEPH as well as the heterogenous comorbidities of the index cohort (acute PE, race, cardiopulmonary disease, cancer, etc.) Registry data suggest that the prevalence of CTEPH is approximately equal in men and women, with an average age of onset in the sixth decade of life. Although CTEPH is thought to be only a direct sequela of chronic thromboembolic disease, a subset of patients diagnosed with CTEPH do not have a history or diagnosis of PE. In past registries, 25% of patients diagnosed with CTEPH did not have a formal diagnosis of acute PE [4–6]. In countries (France, Germany, Italy, Spain, the UK, the USA, and Japan) is expected to reach 37,009 cases, though only 28% may be diagnosed [7]). Underdiagnosis remains a critical issue, with a median diagnostic delay of 14 months from symptom onset. Delays in diagnosis can lead to the progression of the vessel remodeling and distal arteriopathy with worsening right heart dysfunction as a result of elevated pulmonary artery pressure (PAP). Historically, three-year mortality rates as high as 90% have been reported in patients with PAP > 50 mmHg [8,9].

One proposed reason for the underdiagnosis of CTEPH is the geographic differences in CTEPH populations, more specifically due to differences in healthcare systems [10]. For instance, in the United Kingdom, PH care is centralized, and patients must seek care at one of the eight certified PH centers. Once patients are established at one of the eight accredited centers, they are automatically entered into a national PH patient database. In countries where registries are not mandatory, national registries may not be fully representative and are likely underestimating patients. An international physician survey of CTEPH management practice reported that a relatively higher proportion of European patients than in the USA (43% versus 12%, respectively) were diagnosed with CTEPH in expert pulmonary hypertension centers [7]). Further, routine clinical practice between countries may influence the reported prevalence. Right heart catheterization (RHC) remains a gold-standard method for diagnosing and classifying PH. Ventilation/perfusion (V/Q) scans are the preferred initial radiologic test for CTEPH. However, in some countries, RHC or V/Q scans are unavailable or not systematically utilized, leading to misdiagnosis and underdiagnosis [11,12]. Other causes behind the underdiagnosis of CTEPH include a lack of clinical suspicion/underrecognition, a lack of an identifiable causative event and/or a lack of attribution of symptoms to CTEPH [7]. In 2015, the full annual incidence of CTEPH in France, Germany, Italy, Spain, the UK, the US, and Japan was estimated to be 32,636 cases; it was calculated that only 5334 (16%) of these cases were diagnosed. Another challenge is that CTPEH can be misdiagnosed as Group 1 pulmonary arterial hypertension (PAH) due to overlapping symptoms [4,13].

Nevertheless, as underdiagnosed as CTEPH is estimated to be, it is still associated with large societal costs. Societal costs were higher for CTEPH patients when compared to age, sex, and geographic area-matched control groups, which is thought to be most likely due to an overall productivity loss in the CTEPH patient population after diagnosis [14]. Additionally, the main cost drivers were prescribed drugs and hospitalizations for patients. Efforts are being made to reduce diagnostic delays when CTEPH is considered as part of a patient's differential.

3. Affected Populations and Risk Factors

In general, there is a higher prevalence in patients aged 60–70 years, though idiopathic cases occur in younger populations [4,15]. Further, as previously mentioned, CTEPH disproportionately impacts individuals with acute pulmonary embolism (PE), with an estimated prevalence of 0.8–3.8% 2 years after the occurrence of acute PE. Even the symptoms present during the initial presentation of PE have been linked to the development of CTEPH, specifically right ventricular strain manifesting as elevated pulmonary artery systolic pressure. Recurrent events increase the risk of CTEPH 3-fold [6,8]. Other risk factors for the development of CTEPH include splenectomy, infected pacemakers, and ventriculoarterial shunts. Although PE is the primary cause of CTEPH, up to 62% of patients with proximal disease and 49% of those with distal disease have no history of acute PE, which can delay diagnosis in some cases [8,16]. See Figure 1 as a descriptive image of the above-discussed varied factors that lead to the development of CTEPH.

Pedigree analyses have identified families with multiple CTEPH cases, supporting the idea of a dominant inheritance pattern with incomplete penetrance [17]. These findings suggest that, while CTEPH is not a purely Mendelian disorder, genetic predisposition plays a significant role, and ongoing research is focused on identifying specific gene variants and inheritance patterns that contribute to CTEPH risk [18]. Typical hereditary risk factors for acute PE, such as Protein C deficiency, Protein S deficiency, and antithrombin deficiency, were found to be in higher prevalence in CTEPH patients (according to a CTEPH registry in the United States that examined 552 subjects) [19]. Antiphospholipid antibody syndrome, increased von Willebrand factor, and elevated factor VIII levels are known risk factors for developing CTEPH [1,19]. Finally, established comorbidities include chronic inflammatory states, splenectomy, hypothyroidism, thrombolytic therapy, and delayed diagnosis of PE [4,20].

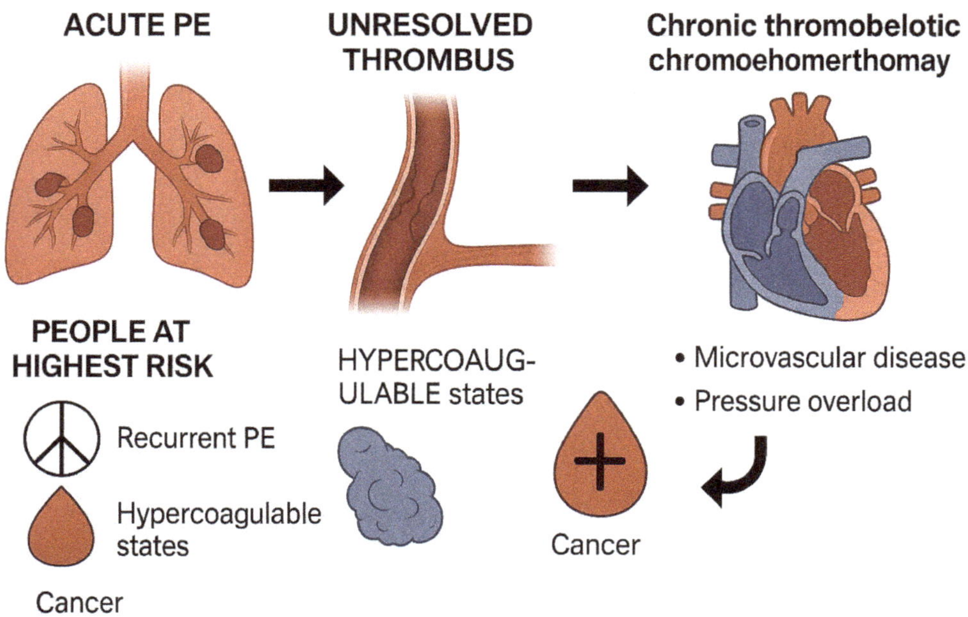

Figure 1. Factors Contributing to the Development of CTEPH.

The most recent epidemiologic data do not support the idea that general lifestyle factors such as diet and exercise increase the risk for chronic thromboembolic pulmonary hypertension (CTEPH). Instead, the evidence shows that certain lifestyle-related conditions—such as prolonged immobility during travel, hospitalization, or illness, obesity and tobacco usage—are established risk factors because they increase the risk of venous thromboembolism (VTE) [12], which can subsequently lead to CTEPH if a pulmonary embolism (PE) is not fully resolved.

4. Pathophysiologic and Molecular Mechanisms

Although the precise pathogenesis of CTEPH is not fully understood, it is thought to be a complication of acute (PE), when there is an incomplete resolution of pulmonary emboli, which then become organized into obstructive fibrotic material with an associated small vessel vasculopathy [1,2,20]. In the diseased vessel, there is an imbalance between clot formation and clot degradation, which initiates a cascade of macrovascular changes and ultimately alters the organization of the thrombus in the patient's pulmonary vasculature. The non-resolved thrombi undergo endothelial-mediated fibrosis via TGF-β/TGFBI signaling, impairing fibrinolysis [13,21]. This remaining fibrotic material leads to systemic dysfunction of the vascular endothelium, increasing PVR with pulmonary vascular remodeling, and a permanent increase in pulmonary artery pressure, eventually leading to right heart failure [1,2,20]. Global dysfunction is also observed in the endothelial cells (ECs) in CTEPH which exhibit dysregulated angiogenesis, promoting aberrant vascular repair [13].

From a microvascular standpoint, a secondary arteriopathy develops in non-occluded vessels, resembling PAH-like changes. Specifically, there is medial hypertrophy of the muscularis layer of the pulmonary arterioles, concentric laminal internal fibroelastosis, and plexiform lesions. This observation of arteriopathy in the non-occluded vessels highlights two crucial points. First, it emphasizes the likelihood that there are some fundamental similarities between PAH and CTEPH on a mechanistic level. Second, the presence of vascular lesions in regions of the lung not affected by thrombotic disease suggests that systemic vascular mediators could provoke the development of these [4,21]. One proposed mechanism is that inflammation itself could affect the normal process of fibrinolysis as levels of C-reactive protein (CRP) and cytokines, such as interleukin-6 (IL-6), are elevated in CTEPH patients. The presence of inflammatory mediators such as CRP and IL-6 drives endothelial dysfunction and smooth muscle proliferation, affecting the process of angiogenesis [4]. A recent single-cell RNA sequencing analysis performed on surgically extracted clots from patients who had undergone pulmonary thrombendarterectomy (PTE) revealed and identified multiple cell types, including macrophages, T cells, and smooth muscle cells (SMCs), that constitute CTEPH thrombus. The identified macrophages were split into two categories, with the larger group characterized by an upregulation of inflammatory signaling through CD4+ and CD8+ T cells. This analysis presents CTEPH as a chronic inflammatory process promoted by the macrophages and T cells that drive smooth muscle cell-mediated vascular remodeling [22]. Historically, genetic or familial connections in CTEPH have not been established. The latest limited research indicates that familial CTEPH has a partially heritable, polygenic basis estimated at a frequency—defined as more than one case in a family—of about 2.2% [17], though this requires confirmation in larger cohorts. Classic hereditary thrombophilias such as Factor V Leiden and Prothrombin G20210A are less common in CTEPH patients than in those with acute PE, indicating that CTEPH may have some distinct genetic underpinnings that differentiate it from acute pulmonary embolism (PE) and deep vein thrombosis (DVT). Recent large-scale genome-wide association studies (GWAS) have identified multiple genetic loci associated with CTEPH, including ABO, FGG, F11, MYH7B, HLA-DRA, F2, TSPAN15, SLC44A2, and F5 [18]. Despite these genetic associations, CTEPH's familial causes currently have a limited role in diagnostics and therefore have not translated to screening practices or clinical predictive algorithms.

While some BMPR2 mutations are found in CTEPH, these mutations are typically associated with PAH [4]. One study has demonstrated that patients with CTEPH had a higher frequency of mutations of known PAH-associated genes [23]. Additionally, only one familial case of CTEPH has been reported, suggesting that distinct pathways lead to the development of CTEPH [24].

5. Current Treatment Paradigm with Advances and Future Directions

Upon diagnosis, all patients with CTEPH should undergo a multidisciplinary review with a team comprised of a surgeon experienced in the CTEPH surgical procedures, a pulmonary hypertension specialist, an interventional cardiologist, and a radiologist to decide the optimal treatment strategy for the patient.

Surgical Evaluation pulmonary thrombendarterectomy (PTE) is the first-line treatment for CTEPH. Per current European Society of Cardiology (ESC) guidelines, pulmonary thrombendarterectomy (PTE) is the overall treatment of choice when a patient is an operable candidate. A crucial consideration for surgery is whether the disease burden can be effectively cleared by surgery. The procedure requires a cardiopulmonary bypass to redirect blood flow from the heart and systemic cooling at 20 °C. The pulmonary arteries are then opened and removed to allow for the complete visualization of the thrombotic material. Next, there is systemic rewarming, and the patient is taken off bypass and is sent to the intensive care unit for post-operative recovery. Currently, in CTEPH-certified centers, perioperative mortality rates are <2.5%. Post-operative pulmonary hypertension is seen in 25% of patients, but overall survival rates and long-term outcomes are favorable. PEA is curative for proximal disease, with 70.6% 10-year survival vs. 41.7% in inoperable cases [15,25].

Balloon Pulmonary Angioplasty (BPA) is an equivalent alternative to surgical treatment in non-operable CTEPH. ESC Guidelines state that balloon pulmonary angioplasty (BPA) is the selected treatment choice when patients are not operable candidates, or when patients have persistent or recurrent PH after PEA [1]. This procedure uses a balloon to open and dilate blood vessels with chronic blockages or narrowing due to embolic burden from CTEPH. Balloon dilation aims to restore adequate pulmonary blood flow and improve hemodynamics in distal disease, achieving 85% 3-year survival [12,26]. While BPA is a surgically less invasive procedure compared to a PEA, it has been associated with more severe post-procedure complications, including hypoxia due to lung injury (18 of 52 patients) and severe hypotension (two of 53 patients) [27]. Overall, BPA offers equivalent survival compared to surgical PTE, and superior functional gains based on six-minute walk testing compared to medical therapy alone. Of note, multiple sessions of BPA are often required to achieve maximal efficacy.

Medical Management of CTEPH: When patients are not eligible for surgical intervention, which occurs in approximately 40% of patients with CTEPH due to unattainable vascular obstruction, disproportionate morphological lesions, and significant comorbidities, medical therapy is the only alternative for the treatment of CTEPH patients [1,28]. In healthy patients, nitric oxide (NO) binds to soluble guanyl-cyclase (sGC), which then results in the production of cyclic guanosine monophosphate (cGMP), which promotes adequate pulmonary blood flow. In patients with PH, endothelial dysfunction impairs NO production and leads to pulmonary vasoconstriction. Riociguat is an oral medication and pulmonary vasodilator that increases the production of sGC and, ultimately, cGMP. cGMP then leads to vasodilation and improves pulmonary circulation [29]. While other pulmonary vasodilators are used in CTEPH management, riociguat is unique due to its FDA approval and demonstrated efficacy in improving exercise capacity and hemodynamics in CTEPH patients [30]. The other concurrent non-surgical management is lifelong anticoagulation, including non-vitamin K antagonist oral anticoagulation (NOACs) or vitamin K antagonists (VKAs) [11]. Additional therapies include supplemental oxygen, diuretics, and pulmonary rehabilitation as part of a strategy to improve functional class and reduce time to clinical worsening [31].

Recent meta-analyses and registry data (2021–2024) reveal significant differences in survival, hemodynamic outcomes, and safety profiles for CTEPH treatments. Among the options of PTE, BPA, and medical management discussed previously, surgical treatment with PTE offers the highest long-term survival (89% at 3 years) but carries surgical risks, especially in low-volume centers (24.4% composite complications vs. 12.1% in high-volume centers) [32]. BPA offers a near-equivalent survival to PTE with lower perioperative mortality [12,33]. Medical therapy alone offered no survival benefit and should be used primarily for inoperable patients or as an adjunctive therapy [12]. Recent studies have shown an overall improvement in survival rates due to multimodal therapies, resulting in a 93% 5-year survival in the current era versus a 68% 5-year survival pre-2000 [28]. Further, combination therapies of sequential PTE/BPA with riociguat reduce residual PH severity [11,26,28]. In general, earlier diagnosis and implementation of these multimodal therapies have reduced the number of late-stage presentations of CTEPH patients. For example, 63% of patients were diagnosed with CTEPH at the New York Heart Association Functioning Class (NYHA FC) III/IV at diagnosis in 2025, versus 72% of patients in 2015 [7,26]. In summary, PTE remains curative for operable CTEPH, while BPA offers comparable long-term survival in selective patients with fewer procedural risks for distal disease. Medical therapy provides palliative benefits with some hemodynamic improvements, but lacks disease-modifying efficacy. Treatment selection must be individualized through multidisciplinary assessment, prioritizing center expertise and anatomic suitability.

6. Recent Research and Future Pathways for Treatment

Biomarker research is underway, evaluating the transforming growth factor–beta induced (TGFBI) protein in distinguishing CTEPH from other types of PH that present with similar clinical features. One study identified endothelial cell phenotype alterations and the induction of TGF-b signaling as a mechanistic contribution to thrombus non-resolution and fibrosis, and identified TGFBI as a specifically up-regulated endothelial marker and a possible causal factor in the pathogenesis of CTEPH [13]. New strategies are currently underway to target known pathways implicated in CTEPH. Single-cell RNA analysis has new approaches for pharmacologically targeting CTEPH as an inflammatory process (similar to atherosclerosis) modulated through macrophage and T cell-driven vascular remodeling [22]. Another pathway suggested by preclinical models is through TGFBI inhibition, which enhances thrombus resolution and helps to decrease CTEPH [13]. This study is promising in that creating biomarker-guided screening, specifically for plasma TGFBI, fibrinogen variants, and microRNA profiles, may enable early diagnosis [4,13].

7. Conclusions

CTEPH has fully evolved from a previously fatal prognosis to a treatable condition, with survival now exceeding 90% at 5 years. Multimodal therapies—surgical endarterectomy, balloon pulmonary angioplasty, and targeted pharmacotherapy—have significantly improved survival and quality of life. Ongoing research into molecular mechanisms and biomarker-driven diagnostics promises earlier identification and more personalized management. Translational research highlights TGF-β/TGFBI as pivotal therapeutic targets, while epidemiologic data underscore the urgency of early detection initiatives. Multidisciplinary care and continued translational investigation are essential to further reduce mortality and optimize outcomes for this complex patient population. Future breakthroughs will hinge on unraveling the molecular drivers of thrombus persistence, and expanding access to multidisciplinary care networks, particularly in underserved populations.

Author Contributions: Conceptualization, L.M.P.; methodology, L.M.P.; software, C.F. and L.M.P.; validation, L.M.P.; resources, C.F. and L.M.P.; writing—original draft preparation, C.F.; writing, review and editing, C.F. and L.M.P.; supervision, L.M.P.; funding acquisition, C.F. All authors have read and agreed to the published version of the manuscript.

Funding: This publication was made possible by an NHLBI-funded predoctoral fellowship to CCF (5T32HL170991-02).

Data Availability Statement: No new data were created or analyzed in this study.

Acknowledgments: Its contents are solely the authors' responsibility and do not necessarily represent the official views of the NHLBI or the NIH, or the Department of Veterans Affairs.

Conflicts of Interest: The authors declare no conflict of interest.

References

1. Humbert, M.; Kovacs, G.; Hoeper, M.M.; Badagliacca, R.; Berger, R.M.F.; Brida, M.; Carlsen, J.; Coats, A.J.S.; Escribano-Subias, P.; Ferrari, P.; et al. 2022 ESC/ERS Guidelines for the diagnosis and treatment of pulmonary hypertension. *Eur. Heart J.* **2022**, *43*, 3618–3731. [CrossRef]
2. Teerapuncharoen, K.; Bag, R. Chronic Thromboembolic Pulmonary Hypertension. *Lung* **2022**, *200*, 283–299. [CrossRef]
3. Simonneau, G.; Dorfmüller, P.; Guignabert, C.; Mercier, O.; Humbert, M. Chronic thromboembolic pulmonary hypertension: The magic of pathophysiology. *Ann. Cardiothorac. Surg.* **2022**, *11*, 106–119. [CrossRef] [PubMed]

4. Medrek, S.; Safdar, Z. Epidemiology and Pathophysiology of Chronic Thromboembolic Pulmonary Hypertension: Risk Factors and Mechanisms. *Methodist DeBakey Cardiovasc. J.* **2016**, *12*, 195–198. [CrossRef] [PubMed]
5. Pepke-Zaba, J.; Delcroix, M.; Lang, I.; Mayer, E.; Jansa, P.; Ambroz, D.; Treacy, C.; D'Armini, A.M.; Morsolini, M.; Snijder, R.; et al. Chronic thromboembolic pulmonary hypertension (CTEPH): Results from an international prospective registry. *Circulation* **2011**, *124*, 1973–1981. [CrossRef] [PubMed]
6. Zhang, M.; Wang, N.; Zhai, Z.; Zhang, M.; Zhou, R.; Liu, Y.; Yang, Y. Incidence and risk factors of chronic thromboembolic pulmonary hypertension after acute pulmonary embolism: A systematic review and meta-analysis of cohort studies. *J. Thorac. Dis.* **2018**, *10*, 4751–4763. [CrossRef]
7. Gall, H.; Hoeper, M.M.; Richter, M.J.; Cacheris, W.; Hinzmann, B.; Mayer, E. An epidemiological analysis of the burden of chronic thromboembolic pulmonary hypertension in the USA, Europe and Japan. *Eur. Respir. Rev. Off. J. Eur. Respir. Soc.* **2017**, *26*, 160121. [CrossRef]
8. D'Armini, A.M. Diagnostic advances and opportunities in chronic thromboembolic pulmonary hypertension. *Eur. Respir. Rev. Off. J. Eur. Respir. Soc.* **2015**, *24*, 253–262. [CrossRef]
9. Verbelen, T.; Godinas, L.; Maleux, G.; Coolen, J.; Claessen, G.; Belge, C.; Meyns, B.; Delcroix, M. Chronic thromboembolic pulmonary hypertension: Diagnosis, operability assessment and patient selection for pulmonary endarterectomy. *Ann. Cardiothorac. Surg.* **2022**, *11*, 82–97. [CrossRef]
10. Leber, L.; Beaudet, A.; Muller, A. Epidemiology of pulmonary arterial hypertension and chronic thromboembolic pulmonary hypertension: Identification of the most accurate estimates from a systematic literature review. *Pulm. Circ.* **2021**, *11*, 2045894020977300. [CrossRef]
11. Kenny, M.; Clarke, M.M.; Pogue, K.T. Overview of Riociguat and Its Role in the Treatment of Pulmonary Hypertension. *J. Pharm. Pract.* **2022**, *35*, 437–444. [CrossRef]
12. Simeone, B.; Maggio, E.; Schirone, L.; Rocco, E.; Sarto, G.; Spadafora, L.; Bernardi, M.; Ambrosio, L.D.; Forte, M.; Vecchio, D.; et al. Chronic thromboembolic pulmonary hypertension: The diagnostic assessment. *Front. Cardiovasc. Med.* **2024**, *11*, 1439402. [CrossRef] [PubMed]
13. Sharma, S.; Antoniak, S. New Insights in the Pathology of Chronic Thromboembolic Pulmonary Hypertension. *JACC. Basic Transl. Sci.* **2024**, *9*, 117–119. [CrossRef] [PubMed]
14. Kjellström, B.; Runheim, H.; Beaudet, A.; Husberg, M.; Ivarsson, B.; Pillai, N.; Levin, L.-A.; Bernfort, L. Societal costs associated to chronic thromboembolic pulmonary hypertension: A study utilizing linked national registries. *Pulm. Circ.* **2023**, *13*, e12254. [CrossRef] [PubMed]
15. Küçükoğlu, M.S.; Sinan, Ü.Y.; Yıldızeli, B. Ten-year outcome of chronic thromboembolic pulmonary hypertension patients in a tertiary center. *Anatol. J. Cardiol.* **2020**, *23*, 105–109. [CrossRef]
16. Condliffe, R.; Kiely, D.G.; Gibbs, J.S.R.; Corris, P.A.; Peacock, A.J.; Jenkins, D.P.; Hodgkins, D.; Goldsmith, K.; Hughes, R.J.; Sheares, K.; et al. Improved outcomes in medically and surgically treated chronic thromboembolic pulmonary hypertension. *Am. J. Respir. Crit. Care Med.* **2008**, *177*, 1122–1127. [CrossRef]
17. Dodson, M.W.; Cirulis, M.M.; Elliott, C.G. Analysis of family histories suggests shared genetic risk for chronic thromboembolic pulmonary hypertension and venous thromboembolism. *Pulm. Circ.* **2022**, *12*, e12170. [CrossRef]
18. Liley, J.; Newnham, M.; Bleda, M.; Bunclark, K.; Auger, W.; Barbera, J.A.; Bogaard, H.; Delcroix, M.; Fernandes, T.M.; Howard, L.; et al. Shared and Distinct Genomics of Chronic Thromboembolic Pulmonary Hypertension and Pulmonary Embolism. *Am. J. Respir. Crit. Care Med.* **2024**, *209*, 1477–1485. [CrossRef]
19. Cannon, J.E.; Jenkins, D.P.; Hoole, S.P. Chronic thromboembolic pulmonary hypertension: A review of risk factors, management and current challenges. *Expert Rev. Cardiovasc. Ther.* **2022**, *20*, 35–43. [CrossRef]
20. Yan, L.; Li, X.; Liu, Z.; Zhao, Z.; Luo, Q.; Zhao, Q.; Jin, Q.; Yu, X.; Zhang, Y. Research progress on the pathogenesis of CTEPH. *Heart Fail. Rev.* **2019**, *24*, 1031–1040. [CrossRef]
21. Olsson, K.M.; Meyer, B.; Hinrichs, J.; Vogel-Claussen, J.; Hoeper, M.M.; Cebotari, S. Chronic thromboembolic pulmonary hypertension. *Dtsch. Arztebl. Int.* **2014**, *111*, 856–862. [CrossRef]
22. Viswanathan, G.; Kirshner, H.F.; Nazo, N.; Ali, S.; Ganapathi, A.; Cumming, I.; Zhuang, Y.; Choi, I.; Warman, A.; Jassal, C.; et al. Single-Cell Analysis Reveals Distinct Immune and Smooth Muscle Cell Populations that Contribute to Chronic Thromboembolic Pulmonary Hypertension. *Am. J. Respir. Crit. Care Med.* **2023**, *207*, 1358–1375. [CrossRef] [PubMed]
23. Xi, Q.; Liu, Z.; Zhao, Z.; Luo, Q.; Huang, Z. High Frequency of Pulmonary Hypertension-Causing Gene Mutation in Chinese Patients with Chronic Thromboembolic Pulmonary Hypertension. *PLoS ONE* **2016**, *11*, e0147396. [CrossRef] [PubMed]
24. Desmarais, J.; Elliott, C.G. Familial Chronic Thromboembolic Pulmonary Hypertension. *Chest* **2016**, *149*, e99–e101. [CrossRef] [PubMed]
25. Brookes, J.D.L.; Li, C.; Chung, S.T.W.; Brookes, E.M.; Williams, M.L.; McNamara, N.; Martin-Suarez, S.; Loforte, A. Pulmonary thromboendarterectomy for chronic thromboembolic pulmonary hypertension: A systematic review. *Ann. Cardiothorac. Surg.* **2022**, *11*, 68–81. [CrossRef]

26. Delcroix, M.; Lang, I.; Pepke-Zaba, J.; Jansa, P.; D'Armini, A.M.; Snijder, R.; Bresser, P.; Torbicki, A.; Mellemkjaer, S.; Lewczuk, J.; et al. Long-Term Outcome of Patients with Chronic Thromboembolic Pulmonary Hypertension: Results From an International Prospective Registry. *Circulation* **2016**, *133*, 859–871. [CrossRef]
27. Jaïs, X.B.; Brenot, P.; Bouvaist, H.; Jevnikar, M.; Canuet, M.; Chabanne, C.; Chaouat, A.; Cottin, V.; De Groote, P.; Favroltet, N.; et al. Balloon pulmonary angioplasty versus riociguat for the treatment of inoperable chronic thromboem-bolic pulmonary hypertension (RACE): A multicentre, phase 3, open-label, randomised controlled trial and ancillary follow-up study. *Lancet Respir. Med.* **2022**, *10*, 961–971. [CrossRef]
28. Kiko, T.; Asano, R.; Yoshikawa, Y.; Endo, H.; Fujisaki, S.; Takano, R.; Akao, M.; Nishi, N.; Hayashi, H.; Ogata, S.; et al. Survival trends in patients with chronic thromboembolic pulmonary hypertension: An observational study. *Eur. Respir. J.* **2025**, 2402268. [CrossRef]
29. Barnikel, M.; Kneidinger, N.; Arnold, P.; Waelde, A.; Behr, J.; Milger, K. Riociguat in Patients with CTEPH and Advanced Age and/or Comorbidities. *J. Clin. Med.* **2022**, *11*, 1084. [CrossRef]
30. Ghofrani, H.-A.; D'Armini, A.M.; Grimminger, F.; Hoeper, M.M.; Jansa, P.; Kim, N.H.; Mayer, E.; Simonneau, G.; Wilkins, M.R.; Fritsch, A.; et al. Riociguat for the treatment of chronic thromboembolic pulmonary hypertension. *N. Engl. J. Med.* **2013**, *369*, 319–329. [CrossRef]
31. Pepke-Zaba, J.; Ghofrani, H.-A.; Hoeper, M.M. Medical management of chronic thromboembolic pulmonary hypertension. *Eur. Respir. Rev. Off. J. Eur. Respir. Soc.* **2017**, *26*, 160107. [CrossRef]
32. Zhang, L.; Bai, Y.; Yan, P.; He, T.; Liu, B.; Wu, S.; Qian, Z.; Li, C.; Cao, Y.; Zhang, M. Balloon pulmonary angioplasty vs. pulmonary endarterectomy in patients with chronic thromboembolic pulmonary hypertension: A systematic review and meta-analysis. *Heart Fail. Rev.* **2021**, *26*, 897–917. [CrossRef]
33. Ismail, N.R.; Makhlouf, H.A.; Hassan, A.; Elshahat, A.; Abdellatif, M.A.; Rezk, A.E.; Mady, A.; Mohamed, R.G.; Elfeky, H.M.; Abdelaziz, A. An in-depth evaluation of the efficacy and safety of various treatment modalities for chronic thromboembolic pulmonary hypertension: A systematic review and network meta-analysis. *Am. Heart J. Plus Cardiol. Res. Pract.* **2024**, *46*, 100466. [CrossRef]

BioChem MDPI

Review

Enhancing Mitochondrial Maturation in iPSC-Derived Cardiomyocytes: Strategies for Metabolic Optimization

Dhienda C. Shahannaz [1,2], Tadahisa Sugiura [2,*] and Brandon E. Ferrell [2]

1 Faculty of Medicine, Universitas Indonesia, Jakarta 10430, Indonesia; dhiendaladdynasrul@gmail.com
2 Department of Cardiothoracic and Vascular Surgery, Montefiore Medical Center/Albert Einstein College of Medicine, Greene Medical Arts Pavilion 5th Floor, 3400 Bainbridge Avenue, Bronx, NY 10467, USA
* Correspondence: tsugiura@montefiore.org; Tel.: +1-718-405-8249

Abstract

Background: Induced pluripotent stem cell-derived cardiomyocytes (iPSC-CMs) hold transformative potential for cardiovascular regenerative medicine, yet their clinical application is hindered by suboptimal mitochondrial maturation and metabolic inefficiencies. This systematic review evaluates targeted strategies for optimizing mitochondrial function, integrating metabolic preconditioning, substrate selection, and pathway modulation to enhance energy production and cellular resilience. Additionally, we examine the role of extracellular matrix stiffness and mechanical stimulation in mitochondrial adaptation, given their influence on metabolism and maturation. **Methods:** A comprehensive analysis of recent advancements in iPSC-CM maturation was conducted, focusing on metabolic interventions that enhance mitochondrial structure and function. Studies employing metabolic preconditioning, lipid and amino acid supplementation, and modulation of key signaling pathways, including PGC-1α, AMPK, and mTOR, were reviewed. Computational modeling approaches predicting optimal metabolic shifts were assessed, alongside insights into reactive oxygen species (ROS) signaling, calcium handling, and the impact of electrical pacing on energy metabolism. **Results:** Evidence indicates that metabolic preconditioning with fatty acids and oxidative phosphorylation enhancers improves mitochondrial architecture, cristae density, and ATP production. Substrate manipulation fosters a shift toward adult-like metabolism, while pathway modulation refines mitochondrial biogenesis. Computational models enhance precision, predicting interventions that best align iPSC-CM metabolism with native cardiomyocytes. The synergy between metabolic and biomechanical cues offers new avenues for accelerating maturation, bridging the gap between in vitro models and functional cardiac tissues. **Conclusions:** Strategic metabolic optimization is essential for overcoming mitochondrial immaturity in iPSC-CMs. By integrating biochemical engineering, predictive modeling, and biomechanical conditioning, a robust framework emerges for advancing iPSC-CM applications in regenerative therapy and disease modeling. These findings pave the way for more physiologically relevant cell models, addressing key translational challenges in cardiovascular medicine.

Keywords: mitochondrial maturation; iPSC-derived cardiomyocytes; oxidative phosphorylation; mitochondrial biogenesis; fatty acid oxidation; energy metabolism; metabolic preconditioning; cardiac regeneration; regenerative medicine

1. Introduction

Cardiovascular disease (CVD) remains the leading cause of death worldwide, with ischemic heart disease alone accounting for 9 million deaths annually [1]. The burden is especially significant in Indonesia [2], Japan [3], and the United States [4], where aging populations and lifestyle-related risk factors contribute to an escalating incidence of heart failure (HF), myocardial infarction (MI), and congenital heart disorders. In Indonesia, CVD causes 36% of all deaths [5], with limited access to advanced cardiac care exacerbating poor outcomes. In Japan, where the aging population is projected to reach 40% by 2040, degenerative cardiac conditions are a growing concern [6]. Meanwhile, in the United States, heart failure affects 6.2 million adults, with associated healthcare costs exceeding USD 30 billion annually [7]. With the escalating burden of cardiovascular diseases worldwide, especially in aging populations, there is a critical need for regenerative therapies that can effectively address the irreversible loss of cardiomyocytes. iPSC-CMs offer a promising solution, but their metabolic immaturity presents significant challenges that must be overcome for clinical success

Despite advances in pharmacological and interventional therapies, cardiomyocyte (CM) loss in cardiac diseases remains irreversible, necessitating regenerative approaches [8]. Induced pluripotent stem cell-derived cardiomyocytes (iPSC-CMs) have emerged as a promising cell source for cardiac repair and disease modeling [9,10]. However, a fundamental challenge remains: iPSC-CMs exhibit metabolic immaturity, underdeveloped mitochondrial architecture, and suboptimal oxidative phosphorylation (OXPHOS) capacity, limiting their ability to mimic the contractile and electrophysiological properties of adult CMs [11–14]. Mitochondrial maturation is crucial for iPSC-CMs to mimic the contractile and electrophysiological properties of adult cardiomyocytes. Without proper mitochondrial maturation, these cells cannot efficiently integrate into host cardiac tissue, limiting their therapeutic potential in clinical applications such as myocardial infarction repair or tissue engineering.

In native human cardiac development, the metabolic landscape undergoes a tightly regulated shift from a glycolytic fetal profile to an oxidative, fatty acid-dependent phenotype in postnatal life [15]. Fetal CMs primarily rely on anaerobic glycolysis and lactate oxidation to support rapid proliferation, with relatively low mitochondrial density and cristae complexity [16,17]. Postnatally, increased oxygen availability and hormonal changes trigger a surge in mitochondrial biogenesis and oxidative capacity, enabling fatty acid oxidation (FAO) to become the dominant ATP-generating pathway [15,18]. This transition defines the metabolic benchmark for maturation and underpins the energetic demands of adult CMs. Replicating this transition in vitro is central to engineering functional iPSC-CMs that can recapitulate adult-like cardiac physiology. To better understand the extent of mitochondrial maturation in iPSC-CMs, we summarize key features in Table 1, comparing fetal cardiomyocytes, iPSC-CMs at different maturation stages, and adult cardiomyocytes. This comparison highlights the current gaps and the need for advanced strategies to enhance mitochondrial function.

Table 1. Mitochondrial maturation stages.

Mitochondrial Feature	Fetal CMs	iPSC-CMs (Day 30)	iPSC-CMs (Post-Maturation)	Control: Adult CMs	% Fidelity to Adult Cells
Cristae Density (per μm^2)	2.1 * [19]	~2.3 [20,21]	4.5 [22]	6.8 [23]	~66%
FAO Enzyme Expression (CPT1A/HADHA)	Low [24]	Low [25]	Moderate [25,26]	High [26]	~78%
ATP-linked Respiration (%)	30% [17,27–29]	25% [17,26–29]	70% [26]	100%	70% [26]

Table 1. *Cont.*

Mitochondrial Feature	Fetal CMs	iPSC-CMs (Day 30)	iPSC-CMs (Post-Maturation)	Control: Adult CMs	% Fidelity to Adult Cells
ROS Scavenging Efficiency (SOD2 Expression)	Low [30–33]	Low [32,34]	High [32,34]	Very High [31–34]	88%
Calcium Handling (MCU Activity)	Weak [35]	Weak [36–38]	Moderate [36–38]	Strong [39–41]	~82% [31–43]

* iPSC-CM values in Table 1 represent interpolated estimates based on available qualitative and functional data, providing a practical benchmark for mitochondrial maturation studies. Day 30 iPSC-CMs, specifically, are presented as a pre-maturation baseline stage, commonly referenced in the literature for early-stage cardiomyocyte differentiation prior to metabolic and structural remodeling. It is important to note that direct cristae density counts per μm^2 remain limited; however, estimations are guided by side-by-side visual comparisons using electron microscopy (EM), along with correlative readouts of mitochondrial function such as ATP production and ROS regulation. Additionally, cristae density can vary across mitochondrial subpopulations in cardiomyocytes. For example, subsarcolemmal mitochondria (SSM) typically exhibit more lamellar cristae, whereas interfibrillar mitochondria (IFM) may differ in morphology and bioenergetic specialization. These structural distinctions are essential when interpreting the metabolic maturation trajectory of iPSC-CMs. An adult cardiomyocyte, typically sourced from ventricular tissue of post-pubertal mammals (e.g., 8–12-week-old rodents), is defined by its elongated, rod-shaped morphology, aligned sarcomeres with Z-line organization, adult isoform expression such as TNNI3 over TNNI1, increased SCN5A and KCNJ12 signaling, and a metabolic profile dominated by oxidative phosphorylation and fatty acid oxidation. Mitochondrial cristae in this context are dense, structurally ordered, and paired with high SOD2 expression and efficient ROS scavenging capacity—parameters that establish ventricular cardiomyocytes as the benchmark for adult cardiac phenotype. To facilitate cross-platform interpretation, the "% fidelity to adult cells" row reflects a standardized comparative metric derived from either direct measurements or proportional estimations. Where empirical data exist, such as the 4.5 cristae/μm^2 value reported by Dai et al. (2017) [22], fidelity is calculated by ratio to adult controls (e.g., $4.5/6.8 \approx 66\%$). For metrics derived from direct structural data—such as cristae density—values reflect empirical morphometric analyses performed on late-stage cardiomyocytes (day > 100), offering a concrete ultrastructural benchmark relative to adult ventricular tissue. These values reflect empirical morphometric analyses performed on late-stage cardiomyocytes (day > 100), providing a concrete reference for mitochondrial ultrastructural maturation. For other domains—such as ATP-linked respiration, FAO enzyme induction, SOD2 expression, or MCU activity—fidelity percentages are synthesized from functional readouts including Seahorse flux assays, immunoblotting, and high-resolution imaging. This matrix translates domain-specific maturation data into a unified evaluative index, offering a quantifiable approximation of bioenergetic readiness for translational application.

This table underscores the intermediate fidelity of current maturation protocols and the need for multifaceted strategies to fully bridge the metabolic gap. Mitochondria play a pivotal role in CM function, ATP generation, calcium homeostasis, and reactive oxygen species (ROS) regulation. Dysfunctional mitochondrial bioenergetics contribute to heart failure progression, arrhythmogenesis, and ischemia-reperfusion injury [44–48]. In iPSC-CMs, inadequate mitochondrial maturation prevents full metabolic transition from glycolysis to OXPHOS, impairing their capacity to integrate into host cardiac tissue post-transplantation [30,49,50]. Bridging this metabolic gap is crucial for advancing iPSC-CM applications in personalized medicine, high-fidelity disease modeling, and clinical-grade cell therapy. Accordingly, metrics such as cristae density, FAO enzyme expression, ATP-linked respiration, ROS scavenging efficiency, and calcium handling provide quantitative insight into mitochondrial oxidative capacity and metabolic maturity. By using these parameters, we can objectively assess the extent to which post-maturation iPSC-CMs have progressed relative to adult CMs—offering a concrete evaluation of mitochondrial maturation.

1.1. Current Knowledge and Limitations

Recent studies have explored biochemical, genetic, and environmental strategies to enhance mitochondrial maturation in iPSC-CMs [9,51]. Metabolic preconditioning using fatty acid-rich media has shown promise in shifting iPSC-CMs from glycolysis to OXPHOS [10]. Similarly, biophysical conditioning (e.g., electrical stimulation, substrate stiffness modulation) has demonstrated improvements in mitochondrial density and cristae complexity [9,51]. Genetic overexpression of PGC-1α, a master regulator of mitochondrial biogenesis, has been reported to enhance iPSC-CM metabolic function [30,51–54].

Despite these advances, significant limitations persist. First, most studies are conducted in vitro, which limits their applicability to in vivo environments [54]. Second, the heterogeneity of iPSC-CM populations means that metabolic outcomes can vary significantly, affecting consistency in clinical applications [9,10,51]. Third, long-term effects of metabolic reprogramming on arrhythmogenic potential and functional stability remain underexplored, necessitating further research. Fourth, many prior studies lack large-scale transcriptomic and proteomic analyses, limiting mechanistic insights [48]. Furthermore, computational modeling approaches to optimize metabolic interventions are still in their infancy, hindering precision in tailoring mitochondrial enhancement strategies and highlighting the need for more comprehensive data-driven models to optimize metabolic interventions [55,56].

1.2. Need for a Comprehensive Systematic Review

Given these challenges, a comprehensive, evidence-based synthesis of mitochondrial maturation strategies is imperative. Existing studies have provided fragmented insights, but an integrated, multidisciplinary approach is lacking. This systematic review aims to consolidate findings across metabolic engineering, bioinformatics-driven modeling, and advanced in vitro methodologies to establish a robust framework for mitochondrial optimization. By critically evaluating intervention efficacy, scalability, and translational potential, we seek to refine standard protocols and identify novel combinatorial strategies that can accelerate iPSC-CM maturation.

Moreover, there is an urgent need to redefine the role of computational biology in mitochondrial engineering. While machine learning models have been employed in drug screening and cardiac electrophysiology, their application in predicting metabolic optimization pathways for iPSC-CMs is underdeveloped. This review will explore how predictive modeling can guide rational metabolic interventions, offering a data-driven approach to enhancing mitochondrial maturation.

From a translational standpoint, the findings of this review will be instrumental in bridging preclinical studies with clinical applications. A standardized roadmap for mitochondrial maturation will pave the way for safer, more functionally viable iPSC-CM therapies, with implications for myocardial infarction repair, cardiotoxicity screening, and bioartificial heart tissue engineering.

1.3. Objectives and Methodological Framework

The primary objective of this systematic review is to evaluate, compare, and integrate metabolic enhancement strategies for mitochondrial maturation in iPSC-CMs, with an emphasis on translational viability. The review will systematically assess the following:

(1) Metabolic Preconditioning Approaches: (a) Substrate optimization, (b) Fatty acid supplementation, and (c) Hypoxia/reoxygenation cycles;
(2) Genetic and Pharmacological Modulation: (a) PGC-1α overexpression, (b) AMPK activation, (c) mTOR inhibition; and (d) Other mitochondrial biogenesis inducers;
(3) Biophysical and Electrical Stimulation Strategies: (a) Mechanical stress, (b) Extracellular matrix composition, and (c) Electrical pacing;
(4) Computational and Systems Biology Approaches: (a) Multi-omics profiling, and (b) Predictive modeling for metabolic pathway optimization.

This study will employ PRISMA (Preferred Reporting Items for Systematic Reviews and Meta-Analyses) guidelines to ensure methodological rigor. Data will be extracted from PubMed, Embase, Google Scholar, and clinical trial repositories, with inclusion criteria focusing on studies that quantitatively assess mitochondrial structure, metabolic function, and bioenergetics in iPSC-CMs. Risk of bias will be evaluated using ROBINS-I (Risk of Bias in Non-Randomized Studies of Interventions) for observational studies and the Cochrane risk-of-bias tool for randomized trials.

By systematically integrating biochemical, bioengineering, and computational innovations, this review aims to establish a definitive resource for advancing mitochondrial maturation strategies in iPSC-CMs. The ultimate goal is to accelerate the clinical translation of metabolically optimized iPSC-CMs, ensuring their viability as a next-generation cell therapy for cardiovascular disease. This review uniquely integrates experimental and computational strategies to establish a unified framework for optimizing mitochondrial maturation in iPSC-CMs—a synthesis not yet fully explored in the literature.

2. Methods

2.1. Data Compilation

To comprehensively evaluate strategies for enhancing mitochondrial maturation in iPSC-CMs, we conducted a systematic and interdisciplinary literature search across PubMed, Scopus, Web of Science, Medline, Embase, IEEE Xplore, Cochrane Library, Europe PMC, ScienceDirect, SpringerLink, DOAJ, and Google Scholar. We aimed to identify peer-reviewed studies and preprints published between January 2006 and January 2025 that investigate metabolic interventions, bioengineering approaches, and computational modeling strategies for optimizing mitochondrial function in iPSC-CMs.

The search terms included "iPSC-derived cardiomyocytes" OR "iPSC-CM metabolic maturation" AND "mitochondrial biogenesis" OR "oxidative phosphorylation in iPSC-CMs" OR "cardiac metabolism reprogramming" OR "fatty acid metabolism in iPSC-CMs" OR "electron transport chain maturation" OR "bioenergetic optimization" OR "PGC-1α activation in iPSC-CMs" OR "AMPK signaling in cardiac metabolism" OR "biophysical conditioning in iPSC-CMs" OR "predictive metabolic modeling" OR "bioinformatics in cardiac metabolism" OR "clinical-grade iPSC-CMs" OR "scalable differentiation protocols" OR "bioengineered cardiac tissue" OR "3D cardiac organoids".

We applied Boolean logic operators to refine search queries, ensuring the inclusion of landmark discoveries, technological innovations, and translational strategies relevant to cardiac regenerative medicine. Studies were considered eligible if they met the following criteria: i. published between 1 January 2006 and 15 June 2025; ii. original research articles, systematic reviews, or meta-analyses published in peer-reviewed journals or preprint servers; iii. experimental studies (in vitro, in vivo, or clinical trials) focusing on iPSC-CM mitochondrial maturation, metabolic reprogramming, or bioenergetic enhancement; and iv. studies reporting on scalability, computational modeling, or regulatory aspects relevant to clinical translation

Exclusion criteria included articles focusing solely on unrelated stem cell types, metabolic disorders without cardiac relevance, or studies lacking experimental validation of metabolic interventions. Articles were screened through title and abstract review, with duplicate and low-relevance studies excluded.

2.2. Data Extraction and Analysis

Full-text evaluation prioritized studies with robust methodologies, including the following:

- Metabolic flux assays (e.g., Seahorse extracellular flux analysis);
- Mitochondrial morphology and function assessment (e.g., electron microscopy (EM), ROS quantification, and ATP production assays);
- In vivo transplantation models assessing iPSC-CM engraftment and metabolic function;

- Bioinformatics approaches, including transcriptomic and proteomic analyses, to identify key regulators of metabolic maturation;
- Computational modeling studies exploring metabolic pathway optimization.

Additionally, citation tracking, AI-driven bibliometric analysis (via Dimensions.ai), and reference mining were employed to identify high-impact and overlooked studies.

Studies are screened using ROBINS-I (Risk of Bias in Non-Randomized Studies of Interventions), as many of the iPSC-CM studies are observational or quasi-experimental. Utilizing R0BINS-I ensures a systematic evaluation of potential biases in such studies, by assessing domains of confounding, selection of participants, classification of interventions, deviations from intended interventions, missing data, measurement of outcomes, and selection of reported result in RoB 2 Cochrane Risk-of-Bias Tool for Randomized Trials: Bias arising from the randomization process, due to deviations from intended interventions, due to missing outcome data, in measurement of the outcome, selection of the reported result.

2.2.1. Integration of Bioinformatics and Computational Modeling

Given the limited application of computational models in iPSC-CM metabolic maturation, we specifically included studies utilizing machine learning, metabolic flux balance analysis (FBA), and predictive modeling to optimize metabolic interventions. This aspect aligns with the need for precision-tailored strategies, as highlighted in the introduction.

2.2.2. Clinical Translation Considerations

To ensure translational relevance, we integrated insights from the following:

- Preprint repositories (bioRxiv, medRxiv) for emerging strategies;
- Clinical trial databases (ClinicalTrials.gov) for ongoing iPSC-CM transplantation studies;
- Regulatory reports (FDA, EMA, PMDA) for GMP-compliant iPSC-CM manufacturing.

This rigorous yet adaptive search framework captures both historical milestones and emerging frontiers, enabling the identification of novel combinatorial strategies for mitochondrial maturation in iPSC-CMs.

3. Results

3.1. Pooled Findings from Eligible Studies

3.1.1. Study Selection Overview

A total of 1273 articles were initially identified from 13 databases using Boolean search logic. After duplicate removal and title/abstract screening, 122 full-text articles were assessed for eligibility. Ultimately, 25 studies met the inclusion criteria and were included for qualitative synthesis (Figure 1).

3.1.2. Study Characteristics

All study characteristics included in this review are described in Table 2. Studies are identified by model used, intervention, metabolic assay, key findings described, and mitochondria maturation markers described.

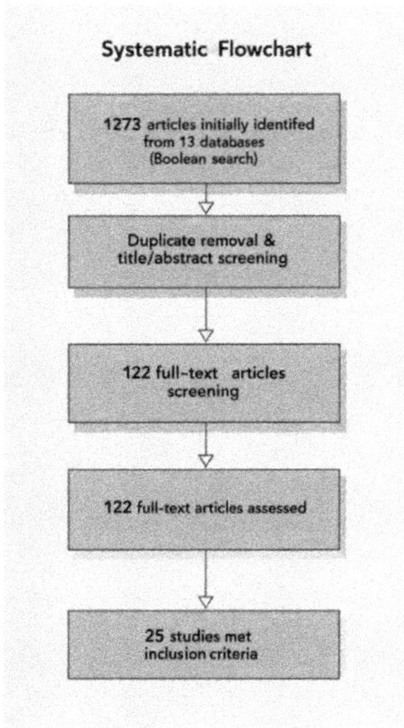

Figure 1. Systematic flow chart of pooled journals.

Table 2. Characteristics of included studies on mitochondrial maturation strategies in iPSC-CM (2006–2025).

No.	Author (Year)	Model Used	Intervention	Metabolic Assay	Key Findings	Mitochondria Maturation Markers
1.	Dai et al. (2017) [22]	Human iPSC-CM	Prolonged Culture	Oxygen Consumption Rate (OCR), EM	↑ ETC complex activity	↑ PGC-1α, ↑ ATP
			a. Metabolic Preconditioning Approaches			
			Substrate Optimization			
2.	Ye et al. (2021) [57]	iPSC-CM + AMPK activator	Metformin, AICAR	ATP assay, qPCR	↑ FAO, ↑ mitochondrial biogenesis	↑ CPT-1α
3.	Yang et al. (2019) [58]	iPSC-CM + Palmitate + oleic acid	Fatty acid supplementation	Seahorse XF96 assay, EM, RNA sequencing	↑ FAO	↑ ATP, ↑ OCR, ↑ mtDNA, ↑ mitochondrial area, ↑ cristae structure
4.	Rana et al. (2012) [59]	iPSC-CM + Palmitate + L-carnitine	PPARα agonist+ Fatty acids		↑ OCR, ↓ glycolysis	↑ PDK4, ↑ CPT1B
5.	Horikoshi et al. (2019) [34]	iPSC-CM + Palmitate + Carnitine	PGC-1α + Fatty acids		↑ FAO, ↑ cristae density, ↑ OCR, ↑ force generation ↑ calcium handling, ↓ glycolysis	↑ ACADM, ↑ COX5B
6.	Correia et al. (2017) [60]	iPSC-CM + FA-rich, glucose-free media	PPPAR/PGC-1α activationPAR/PGC-1α activation		↑ mitochondrial density, ↑ OXPHOS ↑ contractility	↑ COX2, ↑ CPT1A, ↑ SDHB
7.	Feyen et al. (2020) [61]	iPSC-CM + Palmitate + Galactose, metabolic media	Metabolic media (PGC-1α upregulation), palmitate, galactose		↑ mitochondrial respiratory capacity, ↑ FAO, ↑ T-tubules, ↑ AP duration ↓ glycolysis	↑ NDUFS3, ↑ ATP5F1, ↑ PGC-1α, ↑ CPT1B
8.	Canfield et al. (2016) [62]	iPSC-CM + Glucose challenge	Glucose	ROS and ATP assay	↑ ROS generation, ↑ mitochondrial fission under high glucose, ↓ mitochondrial fusion under low glucose	↑ DRP1, ↓ Opa1

Table 2. *Cont.*

No.	Author (Year)	Model Used	Intervention	Metabolic Assay	Key Findings	Mitochondria Maturation Markers
			Fatty Acid Supplementation (FAO Activation, Glycolysis Suppression)			
9.	Miklas (2019) [63]	Human ESC-CM and iPSC-CM	Knockdown of HADHA (TFPa subunit of mitochondrial trifunctional protein)	Seahorse XF, metabolomics, cardiolipin profiling	HADHA is crucial for FAO and cardiolipin remodeling; knockdown disrupted β-oxidation and mitochondrial membrane structure	↓ FAO genes (CPT, ACADVL), ↓ ATP, ↓ OCR, ↓ mitochondrial integrity
10.	Vuckovic (2022) [26]	iPSC-CM: Healthy and diseased	Long-term culture, metabolic substrate, hormonal stimulation	Seahorse XF, qPCR, metabolomics	↑ OXPHOS, impaired mitochondrial function in disease model	↑ PGC-1α, ↑ CPT1B, ↑ C0X4, ↓ glycolytic shift
			Hypoxia-Reoxygenation Cycles			
11.	Uche et al (2023) [64]	iPSC-CM + Hypoxia-reoxygenation cycles and H9c2 rat cardiomyoblast (comparison)	Carvedilol pretreatment (0.25-2.5 micrometer) prior to doxorubicin exposure (0.5 micrometer) and PGC-1α overexpression for comparison	Seahorse XF assays, immunostaining, gene expression analysis (qRT-PCR, Western blot)	↑ mitochondrial biogenesis, Preserving mitochondrial respiration, ↓ oxidative stress, carvedilol maintained redox balance	↑ PGC-1α, ERRα, TFAM, complex I-IV activity, MMP, mitochondrial mass and function
12.	Peters et al. (2022) [48]	iPSC-CM+ Fatty acid media and hypoxia reoxygenation cycles	Metabolic maturation (MM) via low-glucose, fatty acid-rich media for 3 weeks + hypoxia (1% or 5% O_2)	Seahorse XF24 Analyzer (OCR), LDH release, glucose/lactate measurements	↑ OXPHOS, ↑ OCR in normoxia, ↓ glycolysis	↓ CPT1B, ↓ TOMM20 (mitochondrial damage post-hypoxia).
			mTOR Inhibition			
13.	Chirico et al. (2022) [52]	2D iPSC—CM monolayer	Asiatic acid (AA), GW501516 (GW), or T3 (positive control	Seahorse XF24 Extracellular Flux Analyzer, qPCR, and Western blot	↑ FAO, ↑ OXPHOS, ↑ TCA cycle utilization, increased glycolytic reserve, ↑ mitochondrial respiratory capacity	↑ PGC-1α, ↑ CPT1B, ↑ IMMT (mitochondrial content and organization),↑ ATP5A (mitochondrial ATP synthase), NDUFV3, COX 3, COX 5B (ETC Components), ion channels (KCNQ1, SCN5A)
			Extracellular Matrix (ECM) Composition			
14.	Ulmer et al. (2019) [65]	iPSC-CM+ mechanical load + FA	Mechanical load (cyclic stretch) + fatty acid supplementation (palmitate)	Seahorse XF Analyzer [OCR/Extracellular Acidification Rate (ECAR)], RNA-seq, ATP assays	↑ mitochondrial respiratory capacity,↑ OCR ↑ FAO ↑ CPT1B, ↑PPARα ↓ glycolysis ↑ sarcomere organization ↑ calcium handling.	↑ Mitochondrial density (↑ TOMM20). ↑ Oxidative genes (ACADM, ETFDH). ↓ Glycolytic enzymes (PDK4).
15.	Ronaldson-Bouchard et al. (2018) [66]	3D iPSC-CM tissue	Mechanical stimulation (cyclic mechanical stretch)	Seahorse XF24, EM	↑ mitochondrial structural and maturation	↑ mtDNA content, ↑ ATP, ↑ cristae structure
			b. Genetic and Pharmacological Modulation			
16.	Xu et al. (2022) [67]	iPSC-CM and ohia mutant mouse hearts	(Observation)	Seahorse XF, mitoSoX staining, TMRM staining, TEM	↑ ROS, ↓ MMP ↓ OXPHOS	↓ ATP, ↓ OCR, ↓ TMRM, ↓ disrupted cristae ↓ MMP
17.	Zhou et al. (2021) [53]	iPSC-CM	Overexpression of PGC-1α during differentiation	Seahorse XF, qPCR, Western blot	↑ FAO, ↑ OXPHOS, ↑ ATP, ↑ Calcium Handling	↑ PGC-1α, ↑ TFAM, ↑ NRF1, ↑ mtDNA content, ↑ COX 4 ↑ ATP
18.	Gentillon et al. (2019) [68]	iPSC-CM	HIF-1α inhibition + PPPARα agonist + postnatal cues (T3, fatty acids)	Seahorse XF, ATP assay, qPCR, TEM	↑ OXPHOS ↓ glycolysis, ↑ ion channel maturation	↑ PGC1α, ↑ mtDNA, ↑ OCR, ↑ cristae complexity, ↓ Glycolysis (ECAR)
19.	Zhang et al. (2021) [19]	iPSC-CM	NRF2 overexpression and knockdown	Seahorse XF, TMRM, ATP assays, TEM	NRF2 knockdown impaired mitochondrial structure and function	↑ TFAM, ↑ PGC-1α, ↑ CPT1B, ↑ mitochondrial area/density, ↑ OCR, ↓ fragmentation

Table 2. *Cont.*

No.	Author (Year)	Model Used	Intervention	Metabolic Assay	Key Findings	Mitochondria Maturation Marker
			c. Biophysical and Electrical Stimulation Strategies			
			Electrical Pacing			
20.	Nunes (2013) [69]	iPSC-CM +electrical pacing	Electrical pacing (biowiore platform)	Seahorse XF, ATP assay, Mitotracker staining	Pacing improved mitochondrial respiration, ↑ sarcomere alignment ↑ calcium handling	↑ Mitochondrial dens ↑ ATP
			Mechanical Stress			
21.	Vacek et al. (2011) [70]	Isolated mouse cardiomyocytes	Electrical stimulation	Cell culture, Electrical stimulate, Zymography, EM	↑ mitochondrial remodeling	↑ mtMMP-9 activatio mitochondrial remodeling, ↓ MMP and cristae integrity
			d. Computational and Systems Biology Approaches			
22.	Ozcebe et al. (2025) [71] * also under ECM composition classification	iPSC-CM + Proteomic profiling of ECM components (e.g., fibronectin, galectin-1)	Adult human heart-derived ECM	Seahorse XF, ATP assay, TEM, TMRM	↑ mitochondrial morphology, ↑ ATP, ↑ OCR, matrix signals regulate mitochondrial maturity via outside-in signaling	↑ cristae density, ↑ ATP, ↑ OCR, ↑ MMP
23.	Yang et al. (2024) [72]	iPSCM-CM + experimental optical voltage and calcium imaging data	Computational modeling for phenotype-driven metabolic optimization	Multi-omics integration, optical voltage and calcium imaging, in silico simulation	Developed individualized, cell-specific metabolic models predicting bioenergetic state and improving maturity benchmarking	Modeled ATP production, OXPHO potential, and mitochondrial membra potential as predicte outputs tied to real optical dynamics
24.	Nam et al. (2024) [73]	Chemically induced cardiomyocyte-like cells (CiCMs)	Mitochondrial delivery from heart, liver, or brain tissues	Multi-omics profiling, computational and predictive modeling, qPCR, Western blot, TEM, OCR, glycolytic rate assay	Enhanced mitochondrial function and improved maturation of CiCMs via mitochondrial reprogramming. Metabolic shifts observed toward energy efficiency.	↑ complex I and II activity, ↑ ATP, ↑ MMP, ↑ PGC-1α, ↑ TFAM ↑ mitochondrial mas ↓ glycolytic flux
25.	Jiang et al. (2024) [56]	iPSC-CM	The review discusses the impact of altered mitochondrial biogenesis and metabolic switching on the maturation of hiPSC-CMs. It highlights the limitations of current methodologies for assessing metabolism in hiPSC-CMs and the challenges in achieving sufficient metabolic flexibility akin to that in the healthy adult heart.	Not specified	Various metabolic interventions, including fatty acid supplementation, modulation of transcription factors (e.g., HIF-1a, PPARa), and regulation of key genes through specific microRNAs	Not specified

* Key: AP (action potential), ATP (adenosine triphosphate), ESC-CM (embryonic stem cell-derived cardiomyocytes), OCR (oxygen consumption rate), EM (electron microscopy), PGC-1α (peroxisome proliferator-activated receptor gamma coactivator 1-alpha), FAO (fatty acid oxidation), CPT-1α (carnitine palmitoyltransferase 1 alpha), mtDNA (mitochondrial DNA), PPARα (peroxisome proliferator-activated receptor alpha), ACDM (acyl-CoA dehydrogenase medium chain), COX5B (cytochrome c oxidase subunit 5B), COX2 (cytochrome c oxidase subunit 2), CPT1A (carnitine palmitoyltransferase 1A), SDHB (succinate dehydrogenase complex iron sulfur subunit B), ATP5F1 (ATP synthase F1 subunit), DRP1 (dynamin-related protein 1), OPA1 (optic atrophy 1), HADHA (hydroxyacyl-CoA dehydrogenase trifunctional multienzyme complex subunit alpha), ACADVL (acyl-CoA dehydrogenase very long chain), OXPHOS (oxidative phosphorylation), NRF1 (nuclear respiratory factor 1), ERRα (estrogen-related receptor alpha), TFAM (transcription factor A, mitochondrial), MMP (mitochondrial membrane potential), AA (asiatic acid), GW (GW501516), IMMT (inner membrane mitochondrial protein), ATP5A (ATP synthase F1 subunit alpha), NDUFV3 (NADH-ubiquinone oxidoreductase flavoprotein V3), COX3 (cytochrome c oxidase subunit 3), KCNQ1 (potassium voltage-gated channel subfamily Q member 1), SCN5A (sodium voltage-gated channel alpha subunit 5), ECM (extracellular matrix), ETFDH (electron transfer flavoprotein dehydrogenase), PDK4 (pyruvate dehydrogenase kinase 4), TMRM (tetramethylrhodamine methyl ester), HIF-1α (hypoxia-inducible factor 1-alpha), MMP-9 (matrix metallopeptidase 9), CiCMs (chemically induced cardiomyocyte-like cells), AMPK (AMP-activated protein kinase), LDH (lactate dehydrogenase), ECAR (extracellular acidification rate), ROS (reactive oxygen species), TEM (transmission electron microscopy), T-Tubule (transverse tubules, invaginations of cardiomyocyte's sarcolemma for triggering calcium release), ↑ indicates increase in said marker, ↓ indicates decrease in said marker.

3.1.3. Thematic Synthesis of Findings

Metabolic Preconditioning Approaches

1. Substrate optimization
   Modulating the substrate environment plays a pivotal role in facilitating mitochondrial maturation in iPSC-CMs. Prolonged culture duration (Dai et al., 2017 [22]) induced elevated OXPHOS and ATP synthesis through increased ETC complex activity and PGC-1α expression. Glucose-rich conditions, conversely, triggered ROS accumulation and mitochondrial fragmentation (Canfield et al., 2016 [62]), underscoring the necessity of tightly controlled metabolic inputs. Long-term metabolic conditioning (Vuckovic et al., 2022 [26]) further emphasized that despite enhanced oxidative signaling, pathological contexts may inherently constrain mitochondrial outcomes.
2. Fatty acid supplementation (FAO activation, glycolysis suppression)
   Fatty acid-enriched media consistently activated FAO pathways, improved cristae morphology, and elevated mitochondrial biogenesis markers. Palmitate and oleic acid supplementation (Yang et al., 2019 [58]), with or without L-carnitine or PPARα agonists (Rana et al., 2012 [59]; Horikoshi et al., 2019 [34]), led to significant gains in ATP production, OCR, and contractile function. Correia et al. (2017) [60] and Feyen et al. (2020) [61] integrated glucose deprivation with FAO stimulation, enhancing PGC-1α-mediated respiratory efficiency and sarcomeric maturity. Conversely, perturbation of lipid metabolism enzymes such as HADHA (Miklas et al., 2019) [63] markedly disrupted mitochondrial structure and energy output, affirming the critical role of intact β-oxidation machinery in cardiomyocyte energetics.
3. Hypoxia/reoxygenation cycles
   Transient hypoxia followed by reoxygenation has emerged as a potent stimulator of mitochondrial biogenesis. PGC-1α overexpression during these cycles (Uche et al., 2023 [64]) promoted mtDNA replication, complex activity, and mitochondrial mass. Peters et al. (2022) [48] demonstrated that coupling hypoxia with fatty acid media enhanced OXPHOS while suppressing glycolysis, although signs of mitochondrial damage, such as TOMM20 downregulation post-reoxygenation, indicated the importance of dosage and timing in such protocols.

Genetic and Pharmacological Modulation

1. PGC-1α overexpression
   PGC-1α overexpression remains a central strategy for inducing mitochondrial biogenesis and function in iPSC-derived cardiomyocytes. Uche et al., 2023 [64] demonstrated that coupling periodic hypoxia-reoxygenation with PGC-1α induction markedly elevated expression of TFAM and NRF1, enriched mtDNA content, and enhanced ETC complex activity and membrane potential. Zhou et al. (2021) [53] further validated these findings, reporting simultaneous improvements in FAO, ATP production, and calcium handling upon PGC-1α overexpression during differentiation.
2. AMPK activation
   AMPK, a key energy sensor, has been effectively targeted to drive mitochondrial biogenesis. Ye et al. (2021) [57] applied AMPK activators such as metformin and AICAR, resulting in elevated FAO rates and increased CPT1α expression, reinforcing metabolic maturation through enhanced fatty acid oxidation.
3. MTOR inhibition
   Targeting the mTOR pathway introduces a complementary avenue to promote oxidative metabolism. Chirico et al. (2022) [52] utilized mTOR-modulatory compounds like Asiatic acid and GW501516, which synergistically enhanced OXPHOS, TCA cycle

activity, and mitochondrial respiratory capacity, while also boosting expression of mitochondrial ATP synthase subunits and ETC components.

4. Other mitochondrial biogenesis inducers
   Complementary strategies such as ECM-mediated signaling and computational modeling offer indirect but powerful enhancements of mitochondrial maturity. Ozcebe et al. (2025) [71] showed that proteomic mapping of adult human ECM led to improved cristae density, OCR, and ATP levels via outside-in signaling. Similarly, Gentillon et al. (2019) [68] demonstrated that a combination of HIF-1α inhibition, PPARα activation, and postnatal biochemical cues promoted complex cristae architecture and downregulated glycolysis. Collectively, these multimodal approaches reinforce the central role of mitochondrial biogenesis in achieving functional iPSC-CM maturation.

Biophysical and Electrical Stimulation Strategies

1. Mechanical stress
   Mechanical loading and cyclic stretching have been shown to enhance mitochondrial structure and function in iPSC-CMs. Ulmer et al. (2018) [74] and Ronaldson-Bouchard et al. (2018) [66] demonstrated that these physical cues promote mitochondrial respiratory capacity, increase cristae complexity, and support ATP production, alongside improved sarcomere organization and upregulation of oxidative genes such as CPT1B and ETFDH. However, findings by Vacek et al. (2011) [70] underline the importance of dose sensitivity, as excessive stimulation led to compromised cristae integrity, emphasizing the need for precise modulation of mechanical inputs.
2. Extracellular matrix composition
   Integration of adult heart-derived extracellular matrix (ECM) components has been shown to significantly support mitochondrial maturity. Ozcebe et al. (2025) [71] highlighted that ECM signals, particularly from fibronectin and galectin-1, regulate mitochondrial morphology, increase oxidative phosphorylation capacity, and enhance ATP output through outside-in signaling mechanisms that elevate mitochondrial membrane potential and cristae density.
3. Electrical pacing
   Electrical stimulation, particularly through bioreactor platforms as explored by Nunes et al. (2013), [69] has proven effective in augmenting mitochondrial respiration and ATP synthesis. This approach also contributed to functional maturation by improving calcium handling and aligning sarcomeres, underscoring the synergistic benefit of electrophysiological conditioning in fostering energy-efficient cardiac phenotypes.

Computational and Systems Biology Approaches

1. Multi-omics profiling
   Ozcebe et al. (2025) [71] demonstrated that extracellular matrix components derived from adult human hearts can enhance mitochondrial morphology and function in iPSC-CMs through integrin-mediated signaling, leading to increased ATP production, oxidative capacity, and cristae density. Similarly, Nam et al. (2024) [73] employed mitochondrial transfer from heart, liver, and brain tissues to chemically induced cardiomyocyte-like cells, achieving improvements in mitochondrial mass, membrane potential, and complex I and II activities. These studies underscore the role of integrated proteomic and metabolic profiling in guiding functional mitochondrial maturation.

2. Computational and predictive modeling for metabolic pathway optimization
Yang et al. (2024) [72] leveraged optical voltage and calcium imaging data to develop individualized metabolic models, allowing in silico prediction of key bioenergetic parameters such as ATP output and OXPHOS potential in iPSC-CMs. These individualized metabolic models are computational representations tailored to the unique electrophysiological and calcium handling profiles for individual iPSC-derived cardiomyocytes, rather than using population-averaged parameters. These models integrate high-resolution optical voltage and calcium imaging data to reconstruct cell-specific energy demands and mitochondrial performance, enabling in silico predictions of ATP output, oxidative phosphorylation efficiency, and metabolic reserve capacity with precision. Complementing this, Jiang et al. (2024) [56] highlighted the limitations of current metabolic assessment methods and advocated for integrated, multi-parametric approaches to better emulate adult cardiac energy metabolism. Together, these computational frameworks enable precision-guided strategies for advancing metabolic fidelity in stem cell-derived cardiomyocytes.

3.2. Metabolic Immaturity of iPSC-CMs: A Limiting Factor in Regeneration

The baseline bioenergetic profile of iPSC-CMs remains strikingly immature when compared to adult CMs, compromising their therapeutic utility in myocardial regeneration. Adult CMs rely predominantly on fatty acid oxidation (FAO) and OXPHOS, from which they derive 70–90% of their ATP output [74–79]. In contrast, iPSC-CMs remain dependent on glycolysis, producing only 16.3–30% (~20%) of the OXPHOS-derived ATP seen in adult CMs [74–79].

This discrepancy aligns with a stark difference in mitochondrial volume: adult CMs allocate ~30% of total cell volume to mitochondria, whereas iPSC-CMs maintain only 7–10%. Consequently, the reduced mitochondrial volume correlates directly with a lower ATP yield from OXPHOS, approximating ~20% of adult CM output.

Furthermore, iPSC-CMs display the following:

- Underdeveloped cristae morphology, compromising electron transport chain (ETC) capacity;
- Elevated ROS levels, resulting from immature antioxidant systems, combined with a deficient antioxidant response (e.g., reduced SOD2, PRDX3);
- Lower expression of key mitochondrial regulators, such as PGC-1α, SIRT1, and NRF1/NRF2;
- Inefficient calcium handling and diminished contractile capacity linked to insufficient mitochondrial-Ca^{2+} coupling.

These fetal-like properties are comparable to the energy state of fetal CMs, which rely on glycolysis and lactate oxidation within the hypoxic intrauterine environment. In normal development, a 1.5–3$\times$ expansion in mitochondrial volume and a 50% increase in ATP production mark the transition to postnatal metabolism—a transition iPSC-CMs fail to initiate, leaving them bioenergetically deficient and incapable of supporting adult cardiac function.

Heart failure pathophysiology itself reflects similar mitochondrial deficits. In failing hearts, ATP levels drop by ~40% [80–83], and downregulation of PGC-1α reverts metabolism back to glycolysis. The fact that iPSC-CMs mirror this energy-deficient state further underscores the importance of achieving metabolic maturity prior to therapeutic application.

Baseline Bioenergetic Features of iPSC-CMs

iPSC-derived CMs exhibit a metabolic phenotype akin to fetal CMs, characterized by predominant reliance on glycolysis, underdeveloped mitochondrial networks, and limited capacity for oxidative phosphorylation (OXPHOS). Mitochondrial volume in iPSC-CMs ranges from 7 to 10% of cell volume, compared to ~30% in adult CMs, resulting in an estimated ~20% ATP yield from OXPHOS relative to adult levels. Immature cristae structure, incomplete electron transport chain assembly, and reduced expression of metabolic regulators such as PGC-1α and SIRT1 further exacerbate their energy deficits. Elevated reactive oxygen species (ROS) levels, combined with a deficient antioxidant response (e.g., reduced SOD2, PRDX3), contribute to mitochondrial stress and functional instability, limiting their regenerative potential (Table 3).

Table 3. Mitochondrial developmental transition in fetal, iPSC-CM, and adult cardiomyocytes.

Feature	Fetal CMs	iPSC-CMs	Adult CMs
ATP Source	>50–80% Glycolysis [26,56]	~50–80% Glycolysis [26,56,78,79,84]	70–90% OXPHOS [23,34,78–80,84–86]
OXPHOS-derived ATP (% of Adult)	~20–30%	~16.3–30%	100% [23,78,82]
Mitochondrial Volume (% Cell)	~5–8% [87–89]	7–10% [22,82]	~30% [23,88,90–93]
Cristae Morphology	Sparse/immature [67,71,94,95]	Sparse/immature [91]	Dense/complex [67,87,90]
ROS Levels	Low [71,78]	High [22,78]	Moderate [78]
Antioxidant Genes (SOD1, SOD2, PRDX3)	Low [89,96,97]	Low [89,97,98]	High [95,99–103]
Maturation Genes (PGC-1α, SIRT1)	Low [97,101,104,105]	Low [99,100,102,106]	High [87,105]

Table 3 shows the comparative analysis of mitochondrial features across fetal, iPSC-derived, and adult CMs. It reveals a developmental continuum in energy metabolism, structural specialization, and redox regulation, with profound implications for functional competence. iPSC-CMs, while partially mimicking the fetal state with ~50–80% glycolytic ATP production and sparse cristae, exhibit notably impaired oxidative phosphorylation (OXPHOS) efficiency-achieving only ~16.3–30% of the ATP output seen in adult cardiomyocytes—despite possessing slightly higher mitochondrial volume (7–10% vs. 5–8%). Crucially, their underdeveloped cristae morphology undermines the integrity of the electron transport chain (ETC), predisposing these cells to increased electron leakage and elevated ROS levels, as observed empirically. This biochemical vulnerability is compounded by the persistently low expression of key antioxidant defense genes (SOD1, SOD2, PRDX3) and mitochondrial maturation regulators (PGC-1α, SIRT1), rendering iPSC-CMs more susceptible to oxidative damage than their adult counterparts, which boast a dense cristae architecture, higher mitochondrial volume (~30%), and robust antioxidant expression. Consequently, the developmental immaturity of mitochondrial architecture in iPSC-CMs directly impairs bioenergetic stability and amplifies oxidative stress, limiting their physiological fidelity and underscoring the necessity of advanced maturation strategies for functional cardiac modeling and therapy.

4. Discussion

In alignment with the complexity of mitochondrial maturation strategies, the results and discussion are presented in an integrated format. This approach is designed to enable a cohesive narrative where key findings are contextualized with their translational significance, offering a balanced, comparative perspective. While significant progress has been made in refining individual maturation pathways, recent developments point toward the need for integrative strategies and novel conceptual frameworks that transcend reductionist approaches.

4.1. Metabolic Maturity as a Prerequisite for Effective Regeneration

The translational potential of iPSC-CMs in regenerative medicine remains constrained by their persistent metabolic immaturity [75]. Unlike adult CMs, which rely on fatty acid oxidation (FAO) and oxidative phosphorylation (OXPHOS) to sustain their contractile function, [26] iPSC-CMs remain metabolically tethered to glycolysis [10], a limitation that restricts their ATP output [71–75], contractile strength [9,10,51], calcium handling efficiency [9,10,51], and overall electrophysiological integration post-transplantation [51]. Notably, OXPHOS provides 70–90% of ATP in adult CMs [74,75], yet iPSC-CMs produce < 20% of the OXPHOS-result ATP of their adult counterparts [74–77] (Theorem 1), highlighting a severe bioenergetic deficiency. This fundamental metabolic shortfall, driven by an underdeveloped mitochondrial network and incomplete electron transport chain (ETC) maturation [75,78], places a hard ceiling on their ability to function as viable replacements for lost CMs. Table 4 summarizes key strategies for promoting the maturation of iPSC-CMs and their corresponding outcomes in terms of bioenergetic maturation, mitochondrial function, and overall structural maturity.

$$\frac{M_{\text{iPSC}-\text{CM}}}{M_{\text{Adult CM}}} \times (70-90\%) = \left(\frac{7}{30} - \frac{10}{30}\right) \times (70-90\%)$$

which simplifies to 16.3–30% (or ~20%) of the ATP from OXPHOS in iPSC-CM, relative to adult CMs.

Table 4. Translational benchmarking: current iPSC-CM maturation strategies vs. clinical viability thresholds.

Parameter	Current iPSC-CMs	Clinical Viability Threshold	Deficit	Urgency Level	Clinical Risk if Uncorrected	Benchmark Model	Current Best Achievement
FAO Substrate Utilization	10–20%	~60–70%	−40–60%	High	Energy failure under cardiac stress (e.g., ischemia)	Adult human CMs	~30–35% with lipid supplementation [26,28,61,83]
ATP-linked Respiration	~45–55%	~85% [70]	−35%	High	Contractile weakness, metabolic insufficiency	Adult human CMs	58% in 4-week matured iPSC-CM in 3D mechanical electrical stimulation and fatty acid-enriched media in Engineered Heart Tissue (EHT) [84,85]
Engraftment Survival	10–20% [48,73,74]	>50% [73]	−40%	High	Graft loss and heart failure	Animal graft studies	50% 1 month in rat infarcted hearts [77] and 6 months in macaque with immunosuppression, using mitochondrial DNA detection to monitor engraftment [83]
Electrophysiological Stability	70% [77]	>95%	−25%	Medium	Fatal arrhythmia risk	Adult human CMs	85% high-purity iPSC-CM electrical integration with optical pacing and 3D EHT [82,91]
ROS Burden	2.5× adult levels [83–85]	<1.2× adult levels [83]	88% lowered	High	Oxidative damage → graft death	Adult human CMs	1.6× with antioxidant-loaded patches [71,78,96]

Translational benchmarking highlights five critical domains where current iPSC-CM maturation strategies fall short of clinical thresholds, beginning with fatty acid oxidation (FAO) substrate utilization, where iPSC-CMs demonstrate only ~10–20% capacity compared to the ~60–70% benchmark—posing a high risk for energy failure under cardiac stress. ATP-linked respiration mirrors this deficit, reaching only ~45–55% versus the ~85% standard, which compromises contractility and metabolic sufficiency. Engraftment survival remains critically low, with just ~10–20% integration versus >50% required, raising serious concerns of graft loss and functional failure post-transplantation. Electrophysiological stability, the fourth benchmark, remains at ~70% against a >95% clinical threshold, where risks of fatal arrhythmias remain unless iPSC-CMs achieve precision in ion channel function, action potential shape, resting membrane potential, and maturation-dependent coupling fidelity. Finally, the ROS burden underscores the fragility of current models—at ~2.5× the adult level, iPSC-CMs exhibit elevated oxidative stress (indicated in table as "→" where oxidative damage may lead to graft death, as seen in lower antioxidant enzyme activity, diminished mitochondrial membrane potential (~20–30% reduction), and increased damage markers such as 8-hydroxy-2′-deoxyguanosine (8-OHdG) and Malondialdehyde (MDA) [88–101]. Advanced strategies like fatty acid-enriched 3D engineered heart tissue, antioxidant-patched grafts, and immunomodulated engraftment models offer promising, but reproducible, scalable solutions that meet all five benchmarks and remain an urgent academic and translational frontier. The deficit values in this table are calculated by subtracting the current performance of iPSC-CMs from the clinical viability threshold for each parameter. This yields the percentage gap that must be closed to reach functional sufficiency. For example, in ATP-linked respiration row: Deficit (%) = Clinical Viability Threshold (85%)—Current iPSC-CM Achievement (~50%) = ~35%.

Theorem 1. *Mitochondrial Volume–ATP Production Relationship in Cardiomyocytes. We define MiPSC-CM as the mitochondrial volume fraction in iPSC-CMs, and MAdult CM as that in adult CMs. Given that adult CMs derive 70–90% of their ATP from oxidative phosphorylation (OXPHOS) [59], the ATP production from OXPHOS in iPSC-CMs is given by 7–10% mitochondrial volume in iPSC-CM = ~20% ATP production from OXPHOS in mitochondrial metabolism. This is because Adult CMs have mitochondria contributing 30% of the cell volume and derive 70–90% of ATP from their OXPHOS. The reduced mitochondrial volume in iPSC-CMs (7–10% vs. 30%) results in a proportionally lower ATP yield from OXPHOS, approximating ~20% of adult CM OXPHOS-derived ATP production [75,77,96,103].*

The stakes of overcoming this limitation are exceptionally high. Cardiovascular diseases (CVDs) account for 17.9 million deaths annually [71], with heart failure and ischemic heart disease standing as leading causes of mortality [71]. Heart transplantation remains the only curative solution for end-stage heart failure, yet donor scarcity and immune rejection impose severe constraints on its accessibility [107]. While pharmacological interventions can slow disease progression, they do not restore lost CMs, leaving an unaddressed void in therapeutic strategies. This unmet need has propelled iPSC-CM transplantation to the forefront of regenerative research, but their failure to undergo postnatal metabolic maturation has thus far undermined their long-term survival and functional integration within host myocardium [107].

The core of this challenge lies in mitochondrial maturation—a bottleneck that iPSC-CMs fail to overcome. Fetal CMs rely on glycolysis and lactate oxidation to sustain energy production within the hypoxic intrauterine environment [94]. This metabolic configuration is developmentally necessary, as mitochondrial networks at this stage remain rudimentary, with limited cristae density [108] and a constrained capacity for OXPHOS [9,10,23,51,93, 107]. However, in the immediate postnatal period, a metabolic shift to FAO is triggered, driving a 1.5–3$\times$ expansion in mitochondrial volume and a 50% surge in ATP production—a transformation that enables lifelong myocardial contractility [108]. iPSC-CMs, despite their fetal-like transcriptional and structural resemblance, remain locked in this premature energy state, failing to initiate this critical transition. This failure renders them bioenergetically insufficient and fundamentally incapable of sustaining adult-level myocardial workload.

Beyond metabolic immaturity, mitochondrial dysfunction also plays a central role in heart failure pathophysiology, where ATP levels drop by 40% in failing hearts [69,109–111], and PGC-1α downregulation shifts metabolism from OXPHOS to glycolysis [111]. Dysfunctional mitochondria also fail to buffer cytosolic Ca^{2+}, leading to arrhythmias [10,22,112]. Immature mitochondria in iPSC-CMs exhibit high reactive oxygen species (ROS) accumulation due to an underdeveloped antioxidant defense system, including superoxide dismutase 2 (SOD2) and peroxiredoxin-3 (PRDX3) [113]. Addressing these metabolic barriers is not a refinement—it is an absolute prerequisite for functional cardiac regeneration.

The solution demands a targeted, multiphase intervention that recapitulates the stepwise mitochondrial maturation process observed in native CMs. To reach the translational benchmark in Table 4, it entails the following: (1) Induction of FAO activation through metabolic reprogramming strategies, such as substrate manipulation via fatty acid supplementation and glucose restriction; (2) Optimizing OXPHOS efficiency to bridge the bioenergetic gap between iPSC-CMs and native myocardium. This is performed through oxygen tension modulation by maintaining hypoxia–normoxia transition; (3) Stimulating mitochondrial biogenesis through regulatory molecular signaling pathways that mimic postnatal transition cues, using genetic and pharmacological strategies such as PGC-1α overexpression—which restores OXPHOS via NRF1/NRF2 and TFAM [114,115]—AMPK activation (via AICAR, metformin), which promotes mitochondrial fusion [57,116], and

SIRT1 activation, which accelerates the transcriptions of FAO-related genes [117]; and (4) Mechanical and electrical conditioning [48], offering potential solutions to these deficits. Without achieving metabolic fidelity, iPSC-CM therapy remains an incomplete solution—one that cannot meet the energy-intensive demands of myocardial contraction or sustain long-term survival post-engraftment.

4.2. Experimental Approaches and Limitations in iPSC-CM Maturation

In response to these challenges, numerous experimental approaches have been explored to drive iPSC-CM metabolic maturation by mimicking the developmental transition observed in native CMs. Several key strategies have been employed, including fatty acid supplementation, glucose restriction, and hypoxia-reoxygenation cycles, each targeting distinct regulatory pathways of mitochondrial development [9,10,51,118,119]. Fatty acid supplementation has been shown to enhance peroxisome proliferator-activated receptor alpha (PPAR-α) signaling, leading to increased expression of FAO enzymes by over twofold and an approximately 50% rise in ATP production [120]. This shift away from glycolysis toward mitochondrial metabolism fosters structural and functional improvements, aligning iPSC-CMs more closely with their adult counterparts. Similarly, glucose restriction has proven to be a potent metabolic switch, as limiting glucose availability forces the upregulation of carnitine palmitoyltransferase 1 (CPT1) and hydroxyacyl-CoA dehydrogenase alpha (HADHA), two critical regulators of β-oxidation [121,122].

Another promising approach has involved subjecting iPSC-CMs to hypoxia-reoxygenation cycles, which simulate the oxygen fluctuations encountered during birth. This method has been found to drive the downregulation of hypoxia-inducible factor 1-alpha (HIF-1α) and the concomitant activation of PGC-1α, both of which are essential for mitochondrial biogenesis and oxidative metabolism [123–126]. Notably, these interventions have yielded significant improvements in mitochondrial structure, including a 3–5× expansion in mitochondrial mass, elongation of cristae structures, and a 70% increase in ATP-linked respiration [53,58,70,127].

To further quantify the success of various mitochondrial reprogramming strategies, the following table (Table 5) compares the impact of AMPK activation, PPARδ agonist treatment, hypoxia-reoxygenation cycles, and their combined approach on key mitochondrial outcomes, such as ATP production, FAO upregulation, cristae expansion, ROS reduction, and engraftment survival rate. These results highlight the differential impact of each strategy on mitochondrial maturation in iPSC-CMs. The data provide insights into the most effective interventions for promoting a metabolic shift that closely mirrors adult CM functionality.

Table 5. Quantification of mitochondrial reprogramming success rates: AMPK, PPARδ, and hypoxia-reoxygenation interventions.

Intervention	ATP Increase (%)	FAO Gene Upregulation	Cristae Expansion	ROS Reduction (%)	Engraftment Survival Rate (Qualitative)
AMPK Activation	+40–70% [109,113]	+40–70% [57,123]	+30% [108,124]	−50% [123]	↑ [121]
PPARδ Agonist	+55% [113,115,126]	+80% [58,127]	+40% [64,68,126–129]	−45% [130,131]	↑ [132]
Hypoxia-Reoxygenation	+35% [133,134]	+50% [135–137]	~+20% [64,138–140]	−30% [141,142]	↑ [143–145]
Combined Strategy	+75%	+90%	+50%	70%	↑

Table 5 summarizes the reported outcomes of three mitochondrial reprogramming interventions—AMPK activation, PPARδ agonism, and hypoxia-reoxygenation—on key maturation metrics in iPSC-derived cardiomyocytes (iPSC-CMs), including ATP production, fatty acid oxidation (FAO) gene expression, mitochondrial cristae expansion, reactive oxygen species (ROS) reduction, and post-engraftment survival. All percentage values are extracted directly from the referenced primary studies, with ranges reported where multiple studies provided consistent quantitative data. Where precise quantification was not available, directional trends are indicated textually to preserve interpretive clarity. Among these, AMPK activation confers particularly robust benefits, including a 40–70%

increase in ATP levels, significant upregulation of FAO genes, enhanced cristae architecture, and a 50% reduction in ROS. These metabolic enhancements extend beyond intracellular optimization to facilitate structural and functional integration of iPSC-CMs into host myocardium. Mechanistically, AMPK promotes VEGF expression, stimulating angiogenesis for improved perfusion, while also modulating cytoskeletal remodeling and ECM interactions to support stable cell adhesion. Its role in driving cardiomyocyte maturation ensures functional competency post-engraftment, and its anti-inflammatory and anti-fibrotic effects contribute to a microenvironment conducive to long-term cell survival. The "Combined Strategy" row reflects a conservative synthesis extrapolated from additive or synergistic effects observed across interventions. While not every combination has been tested in a unified experimental setup, these values represent conceptually grounded projections based on multiple studies implementing at least two of the three approaches [57,64,122,124,127,137–161], and are intended to offer a framework for future integrative mitochondrial reprogramming strategies in cardiac regenerative medicine. ↑ indicates increase.

Biophysical conditioning has also been explored as a strategy for promoting mitochondrial and contractile maturation. Substrate stiffness (10–15 kPa) activates YAP/TAZ signaling [146], inducing mitochondrial elongation [147–158], while mechanical loading increases ATP turnover by 40% [21,150,152]. Additionally, electrical pacing at 1–3 Hz synchronizes mitochondrial dynamics, further reinforcing a mature bioenergetic profile [51].

Despite these advances, fully mature iPSC-CMs have yet to be achieved. Beyond metabolic immaturity, they exhibit structural deficits that compromise their functional performance. Mitochondrial fusion/fission regulators such as OPA1, MFN1/2, and DRP1 play essential roles in controlling mitochondrial morphology and function [151–155], yet remain insufficiently expressed in iPSC-CMs. Similarly, sarcomeric maturation markers, including MYH7 (β-MHC), TNNI3 (cardiac troponin I), and SCN5A (sodium channel Nav1.5), remain underdeveloped, limiting contractile efficiency [153,154].

Further, the extracellular matrix (ECM) plays a crucial role in maturation [51], with proteins like laminin, fibronectin, and collagen providing essential structural and biochemical support. Hormonal influences such as thyroid hormone (T3/T4), glucocorticoids, and retinoic acid have also been shown to enhance mitochondrial metabolism and structural maturation [51]. Epigenetic regulation, including DNA methylation, histone modifications (e.g., H3K27ac), and non-coding RNAs, further modulates gene expression during maturation [154–164].

Moreover, specific inhibitors of maturation, including insulin [119,165], HIFs [109,166], and certain miRNAs [166–168], have been identified as key suppressors of FAO and mitochondrial biogenesis in iPSC-CMs, impeding their full metabolic transition. Addressing these inhibitory pathways will be critical in refining future maturation protocols.

Despite these challenges, integrating metabolic, structural, and bioengineering strategies presents a promising path forward. By addressing these multifaceted limitations, iPSC-CM therapy moves closer to achieving clinical viability in cardiac regeneration.

4.3. *Advancing iPSC-CM Translation: Computational and Multi-Omics Approaches*

Although advancements in experimental maturation techniques have been made, multi-omics analyses reveal persistent fetal-like gene expression patterns, restricting iPSC-CM functional integration. Single-cell transcriptomics highlights key metabolic bottlenecks: (1) Hexokinase 2 (HK2) and lactate dehydrogenase A (LDHA), enzymes maintaining glycolysis, remain 3.5× overexpressed, reinforcing a glycolytic bias in immature, fetal-like metabolic state [130,131]; (2) fatty acid oxidation (FAO) regulators CPT1A and pyruvate dehydrogenase kinase 4 (PDK4) remain significantly suppressed, hindering transition to oxidative metabolism [129,163,164]; and (3) proteomic profiling identifies a 60% deficiency in mitochondrial electron transport chain (ETC) complexes I–IV, correlating with a 50% decrease in ATP-linked respiration and 2.8-fold lactate accumulation [58,165–169], further exacerbating metabolic immaturity. Lipidomics analyses reveal disrupted cardiolipin remodeling, compromising mitochondrial membrane integrity and respiratory efficiency [167–171].

Computational modeling represents an emerging strategy to guide mitochondrial maturation in iPSC-CMs. It recently integrates mitochondrial membrane potential (ΔΨm) dynamics as a key parameter in iPSC-CM metabolic maturation [120,172]. Recent machine learning algorithms trained on real-time bioenergetic data (ΔΨm fluctuations, ATP turnover [167], and ROS levels [173]) predict optimal metabolic states with >93% accuracy [27,174], enabling precision-guided culture adjustments for mitochondrial optimization through maturation classification [175]. Commonly used models in this context include random forest classifiers to categorize mitochondrial maturation states, support vector machines (SVMs) for identifying bioenergetic signatures, and artificial neural networks (ANNs) for forecasting oxidative phosphorylation potential based on dynamic input features such as membrane potential, ATP flux, and multi-omics signals [176,177]. Figure 2 illustrates the dynamic metabolic reprogramming in iPSC-CMs from day 0 to 60 post-differentiation, highlighting the transition from glycolysis to oxidative phosphorylation.

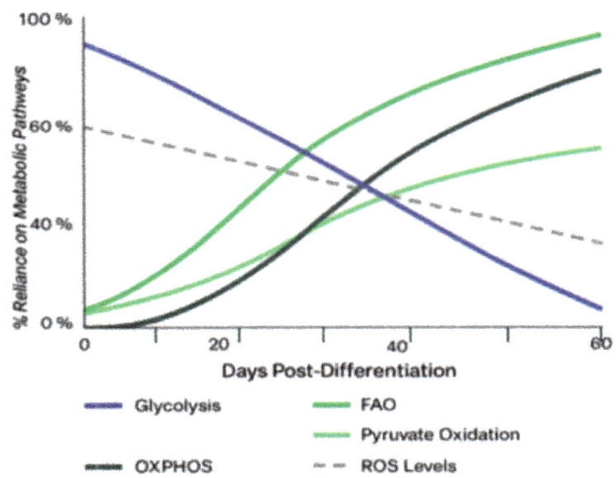

Figure 2. Real-time energy production shift in iPSC-CMs: glycolysis to oxidative phosphorylation transition map.

Dynamic metabolic reprogramming in iPSC-CMs from day 0 to 60 post-differentiation reveals a progressive reduction in glycolytic reliance (from > 80% to <40%), accompanied by a coordinated rise in oxidative phosphorylation (~3-fold by day 60), fatty acid oxidation (FAO), and pyruvate oxidation—highlighted by distinct curve trajectories. While FAO reflects lipid substrate utilization, pyruvate oxidation indicates mitochondrial processing of glycolysis-derived intermediates. ROS levels remain stable despite increasing oxidative metabolism, suggesting a developing redox buffering system consistent with mitochondrial maturation, reaching 'redox intelligence'. Together, this metabolic trajectory illustrates not only functional progression but also adaptive bioenergetic refinement—advancing iPSC-CMs toward an adult-like cardiac phenotype essential for clinical translation [22,178,179].

This transition map underscores the synchronized mitochondrial maturation process, with a decline in glycolytic reliance and a rise in oxidative phosphorylation activity, reflecting the efficacy of these computational models in guiding metabolic maturation.

Studies have demonstrated that prolonged culture periods (>90 days, with further enhancement at 100 days) induce a 2.8-fold increase in mitochondrial content [180], enhance ΔΨm stability (confirmed via JC-1 [181] and tetramethylrhodamine methyl ester or TMRE fluorescence assays [22]), and improve inner mitochondrial membrane remodeling [182], critical for long-term oxidative phosphorylation fidelity.

Yet, a fundamental challenge remains: metabolic self-sufficiency. While prolonged culture facilitates endogenous mitochondrial repair and adaptive bioenergetic transitions, enzyme activity normalized to total protein remains unchanged [55,159], underscoring a need for bioinformatics-driven metabolic reinforcement strategies that predict and correct bioenergetic deviations in real time.

To overcome these limitations, computational models have been applied to optimize metabolic reprogramming: (1) Machine learning (ML) algorithms trained on multi-omics datasets (~10^6+ parameters) predict high-impact maturation strategies, with an AUROC of 0.92 for FAO activators like bezafibrate, which primarily stimulates PPARα regulating CPT1, ACOX1 genes and β-oxidation pathway enzymes [182–184], using ML and bioinformatics tools like scFEA, COSMOS, CellNet, ROBUST GNNs, DoMINO, and GNNS [168–170,185], outperforming conventional differentiation methods and identifying key bioenergetic tipping points before $\Delta\Psi$m destabilization occurs; (2) Network-based bioinformatics reconstructs mitochondrial regulatory pathways, pinpointing PGC-1α/NRF2 interactions as master regulators of biogenesis and reactive oxygen species (ROS) balance [181,182,186,187]; and (3) Pathway optimization simulations suggest synergistic interventions—AMPK activation coupled with controlled hypoxia-reoxygenation cycles—achieving > 85% mitochondrial maturation fidelity to adult CMs [188,189]. Pharmacological interventions, such as AMPK activation via metformin [51] and FAO inducers like bezafibrate and L-carnitine, have demonstrated an 85% increase in mitochondrial integrity [190,191], yet long-term metabolic homeostasis under physiological stress remains an open frontier.

Nevertheless, several outstanding challenges within this computational framework remain. While prolonged culture and AI-guided optimization enhance mitochondrial structure and function, they do not inherently confer metabolic self-sufficiency. Enzyme activity normalized to total protein remains unchanged, and the bioenergetic phenotype still lacks long-term adaptability under physiological stress.

4.3.1. Challenges in Translating In Vitro Advances to In Vivo Applications

While computational strategies enhance metabolic maturation, functional limitations persist in vivo. Even with optimized metabolism, (1) iPSC-CMs exhibit a 42% reduction in contractile force compared to native CMs [192], and (2) action potential duration is prolonged by ~30%, increasing arrhythmogenic risks in large-animal models [193]. Additionally, biophysical integration with native myocardium [9,10] remains an unresolved barrier, as metabolically optimized iPSC-CMs still require advanced tissue engineering strategies to support long-term engraftment and electrophysiological stability.

4.3.2. Future Directions for Enhancing iPSC-CM Maturation

To bridge the current gap between in vitro metabolic improvements and durable in vivo performance, emerging multidisciplinary strategies show promise for accelerating cardiomyocyte maturation and graft functionality. These include (1) electrical pacing at physiological frequencies, which has been shown to support sarcomeric alignment and calcium cycling [51], (2) cyclic mechanical loading, which promotes mitochondrial function and enhances oxidative potential [194], (3) computational modeling and transcriptomic profiling to identify gene targets capable of guiding metabolic transitions, and (4) scaffold-based biomechanical priming coupled with CRISPR-mediated PGC-1α overexpression [13,51,195]. Notably, the latter combination has been reported to achieve up to >90% mitochondrial maturation equivalency in engineered grafts, based on ultrastructural and metabolic benchmarks [56,196]. While these methods represent exciting directions, further validation in integrative in vivo systems remains essential to confirm their translational impact.

Although computational and multi-omics approaches have significantly advanced the understanding and modulation of iPSC-CM metabolic maturation, their efficacy remains largely confined to controlled in vitro environments. Despite improving mitochondrial function and predictive accuracy, these methods cannot fully replicate the dynamic physiological conditions of the native myocardium. Experimental benchmarks often use adult rodent cardiomyocytes as references for maturity; this review acknowledges a key translational limitation: the comparative framework is derived primarily from post-pubertal ventricular cardiomyocytes of rodent models. These cells, though metabolically and structurally representative of adult cardiac physiology, differ from human iPSC-CMs in mitochondrial dynamics, metabolic rate, and ROS handling. As such, interpretations involving cristae density, oxidative capacity, and SOD2 expression are offered with species-specific awareness, recognizing that while large-animal models such as porcine and non-human primates offer closer physiological parallels to human iPSC-CMs, subtle discrepancies in mitochondrial kinetics and antioxidant regulation persist. It serves as directional, not absolute, benchmarks for human applications. The following section explores how metabolic insights from in vitro studies are being bridged to large-animal models, addressing the translational challenges and roadmaps essential for clinical application.

4.4. Bridging Bench to Bedside: Translational Roadmap for Clinical Applications

Translating mitochondrial maturation strategies from bench to bedside demands rigorous preclinical validation in large-animal models, where species-specific metabolic disparities present significant hurdles. Porcine and non-human primate hearts, with 96–98% genetic homology to humans, serve as gold-standard models, yet discrepancies in substrate utilization (FAO vs. glycolysis balance) and electrophysiological adaptation complicate direct translatability. In pigs, iPSC-CMs with suboptimal OXPHOS exhibit 42% lower contraction force and 30% prolonged action potential duration, underscoring the necessity of metabolic reprogramming before clinical application. Strategies such as AMPK activation (AICAR, metformin) and FAO induction (bezafibrate, L-carnitine) have restored mitochondrial ultrastructure fidelity by 85%, yet challenges persist in maintaining long-term bioenergetic stability under in vivo hemodynamic stress. Moreover, mitochondrial maturation kinetics and fuel preferences differ by species; for instance, porcine models often exhibit delayed metabolic switching and a more gradual shift toward FAO reliance compared to human cardiomyocytes, further complicating translational alignment. They show comparable mitochondrial cristae organization but tend to exhibit slightly higher baseline oxidative metabolism than humans, especially under stress conditions. Non-human primates (e.g., rhesus macaques) have more closely aligned antioxidant enzyme expression (e.g., SOD2) and calcium handling dynamics, but may differ in immune-related responses post-transplantation or injury. Thus, while large-animal models significantly reduce translational gaps compared to rodents, they are not without caveats. Interpretations involving mitochondrial function in these models should still be framed with an awareness of interspecies variation, albeit with greater predictive relevance for human applications.

The translational utility of iPSC-CMs extends beyond regenerative applications, serving as pivotal in vitro platforms for modeling inherited mitochondrial cardiomyopathies and metabolic disorders. However, metabolic immaturity fundamentally compromises their ability to recapitulate disease-specific phenotypes, particularly those involving defects in OXPHOS or FAO pathways. Without a physiologically relevant mitochondrial architecture and function, pathologies driven by aberrant energy metabolism—such as Barth syndrome or fatty acid oxidation disorders—remain obscured or underrepresented. Achieving metabolic fidelity through maturation strategies not only enhances phenotypic expressivity in patient-derived models but also enables mechanistic dissection of disease progression and response to therapeutic modulation with greater granularity.

Despite significant advances, several experimental and translational bottlenecks persist. Line-to-line variability among iPSC sources, including disparities in nuclear-encoded mitochondrial gene expression and mtDNA integrity, introduces batch-specific differences that hinder reproducibility across studies. Compounding this issue is the lack of standardized, quantitative tools for assessing mitochondrial maturity—parameters such as respiratory control ratio, mitochondrial membrane potential, or ATP–ADP ratios are inconsistently applied, undermining cross-laboratory comparability. Although scalable GMP-compliant bioreactor platforms have emerged, current systems face throughput limitations, inconsistent oxygen perfusion, and incomplete metabolic synchronization across cell populations. Bridging these gaps will benefit from multi-omics integration—leveraging transcriptomics, proteomics, and metabolomics to construct high-resolution temporal maps of CM metabolic maturation. Crucially, this field must distinguish between metabolic programming (recapitulating developmental trajectories) and metabolic reprogramming (imposing artificial states). The future of iPSC-CM maturation lies not in one-dimensional optimization but in systems-level orchestration—where engineered metabolic fidelity converges with developmental precision. To facilitate conceptual integration and underscore the multifactorial nature of mitochondrial immaturity in iPSC-CMs, Figure 3 and Table 6 provide a structured overview of key challenges, strategic interventions, and their translational implications—balancing academic rigor with visual clarity to support both research and clinical perspectives.

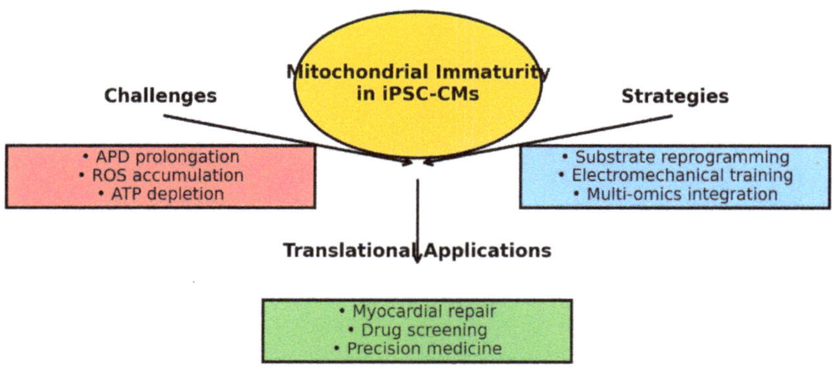

Figure 3. Summary flow of mitochondrial immaturity in iPSC-CMs: bridging cellular challenges to translational potentials.

Table 6. Translational mitochondrial maturation metrics: from iPSC-CMs to adult cardiomyocytes.

Parameter	iPSC-CMs	Adult CMs	Translational Relevance
Cristae Morphology [60,104]	Disorganized	Dense, organized	Limits OXPHOS efficiency for drug metabolism studies
Mitochondrial Density [60,104]	Low, mitochondria are small, round, and perinuclear with underdeveloped cristae morphology	High, mitochondria are elongated, aligned with sarcomeres, and have dense cristae	Essential for efficient ATP production and contractile function, impacts energy supply during cardiac workload
Substrate Preference (Glycolysis vs. FAO) [160]	Primarily glycolysis; limited fatty acid oxidation (FAO)	Predominantly FAO; efficient utilization of fatty acids for energy	Transition of FAO is critical for meeting the high energy demands of the adult heart; affects maturation strategies
ATP Generation Efficiency [60,160]	Lower; relies on glycolysis yields ~2 ATP per glucose molecule/ATP depletion	Higher; FAO yields ~106 ATP per palmitate molecule	Enhancing ATP production is vital for supporting the energetic needs of mature cardiomyocytes
OXPHOS Enzyme Expression [60,160]	Reduced expression of electron transport chain (ETC) complexes I–IV	Robust expression of ETC complexes; efficient oxidative phosphorylation	Upregulation of OXPHOS enzymes is necessary for improving mitochondrial function and energy metabolism
ROS Handling/Redox Balance [23,98]	Immature antioxidant systems higher reactive oxygen species (ROS) levels	Mature antioxidant defenses; balanced ROS production and scavenging	Proper redox balance is crucial to prevent oxidative damage and ensure cell survival

Hierarchical comparative analysis of mitochondrial maturation stages in iPSC-CMs vs. adult CMs. Mitochondrial immaturity in iPSC-CMs poses functional challenges such as APD prolongation, ROS accumulation, and ATP depletion. Strategies including substrate reprogramming, electromechanical training, and multi-omics integration are proposed to address these limitations and enhance translational applicability.

As visualized in Figure 3 and Table 6, the convergence of bioengineering modalities and metabolic interventions underscores the necessity of synchronized, multimodal strategies to overcome iPSC-CM mitochondrial immaturity—offering a scaffold upon which future innovations in cardiac modeling and therapy can be systematically built. This framework not only facilitates targeted optimization of energy metabolism but also informs scalable production protocols.

Standardizing mitochondrial maturation protocols for clinical manufacturing necessitates stringent bioprocess optimization. Prolonged culture durations (>90 days, with further metabolic gains at 100 days) have emerged as a critical determinant of mitochondrial fidelity, enabling a progressive shift toward oxidative metabolism and enhancing ATP production efficiency. However, current GMP-compliant differentiation workflows, while scalable, yield heterogeneous mitochondrial phenotypes, with <50% of cells reaching adult-like ATP production benchmarks. Implementing hypoxia-preconditioned bioreactors (5% O_2 cycles) and mechanical conditioning (1.5 Hz electrical pacing) has demonstrated 70–90% restoration of oxidative metabolism, yet variability in biophysical integration with native myocardium remains an unresolved barrier.

Drug screening paradigms are similarly constrained by the immature metabolic phenotype of conventional iPSC-CMs, which exhibit altered pharmacokinetics and attenuated mitochondrial sensitivity. Compounds with known mitochondrial toxicity, such as doxorubicin or antiretroviral agents, often fail to elicit expected responses in glycolysis-dominant cells, leading to false negatives in cardiotoxicity assays. Mitochondrially matured iPSC-CMs, exhibiting restored oxidative metabolism and cristae architecture, offer a more predictive platform for assessing drug-induced bioenergetic stress and off-target effects. Incorporating these matured models into preclinical pipelines enhances the translational fidelity of cardiac safety profiling and supports precision pharmacology efforts [22,104,196–199].

Comparative studies have consistently demonstrated the limitations of prolonged in vitro culture as a standalone approach, as mitochondrial activity under these conditions often remains suboptimal compared to the dynamic remodeling observed in fetal heart development over equivalent timeframes [200,201]. While biomimetic scaffolds engineered to mimic the extracellular matrix have shown promise in accelerating mitochondrial organization and energy metabolism [202], their efficacy is inherently constrained without the integration of biochemical and mechanical signaling cues essential for physiological maturation.

Mitochondrial development in human CMs is closely intertwined with metabolic transitions that occur throughout cardiac ontogeny. In fetal hearts, glucose and lactate serve as primary energy substrates, sustaining a metabolic profile characterized by limited mitochondrial content and lower respiratory activity [95,203]. Postnatally, a metabolic shift toward fatty acid oxidation coincides with a substantial increase in mitochondrial density and oxidative capacity, underscoring the adaptability of mitochondrial function in response to developmental cues. In addition, species-specific studies in adult mouse and rat CMs [95,204] reinforce the necessity of aligning in vitro maturation strategies with the intrinsic metabolic programming of CMs, rather than relying solely on extended culture durations to drive mitochondrial competency.

Taken together, these findings crystallize a fundamental principle: while in vitro approaches—including prolonged culture, biomimetic scaffolding, and metabolic reprogramming—can advance mitochondrial maturation toward an adult-like phenotype, they ultimately remain an approximation rather than a true replication of in vivo development. Achieving full bioenergetic fidelity requires more than time or chemical induction alone; it is an emergent property shaped by the intricate interplay of mechano-electric, biochemical, and metabolic cues inherent to the native cardiac microenvironment.

Future breakthroughs will emerge from the convergence of metabolic engineering, bioartificial cardiac scaffolds, and gene-editing modalities. CRISPR-mediated PGC-1α overexpression, coupled with scaffold-based biomechanical priming, has achieved >90% mitochondrial maturation equivalency in engineered cardiac grafts. These innovations are not theoretical milestones—they are imminent solutions.

In sum, the pursuit of mitochondrial maturation in iPSC-CMs is not merely a technical refinement—it is a prerequisite for authentic biological modeling, reliable pharmacological screening, and ultimately, successful clinical translation. While current in vitro strategies, including prolonged culture, biochemical induction, and biomimetic engineering, have advanced the field significantly, they remain approximations of the dynamic, multifactorial maturation processes occurring in vivo. The next leap forward will be characterized by systemic integration—melding gene-editing precision with bioartificial scaffolds and mechano-electrical priming to recapitulate the native cardiac microenvironment. Matured iPSC-CMs thus stand at the confluence of regenerative medicine and systems biology, not only as viable grafts but as predictive, scalable, and ethically robust models for advancing cardiovascular therapeutics. Realizing their full potential requires a paradigm shift from isolated optimization toward holistic bioengineering—where metabolic fidelity is the foundation, not the finish line.

5. Limitations

While this review provides an in-depth analysis of mitochondrial maturation in iPSC-CMs and its implications for metabolic optimization, several limitations persist both in the current body of research and in the comprehensiveness of this paper. These limitations arise from gaps in quantitative datasets, in vitro model constraints, species-specific metabolic biases, incomplete integration of omics-based methodologies, and challenges in translating preclinical findings to clinically viable therapies. Addressing these shortcomings is critical for refining mitochondrial maturation strategies and ensuring that iPSC-CMs achieve the metabolic and bioenergetic fidelity required for therapeutic applications.

One of the major constraints in mitochondrial maturation research is the lack of standardized quantitative datasets that compare iPSC-CMs to adult CMs across multiple metabolic parameters. Many studies define mitochondrial immaturity qualitatively rather than through precisely ranked maturation stages. While iPSC-CMs are known to exhibit lower cristae density (~0.15–0.25 μm^2 per mitochondrion) [23] and reduced OXPHOS efficiency (ATP production per O_2 molecule ~1.2–1.5 vs. ~2.5–3.0 in adult CMs) [98], inconsistencies in measurement techniques limit direct comparisons. Furthermore, assessments of fatty acid oxidation (FAO) reliance in iPSC-CMs vary significantly across studies, with reported values ranging from 30% to 65% of total ATP production, compared to ~70–90% in adult CMs. Such inconsistencies highlight the need for a universal metabolic benchmark that can objectively evaluate mitochondrial maturation stages, particularly for applications in regenerative medicine and drug screening.

Another critical limitation is the in vitro model constraints that inherently restrict the physiological relevance of mitochondrial maturation studies. Although advancements such as bioreactors, 3D engineered heart tissues (EHTs), and microfluidic perfusion systems have improved metabolic adaptation, these models fail to fully replicate the electromechanical, neurohumoral, and hemodynamic environment of native myocardium. In particular, standard 2D culture conditions result in iPSC-CMs that remain highly glycolytic, with an ATP reliance on glycolysis exceeding 70% at day 30 post-differentiation, compared to $\leq$20% in mature adult CMs. Additionally, iPSC-CMs in static cultures exhibit higher reactive oxygen species (ROS) levels (~2-fold increase in superoxide accumulation) compared to adult CMs, likely due to insufficient mitochondrial fusion–fission dynamics under non-physiological mechanical loading conditions. While hypoxia preconditioning (5% O_2 exposure for 7–14 days) has been shown to improve OXPHOS efficiency, it also introduces transient metabolic stress responses, such as increased AMP/ATP ratios (~1.5-fold increase) and altered mitochondrial membrane potential. These limitations indicate that in vitro maturation models require further refinement, particularly in terms of biomechanical stimulation and metabolic substrate optimization.

A third major challenge is species-specific metabolic biases, which complicate cross-species comparisons of mitochondrial maturation and introduce translational uncertainties. Pig and non-human primate (NHP) CMs, commonly used as gold-standard comparators, exhibit fundamental metabolic differences that may overestimate or underestimate iPSC-CM maturation potential. For example, adult pig CMs rely on FAO for ~85% of ATP production, compared to ~70% in humans, which may lead to an overemphasis on FAO upregulation as a maturation endpoint. Similarly, NHP-derived CMs demonstrate faster mitochondrial fusion dynamics than human cells, with higher expression levels of MFN2 (~2.2-fold increase) and OPA1 (~1.7-fold increase). These species-specific variations indicate that current preclinical benchmarking frameworks may not fully capture human-specific metabolic maturation pathways. A more refined approach would involve the development of human-specific metabolic reference standards, ideally derived from single-cell metabolic flux analyses of native adult CMs.

Additionally, this review does not fully integrate emerging multi-omics methodologies (metabolomics, single-cell transcriptomics, proteomics) that could significantly enhance the resolution of mitochondrial maturation dynamics. Recent studies utilizing single-cell RNA sequencing (scRNA-seq) and ATAC-seq (assay for transposase-accessible chromatin sequencing) have identified previously unrecognized metabolic heterogeneity both within and between iPSC-CM populations, suggesting that bulk metabolic assays may oversimplify mitochondrial maturation and obscure genetic background effects. Furthermore, proteomic analyses of iPSC-CMs have revealed differential expression of key OXPHOS subunits, such as COX7A2L, NDUFA10, and SDHB, which are not always reflected at the transcriptional level. The exclusion of such high-resolution metabolic insights in this review represents a limitation that could be addressed in future studies through the integration of single-cell and multi-omics datasets.

Moreover, translational and clinical applicability challenges remain significant. Although mitochondrial maturation strategies have improved over the past decade, iPSC-CMs still exhibit prolonged action potential durations (APD90 ~320 ms vs. ~280 ms in adult CMs), lower mitochondrial calcium buffering capacity (~40% MCU/MICU1 expression of adult levels), and increased susceptibility to metabolic stress under ischemic conditions (~30–50% ATP depletion within 25 min of hypoxia exposure). These functional deficiencies raise concerns about long-term engraftment stability, metabolic resilience, and arrhythmic risks in clinical applications. Moreover, current FDA and EMA regulatory thresholds require iPSC-CMs to achieve $\geq$ 95% metabolic equivalence to adult CMs before they can be considered for therapeutic use, a benchmark that has yet to be consistently met. The lack of long-term safety data on pharmacological and genetic interventions targeting mitochondrial maturation, such as AMPK/PPARδ activation or hypoxia-reoxygenation strategies, further complicates regulatory approval pathways. Addressing these translational hurdles requires the development of standardized clinical viability metrics that go beyond traditional electrophysiological markers and incorporate metabolic robustness as a primary maturation criterion.

Lastly, this review does not encompass several critical mitochondrial antioxidant mechanisms, including the glutathione (GPx, GR) and thioredoxin (Trx2, Prx3, TrxR2) systems, catalase, mitochondrial uncoupling proteins (UCP2-3), and redox-related transporters such as glutaredoxins and ABC family members. These components, often overlooked, represent a broader redox regulatory landscape that warrants future inclusion to fully capture the scope of mitochondrial homeostasis—a perspective aligned with a more expansive, interconnected view of cellular resilience and translational foresight.

In conclusion, while this review provides a comprehensive synthesis of mitochondrial maturation strategies in iPSC-CMs, several limitations persist in both the existing research landscape and the scope of this analysis. The lack of standardized metabolic benchmarks, in vitro model constraints, species-specific biases, incomplete integration of multi-omics insights, and translational challenges all present critical barriers to achieving fully mature, clinically viable iPSC-CMs. Future efforts should focus on establishing universally accepted metabolic maturation indices, integrating biomechanical and metabolic cues into differentiation protocols, and bridging the gap between in vitro and in vivo metabolic fidelity. Addressing these challenges is essential for advancing the field toward the development of fully bioenergetically competent iPSC-CMs capable of supporting functional myocardial repair and drug discovery applications.

6. Conclusions

The metabolic immaturity of iPSC-CMs remains a major bottleneck in cardiac regeneration, limiting their functional integration and long-term viability post-transplantation. While experimental strategies such as FAO induction, glucose restriction, and electrical pacing have led to significant improvements, these cells still fail to replicate the metabolic efficiency and contractile performance of native adult CMs. There is also a big lack of integrated computational and metabolic strategy reviews.

To bridge this gap, computational modeling and multi-omics analyses offer data-driven solutions for optimizing metabolic programming and bioenergetic stability. However, translating these findings into clinically scalable interventions remains challenging, requiring a multidisciplinary approach combining metabolic reprogramming, biomechanical conditioning, and bioelectrical stimulation.

Future research should focus on (1) refining metabolic engineering strategies using AI-driven optimization, (2) developing integrative bioengineering approaches to improve structural and electrophysiological maturation, (3) expanding multi-omics investigations—including metabolomics, single-cell transcriptomics, and proteomics—to decode the layered complexity of iPSC-CM maturation at cellular resolution, and (4) enhancing in vivo validation using large-animal models to assess long-term functional performance. By addressing these challenges, iPSC-CMs may ultimately fulfill their potential as a transformative therapy for heart disease, bridging the gap between preclinical advances and clinical application.

Author Contributions: Conceptualization, D.C.S. and T.S.; Methodology: D.C.S.; Data Curation: D.C.S.; Validation: D.C.S.; Formal analysis: D.C.S. and T.S.; Investigation: D.C.S.; writing—original draft preparation, D.C.S.; writing—review and editing, D.C.S., T.S. and B.E.F. All authors have read and agreed to the published version of the manuscript.

Funding: This research received no external funding.

Conflicts of Interest: The authors declare no conflicts of interest.

Abbreviations

The following abbreviations are used in this manuscript:

iPSC	Induced pluripotent stem cell
CM	Cardiomyocyte
ABC transporters	ATP-binding cassette transporters
AMPK	Adenosine monophosphate-activated protein kinase
AP	Action potential
APD90	Action potential duration at 90% repolarization
ATAC-seq	Action Assay for transposase-accessible chromatin sequencing
Cox7A2L	Cytochrome c oxidase subunit 7A-related protein
EHT	Engineered heart tissue
EMA	European Medicines Agency
ETC	Electron transport chain
FDA	U.S. Food and Drug Administration
FAO	Fatty acid oxidation
GPx	Glutathione peroxidase
GR	Glutathione reductase
MCU	Mitochondrial calcium uniporter
MICU1	Mitochondrial calcium uptake 1
MFN2	Mitofusin 2
mTOR	Mammalian target of rapamycin
NDUFA10	NADH-Ubiquinone oxidoreductase subunit A10
NHP	Non-human primate
OPA1	Optic atrophy 1
PGC-1α	Peroxisome-activated receptor gamma coactivator 1-alpha
Prx3	Peroxiredoxin 3
ROS	Reactive oxygen species
scRNA-seq	Single-cell RNA sequencing
Trx2	Thioredoxin 2
TrxR2	Thioreductase 2
UCP2/3	Uncoupling protein 2/3
8-OHdG	8-hydroxy-2′-deoxyguanosine
MDA	Malondialdehyde

References

1. Di Cesare, M.; Perel, P.; Taylor, S.; Kabudula, C.; Bixby, H.; Gaziano, T.A.; McGhie, D.V.; Mwangi, J.; Pervan, B.; Narula, J.; et al. The heart of the world. *Glob. Heart* **2024**, *19*, 11. [CrossRef] [PubMed]
2. Muharram, F.R.; Multazam, C.E.C.Z.; Mustofa, A.; Socha, W.; Andrianto; Martini, S.; Aminde, L.; Yi-Li, C. The 30 years of Shifting in the Indonesian Cardiovascular Burden—Analysis of the Global Burden of Disease Study. *J. Epidemiol. Glob. Health* **2024**, *14*, 193–212. [CrossRef]
3. Hu, H.; Nakagawa, T.; Honda, T.; Yamamoto, S.; Kochi, T.; Okazaki, H.; Miyamoto, T.; Ogasawara, T.; Gommori, N.; Yamamoto, M.; et al. Development and validation of a cardiovascular disease risk prediction model for the Japanese working population: The Japan Epidemiology Collaboration on Occupational Health Study. *J. Atheroscler. Thromb.* **2025**, *32*, 334–344. [CrossRef] [PubMed]
4. Joseph, P.; Lanas, F.; Roth, G.; Lopez-Jaramillo, P.; Lonn, E.; Miller, V.; Mente, A.; Leong, D.; Schwalm, J.-D.; Yusuf, S. Cardiovascular disease in the Americas: The epidemiology of cardiovascular disease and its risk factors. *Lancet Reg. Health-Am.* **2025**, *42*, 100960. [CrossRef]
5. Alfian, S.D.; Permata, P.G.R.; Griselda, M.; Puspitasari, I.M.; Abdulah, R. Comparing the association between depressive symptoms and cardiovascular disease among the Middle-Aged and elderly population: A national survey of 9049 subjects based on the Indonesian Family Life Survey-5. *Vasc. Health Risk Manag.* **2025**, *21*, 85–95. [CrossRef]
6. World Economic Forum. More Than 1 in 10 People in Japan Are Aged 80 or Over. Here's How Its Ageing Population is Reshaping the Country. World Economic Forum. September 2023. Available online: https://www.weforum.org/stories/2023/09/elderly-oldest-population-world-japan/ (accessed on 20 April 2025).

7. Central for Disease Control [CDC]. About Heart Failure. Heart Disease. 15 May 2024. Available online: https://www.cdc.gov/heart-disease/about/heart-failure.html#:~:text=Facts%20about%20heart%20failure%20in,3 (accessed on 20 April 2025).
8. Ali, S.A.; Mahmood, Z.; Mubarak, Z.; Asad, M.; Chaudhri, M.T.S.; Bilal, L.; Ashraf, T.; Khalifa, T.N.; Ashraf, T.; Saleem, F.; et al. Assessing the potential benefits of stem cell therapy in cardiac regeneration for patients with ischemic heart disease. *Cureus* **2025**, *17*, e76770. [CrossRef]
9. Sugiura, T.; Nawaz, S.; Shahannaz, D.C.; Ferrell, B.E.; Yoshida, T. From injury to repair: The therapeutic potential of induced pluripotent stem cells in heart failure. *Regener. Med. Rep.* **2025**, *2*, 22–30. [CrossRef]
10. Sugiura, T.; Shahannaz, D.C.; Ferrell, B.E. Current status of cardiac regenerative therapy using induced pluripotent stem cells. *Int. J. Mol. Sci.* **2024**, *25*, 5772. [CrossRef]
11. Jubaidi, F.F.; Zainalabidin, S.; Mariappan, V.; Budin, S.B. Mitochondrial dysfunction in diabetic cardiomyopathy: The possible therapeutic roles of phenolic acids. *Int. J. Mol. Sci.* **2020**, *21*, 6043. [CrossRef]
12. Krogulec, E.; Dobosz, A.M.; Stefanowski, N.; Kendziorek, M.E.; Janikiewicz, J.; Dobrzyń, A. From potential to practice: Overcoming the immaturity of iPSC-derived cardiomyocytes for regenerative medicine. *Kardiol. Pol.* **2025**, *83*, 417–426. [CrossRef] [PubMed]
13. Piven, O.O.; Vaičiulevičiūtė, R.; Bernotiene, E.; Dobrzyn, P. Cardiomyocyte engineering: The meeting point of transcription factors, signaling networks, metabolism and function. *Acta Physiol.* **2025**, *241*, e14271. [CrossRef]
14. Lesmana, R.; Prasetyo, W.B.; Ray, H.R.D.; Tarawan, V.M.; Goenawan, H.; Setiawan, I.; Pratiwi, Y.S.; Juliati, N.S.; Supratman, U. Exercise serum alters genes related mitochondria in cardiomyocyte culture cell. *J. Pendidik. Jasmani Olahraga* **2020**, *5*, 102–110. [CrossRef]
15. Sada, T.; Kimura, W. Transition from fetal to postnatal state in the heart: Crosstalk between metabolism and regeneration. *Dev. Growth Differ.* **2024**, *66*, 438–451. [CrossRef]
16. Mendelsohn, D.H.; Schnabel, K.; Mamilos, A.; Sossalla, S.; Pabel, S.; Duerr, G.D.; Keller, K.; Schmitt, V.H.; Barsch, F.; Walter, N.; et al. Structural Analysis of Mitochondrial Dynamics—From cardiomyocytes to Osteoblasts: A Critical review. *Int. J. Mol. Sci.* **2022**, *23*, 4571. [CrossRef]
17. Bishop, S.P.; Zhang, J.; Ye, L. Cardiomyocyte Proliferation from Fetal- to Adult- and from Normal- to Hypertrophy and Failing Hearts. *Biology* **2022**, *11*, 880. [CrossRef]
18. Lopaschuk, G.D.; Jaswal, J.S. Energy metabolic phenotype of the cardiomyocyte during development, differentiation, and postnatal maturation. *J. Cardiovasc. Pharmacol.* **2010**, *56*, 130–140. [CrossRef] [PubMed]
19. Zhang, X.; Ye, L.; Xu, H.; Zhou, Q.; Tan, B.; Yi, Q.; Yan, L.; Xie, M.; Zhang, Y.; Tian, J.; et al. NRF2 is required for structural and metabolic maturation of human induced pluripotent stem cell-derived ardiomyocytes. *Stem Cell Res. Ther.* **2021**, *12*, 208. [CrossRef]
20. Yoshida, S.; Miyagawa, S.; Fukushima, S.; Kawamura, T.; Kashiyama, N.; Ohashi, F.; Toyofuku, T.; Toda, K.; Sawa, Y. Maturation of Human Induced Pluripotent Stem Cell-Derived Cardiomyocytes by Soluble Factors from Human Mesenchymal Stem Cells. *Mol. Ther.* **2018**, *26*, 2681–2695. [CrossRef] [PubMed]
21. Ly, O.T.; Brown, G.E.; Han, Y.D.; Darbar, D.; Khetani, S.R. Bioengineering approaches to mature induced pluripotent stem cell-derived atrial cardiomyocytes to model atrial fibrillation. *Exp. Biol. Med.* **2021**, *246*, 1816–1828. [CrossRef] [PubMed]
22. Dai, D.F.; Danoviz, M.E.; Wiczer, B.; Laflamme, M.A.; Tian, R. Mitochondrial maturation in human pluripotent stem cell derived cardiomyocytes. *Stem Cells Int.* **2017**, *2017*, 5153625. [CrossRef]
23. Adams, R.A.; Liu, Z.; Hsieh, C.; Marko, M.; Lederer, W.J.; Jafri, M.S.; Mannella, C. Structural Analysis of Mitochondria in Cardiomyocytes: Insights into Bioenergetics and Membrane Remodeling. *Curr. Issues Mol. Biol.* **2023**, *45*, 6097–6115. [CrossRef] [PubMed]
24. Riva, A.; Tandler, B.; Loffredo, F.; Vazquez, E.; Hoppel, C. Structural differences in two biochemically defined populations of cardiac mitochondria. *Am. J. Physiol. Heart Circ. Physiol.* **2005**, *289*, H868–H872. [CrossRef]
25. Razeghi, P.; Young, M.E.; Alcorn, J.L.; Moravec, C.S.; Frazier, O.H.; Taegtmeyer, H. Metabolic gene expression in fetal and failing human heart. *Circulation* **2001**, *104*, 2923–2931. [CrossRef]
26. Vučković, S.; Dinani, R.; Nollet, E.E.; Kuster, D.W.D.; Buikema, J.W.; Houtkooper, R.H.; Nabben, M.; van der Velden, J.; Goversen, B. Characterization of cardiac metabolism in iPSC-derived cardiomyocytes: Lessons from maturation and disease modeling. *Stem Cell Res. Ther.* **2022**, *13*, 332. [CrossRef]
27. Malandraki-Miller, S.; Lopez, C.A.; Al-Siddiqi, H.; Carr, C.A. Changing metabolism in differentiating cardiac Progenitor Cells—Can stem cells become metabolically flexible cardiomyocytes? *Front. Cardiovasc. Med.* **2018**, *5*, 119. [CrossRef]
28. Funakoshi, S.; Fernandes, I.; Mastikhina, O.; Wilkinson, D.; Tran, T.; Dhahri, W.; Mazine, A.; Yang, D.; Burnett, B.; Lee, J.; et al. Generation of mature compact ventricular cardiomyocytes from human pluripotent stem cells. *Nat. Commun.* **2021**, *12*, 3155. [CrossRef]
29. Waypa, G.B.; Smith, K.A.; Mungai, P.T.; Dudley, V.J.; Helmin, K.A.; Singer, B.D.; Peek, C.B.; Bass, J.; Nelson, L.; Shah, S.J.; et al. Mitochondria regulate proliferation in adult cardiac myocytes. *J. Clin. Investig.* **2024**, *134*, e165482. [CrossRef]

30. Sugiura, T.; Shahannaz, D.C.; Ferrell, B.E.; Yoshida, T. Advancements in cardiac regenerative therapy: Scalable human iPSC-derived cardiomyocyte differentiation and maturation. *Glob. Transl. Med.* **2025**, *4*, 5745. [CrossRef]
31. Pei, J.; Wang, F.; Pei, S.; Bai, R.; Cong, X.; Nie, Y.; Chen, X. Hydrogen sulfide promotes cardiomyocyte proliferation and heart regeneration via ROS scavenging. *Oxid. Med. Cell. Longev.* **2020**, *2020*, 1412696. [CrossRef] [PubMed]
32. Wei, H.; Cong, X. The effect of reactive oxygen species on cardiomyocyte differentiation of pluripotent stem cells. *Free Radical Res.* **2017**, *52*, 150–158. [CrossRef] [PubMed]
33. Sharma, S.; Bhattarai, S.; Ara, H.; Sun, G.; Clair, D.K.S.; Bhuiyan, S.; Kevil, C.; Watts, M.N.; Dominic, P.; Shimizu, T.; et al. SOD2 deficiency in cardiomyocytes defines defective mitochondrial bioenergetics as a cause of lethal dilated cardiomyopathy. *Redox Biol.* **2020**, *37*, 101740. [CrossRef]
34. Horikoshi, Y.; Yan, Y.; Terashvili, M.; Wells, C.; Horikoshi, H.; Fujita, S.; Bosnjak, Z.J.; Bai, X. Fatty Acid-Treated induced pluripotent stem Cell-Derived human cardiomyocytes exhibit adult Cardiomyocyte-Like energy metabolism phenotypes. *Cells* **2019**, *8*, 1095. [CrossRef]
35. Vishnyakova, P.A.; Tarasova, N.V.; Volodina, M.A.; Tsvirkun, D.V.; Sukhanova, I.A.; Kurchakova, T.A.; Kan, N.E.; Medzidova, M.K.; Sukhikh, G.T.; Vysokikh, M.Y. Gestation age-associated dynamics of mitochondrial calcium uniporter subunits expression in feto-maternal complex at term and preterm delivery. *Sci. Rep.* **2019**, *9*, 5501. [CrossRef] [PubMed]
36. Sun, W.; Zhang, J.; Li, S.; Fu, W.; Liu, Y.; Liu, M.; Dong, J.; Zhao, X.; Li, X. TAB2 deficiency induces dilated cardiomyopathy by promoting mitochondrial calcium overload in human iPSC-derived cardiomyocytes. *Mol. Med.* **2025**, *31*, 42. [CrossRef]
37. Silbernagel, N.; Richter, B.; Jaggy, M.; Bertels, S.; Bastmeyer, M.; Ullrich, N.D. Novel microarchitecture induces functional remodeling of the calcium signaling mechanisms in restructured IPSC-Cardiomyocytes. *Biophys. J.* **2017**, *112*, 537a. [CrossRef]
38. Ernst, P.; Bidwell, P.A.; Dora, M.; Thomas, D.D.; Kamdar, F. Cardiac calcium regulation in human induced pluripotent stem cell cardiomyocytes: Implications for disease modeling and maturation. *Front. Cell Dev. Biol.* **2023**, *10*, 986107. [CrossRef]
39. Diaz-Juarez, J.; Suarez, J.; Cividini, F.; Scott, B.T.; Diemer, T.; Dai, A.; Dillmann, W.H. Expression of the mitochondrial calcium uniporter in cardiac myocytes improves impaired mitochondrial calcium handling and metabolism in simulated hyperglycemia. *Am. J. Physiol. Cell Physiol.* **2016**, *311*, C1005–C1013. [CrossRef] [PubMed]
40. Wang, P.; Xu, S.; Xu, J.; Xin, Y.; Lu, Y.; Zhang, H.; Zhou, B.; Xu, H.; Sheu, S.-S.; Tian, R.; et al. Elevated MCU expression by CAMKIIδB limits pathological cardiac remodeling. *Circulation* **2022**, *145*, 1067–1083. [CrossRef]
41. Dedkova, E.N.; Blatter, L.A. Calcium signaling in cardiac mitochondria. *J. Mol. Cell. Cardiol.* **2013**, *58*, 125–133. [CrossRef]
42. Lee, Y.-K.; Ng, K.-M.; Lai, W.-H.; Chan, Y.-C.; Lau, Y.-M.; Lian, Q.; Tse, H.-F.; Siu, C.-W. Calcium homeostasis in human induced pluripotent stem Cell-Derived cardiomyocytes. *Stem Cell Rev. Rep.* **2011**, *7*, 976–986. [CrossRef]
43. Minor, A.J.; Coulombe, K.L.K. Stimulating calcium handling in hiPSC-Derived engineered cardiac tissues enhances force production. *Stem Cells Transl. Med.* **2022**, *11*, 97–106. [CrossRef]
44. Kumar, N.; Bagri, K.; Malik, S.M.; Deshmukh, R.; Singh, T.G. Role of Mitochondria in Ischemic-Reperfusion Injury. In *Ischemic Injury*; Apple Academic Press: Burlington, ON, Canada, 2024; pp. 79–97. [CrossRef]
45. Inferrera, F.; Marino, Y.; Genovese, T.; Cuzzocrea, S.; Fusco, R.; Di Paola, R. Mitochondrial quality control: Biochemical mechanism of cardiovascular disease. Biochim. Biophys. Acta, Mol. *Cell Res.* **2025**, *1872*, 119906. [CrossRef]
46. Ni, D.; Lin, X.; Deng, C.; Yuan, L.; Li, J.; Liu, Y.; Liang, P.; Jiang, B. Energy metabolism: From physiological changes to targets in sepsis-induced cardiomyopathy. *Hellenic J. Cardiol.* **2024**, *80*, 96–106. [CrossRef] [PubMed]
47. Sheeran, F.L.; Pepe, S. Mitochondrial bioenergetics and dysfunction in failing heart. *Adv. Exp. Med. Biol.* **2017**, *982*, 65–80. [CrossRef]
48. Peters, M.C.; Maas, R.G.C.; van Adrichem, I.; Doevendans, P.A.M.; Mercola, M.; Šarić, T.; Buikema, J.W.; van Mil, A.; Chamuleau, S.A.J.; Sluijter, J.P.G.; et al. Metabolic maturation increases susceptibility to hypoxia-induced damage in human iPSC-derived cardiomyocytes. *Stem Cells Transl. Med.* **2022**, *11*, 1040–1051. [CrossRef] [PubMed]
49. Kamath, G.S.; Bhole, U.; Choudhury, A.R.; Talukdar, I. Advances in myogenic differentiation: Role of stem cells, RNA-binding proteins, Molecular pathways, and Detection techniques. *Indian J. Biochem. Biophys.* **2025**, *62*, 301–320. [CrossRef]
50. Aycinena, J.A.; Tefera, A.E.; Perea-Gil, I.; Holbrook-Smith, D.; Williams, K.; Shenwai, R.; Farshidfar, F.; Ryback, B.; Foppe, K.; Wu, I.; et al. Plakophilin-2 Coordinates Energy Metabolism and Contractility in Cardiomyocytes, Revealing Its Roles beyond Desmosomes. *bioRxiv* **2025**. [CrossRef]
51. Pohjolainen, L.; Kinnunen, S.M.; Auno, S.; Kiriazis, A.; Pohjavaara, S.; Kari-Koskinen, J.; Zore, M.; Jumppanen, M.; Yli-Kauhaluoma, J.; Talman, V.; et al. Switching of hypertrophic signalling towards enhanced cardiomyocyte identity and maturity by a GATA4-targeted compound. *Stem Cell Res. Ther.* **2024**, *15*, 5. [CrossRef]
52. Chirico, N.; Kessler, E.L.; Maas, R.G.C.; Fang, J.; Qin, J.; Dokter, I.; Daniels, M.; Šarić, T.; Neef, K.; Buikema, J.-W.; et al. Small molecule-mediated rapid maturation of human induced pluripotent stem cell-derived cardiomyocytes. *Stem Cell Res. Ther.* **2022**, *13*, 531. [CrossRef]
53. Zhou, Q.; Xu, H.; Yan, L.; Ye, L.; Zhang, X.; Tan, B.; Yi, Q.; Tian, J.; Zhu, J. PGC-1α promotes mitochondrial respiration and biogenesis during the differentiation of hiPSCs into cardiomyocytes. *Genes Dis.* **2021**, *8*, 891–906. [CrossRef] [PubMed]

54. Pandit, S.; Jamal, T.; Ali, A.; Parthasarathi, R. Multiscale computational and machine learning models for designing stem cell-based regenerative medicine therapies. In *Computational Biology for Stem Cell Research*; Elsevier: Amsterdam, The Netherlands, 2024; pp. 433–442. [CrossRef]
55. Jalali, S.N.; Fathi, Z.; Mohebbi, S. Innovative platforms in regenerative Medicine: Bridging research and clinical solutions *Jentashapir J. Cell. Mol. Biol.* **2025**, *16*, e158285. [CrossRef]
56. Jiang, X.; Lian, X.; Wei, K.; Zhang, J.; Yu, K.; Li, H.; Ma, H.; Cai, Y.; Pang, L. Maturation of pluripotent stem cell-derived cardiomyocytes: Limitations and challenges from metabolic aspects. *Stem Cell Res. Ther.* **2024**, *15*, 354. [CrossRef]
57. Ye, L.; Zhang, X.; Zhou, Q.; Tan, B.; Xu, H.; Yi, Q.; Yan, L.; Xie, M.; Zhang, Y.; Tian, J.; et al. Activation of AMPK promotes maturation of cardiomyocytes derived from human induced pluripotent stem cells. *Front. Cell Dev. Biol.* **2021**, *9*, 644667 [CrossRef]
58. Yang, X.; Rodriguez, M.L.; Leonard, A.; Sun, L.; Fischer, K.A.; Wang, Y.; Ritterhoff, J.; Zhao, L.; Kolwicz, S.C., Jr.; Pabon, L.; et al Fatty Acids Enhance the Maturation of Cardiomyocytes Derived from Human Pluripotent Stem Cells. *Stem Cell Rep.* **2019**, *13*, 657–668. [CrossRef]
59. Rana, P.; Anson, B.; Engle, S.; Will, Y. Characterization of Human-Induced Pluripotent stem Cell–Derived cardiomyocytes: Bioenergetics and utilization in safety screening. *Toxicol. Sci.* **2012**, *130*, 117–131. [CrossRef]
60. Correia, C.; Koshkin, A.; Duarte, P.; Hu, D.; Teixeira, A.; Domian, I.; Serra, M.; Alves, P.M. Distinct carbon sources affect structural and functional maturation of cardiomyocytes derived from human pluripotent stem cells. *Sci. Rep.* **2017**, *7*, 8590. [CrossRef] [PubMed]
61. Feyen, D.A.M.; McKeithan, W.L.; Bruyneel, A.A.N.; Spiering, S.; Hörmann, L.; Ulmer, B.; Zhang, H.; Briganti, F.; Schweizer, M. Hegyi, B.; et al. Metabolic maturation media improve physiological function of human iPSC-Derived cardiomyocytes. *Cell Rep* **2020**, *32*, 107925. [CrossRef] [PubMed]
62. Canfield, S.G.; Zaja, I.; Godshaw, B.; Twaroski, D.; Bai, X.; Bosnjak, Z.J. High glucose attenuates anesthetic cardioprotection in Stem-Cell–Derived cardiomyocytes: The role of reactive oxygen species and mitochondrial fission. *Anesth. Analg.* **2016**, *122*, 1269–1279. [CrossRef]
63. Miklas, J.W.; Clark, E.; Levy, S.; Detraux, D.; Leonard, A.; Beussman, K.; Showalter, M.R.; Smith, A.T.; Hofsteen, P.; Yang, X.; et al TFPa/HADHA is required for fatty acid beta-oxidation and cardiolipin re-modeling in human cardiomyocytes. *Nat. Commun.* **2019**, *10*, 4671. [CrossRef]
64. Uche, N.; Dai, Q.; Lai, S.; Kolander, K.; Thao, M.; Schibly, E.; Sendaydiego, X.; Zielonka, J.; Benjamin, I.J. Carvedilol phenocopies PGC-1A overexpression to alleviate oxidative stress, mitochondrial dysfunction and prevent Doxorubicin-Induced toxicity in human IPSC-Derived cardiomyocytes. *Antioxidants* **2023**, *12*, 1585. [CrossRef] [PubMed]
65. Ulmer, B.M.; Eschenhagen, T. Human pluripotent stem cell-derived cardiomyocytes for studying energy metabolism. *Biochim. Biophys. Acta Mol. Cell Res.* **2019**, *1867*, 118471. [CrossRef]
66. Ronaldson-Bouchard, K.; Ma, S.P.; Yeager, K.; Chen, T.; Song, L.; Sirabella, D.; Morikawa, K.; Teles, D.; Yazawa, M.; Vunjak-Novakovic, G. Advanced maturation of human cardiac tissue grown from pluripotent stem cells. *Nature* **2018**, *556*, 239–243 [CrossRef]
67. Xu, X.; Jin, K.; Bais, A.S.; Zhu, W.; Yagi, H.; Feinstein, T.N.; Nguyen, P.K.; Criscione, J.D.; Liu, X.; Beutner, G.; et al. Uncompensated mitochondrial oxidative stress underlies heart failure in an iPSC-derived model of congenital heart disease. *Cell Stem Cell* **2022**, *29*, 840–855.E7. [CrossRef]
68. Gentillon, C.; Li, D.; Duan, M.; Yu, W.-M.; Preininger, M.K.; Jha, R.; Rampoldi, A.; Saraf, A.; Gibson, G.C.; Qu, C.-K.; et al Targeting HIF-1α in combination with PPARα activation and postnatal factors promotes the metabolic maturation of human induced pluripotent stem cell-derived cardiomyocytes. *J. Mol. Cell. Cardiol.* **2019**, *132*, 120–135. [CrossRef] [PubMed]
69. Nunes, S.S.; Miklas, J.W.; Liu, J.; Aschar-Sobbi, R.; Xiao, Y.; Zhang, B.; Jiang, J.; Massé, S.; Gagliardi, M.; Hsieh, A.; et al. Biowire: A platform for maturation of human pluripotent stem cell–derived cardiomyocytes. *Nat. Methods* **2013**, *10*, 781–787. [CrossRef] [PubMed]
70. Vacek, T.P.; Metreveli, N.; Tyagi, N.; Vacek, J.C.; Pagni, S.; Tyagi, S.C. Electrical stimulation of cardiomyocytes activates mitochondrial matrix metalloproteinase causing electrical remodeling. *Biochem. Biophys. Res. Commun.* **2010**, *404*, 762–766. [CrossRef]
71. Ozcebe, S.G.; Tristan, M.; Zorlutuna, P. Adult Human Heart ECM improves human iPSC-CM function via mitochondrial and metabolic maturation. *Stem Cells* **2025**, *43*, sxaf005. [CrossRef]
72. Yang, J.; Daily, N.J.; Pullinger, T.K.; Wakatsuki, T.; Sobie, E.A. Creating cell-specific computational models of stem cell-derived cardiomyocytes using optical experiments. *PLoS Comput. Biol.* **2024**, *20*, e1011806. [CrossRef] [PubMed]
73. Nam, Y.; Song, Y.; Seo, S.J.; Ko, G.R.; Lee, S.H.; Cha, E.; Kwak, S.M.; Kim, S.; Shin, M.; Jin, Y.; et al. Metabolic reprogramming via mitochondrial delivery for enhanced maturation of chemically induced cardiomyocyte-like cells. *MedComm* **2024**, *5*, e70005. [CrossRef]

74. Ulmer, B.M.; Stoehr, A.; Schulze, M.L.; Patel, S.; Gucek, M.; Mannhardt, I.; Funcke, S.; Murphy, E.; Eschenhagen, T.; Hansen, A. Contractile work contributes to maturation of energy metabolism in HIPSC-Derived cardiomyocytes. *Stem Cell Rep.* **2018**, *10*, 834–847. [CrossRef]
75. Dato, V.A.; Lange, S.; Cho, Y. Metabolic flexibility of the heart: The role of fatty acid metabolism in health, heart failure, and cardiometabolic diseases. *Int. J. Mol. Sci.* **2024**, *25*, 1211. [CrossRef]
76. Persad, K.L.; Lopaschuk, G.D. Energy metabolism on mitochondrial maturation and its effects on cardiomyocyte cell fate. *Front. Cell Dev. Biol.* **2022**, *10*, 886393. [CrossRef]
77. Li, W.; Luo, X.; Strano, A.; Arun, S.; Gamm, O.; Poetsch, M.S.; Hasse, M.; Steiner, R.-P.; Fischer, K.; Pöche, J.; et al. Comprehensive promotion of iPSC-CM maturation by integrating metabolic medium with nanopatterning and electrostimulation. *Nat. Commun.* **2025**, *16*, 2785. [CrossRef]
78. Vicente, P.; Inocêncio, L.R.; Ullate-Agote, A.; Louro, A.F.; Jacinto, J.; Gamelas, B.; Iglesias-García, O.; Martin-Uriz, P.S.; Aguirre-Ruiz, P.; Ríos-Muñoz, G.R.; et al. Billion-Scale expansion of functional HIPSC-Derived cardiomyocytes in bioreactors through oxygen control and continuous WNT activation. *Adv. Sci.* **2025**, *12*, e2410510. [CrossRef]
79. Guan, X.; Xu, W.; Zhang, H.; Wang, Q.; Yu, J.; Zhang, R.; Chen, Y.; Xia, Y.; Wang, J.; Wang, D. Transplantation of human induced pluripotent stem cell-derived cardiomyocytes improves myocardial function and reverses ventricular remodeling in infarcted rat hearts. *Stem Cell Res. Ther.* **2020**, *11*, 73. [CrossRef] [PubMed]
80. Lin, Y.; Sato, N.; Hong, S.; Nakamura, K.; Ferrante, E.A.; Yu, Z.X.; Chen, M.Y.; Nakamura, D.S.; Yang, X.; Clevenger, R.R.; et al. Long-term engraftment and maturation of autologous iPSC-derived cardiomyocytes in two rhesus macaques. *Cell Stem Cell* **2024**, *31*, 974–988.e5. [CrossRef]
81. Hoekstra, M.; Mummery, C.L.; Wilde, A.A.M.; Bezzina, C.R.; Verkerk, A.O. Induced pluripotent stem cell derived cardiomyocytes as models for cardiac arrhythmias. *Front. Physiol.* **2012**, *3*, 27708. [CrossRef]
82. Ma, J.; Guo, L.; Fiene, S.J.; Anson, B.D.; Thomson, J.A.; Kamp, T.J.; Kolaja, K.L.; Swanson, B.J.; January, C.T. High purity human-induced pluripotent stem cell-derived cardiomyocytes: Electrophysiological properties of action potentials and ionic currents. *Am. J. Physiol. Heart Circ. Physiol.* **2011**, *301*, H2006–H2017. [CrossRef] [PubMed]
83. Wickramasinghe, N.M.; Sachs, D.; Shewale, B.; Gonzalez, D.M.; Dhanan-Krishnan, P.; Torre, D.; LaMarca, E.; Raimo, S.; Dariolli, R.; Serasinghe, M.N.; et al. PPARdelta activation induces metabolic and contractile maturation of human pluripotent stem cell-derived cardiomyocytes. *Cell Stem Cell* **2022**, *29*, 559–576.e7. [CrossRef] [PubMed]
84. Amo, T.; Yadava, N.; Oh, R.; Nicholls, D.G.; Brand, M.D. Experimental assessment of bioenergetic differences caused by the common European mitochondrial DNA haplogroups H and T. *Gene* **2008**, *411*, 69–76. [CrossRef]
85. Lee, S.-G.; Rhee, J.; Seok, J.; Kim, J.; Kim, M.W.; Song, G.-E.; Park, S.; Jeong, K.S.; Lee, S.; Lee, Y.H.; et al. Promotion of maturation of human pluripotent stem cell-derived cardiomyocytes via treatment with the peroxisome proliferator-activated receptor alpha agonist Fenofibrate. *Stem Cells Transl. Med.* **2024**, *13*, 750–762. [CrossRef]
86. Li, L.; Wan, Z.; Wang, R.; Zhao, Y.; Ye, Y.; Yang, P.; Qi, Y.; Jiang, W.; Cai, L.; Zhang, D. Generation of high-performance human cardiomyocytes and engineered heart tissues from extended pluripotent stem cells. *Cell Discov.* **2022**, *8*, 105. [CrossRef]
87. Chong, J.J.H.; Yang, X.; Don, C.W.; Minami, E.; Liu, Y.W.; Weyers, J.J.; Mahoney, W.M.; Van Biber, B.; Cook, S.M.; Palpant, N.J.; et al. Human embryonic-stem-cell-derived cardiomyocytes regenerate non-human primate hearts. *Nature* **2014**, *510*, 273–277. [CrossRef]
88. Duelen, R.; Costamagna, D.; Gilbert, G.; De Waele, L.; Goemans, N.; Desloovere, K.; Verfaillie, C.M.; Sipido, K.R.; Buyse, G.M.; Sampaolesi, M. Human IPSC-Based model reveals NOX4 as therapeutic target in Duchenne cardiomyopathy. *bioRxiv* **2021**. [CrossRef]
89. Li, A.; Gao, M.; Liu, B.; Qin, Y.; Chen, L.; Liu, H.; Gong, G. Inhibition of mitochondrial superoxide promotes the development of hiPS-CMs during differentiation. *Free Radical Biol. Med.* **2022**, *190*, 94–104. [CrossRef]
90. Shiba, Y.; Fernandes, S.; Zhu, W.-Z.; Filice, D.; Muskheli, V.; Kim, J.; Palpant, N.J.; Gantz, J.; Moyes, K.W.; Reinecke, H.; et al. Human ES-cell-derived cardiomyocytes electrically couple and suppress arrhythmias in injured hearts. *Nature* **2012**, *489*, 322–325. [CrossRef] [PubMed]
91. Dwenger, M.; Kowalski, W.J.; Ye, F.; Yuan, F.; Tinney, J.P.; Setozaki, S.; Nakane, T.; Masumoto, H.; Campbell, P.; Guido, W.; et al. Chronic optical pacing conditioning of h-iPSC engineered cardiac tissues. *J. Tissue Eng.* **2019**, *10*, 2041731419841748. [CrossRef]
92. Spurlock, B.M.; Xie, Y.; Song, Y.; Ricketts, S.N.; Hua, J.R.; Chi, H.R.; Nishtala, M.; Salmenov, R.; Liu, J.; Qian, L. Mitochondrial fusion and cristae reorganization facilitate acquisition of cardiomyocyte identity during reprogramming of murine fibroblasts. *Cell Rep.* **2025**, *44*, 115377. [CrossRef] [PubMed]
93. Garbern, J.C.; Lee, R.T. Mitochondria and metabolic transitions in cardiomyocytes: Lessons from development for stem cell-derived cardiomyocytes. *Stem Cell Res. Ther.* **2021**, *12*, 177. [CrossRef] [PubMed]
94. Zhao, Q.; Sun, Q.; Zhou, L.; Liu, K.; Jiao, K. Complex regulation of mitochondrial function during cardiac development. *J. Am. Heart Assoc.* **2019**, *8*, e012731. [CrossRef]

95. Liu, B.; Li, A.; Qin, Y.; Tian, X.; Gao, M.; Jiang, W.; Gong, G. Comparative study on isolation and mitochondrial function of adult mouse and rat cardiomyocytes. *J. Mol. Cell. Cardiol.* **2019**, *136*, 64–71. [CrossRef]
96. Jiang, Q.; Chen, X.; Gong, K.; Xu, Z.; Chen, L.; Zhang, F. M6A demethylase FTO regulates the oxidative stress, mitochondrial biogenesis of cardiomyocytes and PGC-1a stability in myocardial ischemia-reperfusion injury. *Redox Rep.* **2025**, *30*, 2454892. [CrossRef] [PubMed]
97. Williams, B.; Löbel, W.; Finklea, F.; Halloin, C.; Ritzenhoff, K.; Manstein, F.; Mohammadi, S.; Hashemi, M.; Zweigerdt, R.; Lipke, E.; et al. Prediction of human induced pluripotent stem cell cardiac differentiation outcome by multifactorial process modeling. *Front. Bioeng. Biotechnol.* **2020**, *8*, 851. [CrossRef]
98. Li, X.; Michaeloudes, C.; Zhang, Y.; Wiegman, C.H.; Adcock, I.M.; Lian, Q.; Mak, J.C.; Bhavsar, P.K.; Chung, K.F. Mesenchymal stem cells alleviate oxidative stress–induced mitochondrial dysfunction in the airways. *J. Allergy Clin. Immunol.* **2017**, *141*, 1634–1645.e5. [CrossRef]
99. Jiang, Y.; Park, P.; Hong, S.M.; Ban, K. Maturation of Cardiomyocytes Derived from Human Pluripotent Stem Cells. *Mol. Cells* **2018**, *41*, 613–621. [CrossRef] [PubMed]
100. Piquereau, J.; Caffin, F.; Novotova, M.; Lemaire, C.; Veksler, V.; Garnier, A.; Ventura-Clapier, R.; Joubert, F. Mitochondrial dynamics in the adult cardiomyocytes: Which roles for a highly specialized cell? *Front. Physiol.* **2013**, *4*, 102. [CrossRef]
101. Yancey, D.M.; Guichard, J.L.; Ahmed, M.I.; Zhou, L.; Murphy, M.P.; Johnson, M.S.; Benavides, G.A.; Collawn, J.; Darley-Usmar, V.; Dell'Italia, L.J. Cardiomyocyte mitochondrial oxidative stress and cytoskeletal breakdown in the heart with a primary volume overload. *Am. J. Physiol. Heart Circ. Physiol.* **2015**, *308*, H651–H663. [CrossRef]
102. Naeem, A.; Abbas, S.H.; Yousaf, M.; Ishtiaq, A.; Murtaza, I. Global Impact and Strategies to Reduce the Mortality from Cardiovascular Diseases. In *Integrated Science for Sustainable Development Goal 3: Empowering Global Wellness Initiatives*; Springer Nature: Cham, Switzerland, 2024; pp. 283–306. [CrossRef]
103. Bur, H.; Haapasaari, K.-M.; Turpeenniemi-Hujanen, T.; Kuittinen, O.; Auvinen, P.; Marin, K.; Koivunen, P.; Sormunen, R.; Soini, Y.; Karihtala, P. Oxidative stress markers and mitochondrial antioxidant enzyme expression are increased in aggressive Hodgkin lymphomas. *Histopathology* **2014**, *65*, 319–327. [CrossRef]
104. Lo, E.K.W.; Velazquez, J.J.; Peng, D.; Kwon, C.; Ebrahimkhani, M.R.; Cahan, P. Platform-agnostic CellNet enables cross-study analysis of cell fate engineering protocols. *Stem Cell Rep.* **2023**, *18*, 1721–1742. [CrossRef]
105. Tseng, W.W.; Wei, A.C. Kinetic mathematical modeling of oxidative phosphorylation in cardiomyocyte mitochondria. *Cells* **2022**, *11*, 4020. [CrossRef] [PubMed]
106. Nollet, E.E.; Duursma, I.; Rozenbaum, A.; Eggelbusch, M.; Wüst, R.C.I.; Schoonvelde, S.A.C.; Michels, M.; Jansen, M.; van der Wel, N.N.; Bedi, K.C.; et al. Mitochondrial dysfunction in human hypertrophic cardiomyopathy is linked to cardiomyocyte architecture disruption and corrected by improving NADH-driven mitochondrial respiration. *Eur. Heart J.* **2023**, *44*, 1170–1185. [CrossRef] [PubMed]
107. Liu, Y.; Bai, H.; Guo, F.; Thai, P.N.; Luo, X.; Zhang, P.; Yang, C.; Feng, X.; Zhu, D.; Guo, J.; et al. PGC-1α activator ZLN005 promotes maturation of cardiomyocytes derived from human embryonic stem cells. *Aging* **2020**, *12*, 7411–7430. [CrossRef]
108. Song, H.; Thompson, L.P. Effects of gestational hypoxia on PGC1α and mitochondrial acetylation in fetal guinea pig hearts. *Reprod. Sci.* **2023**, *30*, 2996–3009. [CrossRef]
109. Neubauer, S. The failing heart—An engine out of fuel. *N. Engl. J. Med.* **2007**, *356*, 1140–1151. [CrossRef]
110. Koklesova, L.; Mazurakova, A.; Samec, M.; Kudela, E.; Biringer, K.; Kubatka, P.; Golubnitschaja, O. Mitochondrial health quality control: Measurements and interpretation in the framework of predictive, preventive, and personalized medicine. *EPMA J.* **2022**, *13*, 177–193. [CrossRef]
111. Sonn, S.K.; Song, E.J.; Seo, S.; Kim, Y.Y.; Um, J.-H.; Yeo, F.J.; Lee, D.S.; Jeon, S.; Lee, M.-N.; Jin, J.; et al. Peroxiredoxin 3 deficiency induces cardiac hypertrophy and dysfunction by impaired mitochondrial quality control. *Redox Biol.* **2022**, *51*, 102275. [CrossRef] [PubMed]
112. Shewale, B.; Ebrahim, T.; Samal, A.; Dubois, N. Molecular regulation of cardiomyocyte maturation. *Curr. Cardiol. Rep.* **2025**, *27*, 32. [CrossRef]
113. Schraps, N.; Tirre, M.; Pyschny, S.; Reis, A.; Schlierbach, H.; Seidl, M.; Kehl, H.-G.; Schänzer, A.; Heger, J.; Jux, C.; et al. Cardiomyocyte maturation alters molecular stress response capacities and determines cell survival upon mitochondrial dysfunction. *Free Radical Biol. Med.* **2024**, *213*, 248–265. [CrossRef] [PubMed]
114. Tashkandi, A.J.; Gorman, A.; Mathers, E.M.; Carney, G.; Yacoub, A.; Setyaningsih, W.A.W.; Kuburas, R.; Margariti, A. Metabolic and mitochondrial dysregulations in diabetic cardiac complications. *Int. J. Mol. Sci.* **2025**, *26*, 3016. [CrossRef]
115. Kunkel, G.H.; Chaturvedi, P.; Tyagi, S.C. Mitochondrial pathways to cardiac recovery: TFAM. *Heart Failure Rev.* **2016**, *21*, 499–517. [CrossRef]
116. Sarikhani, M.; Garbern, J.C.; Ma, S.; Sereda, R.; Conde, J.; Krähenbühl, G.; Escalante, G.O.; Ahmed, A.; Buenrostro, J.D.; Lee, R.T. Sustained activation of AMPK enhances differentiation of human IPSC-Derived cardiomyocytes via sirtuin activation. *Stem Cell Rep.* **2020**, *15*, 498–514. [CrossRef]

117. Peng, B.; Wang, Y.; Zhang, H. Mitonuclear communication in stem cell function. *Cell Prolif.* **2024**, *58*, e13796. [CrossRef] [PubMed]
118. Snyder, C.A.; Dwyer, K.D.; Coulombe, K.L.K. Advancing Human iPSC-Derived Cardiomyocyte Hypoxia Resistance for Cardiac Regenerative Therapies through a Systematic Assessment of In Vitro Conditioning. *Int. J. Mol. Sci.* **2024**, *25*, 9627. [CrossRef]
119. Häkli, M.; Kreutzer, J.; Mäki, A.-J.; Välimäki, H.; Lappi, H.; Huhtala, H.; Kallio, P.; Aalto-Setälä, K.; Pekkanen-Mattila, M. Human induced pluripotent stem cell-based platform for modeling cardiac ischemia. *Sci. Rep.* **2021**, *11*, 4153. [CrossRef] [PubMed]
120. Lopez, C.A.; Al-Siddiqi, H.H.A.A.; Purnama, U.; Iftekhar, S.; Bruyneel, A.A.N.; Kerr, M.; Nazir, R.; Fialho, M.d.L.S.; Malandraki-Miller, S.; Alonaizan, R.; et al. Physiological and pharmacological stimulation for in vitro maturation of substrate metabolism in human induced pluripotent stem cell-derived cardiomyocytes. *Sci. Rep.* **2021**, *11*, 7802. [CrossRef]
121. Li, D.; Armand, L.C.; Sun, F.; Hwang, H.; Wolfson, D.; Rampoldi, A.; Liu, R.; Forghani, P.; Hu, X.; Yu, W.-M.; et al. AMPK activator-treated human cardiac spheres enhance maturation and enable pathological modeling. *Stem Cell Res. Ther.* **2023**, *14*, 322. [CrossRef]
122. Detraux, D.; Renard, P. Succinate as a new actor in pluripotency and early development? *Metabolites* **2022**, *12*, 651. [CrossRef]
123. Nit, K.; Tyszka-Czochara, M.; Bobis-Wozowicz, S. Oxygen as a master regulator of human pluripotent stem cell function and metabolism. *J. Pers. Med.* **2021**, *11*, 905. [CrossRef]
124. Giardinelli, S.; Meliota, G.; Mentino, D.; D'Amato, G.; Faienza, M.F. Molecular basis of cardiomyopathies in Type 2 Diabetes. *Int. J. Mol. Sci.* **2024**, *25*, 8280. [CrossRef] [PubMed]
125. Hamledari, H.; Asghari, P.; Jayousi, F.; Aguirre, A.; Maaref, Y.; Barszczewski, T.; Ser, T.; Moore, E.; Wasserman, W.; Geltink, R.K.; et al. Using human induced pluripotent stem cell-derived cardiomyocytes to understand the mechanisms driving cardiomyocyte maturation. *Front. Cardiovasc. Med.* **2022**, *9*, 967659. [CrossRef] [PubMed]
126. Lamberto, F.; Peral-Sanchez, I.; Muenthaisong, S.; Zana, M.; Willaime-Morawek, S.; Dinnyés, A. Environmental Alterations during Embryonic Development: Studying the Impact of Stressors on Pluripotent Stem Cell-Derived Cardiomyocytes. *Genes* **2021**, *12*, 1564. [CrossRef]
127. Xiang, H.; Xu, H.; Tan, B.; Yi, Q.; Zhang, X.; Wang, R.; Chen, T.; Xie, Q.; Tian, J.; Zhu, J. AKAP1 Regulates Mitochondrial Dynamics during the Fatty-Acid-Promoted Maturation of Human-Induced Pluripotent Stem Cell-Derived Cardiomyocytes as Indicated by Proteomics Sequencing. *Int. J. Mol. Sci.* **2023**, *24*, 8112. [CrossRef]
128. Hu, X.; Yu, S.P.; Fraser, J.L.; Lu, Z.; Ogle, M.E.; Wang, J.-A.; Wei, L. Transplantation of hypoxia-preconditioned mesenchymal stem cells improves infarcted heart function via enhanced survival of implanted cells and angiogenesis. *J. Thorac. Cardiovasc. Surg.* **2008**, *135*, 799–808. [CrossRef] [PubMed]
129. Deogharia, M.; Venegas-Zamora, L.; Agrawal, A.; Shi, M.; Jain, A.K.; McHugh, K.J.; Altamirano, F.; Marian, A.J.; Gurha, P. Histone demethylase KDM5 regulates cardiomyocyte maturation by promoting fatty acid oxidation, oxidative phosphorylation, and myofibrillar organization. *Cardiovasc. Res.* **2024**, *120*, 630–643. [CrossRef]
130. Herzig, S.; Shaw, R.J. AMPK: Guardian of metabolism and mitochondrial homeostasis. *Nat. Rev. Mol. Cell Biol.* **2017**, *19*, 121–135. [CrossRef] [PubMed]
131. Drake, J.C.; Wilson, R.J.; Laker, R.C.; Guan, Y.; Spaulding, H.R.; Nichenko, A.S.; Shen, W.; Shang, H.; Dorn, M.V.; Huang, K.; et al. Mitochondria-localized AMPK responds to local energetics and contributes to exercise and energetic stress-induced mitophagy. *Proc. Natl. Acad. Sci. USA* **2021**, *118*, e2025932118. [CrossRef] [PubMed]
132. Jain, M.R.; Giri, S.R.; Trivedi, C.; Bhoi, B.; Rath, A.; Vanage, G.; Vyas, P.; Ranvir, R.; Patel, P.R. Saroglitazar, a novel PPARα/γ agonist with predominant PPARα activity, shows lipid-lowering and insulin-sensitizing effects in preclinical models. *Pharmacol. Res. Perspect.* **2015**, *3*, e00136. [CrossRef]
133. Lee, Y.J.; Zhao, Z.S.; Kim, S.K.; Kim, H.J.; Shim, W.S.; Ahn, C.W.; Lee, H.C.; Cha, B.S. Effects of PPAR-α and -γ Agonists on Fatty Acid Metabolism of Muscle Cells in Hyperlipidemic and Hyperglycemic Conditions. *J. Korean Diabetes Assoc.* **2006**, *30*, 324. [CrossRef]
134. Warren, J.S.; Oka, S.I.; Zablocki, D.; Sadoshima, J. Metabolic reprogramming via PPARα signaling in cardiac hypertrophy and failure: From metabolomics to epigenetics. *Am. J. Physiol. Heart Circ. Physiol.* **2017**, *313*, H584–H596. [CrossRef]
135. Finck, B.N.; Lehman, J.J.; Leone, T.C.; Welch, M.J.; Bennett, M.J.; Kovacs, A.; Han, X.; Gross, R.W.; Kozak, R.; Lopaschuk, G.D.; et al. The cardiac phenotype induced by PPARα overexpression mimics that caused by diabetes mellitus. *J. Clin. Investig.* **2002**, *109*, 121–130. [CrossRef]
136. Lee, T.W.; Bai, K.J.; Lee, T.I.; Chao, T.F.; Kao, Y.H.; Chen, Y.J. PPARs modulate cardiac metabolism and mitochondrial function in diabetes. *J. Biomed. Sci.* **2017**, *24*, 5. [CrossRef]
137. Wang, X.; Zhu, X.X.; Jiao, S.Y.; Qi, D.; Yu, B.Q.; Xie, G.M.; Liu, Y.; Song, Y.T.; Xu, Q.; Xu, Q.B.; et al. Cardiomyocyte peroxisome proliferator-activated receptor α is essential for energy metabolism and extracellular matrix homeostasis during pressure overload-induced cardiac remodeling. *Acta Pharmacol. Sin.* **2021**, *43*, 1231–1242. [CrossRef]
138. Beard, D.A.; Marzban, B.; Li, O.Y.; Campbell, K.S.; Janssen, P.M.; Chesler, N.C.; Baker, A.J. Reduced cardiac muscle power with low ATP simulating heart failure. *Biophys. J.* **2022**, *121*, 3213–3223. [CrossRef]

139. Zolezzi, J.M.; Silva-Alvarez, C.; Ordenes, D.; Godoy, J.A.; Carvajal, F.J.; Santos, M.J.; Inestrosa, N.C.; Ferreira, S.T. Peroxisome Proliferator-Activated Receptor (PPAR) γ and PPARα Agonists Modulate Mitochondrial Fusion-Fission Dynamics: Relevance to Reactive Oxygen Species (ROS)-Related Neurodegenerative Disorders? *PLoS ONE* **2013**, *8*, e64019. [CrossRef] [PubMed]
140. Birket, M.J.; Casini, S.; Kosmidis, G.; Elliott, D.A.; Gerencser, A.A.; Baartscheer, A.; Schumacher, C.; Mastroberardino, P.G.; Elefanty A.G.; Stanley, E.G.; et al. PGC-1α and reactive oxygen species regulate human embryonic stem Cell-Derived cardiomyocyte function. *Stem Cell Rep.* **2013**, *1*, 560–574. [CrossRef]
141. Lee, J.; Lee, J.; Cho, Y.S. Peroxisome Proliferator-Activated Receptor α Agonist and Its Target Nanog Cooperate to Induce Pluripotency. *J. Clin. Med.* **2018**, *7*, 488. [CrossRef]
142. Portman, M.A.; Standaert, T.A.; Ning, X.H. Relation of myocardial oxygen consumption and function to high energy phosphate utilization during graded hypoxia and reoxygenation in sheep in vivo. *J. Clin. Investig.* **1995**, *95*, 2134–2142. [CrossRef] [PubMed]
143. Cao, S.; Liu, Y.; Sun, W.; Zhao, L.; Zhang, L.; Liu, X.; Yu, T. Genome-Wide expression profiling of Anoxia/Reoxygenation in rat cardiomyocytes uncovers the role of MitoKATP in energy homeostasis. *Oxid. Med. Cell. Longev.* **2015**, *2015*, 756576. [CrossRef] [PubMed]
144. Zhu, S.C.; Chen, C.; Wu, Y.N.; Ahmed, M.; Kitmitto, A.; Greenstein, A.S.; Kim, S.J.; Shao, Y.F.; Zhang, Y.H. Cardiac complex II activity is enhanced by fat and mediates greater mitochondrial oxygen consumption following hypoxic re-oxygenation. *Pflug Arch. Eur. J. Physiol.* **2020**, *472*, 367–374. [CrossRef]
145. Morash, A.J.; Kotwica, A.O.; Murray, A.J. Tissue-specific changes in fatty acid oxidation in hypoxic heart and skeletal muscle. *Am J. Physiol. Regul. Integr. Comp. Physiol.* **2013**, *305*, R534–R541. [CrossRef]
146. Querceto, S.; Santoro, R.; Gowran, A.; Grandinetti, B.; Pompilio, G.; Regnier, M.; Tesi, C.; Poggesi, C.; Ferrantini, C.; Pioner, J.M The harder the climb the better the view: The impact of substrate stiffness on cardiomyocyte fate. *J. Mol. Cell. Cardiol.* **2022**, *166* 36–49. [CrossRef] [PubMed]
147. Hwang, J.-H.; Kim, K.M.; Oh, H.T.; Yoo, G.D.; Jeong, M.G.; Lee, H.; Park, J.; Jeong, K.; Kim, Y.K.; Ko, Y.-G.; et al. TAZ links exercise to mitochondrial biogenesis via mitochondrial transcription factor A. *Nat. Commun.* **2022**, *13*, 653. [CrossRef] [PubMed]
148. Huang, S.; Wang, X.; Wu, X.; Yu, J.; Li, J.; Huang, X.; Zhu, C.; Ge, H. Yap regulates mitochondrial structural remodeling during myoblast differentiation. *Am. J. Physiol. Cell Physiol.* **2018**, *315*, C474–C484. [CrossRef]
149. Vinarsky, V.; Pagliari, S.; Martino, F.; Mazzotti, C.; Jirakova, K.; Garlikova, Z.; Iuri, E.D.; Kytyr, D.; Benzoni, P.; Arici, M.; et al YAP1 contributes to the development of contractile force and sarcomere maturation in human pluripotent stem cell derived cardiomyocytes. *bioRxiv* **2024**. [CrossRef]
150. Körner, A.; Mosqueira, M.; Hecker, M.; Ullrich, N.D. Substrate stiffness influences structural and functional remodeling in induced pluripotent stem Cell-Derived cardiomyocytes. *Front. Physiol.* **2021**, *12*, 710619. [CrossRef]
151. Xiong, Q.; Ye, L.; Zhang, P.; Lepley, M.; Tian, J.; Li, J.; Zhang, L.; Swingen, C.; Vaughan, J.T.; Kaufman, D.S.; et al. Functional consequences of human induced pluripotent stem cell therapy. *Circulation* **2013**, *127*, 997–1008. [CrossRef]
152. Wai, T. Is mitochondrial morphology important for cellular physiology? *Trends Endocrinol. Metab.* **2024**, *35*, 854–871. [CrossRef]
153. Kawano, I.; Bazila, B.; Ježek, P.; Dlasková, A. Mitochondrial dynamics and cristae shape changes during metabolic reprogramming *Antioxid. Redox Signal.* **2023**, *39*, 684–707. [CrossRef]
154. Seo, B.J.; Yoon, S.H.; Do, J.T. Mitochondrial dynamics in stem cells and differentiation. *Int. J. Mol. Sci.* **2018**, *19*, 3893. [CrossRef]
155. Hoque, A.; Sivakumaran, P.; Bond, S.T.; Ling, N.X.Y.; Kong, A.M.; Scott, J.W.; Bandara, N.; Hernández, D.; Liu, G.-S.; Wong, R.C.B.; et al. Mitochondrial fission protein Drp1 inhibition promotes cardiac mesodermal differentiation of human pluripotent stem cells. *Cell Death Discov.* **2018**, *4*, 39. [CrossRef] [PubMed]
156. Butler, D.; Reyes, D.R. Heart-on-a-chip systems: Disease modeling and drug screening applications. *Lab Chip* **2024**, *24*, 1494–1528 [CrossRef]
157. Tsui, J.H.; Leonard, A.; Camp, N.D.; Long, J.T.; Nawas, Z.Y.; Chavanachat, R.; Smith, A.S.; Choi, J.S.; Dong, Z.; Ahn, E.H.; et al Tunable electroconductive decellularized extracellular matrix hydrogels for engineering human cardiac microphysiological systems. *Biomaterials* **2021**, *272*, 120764. [CrossRef]
158. Barnes, A.M.; Holmstoen, T.B.; Bonham, A.J.; Rowland, T.J. Differentiating human pluripotent stem cells to cardiomyocytes using purified extracellular matrix proteins. *Bioengineering* **2022**, *9*, 720. [CrossRef]
159. Smandri, A.; Al-Masawa, M.E.; Hwei, N.M.; Fauzi, M.B. ECM-derived biomaterials for regulating tissue multicellularity and maturation. *iScience* **2024**, *27*, 109141. [CrossRef]
160. Zhu, D.; Chen, Z.; Guo, K.; Xie, Q.; Zou, Y.; Mou, Q.; Zhou, Z.; Jin, G. Enhanced viability and functional maturity of iPSC-derived islet organoids by collagen-VI-enriched ECM scaffolds. *Cell Stem Cell* **2025**, *32*, 547–563.e7. [CrossRef] [PubMed]
161. Zare, A.; Salehpour, A.; Khoradmehr, A.; Bakhshalizadeh, S.; Najafzadeh, V.; Almasi-Turk, S.; Mahdipour, M.; Shirazi, R.; Tamadon, A. Epigenetic Modification Factors and microRNAs Network Associated with Differentiation of Embryonic Stem Cells and Induced Pluripotent Stem Cells toward Cardiomyocytes: A Review. *Life* **2023**, *13*, 569. [CrossRef] [PubMed]
162. Al-Awar, A.; Hussain, S. Interplay of Reactive Oxygen Species (ROS) and Epigenetic Remodelling in Cardiovascular Diseases Pathogenesis: A Contemporary perspective. *Front. Biosci.-Landmark* **2024**, *29*, 398. [CrossRef]

163. George, R.M.; Firulli, A.B. Epigenetics and heart development. *Front. Cell Dev. Biol.* **2021**, *9*, 637996. [CrossRef] [PubMed]
164. Hajdarpašić, A.; Tukker, M.; Rijdt, W.T.; Mohamedhoesein, S.; Meijers, W.C.; Caliskan, K. Epigenetics of cardiomyopathies: The next frontier. *Heart Fail. Rev.* **2024**, *30*, 257–270. [CrossRef]
165. Sinenko, S.A.; Tomilin, A.N. Metabolic control of induced pluripotency. *Front. Cell Dev. Biol.* **2024**, *11*, 1328522. [CrossRef]
166. Queiroz, A.L.; Lessard, S.J.; Ouchida, A.T.; Araujo, H.N.; Gonçalves, D.A.; Guimarães, D.S.P.S.F.; Teodoro, B.G.; So, K.; Espreafico, E.M.; Hirshman, M.F.; et al. The MicroRNA miR-696 is regulated by SNARK and reduces mitochondrial activity in mouse skeletal muscle through Pgc1α inhibition. *Mol. Metab.* **2021**, *51*, 101226. [CrossRef]
167. Zhang, X.; Ji, R.; Liao, X.; Castillero, E.; Kennel, P.J.; Brunjes, D.L.; Franz, M.; Möbius-Winkler, S.; Drosatos, K.; George, I.; et al. MicroRNA-195 regulates metabolism in failing myocardium via alterations in sirtuin 3 expression and mitochondrial protein acetylation. *Circulation* **2018**, *137*, 2052–2067. [CrossRef]
168. Hu, D.; Linders, A.; Yamak, A.; Correia, C.; Kijlstra, J.D.; Garakani, A.; Xiao, L.; Milan, D.J.; van der Meer, P.; Serra, M.; et al. Metabolic maturation of human pluripotent stem Cell-Derived cardiomyocytes by inhibition of HIF1α and LDHA. *Circ. Res.* **2018**, *123*, 1066–1079. [CrossRef]
169. Kumar, A.; Choudhary, A.; Munshi, A. Epigenetic reprogramming of mtDNA and its etiology in mitochondrial diseases. *J. Physiol. Biochem.* **2024**, *80*, 727–741. [CrossRef]
170. Soma, Y.; Tohyama, S.; Kubo, A.; Yamasaki, T.; Kabasawa, N.; Haga, K.; Tani, H.; Morita-Umei, Y.; Umei, T.C.; Sekine, O.; et al. Metabolic changes of human induced pluripotent stem cell-derived cardiomyocytes and teratomas after transplantation. *iScience* **2024**, *27*, 111234. [CrossRef]
171. Pettersen, I.K.N.; Tusubira, D.; Ashrafi, H.; Dyrstad, S.E.; Hansen, L.; Liu, X.-Z.; Nilsson, L.I.H.; Løvsletten, N.G.; Berge, K.; Wergedahl, H.; et al. Upregulated PDK4 expression is a sensitive marker of increased fatty acid oxidation. *Mitochondrion* **2019**, *49*, 97–110. [CrossRef] [PubMed]
172. Lin, Z.; Liu, F.; Shi, P.; Song, A.; Huang, Z.; Zou, D.; Chen, Q.; Li, J.; Gao, X. Fatty acid oxidation promotes reprogramming by enhancing oxidative phosphorylation and inhibiting protein kinase C. *Stem Cell Res. Ther.* **2018**, *9*, 47. [CrossRef] [PubMed]
173. Meshrkey, F.; Scheulin, K.M.; Littlejohn, C.M.; Stabach, J.; Saikia, B.; Thorat, V.; Huang, Y.; LaFramboise, T.; Lesnefsky, E.J.; Rao, R.R.; et al. Induced pluripotent stem cells derived from patients carrying mitochondrial mutations exhibit altered bioenergetics and aberrant differentiation potential. *Stem Cell Res. Ther.* **2023**, *14*, 320. [CrossRef] [PubMed]
174. El-Hattab, A.W.; Scaglia, F. Mitochondrial cytopathies. *Cell Calcium* **2016**, *60*, 199–206. [CrossRef] [PubMed]
175. Venkatraman, K.; Budin, I. Cardiolipin remodeling maintains the inner mitochondrial membrane in cells with saturated lipidomes. *J. Lipid Res.* **2024**, *65*, 100601. [CrossRef]
176. Wu, Z.; Park, J.; Steiner, P.R.; Zhu, B.; Zhang, J.X.J. Generative adversarial network model to classify human induced pluripotent stem cell-cardiomyocytes based on maturation level. *Sci. Rep.* **2024**, *14*, 27016. [CrossRef] [PubMed]
177. Kalkunte, N.; Cisneros, J.; Castillo, E.; Zoldan, J. A review on machine learning approaches in cardiac tissue engineering. *Front. Biomater. Sci.* **2024**, *3*, 1358508. [CrossRef]
178. Liang, J.; Wu, M.; Chen, C.; Mai, M.; Huang, J.; Zhu, P. Roles of reactive oxygen species in cardiac differentiation, reprogramming, and regenerative therapies. *Oxid. Med. Cell. Longev.* **2020**, *2020*, 2102841. [CrossRef] [PubMed]
179. Momtahan, N.; Crosby, C.O.; Zoldan, J. The role of reactive oxygen species in in vitro cardiac maturation. *Trends Mol. Med.* **2019**, *25*, 482–493. [CrossRef]
180. Lewandowski, J.; Rozwadowska, N.; Kolanowski, T.J.; Malcher, A.; Zimna, A.; Rugowska, A.; Fiedorowicz, K.; Łabędź, W.; Kubaszewski, Ł.; Chojnacka, K.; et al. The impact of in vitro cell culture duration on the maturation of human cardiomyocytes derived from induced pluripotent stem cells of myogenic origin. *Cell Transplant.* **2018**, *27*, 1047–1067. [CrossRef]
181. Huebsch, N.; Charrez, B.; Neiman, G.; Siemons, B.; Boggess, S.C.; Wall, S.; Charwat, V.; Jæger, K.H.; Cleres, D.; Telle, Å.; et al. Metabolically driven maturation of human-induced-pluripotent-stem-cell-derived cardiac microtissues on microfluidic chips. *Nat. Biomed. Eng.* **2022**, *6*, 372–388. [CrossRef]
182. Theodorakis, N.; Feretzakis, G.; Tzelves, L.; Paxinou, E.; Hitas, C.; Vamvakou, G.; Verykios, V.S.; Nikolaou, M. Integrating Machine Learning with Multi-Omics Technologies in Geroscience: Towards Personalized Medicine. *J. Pers. Med.* **2024**, *14*, 931. [CrossRef]
183. Vázquez-Carrada, M.; Vilchis-Landeros, M.M.; Vázquez-Meza, H.; Uribe-Ramírez, D.; Matuz-Mares, D. A new perspective on the role of alterations in mitochondrial proteins involved in ATP synthesis and mobilization in cardiomyopathies. *Int. J. Mol. Sci.* **2025**, *26*, 2768. [CrossRef]
184. Wong, S.; Bertram, K.R.; Raghuram, N.; Knight, T.; Hughes, A.L. Alterations in lipid saturation trigger remodeling of the outer mitochondrial membrane. *bioRxiv* **2025**. [CrossRef]
185. Rohani, L.; Machiraju, P.; Sabouny, R.; Meng, G.; Liu, S.; Zhao, T.; Iqbal, F.; Wang, X.; Ravandi, A.; Wu, J.C.; et al. Reversible mitochondrial fragmentation in IPSC-Derived cardiomyocytes from children with DCMA, a mitochondrial cardiomyopathy. *Can. J. Cardiol.* **2019**, *36*, 554–563. [CrossRef]

186. Ghai, M.K.; Khatri, A.; Kumar, K.; Thakur, I.S. Multi-omics and advance technologies in biodegradation of emerging contaminants and eco-estrogens in environmental waste. *Total Environ. Adv.* **2024**, *11*, 200113. [CrossRef]
187. Leventhal, M.J.; Zanella, C.A.; Kang, B.; Peng, J.; Gritsch, D.; Liao, Z.; Bukhari, H.; Wang, T.; Pao, P.-C.; Danquah, S.; et al. An integrative systems-biology approach defines mechanisms of Alzheimer's disease neurodegeneration. *Nat. Commun.* **2025**, *16*. [CrossRef]
188. He, H.; Qin, G.; Xue, M.; Feng, Z.; Mao, J.; Tao, W.; Chen, H.; Wang, X.; Yu, D.; Huang, F. Mitochondrial nanoprobe for precise cellular and drug analysis via graph Neural network. *Chem. Eng. J.* **2024**, *493*, 152709. [CrossRef]
189. Grabowska, M.E.; Chun, B.; Moya, R.; Saucerman, J.J. Computational model of cardiomyocyte apoptosis identifies mechanisms of tyrosine kinase inhibitor-induced cardiotoxicity. *J. Mol. Cell. Cardiol.* **2021**, *155*, 66–77. [CrossRef]
190. Kar, D.; Bandyopadhyay, A. Targeting Peroxisome Proliferator Activated Receptor α (PPAR α) for the Prevention of Mitochondrial Impairment and Hypertrophy in Cardiomyocytes. *Cell. Physiol. Biochem.* **2018**, *49*, 245–259. [CrossRef]
191. Sadria, M.; Layton, A.T. Interactions among mTORC, AMPK and SIRT: A computational model for cell energy balance and metabolism. Cell Commun. *Signaling* **2021**, *19*, 57. [CrossRef]
192. Krogh-Madsen, T.; Sobie, E.A.; Christini, D.J. Improving cardiomyocyte model fidelity and utility via dynamic electrophysiology protocols and optimization algorithms. *J. Physiol.* **2015**, *594*, 2525–2536. [CrossRef] [PubMed]
193. Zhang, Y.; Fu, Y.; Jiang, T.; Liu, B.; Sun, H.; Zhang, Y.; Fan, B.; Li, X.; Qin, X.; Zheng, Q. Enhancing fatty acids oxidation via L-Carnitine attenuates Obesity-Related atrial fibrillation and structural remodeling by activating AMPK signaling and alleviating cardiac lipotoxicity. *Front. Pharmacol.* **2021**, *12*, 771940. [CrossRef] [PubMed]
194. Chegary, M.; Brinke, H.T.; Doolaard, M.; Ijlst, L.; Wijburg, F.A.; Wanders, R.J.; Houten, S.M. Characterization of l-aminocarnitine, an inhibitor of fatty acid oxidation. *Mol. Genet. Metab.* **2008**, *93*, 403–410. [CrossRef] [PubMed]
195. Wheelwright, M.; Win, Z.; Mikkila, J.L.; Amen, K.Y.; Alford, P.W.; Metzger, J.M. Investigation of human iPSC-derived cardiac myocyte functional maturation by single cell traction force microscopy. *PLoS ONE* **2018**, *13*, e0194909. [CrossRef]
196. Joshi, J.; Albers, C.; Smole, N.; Guo, S.; Smith, S.A. Human induced pluripotent stem cell-derived cardiomyocytes (iPSC-CMs) for modeling cardiac arrhythmias: Strengths, challenges and potential solutions. *Front. Physiol.* **2024**, *15*, 1475152. [CrossRef]
197. Alberts, B.; Johnson, A.; Lewis, J.; Raff, M.; Roberts, K.; Walter, P. The Mitochondrion. Molecular Biology of the Cell-NCBI Bookshelf. 2002. Available online: https://www.ncbi.nlm.nih.gov/books/NBK26894/#:~:text=A%20Chemiosmotic%20Process%20Converts%20Oxidation,(Figure%2014%2D11) (accessed on 22 April 2025).
198. Ren, H.; Hu, W.; Jiang, T.; Yao, Q.; Qi, Y.; Huang, K. Mechanical stress induced mitochondrial dysfunction in cardiovascular diseases: Novel mechanisms and therapeutic targets. *Biomed. Pharmacother.* **2024**, *174*, 116545. [CrossRef] [PubMed]
199. Álvarez, M.; Ruiz-Sala, P.; Pérez, B.; Desviat, L.R.; Richard, E. Dysregulated Cell Homeostasis and miRNAs in Human iPSC-Derived Cardiomyocytes from a Propionic Acidemia Patient with Cardiomyopathy. *Int. J. Mol. Sci.* **2023**, *24*, 2182. [CrossRef] [PubMed]
200. Urzì, O.; Gasparro, R.; Costanzo, E.; De Luca, A.; Giavaresi, G.; Fontana, S.; Alessandro, R. Three-Dimensional Cell Cultures: The Bridge between In Vitro and In Vivo Models. *Int. J. Mol. Sci.* **2023**, *24*, 12046. [CrossRef] [PubMed]
201. Afzal, J.; Liu, Y.; Du, W.; Suhail, Y.; Zong, P.; Feng, J.; Ajeti, V.; Sayyad, W.A.; Nikolaus, J.; Yankova, M.; et al. Cardiac ultrastructure inspired matrix induces advanced metabolic and functional maturation of differentiated human cardiomyocytes. *Cell Rep.* **2022**, *40*, 111146. [CrossRef]
202. Stoll, S.; Leimena, C.; Qiu, H. Mitochondria and heart disease. In *Mitochondrial Diseases*; InTechOpen: Rijeka, Croatia, 2018. [CrossRef]
203. Wolk, A. Relationship Between Cardiac Contractile Performance and Mitochondrial Respiratory Capacity in Patients Undergoing Coronary Artery Bypass Grafting Surgery. 2021. Available online: https://dabar.srce.hr/en/islandora/object/mefst%3A1066 (accessed on 24 April 2025).
204. Yücel, D.; Garay, B.I.; Perlingeiro, R.C.R.; Van Berlo, J.H. Stimulation of cardiomyocyte proliferation is dependent on species and level of maturation. *Front. Cell Dev. Biol.* **2022**, *10*, 806564. [CrossRef]

BioChem

MDPI

Review

Cancer Therapy-Related Left Ventricular Dysfunction: Are There New Gatekeepers?

Mariagrazia Piscione [1,*], Maria Carmela Di Marcantonio [2], Barbara Pala [3,*] and Gabriella Mincione [2]

1 Fondazione Policlinico, Campus Bio-Medico University of Rome, Via Alvaro del Portillo, 200, 00128 Rome, Italy
2 Department of Innovative Technologies in Medicine & Dentistry, University "G. d'Annunzio" of Chieti-Pescara, Via dei Vestini 29, 66100 Chieti, Italy; dimarcantonio@unich.it (M.C.D.M.); gabriella.mincione@unich.it (G.M.)
3 School of Applied Medical-Surgical Sciences, University of Tor Vergata, Via Montpellier 1, 20, 00133 Rome, Italy
* Correspondence: mariagrazia.piscione@unicampus.it (M.P.); barbara.pala@uniroma1.it (B.P.)

Abstract

The growing success of oncologic therapies has led to a significant improvement in patient survival; however, this has been accompanied by an increasing incidence of cardiovascular adverse events, particularly cancer therapy-related cardiac dysfunction (CTRCD). Among these, left ventricular impairment represents a major concern due to its potential to compromise both cardiac and oncologic outcomes. This review provides an in-depth overview of the cardiotoxic adverse events associated with several classes of anticancer agents. Particular focus is given to the molecular mechanisms involved in myocardial injury, such as oxidative stress, mitochondrial dysfunction, calcium dysregulation, endothelial reticulum stress, autophagy, and apoptosis. In parallel, established and emerging cardioprotective strategies, from conventional to newer therapeutic approaches, are explored. The role of advanced imaging modalities, as well as cardiac biomarkers, is discussed in the context of early detection and monitoring of subclinical cardiac injury. Finally, the integration of pharmacogenomics and epigenetics is considered as a promising avenue to personalize risk stratification and preventive therapy. By elucidating the complex interplay between cancer treatments and cardiovascular health, this review underscores the importance of a multidisciplinary, precision medicine approach to optimizing the care of patients undergoing potentially cardiotoxic therapies.

Keywords: cardio-oncology; cancer therapy-related cardiac dysfunction; tailored approach; left ventricular impairment

1. Introduction

In recent years, cancer survival has markedly increased due to the plural progress in early diagnoses, improved surgical approaches, and new therapies [1]. As a matter of fact, updated detection techniques, precision oncology, and the emergence of new anticancer drugs have significantly improved the efficacy of cancer treatment protocols [1–3]. Nonetheless, these advancements have simultaneously increased the potential risks of cancer therapy-related cardiac dysfunction (CTRCD); treatment-induced hypertension; vasospastic and thromboembolic ischemia; and rhythm disorders, including life-threatening arrhythmias [4]. In addition, chest radiation is also considered responsible for pericardial and myocardial diseases, valve heart disease and prosthesis dysfunction, and myocardial

 https://doi.org/10.3390/biochem5030025

fibrosis [4,5]. The prevalence of cardiovascular damage generated by cancer treatments significantly varies according to the type and the duration of therapy, and a patient's preexisting comorbidities [4]. Surprisingly, according to a comprehensive review considering breast cancer survivors in the United States, treated women faced a markedly elevated risk of mortality due to cardiovascular disease, exceeding the probability of death from the primary cancer or its recurrence [6,7]. That is the reason why, over the decades, the acute and long-term cardiovascular consequences of cancer therapies have been deeply studied by a new discipline: cardio-oncology [4]. Nowadays, precision cardio-oncology plays a pivotal role, integrating molecular profiling and risk stratification to tailor surveillance and cardioprotective strategies [8]. Indeed, precision cardio-oncology, together with advanced imaging techniques—such as speckle-tracking echocardiography and cardiac magnetic resonance (CMR)—and the use of biomarkers have a proven efficacy for detecting subclinical myocardial dysfunction before irreversible damage occurs [9]. These are now considered essential to comprehensive cancer care, since they can guide cardiovascular assessments before, during, and after treatment, with the goal of integrating both oncologic efficacy and cardiovascular safety into standard clinical practice, emphasizing early intervention and continuous monitoring [4]. By contrast, a research field still in its early stages that deserves further development concerns the use of novel pharmacological strategies for both the prevention of chemotherapy-induced damage and the improvement of ventricular contractility at the end of therapeutic cycles [4].

This review aims to summarize the molecular mechanism of damage of the most dangerous drugs that cause left ventricular (LV) impairment. Furthermore, innovative approaches for evaluating and managing the risk factors associated with cardiovascular disease (CVD) in cancer-treated patients are described. Last, but not least, the role of new protective drugs for both asymptomatic and symptomatic CTRCD is discussed, with a focus on new technologies that provide individualized approaches to CTRCD prevention, such as genomics and gene therapy. According to the guidelines, this review emphasizes the need for a collaborative, multidisciplinary approach involving dedicated cardio-oncology teams, which are essential for identifying patients at risk early, tailoring monitoring strategies, and implementing preventive measures to reduce the burden of CVD.

2. Cancer Therapy and Cardiotoxicity: An Overview

Cardiotoxicity is a well-known adverse event of several cancer therapies (Table 1), and it represents a leading cause of morbidity among cancer survivors [4].

Anthracyclines, which are commonly used to treat solid tumors and hematologic malignancies, have cumulative, dose-dependent, and progressive cardiotoxic effects [10]. Doxorubicin (DOXO), daunorubicin, epirubicin, and idarubicin are derived from *Streptomyces peucetius* var. *caesius* [10]. Their clinical utility is hindered by the potential for toxicity, which can appear up to ten years after the cessation of chemotherapy [10]. Their acute side effects include hematological suppression, nausea, vomiting, and alopecia; however, the most feared one is cardiotoxicity. Acute toxicity is clinically characterized by a wide variety of signs, from asymptomatic electrocardiographic abnormalities to pericarditis and decompensated biventricular heart failure (HF) [11]. The likelihood of developing HF is predominantly dose-dependent, but cardiotoxicity may also manifest at low dosages and be due to individual vulnerabilities [11]. Age, the concurrent administration of additional cardiotoxic agents, mediastinal radiotherapy, and the cumulative dosage of DOXO are risk factors associated with an elevated risk of DOXO-induced HF [12–15]. Beyond these, gender differences have been identified as a risk factor for the harmful effects of DOXO [11]. Lipshultz, S.E. et al. showed that women experience more severe cardiotoxicity with greater contractility depression [12].

Even though the mechanisms of damage are not clearly and completely known, they can be grouped into genomic instability, oxidative stress, mitochondrial dysfunction, and activation of apoptotic pathways in cardiomyocytes [9], and are summarized in Figure 1.

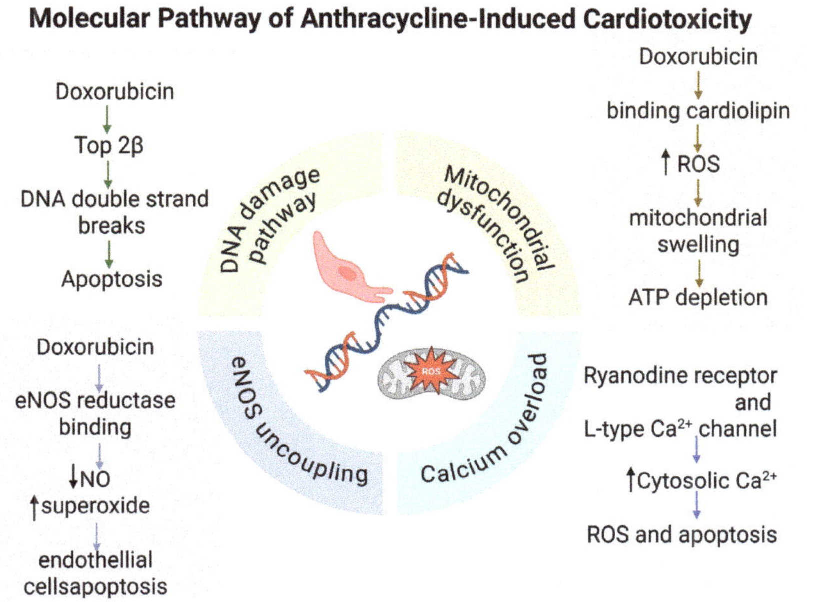

Figure 1. Summarized molecular mechanisms of anthracycline-induced cardiotoxicity (created in BioRender.com on 30 May 2025). Abbreviations: ATP: adenosine tri-phosphate; Ca^{2+}: calcium ion; DNA: deoxyribonucleic acid; eNOS: endothelial nitric oxide synthase; HSF-1: heat shock factor 1; Hsp 25: heat shock protein 25; p53: protein 53; Top2β: topoisomerase 2β; ROS: oxygen reactive species.

First of all, anthracyclines inhibit topoisomerase 2β (Top2β), which prevents DNA repair [16,17]. Two Top2 enzymes exist: Top2α and Top2β [18]. Top2α, a recognized marker of cellular proliferation, is overexpressed in malignancies but is undetectable in quiescent tissues, and in the past it was thought to be responsible for anthracycline toxicity [18]. This notion remained accepted until Zhang et al., in 2012, identified topoisomerase 2β (Top2β) as a critical mediator of DOXO-related cardiac injury [17]. In particular, the researchers showed in murine models that the interaction between DOXO and Top2β in cardiomyocytes led to deoxyribonucleic acid (DNA) double-strand breaks [18]. Using a cardiomyocyte-specific deletion of Top2β, the authors demonstrated that the absence of this isoform completely removed the cardiotoxic effects of DOXO without altering its anticancer efficacy. These findings not only elucidated the molecular mechanism underlying cardiac harm, but also highlighted Top2β as a potential therapeutic target for selective cardioprotection during anthracycline therapy [17].

Another mechanism through which DOXO damages the myocardium is the stimulation of oxygen reactive species (ROS), considering that the heart is very susceptible to oxidative stress due to its high oxygen demand, dense mitochondrial content, limited antioxidant defenses, and vulnerability to injury during ischemia/reperfusion [18]. Mitochondria are severely and progressively involved since the cationic molecule of DOXO is sequestered in the mitochondrial inner membrane by establishing a nearly irreversible combination with cardiolipin, which is an essential phospholipid found mainly in the mitochondrial inner membrane, where it plays a critical role in oxidative phosphorylation [19].

The proteins of the electron transport chain need an optimal functioning of cardiolipin, but DOXO disrupts the cardiolipin–protein interface, leading to increased superoxide (O_2^-) production [20]. It is highly probable that ROS production interferes with mitochondria and consequently cellular metabolism, as mitochondria generate over 90% of the adenosine tri-phosphate (ATP) consumed by cardiomyocytes [21]. This functional impairment results in ultrastructural pathological alterations, including mitochondrial swelling and the formation of myelin patterns inside these organelles [21]. However, it has to be considered that the doses used to treat malignancies are lower than those used in in vitro studies, so that DOXO-related side effects are generally less severe in vivo [22].

Moreover, several studies have highlighted the central role of endothelial nitric oxide synthase (eNOS) in DOXO-induced cardiotoxicity, which not only involves cardiomyocytes but also the cardiac vasculature [22]. Vasquez-Vivar et al. demonstrated that DOXO binds to the reductase domain of eNOS, leading to the generation of O_2^- via the formation of a semiquinone radical, which is an unstable and reactive molecule [22]. This reaction does not involve the oxygenase domain, which is commonly responsible for the production of nitric oxide (NO) that is protective of the endothelium [23]. By contrast, the redox activation of DOXO by eNOS is associated with apoptosis and endothelial dysfunction, shifting the enzyme function from NO production to ROS generation [24]. Murine models have shown that the genetic deletion of eNOS protects against DOXO-induced cardiac dysfunction, while eNOS overexpression worsens it [22,25].

Another pathogenic mechanism involves the dysregulation of intracellular calcium; actually, it has been demonstrated that DOXO-induced cardiotoxicity is associated with elevated intracellular calcium levels, while a calcium ion chelator can decrease DOXO-induced ROS production and apoptosis [26]. According to Saeki et al., DOXO can bind to the ryanodine channel in the sarcoplasmic reticulum, regardless of the channel's state. DOXO inhibits the sodium–calcium exchanger channel in the sarcolemma, which expels calcium from cells and increases the activity of the L-type calcium channel, increasing intracellular calcium levels and promoting the development of DOXO-induced cardiomyopathy [27]. It is widely established that DOXO-induced oxidative stress stimulates apoptotic signaling, resulting in cardiomyocyte death, and that both extrinsic and intrinsic apoptotic mechanisms are involved [28,29].

Furthermore, DOXO has also been shown to trigger apoptosis by mechanisms other than ROS generation and oxidative stress, but this is confounded by the fact that apoptosis itself produces free radicals. Research studies have proved that oxidative stress stimulates heat shock factor 1 (HSF-1), which is a "sensor" protein that becomes activated when a cell is under stress [28]. Then, HSF-1 translocates to the nucleus and induces the expression of specific protective proteins known as heat shock proteins (HSPs), in particular Hsp25 (heat shock protein 25), which stabilizes protein 53 (p53), a key regulatory protein that monitors cellular damage and determines whether to initiate repair processes or to trigger cell elimination. When p53 is not degraded, it promotes the expression of pro-apoptotic proteins, which drive cell apoptosis [28]. The interplay of HSF-1, Hsp 25, and p53 provides an example of the difficulty in distinguishing between oxidative stress and apoptosis, since this cascade illustrates how a protective response can paradoxically lead to unjustified cell death and myocardial tissue loss. The close interconnection between oxidative stress and apoptosis complicates the distinction between cause and consequence, as each process can amplify the other [28].

Trastuzumab, a humanized monoclonal antibody which targets human epidermal growth factor receptor 2 (HER2), is approved for treating HER2-positive breast cancer and metastatic gastric cancer [30]. It suppresses tumor growth by binding to domain IV of the extracellular domain of HER2. It activates antibody-dependent cell-mediated cytotoxicity and inhibits HER2 extracellular domain cleavage [30]. The latter prevents cancer cells from releasing a truncated, active form of the HER2 receptor, which can promote tumor growth even in the absence of external signals. Trastuzumab blocks oncogenic cellular signaling, and downregulates angiogenesis and DNA repair pathways [30].

The specific molecular mechanisms of trastuzumab-induced cardiotoxicity are still unclear, despite extensive research [31]. In general, trastuzumab-induced cardiac dysfunction is considered milder and more reversible than anthracycline-related cardiomyopathy due to the absence of ultrastructural changes in the cardiomyocytes, since patients treated with trastuzumab do not experience primary myocyte injury [32]. Nonetheless, therapy with this monoclonal antibody has been linked to both short-term and long-term adverse events, highlighting the importance of prolonged cardiac follow-up [33]. In fact, its transient toxicity has been questioned by some in vivo studies, which have shown that it causes ultrastructural changes in mouse cardiac tissues, as observed by electron microscopy [34]. Indeed, treatment with it has altered the gene expression profiles related to cardiac and mitochondrial activities, as well as DNA repair [34]. The mechanism of myocardial injury consists of suppressed autophagy in cardiomyocytes, leading to the accumulation of toxic ROS [35]. Trastuzumab alters HER signaling by phosphorylating HER1 and HER2 at 845 and 1248 sites, respectively, and activating the autophagy-inhibitory signaling cascade constituted of extracellular signal-regulated kinase/mammalian target of rapamycin/Unc-51-like autophagy-activating kinase 1 (Erk/mTOR/Ulk 1). This compromises cardiomyocytes' ability to eliminate toxic cellular substrates, resulting in cardiotoxicity [35]. Different from trastuzumab, pertuzumab, an anti-HER2 monoclonal antibody with a distinct epitope, has not affected the HER2 signaling cascade in in vitro models [36]. Actually, pertuzumab targets the dimerization domain (domain II) of the HER2 receptor, thereby preventing the heterodimerization of HER2 with other members of the endothelial growth factor receptor (EGFR) family, particularly HER3 [36]. This blockade effectively inhibits downstream signaling pathways, which are critical for tumor cell proliferation and survival. Pertuzumab is typically used in combination with trastuzumab and chemotherapy, taking advantage of their complementary mechanisms of action to enhance antitumor efficacy [36]. By contrast, the mechanisms of cardiac toxicity associated with the combination of anthracycline and trastuzumab are still unknown [35]. In a recent paper, it was demonstrated that Trastuzumab reduced Top2β, a key target in DOXO-induced cardiotoxicity, and that the combined use of this drug with DOXO appeared to exert a protective effect [35]. Conversely, trastuzumab and DOXO treatment significantly increased apoptosis, inhibited cell growth, and promoted production of ROS and nitrative species in human cardiomyocytes [35].

Proteasome inhibitors (PIs), bortezomib, carfilzomib, and the oral ixazomib, together with immunomodulatory drugs, monoclonal antibodies, and chemotherapy, are used to treat patients with plasma cell dyscrasias [37]. Proteostasis is essential for maintaining the integrity of the cellular proteome, ensuring cell function and viability [38]. The ubiquitin proteasome (UPP) is the primary proteolytic mechanism in mammalian cells, which degrading up to 80% to 90% of intracellular proteins. The UPP consists of three components: ubiquitin, ubiquitin-conjugating enzymes, and the proteasome [39]. The proteasome removes both short-lived ubiquitinated proteins and non-repairable polypeptides, and decreased proteasome activity leads to increased proteome instability, which can cause severe proteotoxic stress [39]. Inhibition of proteasomes resulted in both reduced angiogenesis and increased cancer cell death in multiple myeloma (MM) cells [40]. At the same time, cardiomyocytes, which are highly dependent on proteasomal activity for their contractile function, are particularly vulnerable to PIs [4]. A recent study demonstrated the molecular mechanism of heart toxicity due to carfilzomib [41]. In Drosophila, the genetic suppression of the proteasome altered heart function and decreased cell lifespan [41]. In addition, carfilzomib may cause LV dysfunction in mice via increasing phosphoprotein phosphatase 2A (PP2A) activity and disrupting autophagy by inhibiting AMP-activated protein kinase a (AMPKa) and its downstream pathways [41]. Furthermore, reduced UPP activity also depends on a patient's fragility [42]. A study proved that PIs in old age lead to increased proteome instability [42]. Additionally, the elderly are more likely to have traditional cardiovascular risk factors, and patients with MM might have been exposed to potentially cardiotoxic therapies like anthracyclines or chest radiotherapy, which can increase the risk of developing HF [42]. Carfilzomib has been associated with a higher incidence of cardiovascular events compared to other PIs, likely due to its irreversible inhibition of proteasomal activity [43]. A systematic review and meta-analysis of 24 clinical trials reported an 18.1% incidence of cardiac adverse events and a 4.1% incidence of HF in patients treated with this drug [44]. Notably, in a prospective study, the patients receiving PIs developed signs and symptoms of HF despite having a normal or near-normal left ventricular ejection fraction (LVEF), suggesting a phenotype consistent with HF with preserved ejection fraction (HFpEF) [45]. Hypertension, a key driver in HFpEF pathogenesis, has been frequently observed during carfilzomib therapy and is considered one of the most common cardiovascular adverse events associated with its use [45]. Although the data remain limited, several risk factors for carfilzomib-related cardiotoxicity have been identified, including pre-existing cardiovascular disease (CVD), age over 75, obesity, use of a twice-weekly dosing schedule, concurrent administration of immunomodulatory drugs, doses exceeding 45 mg/m^2, and the presence of chronic obstructive pulmonary disease [45]. The pathophysiological mechanisms appear to be multifactorial, with vascular dysfunction playing a central role [43]. Carfilzomib-induced endothelial damage leads to reduced NO bioavailability, increased vascular stiffness, and elevated systemic vascular resistance, ultimately resulting in persistent hypertension with LV remodeling and myocardial stiffening [43]. Additionally, this drug has been shown to promote pro-inflammatory signaling and oxidative stress, both of which contribute to the activation of fibrotic pathways and extracellular matrix deposition in the myocardium, which impairs diastolic relaxation [43,44]. Together, these mechanisms provide a plausible explanation for the HFpEF phenotype observed in patients treated with carfilzomib, highlighting the need for careful blood pressure control and the early recognition of subclinical diastolic dysfunction during therapy [43]. In contrast, bortezomib appears to have a more favorable cardiovascular safety profile [46]. A retrospective observational cohort study involving 1790 patients with MM found no significant increase in the risk of hospitalization for HF in bortezomib-treated patients compared to those receiving lenalidomide, following propensity score matching [46].

Small-molecule tyrosine kinase inhibitors (TKIs) targeting the Breakpoint Cluster Region–Abelson Leukemia (BCR-ABL) have been shown to effectively treat blood cancer and advanced solid tumors [47]. TKIs are thought to be safer than classic chemotherapy medications like DOXO [48]. However, since TKIs are often administered for long periods without a maximum dosage limit, some of them continue to cause significant cardiac side effects [48]. In a meta-analysis of 6935 patients, sunitinib was found to be related to 4.1% of congestive HF [49]. The mechanisms of LV dysfunction are many and interrelated. Recent findings, including those by Wang et al., have highlighted a common molecular mechanism underlying the cardiotoxic effects of multiple TKIs: the induction of ER stress [47]. In their study, human induced pluripotent stem cell-derived cardiomyocytes (hiPSC-CMs) were used to model the cardiac cellular response to eight different TKIs. All the compounds were found to activate the inositol-requiring enzyme 1 alpha (IRE1α) pathway, leading to the splicing of X-box binding protein 1 (XBP1), a key branch of the unfolded protein response [47]. This ER stress response was preceded by a transient but significant increase in ROS production, followed by lipid peroxidation—a typical manifestation of oxidative damage [47]. The downstream consequences of ER stress included the upregulation of pro-inflammatory genes and the reactivation of fetal gene programs, both of which are considered molecular signatures of pathological cardiac remodeling [47]. These findings provide mechanistic insight into how TKIs disrupt cardiac homeostasis and point to ER stress and oxidative imbalance as key contributors to their cardiotoxic potential [47].

Mitogen-activated protein kinase kinase (MEK) inhibitors target the RAS sarcoma–rapid accelerated fibrosarcoma–mitogen-activated protein kinase (Ras-Raf-MAP kinase) pathway and are used to treat metastatic melanoma [50]. They impede the downstream target of BRAF. The main cardiovascular adverse event is LV dysfunction, especially when combined with a BRAF inhibitor [50]. Interrupting or reducing the doses of this drug can lead to improved LV function [50]. The incidence of HF in patients treated with BRAF and MEK inhibitors in clinical trials, with a median follow-up of 9 to 16 months, has been reported to be 2% to 12% [51]. However, this could be an underestimate for a variety of reasons, one of the most important being that patients with severe underlying cardiovascular morbidity are often excluded from clinical studies [50].

Immune checkpoint inhibitors (ICIs) are monoclonal antibodies that prevent immune checkpoints from interacting with their ligands [52]. They have been evaluated in both pre-clinical and clinical trials [53]. Despite their efficacy against cancer, ICIs are associated with immune-related adverse events caused by off-target immune system hyperactivation and dysregulation, resulting in damage to various organ systems with clinical consequences, such as colitis, hepatitis, dermatitis, thyroiditis, myocarditis, or hypophysitis [53]. Although the rate of cardiac toxicity found in clinical trials following ICI treatments is rather rare, 3.1% in monotherapies and 5.8% in dual/combination therapies, it can be fatal [53]. Rubio-Infante et al. conducted a systematic analysis to elucidate the potential immune mechanisms contributing to cardiac toxicity associated with ICIs [54]. The study identified three primary pathways: T cell recruitment and activation, autoantibody-mediated cardiotoxicity, and inflammatory cytokine release [55]. ICIs alter peripheral immune tolerance, leading to the activation and infiltration of cluster of differentiation 4+ (CD4+) T cells and cluster of differentiation 8+ (CD8+) T cells into cardiac tissue [55]. This immune response can result in myocardial inflammation and damage [55]. Moreover, the pro-inflammatory environment induced by ICIs may activate autoreactive B cells, leading to the production of autoantibodies that target cardiac antigens [56]. These autoantibodies can mediate cardiotoxicity through mechanisms such as complement-dependent cytotoxicity and antibody-dependent cellular cytotoxicity. ICIs can also promote the release of pro-inflammatory cytokines, contributing to systemic inflammation and potential cardiac tissue damage [57]. In synthesis, myocardial injury is primarily driven by an aberrant activation of cytotoxic T lymphocytes, which can infiltrate cardiac tissue and induce direct cytotoxicity through the release of perforin, granzymes, and interferon-γ [57]. The heart naturally expresses Programmed Death Ligand-1 (PD-L1) as a protective mechanism to prevent autoimmunity; the blockade of this pathway renders cardiac tissue more susceptible to immune-mediated attack [57]. Additionally, antigenic cross-reactivity between tumor and myocardial epitopes (molecular mimicry) may trigger unintended autoimmune responses. The systemic release of pro-inflammatory cytokines, such as tumor necrosis-α (TNF-α) and interleukin-6 (IL-6), along with the activation of intracellular signaling pathways, contribute to endothelial dysfunction, myocardial fibrosis, and ventricular remodeling—key components in the pathogenesis of ICI-induced HF [57].

Table 1. Summary of the main antitumor treatments and their possible mechanisms of cardiotoxicity.

Treatments	Mechanism of Cardiotoxicity	CRTCD*/LV Systolic Dysfunction	References
Anthracyclines	Top2β inhibition: DNA breaks; ROS production; mitochondrial dysfunction; calcium dysregulation; apoptosis; eNOS uncoupling.	Cumulative and dose-dependent; occurs years after therapy.	[17,22,24]
HER2-Targeted Therapies	Oxidative stress; suppression of autophagy; mitochondrial gene expression alteration.	Typically reversible; increased risk when combined with anthracyclines.	[7,31,32]
Proteasome Inhibitors	UPP inhibition: proteotoxic stress; autophagy suppression; vascular dysfunction; ROS and inflammation.	HFpEF, subclinical diastolic dysfunction; higher risk in elderly or those with comorbidities.	[39–41]
Tyrosine Kinase Inhibitors	ER stress; ROS burst and lipid peroxidation; upregulation of pro-inflammatory and fetal gene programs.	LV dysfunction; cardiac remodeling associated with long-term use.	[48–50]
MEK/BRAF Inhibitors	MAPK pathway inhibition: LV dysfunction; reversibility upon dose reduction.	LV dysfunction; underreported due to patient selection.	[51,52]
Immune Checkpoint Inhibitors	T CD4+ and CD8+ activation; autoantibody generation; cytokine storm; PD-L1 pathway suppression; molecular mimicry: myocarditis, fibrosis, HF.	Myocarditis, HF; rare but often severe or fatal.	[55,56]

* Abbreviations: CRTCD: cancer therapy-related cardiac dysfunction; CD4+: cluster of differentiation 4+; CD8+: cluster of differentiation 8+; DNA: deoxyribonucleic acid; ER: endoplasmic reticulum; eNOS: endothelial nitric oxide synthase; HER2: human epidermal growth factor receptor 2; HF: heart failure; HFpEF: heart failure with preserved ejection fraction; LV: left ventricle; MAPK: mitogen-activated protein kinase; MEK/BRAF: mitogen-activated protein kinase/v-Raf murine sarcoma viral oncogene homolog B1; PD-L1: Programmed Death Ligand-1; ROS: reactive oxygen species; Top2β: topoisomerase 2β; UPP: ubiquitin proteasome.

3. The "Old Gatekeepers": Early Detection, Biomarkers, and Risk Stratification of CVDs

In a cardio-oncology setting, cardiovascular adverse effects can develop insidiously during or after cancer therapy, often progressing to irreversible myocardial damage before symptoms appear. The early detection of subclinical injury is therefore crucial to enable timely intervention and prevent long-term sequelae. Biomarkers and advanced cardiac imaging provide sensitive and specific means to identify these early changes, while validated risk stratification tools help individualize the surveillance intensity and therapeutic strategies. This combination forms the foundation for effective prevention, diagnosis, and management in contemporary cardio-oncology practice. Although this review primarily focuses on the molecular mechanisms underlying the cardiotoxic effects of major chemotherapeutic agents and new drugs, it is essential to briefly highlight the critical role of biomarkers, advanced imaging, and risk stratification tools in clinical practice [4] (Figure 2).

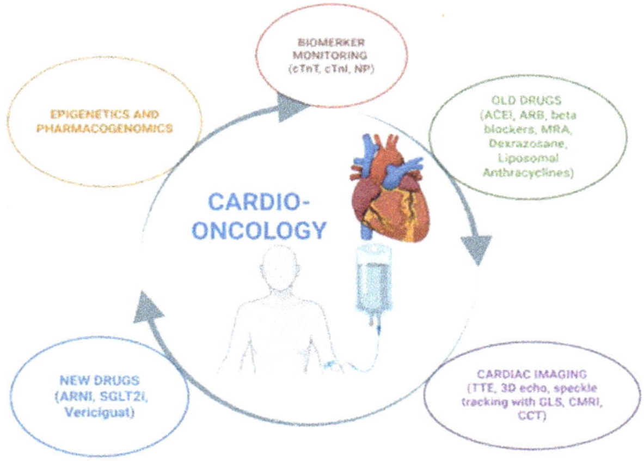

Figure 2. Schematic illustration of the phases of risk stratification, primary prevention, monitoring, and treatment of CTRCD (created in BioRender.com on 30 May 2025). Abbreviations: ACEIs: angiotensin-converting enzyme inhibitors; ARBs: angiotensin receptor blockers; ARNI: angiotensin receptor-neprilysin inhibitor; CCT: cardiac coronary CT; CMRI: cardiac magnetic resonance imaging; cTnI cardiac troponin I; cTnT: cardiac troponin T; GLS: global longitudinal strain; MRA: mineralcorticoid receptor antagonist; SGLT2i: sodium–glucose cotransporter inhibitor; TTE: trans-thoracic echocardiogram.

The 2022 ESC guidelines recommend measuring cardiac biomarkers (cardiac troponin I (cTnI) or cardiac troponin T (cTnT)) and natriuretic peptides (NPs), including B-type natriuretic peptide (BNP) and N-terminal pro-BNP (NT-proBNP), to assess the cardiovascular risk of patients receiving anthracyclines, HER2-targeted treatments, and vascular endothelial growth factor (VEGF) inhibitors [4,58]. Monitoring biomarkers during treatment can help identify patients who may benefit from cardioprotective therapies. Therefore, it is essential to establish the baseline levels of NPs and/or cTn for patients at high risk of CTRCD and to track their changes throughout treatment [58]. However, there is no universally agreed threshold value for these in cancer patients because of confounding factors, such as age, renal function, and comorbidities, so their alterations should be evaluated taking into account patients' clinical profiles [58]. A 2020 meta-analysis found that elevated troponin levels following chemotherapy increased the likelihood of LV dysfunction by sevenfold, with a 93% negative predictive value [59]. In contrast, a BNP and NT-proBNP levels increase following cancer treatment had an inconsistent prognostic value [59].

In addition to biomarkers evaluation, cardiovascular imaging plays a crucial role in detecting the early signs of cardiac dysfunction, assessing pre-existing heart conditions, and monitoring changes during and after cancer treatment [4,60]. Transthoracic echocardiography (TTE) is essential for diagnosing LV failure, and the LV ejection fraction (EF) is the key measurement [60]. Three-dimensional echocardiography is now considered the best method to measure the LVEF and chamber volumes due to its higher accuracy and lower interobserver variability (5–6%) [60,61]. If 3D is not accessible, the global longitudinal strain (GLS), derived from speckle tracking, is a more sensitive and reproducible measure of LV systolic function, since it can detect subclinical dysfunction, corresponding with early molecular alterations, such as mitochondrial dysfunction, oxidative stress, and cardiomyocyte apoptosis [62]. A $\geq$15% decline in the GLS from baseline indicates subclinical cardiotoxicity and an increased risk of eventual LV dysfunction [62]. It is important to note that the GLS may exhibit slight variations depending on the equipment used and on chemotherapy-induced changes in the loading conditions [62]. Myocardial work (global work index, global constructive work, global wasted work, and global work efficiency) is a promising early cardiotoxicity marker, especially with anthracycline or anthracycline/trastuzumab therapy [63]. Although further research is needed, it is established that the GLS provides a non-invasive insight into molecular cardiotoxicity, reinforcing its role in risk stratification and the early monitoring of at-risk patients [4].

CMR represents an interesting alternative to TTE for the detection of CTRCD [64]. In addition to its superior spatial resolution, CMR offers unparalleled accuracy in quantifying both left and right ventricular function, an aspect of particular relevance in patients treated with HER2-targeted agents [64]. Crucially, CMR allows for myocardial tissue characterization via T1 and T2 mapping sequences, thereby enabling the detection of diffuse fibrosis, interstitial expansion, and myocardial edema—hallmarks of early molecular injury [65]. Among the most promising tools derived from CMR is the GLS obtained through feature tracking (FT). This parameter has been demonstrated to detect subclinical myocardial dysfunction well before an overt reduction in the LVEF, reflecting the early reduction in the EF linked to mitochondrial dysfunction, oxidative stress, and apoptosis of longitudinally oriented subendocardial fibers—structures especially vulnerable to cytotoxic insult [65]. In this regard, the retrospective analysis by Romano et al., considering 470 patients with reduced LVEF and CMR evaluation using both cine and late gadolinium enhancement (LGE), revealed that FT-derived GLS independently predicted all-cause mortality in both ischemic and non-ischemic cardiomyopathy. Remarkably, the GLS has added incremental prognostic value beyond that of established markers, such as the LVEF and LGE burden [65]. This finding underscores the vulnerability of the subendocardial longitudinal fibers, whose dysfunction likely represents a downstream consequence of cumulative molecular damage, including ROS generation, calcium dysregulation, and mitochondrial injury [65]. Moreover, CMR has proven capable of detecting early myocardial edema through T2-weighted imaging in patients receiving anthracyclines, suggesting its utility in recognizing the inflammatory and metabolic alterations that precede irreversible structural damage [66,67]. These features make CMR not only a diagnostic modality but also a non-invasive molecular "lens" through which cardiotoxicity can be visualized and understood in its earliest and most reversible stages [66].

Cardiac computed tomography (CCT) can examine heart chamber function by providing detailed volumetric and structural data for both ventricles using retrospective multiphase ECG-gated imaging [68]. CCT is a reliable alternative to CMR for functional and morphological analyses, and it is highly sensitive in excluding obstructive CAD in patients with a low-to-intermediate pre-test risk of CAD (high negative predictive value) [69]. Quantifying the coronary artery calcium score with a non-contrast cardiac CT has been proposed to estimate the cardiovascular risk in cancer patients and survivors [69]. Recent research suggests that the CCT-derived extracellular volume (ECV) could be a biomarker for monitoring anthracycline-induced cardiotoxicity in breast cancer patients [70,71]. A proactive, multimodal strategy combining biomarkers and imaging is essential to optimize cardio-oncologic care.

4. The "Old Gatekeepers": Drugs Which Protect Against Ventricular Dysfunction

The 2022 ESC cardio-oncology guidelines recommend angiotensin-converting enzyme inhibitors (ACEIs) or angiotensin receptor blockers (ARBs), along with beta-blockers, for preventing cardiotoxicity in high- or very-high-risk patients undergoing anthracycline, anti-HER-2, or other targeted anticancer therapies that may cause LV systolic dysfunction, according to the Heart Failure Association–International Cardio-Oncology Society score (HFA-ICOS score) [4] (Table 2).

Table 2. This tables summarizes the main drugs used to treat or prevent left ventricular dysfunction.

Drug/Class	Mechanism of Protection	Key Evidence
ACE Inhibitors	↓ ROS, ↑ SOD, mitochondrial support, PPARα modulation, renin–angiotensin system balance.	[4,72–78]
Angiotensin Receptor Blockers	Antioxidant effects.	[4]
Beta-Blockers (Carvedilol)	β-arrestin-biased signaling, ↓ apoptosis, mitochondrial respiration preservation.	[79–81]
Mineralcorticoid Receptoragonists	Aldosterone receptor blockade, antioxidant effects.	[82]
Dexrazoxane	Iron chelation, Top2β inhibition, prevention of DNA damage, and ROS generation.	[83]
Liposomal Anthracyclines	↓ cardiac exposure to anthracycline, prolonged circulation, ↓ ROS, and apoptosis.	[84,85]
Statins	↓ oxidative stress, ↑ SOD2, ↓ heart inflammation as evidenced by ↓ TNF-α expression.	[86–89]
Angiotensin Receptor-Neprilysin Inhibitors	AMPKα–mTORC1 modulation, ↓ oxidative stress and inflammation, ↓ autophagy.	[86]
Sodium–Glucose Cotransporter Inhibitors	↓ ER stress, ↑ autophagy/ketogenesis, antioxidant/anti-inflammatory effects, ↑ survival.	[87–90]
Soluble Guanylate Cyclase Stimulators	↑ cGMP, ↓ inflammation, mitochondrial support, vasodilation.	[91,92]

Abbreviations: ↓: reduction, ↑: increase, ACE: angiotensin-converting enzyme; AMPKα–mTORC1: AMP-activated protein kinaseα–mechanistic target of rapamycin complex1; cGMP: cyclic guanosine monophosphate; ER: endoplasmic reticulum; PPARα: Peroxisome Proliferator-Activated Receptor; ROS: oxygen reactive species; SOD: superoxide dismutase; TNF-alpha: tumor necrosis factor-alpha.

4.1. *Angiotensin-Converting Enzyme Inhibitors and Angiotensin Receptor Blockers*

ACEIs and ARBs reduce oxidative stress but enhance mitochondrial activity and cardiomyocyte metabolism. In a study, captopril and enalapril were tested for their antioxidant protection against DOXO (adriamycin)-induced cardiac and liver toxicity. Rats were given captopril (10 mg/kg) or enalapril (2 mg/kg) intragastrically (i.g.) daily for 7 days before receiving a single i.p. injection of DOXO (15 mg/kg). After 30 h, DOXO killed the animals [72]. Indeed, it increased thiobarbituric acid reactive substances (TBARS), an indicator of lipid peroxidation, and inhibited superoxide dismutase (SOD) in the heart and liver tissues, indicating acute cardiac toxicity [72]. Nonetheless, it was demonstrated how ACEI pretreatment significantly lowered heart and liver TBARS concentrations and improved cardiac and hepatic SOD activity inhibition [72]. These findings suggest that captopril and enalapril may protect the heart from DOXO-induced acute oxidative damage by limiting

ROS production and reducing oxidative stress [72]. Furthermore, Vaynblant et al. evaluated the protective effects of enalapril against DOXO-induced HF in a canine model [73]. Seventeen dogs were treated with DOXO over four weeks, with one group receiving concurrent enalapril therapy [73]. The results showed a marked improvement in survival (100% vs. 36%), as well as significant improvement in the LVEF in the enalapril-treated group [73]. The experimental data from a rat model of daunorubicin-induced cardiotoxicity suggest that ACEIs may exert early cardioprotective effects, improving myocardial contractility and normalizing QT prolongation, even in the absence of a full recovery of the LVEF [74]. These findings support the potential for early neurohormonal modulation to prevent structural deterioration during anthracycline therapy [74]. Other preclinical studies have maintained that anthracyclines alter the renin–angiotensin–aldosterone system (RAAS), leading to increased levels of angiotensin II and overexpression of angiotensin type 1 receptor, thereby contributing to cardiotoxicity [75]. The clinical evidence supports these findings: both retrospective and prospective studies have demonstrated that ACEIs reduce the likelihood of LVEF decline in this setting [76,77]. Moreover, starting an ACEI medication within six months of chemotherapy improves the results [78]. Furthermore, HER-2 inhibition, as occurs with trastuzumab and other targeted therapies, can disrupt the myocardial signaling pathways that normally limit activation of the RAAS. This disruption has been associated with increased angiotensin II production and oxidative stress via NADPH oxidase activation [77]. These effects may contribute to myocardial injury, providing a mechanistic rationale for considering ACEIs to counteract neurohormonal activation and oxidative damage in this setting. So, according to recent studies, ACEIs may also help manage the cardiotoxicity of HER2 inhibitors [77].

4.2. Beta-Blockers

Beta-blockers have cardioprotective functions, including antioxidant and antiapoptotic properties [79]. β_1-adrenergic receptors (β_1ARs) are key regulators of cardiac function and are therapeutic targets in heart disease, mediating β-arrestin-dependent pathways [79]. Carvedilol, a β-arrestin-biased β-blocker, selectively activates these non-canonical signals [79]. To test the effects of carvedilol, Santos et al. extracted heart and liver mitochondria from rats treated for 7 weeks with DOXO (2 mg/kg subcutaneously per week), carvedilol (1 mg/kg intraperitoneally per week), or their combination [47]. The mitochondrial function was assessed by measuring state 3 respiration—the rate of oxygen consumption upon the addition of adenosine diphosphate (ADP)—which reflects active engagement in ATP synthesis and is a key indicator of cellular metabolic capacity [47]. Moreover, they investigated the respiratory control ratio (RCR), which is the ratio between state 3 respiration (ADP-stimulated, active respiration) and state 4 respiration (resting, basal respiration) and reflects how effectively mitochondria couple oxygen consumption to ATP production [47]. The heart mitochondria isolated from DOXO-treated rats had lower rates of state 3 respiration (336 +/− 26 versus 425 +/− 53 natom O_2/min/mg protein) and a lower RCR (4.3 +/− 0.6 versus 5.8 +/− 0.4) than the cardiac mitochondria isolated from the saline-treated rats [47]. The DOXO treatment reduced the mitochondrial calcium-loading capacity and NADH-dehydrogenase activity. Moreover, it was observed that carvedilol coadministration reduced cellular vacuolization in cardiac myocytes and avoided DOXO's inhibitory effect on mitochondrial respiration in both the heart and liver [47]. In conclusion, carvedilol gives protection against both structural and functional cardiac tissue damage and provides significant clinical benefit in reducing the dose-limiting mitochondrial dysfunction that accompanies long-term DOXO therapy in cancer patients [80]. In addition, in clinical investigations, such as the CarDHA study, it was demonstrated that carvedilol was necessary to minimize DOXO-induced cardiotoxicity and preserve LVEF. However, its oral intake does not necessarily result in a significant reduction in HF hospitalizations or cardiac fatalities [81].

4.3. Mineralcorticoid Receptor Antagonist

Spironolactone is a mineralcorticoid receptor antagonist (MRA) that inhibits the final phase of the RAAS pathway, which has shown antioxidative effects and is being studied for possible benefits; however, the research is limited [82].

4.4. Dexrazoxane and Liposomal Anthracyclines

The use of dexrazoxane and liposomal anthracyclines in individuals at high or very high risk of cardiotoxicity who require anthracycline therapy is strongly recommended [4].

Dexrazoxane is the only clinically approved cardioprotective agent for preventing anthracycline-induced cardiotoxicity, and its efficacy relies on a dual mechanism of action. First, it acts as an iron chelator, preventing the formation of anthracycline–iron complexes that drive the production of ROS. Second, it inhibits Top2β, a key mediator of DOXO-induced DNA damage in cardiac cells [83]. By competitively binding to Top2β, dexrazoxane prevents the formation of DOXO–Top2β complexes, avoiding double-strand DNA breaks and the subsequent activation of pro-apoptotic pathways [83]. These complementary mechanisms help preserve myocardial integrity without compromising the antitumor efficacy of anthracyclines [83].

Liposomal anthracyclines, particularly pegylated liposomal DOXO, have been developed to reduce the cardiotoxic effects of conventional anthracyclines [84]. Liposomal formulations of DOXO modify drug distribution, leading to lower peak plasma concentrations and reduced myocardial exposure. By preferentially targeting tumor tissue while limiting cardiac uptake, they have been associated with a lower incidence of anthracycline-induced cardiotoxicity [84,85]. As a result, cardiac exposure to DOXO is significantly decreased, leading to lower levels of ROS, reduced mitochondrial damage, and less induction of apoptosis in cardiomyocytes [85].

4.5. Statins

Even though statins are mainly considered for acute and chronic ischemic diseases, emerging evidence suggests that these drugs may offer cardioprotective effects against chemotherapy-induced cardiotoxicity, particularly in patients receiving anthracyclines and trastuzumab. Preclinical and early clinical studies first highlighted this potential, such as the pilot study by Acar et al., which demonstrated that atorvastatin significantly reduced the development of anthracycline-induced cardiomyopathy, possibly via anti-inflammatory and antioxidant mechanisms in addition to their ability to decrease cholesterol [86]. Moreover, Riad et al. demonstrated in an animal model that pretreatment with fluvastatin reduced the risk of chemotherapy-induced toxicity; actually, in fluvastatin-treated animals, they found lower oxidative stress, increased expression of the antioxidative enzyme SOD2, and decreased heart inflammation as evidenced by lower TNF-alpha expression [87]. More recently, a large propensity score-matched cohort study by Abdel-Qadir et al. confirmed that prior or concurrent statin use was associated with a lower incidence of HF in women undergoing anthracycline- or trastuzumab-based therapy for early breast cancer [88]. Supporting these findings, a meta-analysis by Shahid et al. concluded that statins significantly decreased the risk of LV dysfunction and clinical HF in cancer patients exposed to cardiotoxic agents [89]. Together, these findings support the hypothesis that statins could represent a promising strategy for cardioprotection in a cardio-oncology setting, warranting further validation through prospective randomized trials.

5. The "New Gatekeepers": Drugs Which Protect Against Ventricular Dysfunction

In the evolving landscape of cardio-oncology, the term "new gatekeepers" refers to emerging cardioprotective agents that extend beyond the established classes traditionally used to mitigate cancer therapy-related cardiac dysfunction (CTRCD), such as ACE inhibitors, ARBs, and β-blockers. While these older agents primarily target neurohormonal activation and hemodynamic stress, the novel agents under this category act through additional or complementary mechanisms, often directly modulating oxidative stress, inflammation, and cardiomyocyte survival pathways.

5.1. Angiotensin Receptor-Neprilysin Inhibitors

Angiotensin receptor-neprilysin inhibitors (ARNIs) have prevented anthracycline-induced CTRCD by regulating the AMPKα–mechanistic target of rapamycin complex1 (mTORC1) signaling pathway in mice models, reducing myocardial oxidative stress and inflammation, and modulating dysregulated autophagy in cardiomyocytes [90,91]. The SARAH trial was presented in November 2024 at the American Heart Association meeting [92]. The study assessed the effectiveness of ARNIs at preventing anthracycline-induced CRTCD. From March 2022 to August 2024, 114 high-risk patients with raised high-sensitivity cTnI levels after any dose of anthracyclines were treated at a cancer hospital in Brazil. The majority of the patients were women with an average age of 52 [90]. The participants were randomly assigned to receive an ARNI or a placebo for 6 months, beginning with a low dose and gradually increasing it. The treatment with an ARNI resulted in a reduced percentage of patients experiencing a $\geq$15% drop in the GLS at 6 months compared to placebo (7.1% vs. 25.0%; HR, 0.23; 95% CI, 0.07–0.75; $p = 0.015$), lowering the risk of subclinical cardiotoxicity [90].

5.2. Sodium–Glucose Cotransporter Inhibitors-2

Also, sodium–glucose cotransporter inhibitors-2 (SGLT2is) have shown promise in cardio-oncology for preventing CTRCD. Preclinical research, including animal models, has demonstrated that SGLT2is can improve LV dysfunction in patients with CRTCD due to anthracyclines [93]. This protective impact is achieved through various mechanisms, including anti-inflammatory and antioxidant effects; a reduction in ER stress; and increases in ketogenesis, cell metabolism, autophagy, and ferroptosis [93]. Clinical research suggests that SGLT2is can benefit cancer patients undergoing cardiotoxic treatments, since they are associated with decreased cardiac event rates and better survival outcomes, according to retrospective observational studies [94]. The EMPACARD-PILOT trial, a prospective case–control study, assessed empagliflozin (10 mg/day) in high-cardiotoxic-risk breast cancer patients receiving anthracycline therapy. The trial demonstrated that empagliflozin significantly reduced the incidence of CTRCD (6.5% vs. 35.5%, $p = 0.005$) and improved the LVEF over six months, although there were no significant differences in biomarkers, such as NT-proBNP, or hospitalization rates [93,95].

5.3. Soluble Guanylate Cyclase Stimulators

Vericiguat stimulates soluble guanylate cyclase (sGC), an enzyme that synthesizes cyclic guanosine monophosphate (cGMP), which has vasodilating, antiproliferative, anti-inflammatory, and antifibrotic effects [96]. Vericiguat has been found to effectively prevent DOXO-induced apoptosis and inflammation in cellular models; moreover, its cardioprotective effects are attributed to its ability to lower inflammation by regulating critical pathways, such the NOD-like receptor protein 3 (NLRP3) inflammasome and cytokine production [96]. It also prevents mitochondrial dysfunction by balancing calcium levels and promoting mitochondrial biogenesis, essential for heart cell survival [96,97]. In addition, this drug increases cGMP levels, which improves vasodilation and lowers the afterload, resulting in improved LVEF [97].

6. Future Direction: Pharmacogenomics and Epigenetics

Pharmacogenomics can be defined as how an individual's genetic makeup affects their responses to drugs, considering both the therapeutic efficacy and the risk of adverse reactions to treatments. It aims to optimize therapies based on genetic profiles, thus advancing the goals of precision medicine [98]. Genetic variations affecting drug metabolism and transport are key factors in susceptibility to chemotherapy-induced cardiotoxicity [98]. Variants in the ATP-binding cassette (ABC) genes, which control drug excretion across cell membranes, can vary the clearance and accumulation of drugs in myocardial cells, raising the risk of cardiac toxicity in some patients [98]. In Muckiene et al.'s study, a cohort of 71 individuals treated with DOXO-based chemotherapy had single-nucleotide polymorphisms (SNPs) in their ABC transporter genes evaluated for their association with cardiac dysfunction. The authors identified the ABCC1 rs4148350 TG genotype as significantly associated with an increased risk of anthracycline-induced cardiotoxicity (OR = 8.0; $p = 0.019$), as defined by a $\geq$10% reduction in the LVEF. These findings suggest that genetic variants in ABC transporters, particularly ABCC1 rs4148350, may serve as predictive biomarkers for cardiotoxicity risk stratification in breast cancer patients undergoing anthracycline therapy [99]. Conversely, variants in soluble carrier transporters (SLCs) can improve drug excretion and lower the risk of anthracycline-induced cardiotoxicity [100].

Moreover, chemotherapy-induced cardiac damage is heavily influenced by epigenetic factors [101]. Epigenetics refers to heritable changes in gene expression that occur without alterations to the DNA sequence itself, but are instead regulated by mechanisms such as DNA methylation, histone modification, and non-coding ribonucleic acids (RNAs) [101]. The methylated DNA of circulating and damaged myocytes may serve as a biomarker for cardiotoxicity. The overexpression of the Methyltransferase-Like 4 (METTL4) gene, resulting in increased mitochondrial DNA methylation, has been associated with mitochondrial malfunction and HF [101]. Integrating genetic and epigenetic profiling is a viable technique for stratifying oncology patients based on their cardiotoxicity risk. This technique could improve patient outcomes by detecting cardiac diseases early, guiding prompt therapies, and tailoring oncological therapy accordingly.

7. Future Direction: Gene Therapy

Gene therapy is a promising area in the fight against CTRCD since it may provide long-term and effective myocardial protection. Clinical investigations have focused on reducing the influence of autophagy, a key mechanism in DOXO-induced HF [97]. Excessive autophagy has been associated with cardiomyocyte death, with studies showing an upregulation of autophagic regulatory proteins, such as MAP1LC3 (Microtubule-Associated Protein 1 Light Chain 3), following DOXO therapy [102].

Xiaofan Sun et al. investigated a new gene therapy for autophagy [103]. The study showed that inhibiting autophagy-related protein 7 (Atg7), a key regulator of autophagy, could mitigate the course of DOXO-induced cardiotoxicity [103].

A Western blot analysis of cardiac tissue following tamoxifen injection confirmed the deletion of Atg7 in a mouse model. Both wild-type and Atg7-knockout mice received weekly DOXO injections for four weeks [103]. In the wild-type mice, the LVEF decreased to roughly 40%, while the Atg7-knockout mice maintained their systolic function at around 55% [103]. Moreover, a histological analysis confirmed that the Atg7-deleted animals had less autophagosome and vacuole production in their cardiomyocytes compared to the wild-type mice, highlighting the importance of autophagy in cardiotoxicity [103].

Targeting autophagy with gene therapy is a promising technique to reduce DOXO-induced cardiotoxicity. Further research into these approaches may lead to the development of new therapies that could improve cardiac outcomes in cancer patients undergoing chemotherapy.

8. Conclusions

Cancer therapy-related cardiac dysfunction represents a significant clinical challenge in the era of precision oncology. As survival rates increase, the burden of cardiovascular adverse events becomes more apparent. Consistent with international best practices, integrating cardio-oncology into routine cancer care demands a comprehensive, personalized approach that combines clinical evaluation, imaging, biomarkers, and cutting-edge technologies. This review highlights the complex interplay of molecular mechanisms—particularly oxidative stress, ER stress, mitochondrial dysfunction, and apoptosis—that underlie the cardiotoxicity of several modern oncologic therapies. Established cardioprotective agents, including ACEIs and beta-blockers, remain the cornerstone of prevention and treatment. However, newer pharmacological strategies, such as ARNIs, SGLT2 inhibitors, and vericiguat, offer additional tools with which to preserve cardiac function. In particular, the integration of biomarkers, advanced imaging techniques (such as GLS and cardiac MRI), and individualized risk stratification is transforming cardio-oncology into a precision-based discipline.

The emerging insights from pharmacogenomics and epigenetics further support a personalized approach, enabling the identification of patients at higher risk of cardiotoxicity and the tailoring of therapeutic regimens. Gene therapy targeting autophagy regulation also holds promise, although it remains at an early stage.

In conclusion, the field of cardio-oncology is rapidly evolving. Continued translational research, combined with multidisciplinary collaboration, is essential to mitigate cardiotoxic risks and ensure that the benefits of cancer therapy are not outweighed by preventable cardiovascular harm.

Author Contributions: Conceptualization, M.P. and B.P.; methodology, M.C.D.M.; investigation, G.M.; writing—original draft preparation, M.P.; writing—review and editing, M.P. and B.P.; supervision, G.M. All authors have read and agreed to the published version of the manuscript.

Funding: This research received no external funding.

Informed Consent Statement: Not applicable.

Conflicts of Interest: The authors declare no conflicts of interest.

Abbreviations

The following abbreviations are used in this manuscript:

ACB	ATP-binding cassette
ACEIs	angiotensin-converting enzyme inhibitors
ADP	adenosine diphosphate
AMPKa	AMP-activated protein kinase a
ARBs	angiotensin receptor blockers
ARNI	angiotensin receptor-neprilysin inhibitor
Atg7	autophagy-related protein 7
ATP	adenosine tri-phosphate
β_1ARs	β_1-adrenergic receptors

BCR-ABL	breakpoint cluster region–Abelson leukemia
BNP	B-type natriuretic peptide
BRAF	v-Raf murine sarcoma viral oncogene homolog B1
CCT	cardiac computed tomography
CD4+	cluster of differentiation 4+
CD8+	cluster of differentiation 8+
cGMP	cyclic guanosine monophosphate
CMR	cardiac magnetic resonance
cTnI	cardiac troponin I
cTnT	cardiac troponin T
CTRCD	cancer therapy-related cardiac dysfunction
DNA	deoxyribonucleic acid
ECV	extracellular volume
EF	ejection fraction
eNOS	endothelial nitric oxide synthase
EGFR	endothelial growth factor receptor
ER	endoplasmic reticulum
ERK	extracellular signal-regulated kinase
FT	feature tracking
GLS	global longitudinal strain
HER2	human epidermal growth factor receptor 2
HF	heart failure
HFA-ICOS	Heart Failure Association–International Cardio-Oncology Society
HFpEF	heart failure with preserved ejection fraction
hiPSC-CMs	induced pluripotent stem cell-derived cardiomyocytes
HSF-1	heat shock factor 1
Hsp 25	heat shock protein 25
ICIs	immune checkpoint inhibitors
IL-6	interleukin-6
IRE1α-	inositol-requiring transmembrane kinase/endoribonuclease 1α-spliced
LV	left ventricle
MAP1LC3	microtubule-associated protein 1 light chain 3
MAPK	mitogen-activated protein kinase
MEK	mitogen-activated protein kinase kinase
MM	multiple myeloma
MRA	mineralcorticoid receptor antagonist
MRI	magnetic resonance imaging
mTOR	mechanistic target of rapamycin
mTORC1	mechanistic target of rapamycin complex1
NADPH	nicotinamide adenine dinucleotide phosphate
NLP3	NOD-like receptor protein 3
NO	nitric oxide
NT-ptoBNP	N-terminal pro-BNP
O_2^-	superoxide
p53	protein 53
PDL-1	programmed death ligand-1
PIs	proteasome inhibitors
PP2A	phosphoprotein phosphatase 2A
PPRA	peroxisome proliferator-activated receptor
RAAS	renin–angiotensin aldosterone system
RAF	rapid accelerated fibrosarcoma
RV	right ventricle
RCR	respiratory control ratio
RNA	ribonucleic acid

ROS	oxygen reactive species
sGC	soluble guanylate cyclase
SGLT2 inhibitors	sodium–glucose cotransporter inhibitors-2
SLC	soluble carrier transporters
SOD	superoxide dismutase
TBARS	thiobarbituric acid reactive substances
TKIs	tyrosine kinase inhibitors
TNFα	tumor necrosis factor-α
Top2β	topoisomerase 2β
TTE	trans-thoracic echocardiogram
UlK1	Unc-51-like autophagy-activating kinase 1
UPP	ubiquitin proteasome
VEGF	vascular endothelial growth factor
XBP1	X-box binding protein 1

References

1. Haj-Yehia, E.; Michel, L.; Mincu, R.I.; Rassaf, T.; Totzeck, M. Prevention of cancer-therapy related cardiac dysfunction. *Curr. Heart Fail. Rep.* **2025**, *22*, 9. [CrossRef]
2. Desai, A.; Reddy, N.K.; Subbiah, V. Top advances of the year: Precision oncology. *Cancer* **2023**, *129*, 1634–1642. [CrossRef] [PubMed]
3. Fitzgerald, R.C.; Antoniou, A.C.; Fruk, L.; Rosenfeld, N. The future of early cancer detection. *Nat. Med.* **2022**, *28*, 666–677 [CrossRef]
4. Lyon, A.R.; López-Fernández, T.; Couch, L.S.; Asteggiano, R.; Aznar, M.C.; Bergler-Klein, J.; Boriani, G.; Cardinale, D.; Cordoba, R. Cosyns, B.; et al. 2022 ESC Guidelines on cardio-oncology developed in collaboration with the European Hematology Association (EHA), the European Society for Therapeutic Radiology and Oncology (ESTRO) and the International Cardio-Oncology Society (IC-OS). *Eur. Heart J. Cardiovasc. Imaging* **2022**, *43*, e333–e465, Erratum in *Eur. Heart J. Cardiovasc. Imaging* **2023**, *44*, e98..
5. Koene, R.J.; Prizment, A.E.; Blaes, A.; Konety, S.H. Shared Risk Factors in Cardiovascular Disease and Cancer. *Circulation* **2016**, *133*, 1104–1114. [CrossRef]
6. Bodai, B.I.; Tuso, P. Breast cancer survivorship: A comprehensive review of long-term medical issues and lifestyle recommendations. *Perm. J.* **2015**, *19*, 48–79. [CrossRef]
7. Lenneman, C.G.; Sawyer, D.B. Cardio-oncology: An update on cardiotoxicity of cancer-related treatment. *Circ. Res.* **2016**, *118*, 1008–1020. [CrossRef] [PubMed]
8. Alvarez-Cardona, J.A.; Ray, J.; Carver, J.; Zaha, V.; Cheng, R.; Yang, E.; Mitchell, J.D.; Stockerl-Goldstein, K.; Kondapalli, L.; Dent, S.; et al. Cardio-Oncology Leadership Council. Cardio-Oncology Education and Training: JACC Council Perspectives. *J. Am. Coll. Cardiol.* **2020**, *19*, 2267–2281. [CrossRef] [PubMed] [PubMed Central]
9. Contaldi, C.; D'Aniello, C.; Panico, D.; Zito, A.; Calabrò, P.; Di Lorenzo, E.; Golino, P.; Montesarchio, V. Cancer-Therapy-Related Cardiac Dysfunction: Latest Advances in Prevention and Treatment. *Life* **2025**, *15*, 471. [CrossRef]
10. Octavia, Y.; Tocchetti, C.G.; Gabrielson, K.L.; Janssens, S.; Crijns, H.J.; Moens, A.L. Doxorubicin-induced cardiomyopathy: From molecular mechanisms to therapeutic strategies. *J. Mol. Cell. Cardiol.* **2012**, *52*, 1213–1225. [CrossRef]
11. Jain, D. Cardiotoxicity of doxorubicin and other anthracycline derivatives. *J. Nucl. Cardiol.* **2000**, *7*, 53–62. [CrossRef]
12. Lipshultz, S.E.; Lipsitz, S.R.; Mone, S.M.; Goorin, A.M.; Sallan, S.E.; Sanders, S.P.; Orav, E.J.; Gelber, R.D.; Colan, S.D. Female sex and higher drug dose as risk factors for late cardiotoxic effects of doxorubicin therapy for childhood cancer. *N. Engl. J. Med.* **1995**, *332*, 1738–1744. [CrossRef] [PubMed]
13. Capranico, G.; Tinelli, S.; Austin, C.A.; Fisher, M.L.; Zunino, F. Different patterns of gene expression of topoisomerase II isoforms in differentiated tissues during murine development. *Biochim. Biophys. Acta* **1992**, *1132*, 43–48. [CrossRef]
14. Biancaniello, T.; Meyer, R.A.; Wong, K.Y.; Sager, C.; Kaplan, S. Doxorubicin cardiotoxicity in children. *J. Pediatr.* **1980**, *97*, 45–50. [CrossRef]
15. Hershman, D.L.; McBride, R.B.; Eisenberger, A.; Tsai, W.Y.; Grann, V.R.; Jacobson, J.S. Doxorubicin, cardiac risk factors, and cardiac toxicity in elderly patients with diffuse B-cell non-Hodgkin's lymphoma. *J. Clin. Oncol.* **2008**, *26*, 3159–3165. [CrossRef] [PubMed]
16. Martin, E.; Thougaard, A.V.; Grauslund, M.; Jensen, P.B.; Bjorkling, F.; Hasinoff, B.B.; Tjørnelund, J.; Sehested, M.; Jensen, L.H. Evaluation of the topoisomerase II-inactive bisdioxopiperazine ICRF-161 as a protectant against doxorubicin-induced cardiomyopathy. *Toxicology* **2008**, *255*, 72–79. [CrossRef]

17. Zhang, S.; Liu, X.; Bawa-Khalfe, T.; Lu, L.-S.; Lyu, Y.L.; Liu, L.F.; Yeh, E.T.H. Identification of the molecular basis of doxorubicin-induced cardiotoxicity. *Nat. Med.* **2012**, *18*, 1639–1642. [CrossRef]
18. Doroshow, J.H.; Locker, G.Y.; Myers, C.E. Enzymatic defenses of the mouse heart against reactive oxygen metabolites: Alterations produced by doxorubicin. *J. Clin. Investig.* **1980**, *65*, 128–135. [CrossRef]
19. Schlame, M.; Rua, D.; Greenberg, M.L. The biosynthesis and functional role of cardiolipin. *Prog. Lipid Res.* **2000**, *39*, 257–288. [CrossRef] [PubMed]
20. Ventura-Clapier, R.; Garnier, A.; Veksler, V. Energy metabolism in heart failure. *J. Physiol.* **2004**, *555*, 1–13. [CrossRef]
21. Pecoraro, M.; Pala, B.; Di Marcantonio, M.C.; Muraro, R.; Marzocco, S.; Pinto, A.; Mincione, G.; Popolo, A. Doxorubicin-induced oxidative and nitrosative stress: Mitochondrial connexin 43 is at the crossroads. *Int. J. Mol. Med.* **2020**, *46*, 1197–1209. [CrossRef] [PubMed]
22. Vásquez-Vivar, J.; Martasek, P.; Hogg, N.; Masters, B.S.; Pritchard, K.A., Jr.; Kalyanaraman, B. Endothelial nitric oxide synthase-dependent superoxide generation from adriamycin. *Biochemistry* **1997**, *36*, 11293–11297. [CrossRef] [PubMed]
23. Kalivendi, S.V.; Kotamraju, S.; Zhao, H.; Joseph, J.; Kalyanaraman, B. Doxorubicin-induced apoptosis is associated with increased transcription of endothelial nitric-oxide synthase: Effect of antiapoptotic antioxidants and calcium. *J. Biol. Chem.* **2001**, *276*, 47266–47276. [CrossRef] [PubMed]
24. Deng, S.; Kruger, A.; Kleschyov, A.; Kalinowski, L.; Daiber, A.; Wojnowski, L. Gp91phox-containing NAD(P)H oxidase increases superoxide formation by doxorubicin and NADPH. *Free. Radic. Biol. Med.* **2007**, *42*, 466–473. [CrossRef]
25. Minotti, G.; Menna, P.; Salvatorelli, E.; Cairo, G.; Gianni, L. Anthracyclines: Molecular advances and pharmacologic developments in antitumor activity and cardiotoxicity. *Pharmacol. Rev.* **2004**, *56*, 185–229. [CrossRef]
26. Boudoulas, K.D.; Triposkiadis, F.; Gumina, R.; Addison, D.; Iliescu, C.; Boudoulas, H. Cardiovascular Disease, Cancer, and Multimorbidity Interactions: Clinical Implications. *Cardiology* **2022**, *147*, 196–206. [CrossRef]
27. Kalivendi, S.V.; Konorev, E.A.; Cunningham, S.; Vanamala, S.K.; Kaji, E.H.; Joseph, J.; Kalyanaraman, B. Doxorubicin activates nuclear factor of activated T-lymphocytes and Fas ligand transcription: Role of mitochondrial reactive oxygen species and calcium. *Biochem. J.* **2005**, *389*, 527–539. [CrossRef]
28. Saeki, K.; Obi, I.; Ogiku, N.; Shigekawa, M.; Imagawa, T.; Matsumoto, T. Doxorubicin directly binds to the cardiac-type ryanodine receptor. *Life Sci.* **2002**, *70*, 2377–2389. [CrossRef]
29. Vedam, K.; Nishijima, Y.; Druhan, L.J.; Khan, M.; Moldovan, N.I.; Zweier, J.L.; Ilangovan, G. Role of heat shock factor-1 activation in the doxorubicin-induced heart failure in mice. *Am. J. Physiol. Heart Circ. Physiol.* **2010**, *298*, H1832–H1841. [CrossRef]
30. Kim, S.-C.; Stice, J.P.; Chen, L.; Jung, J.S.; Gupta, S.; Wang, Y.; Baumgarten, G.; Trial, J.; Knowlton, A.A. Extracellular heat shock protein 60, cardiac myocytes, and apoptosis. *Circ. Res.* **2009**, *105*, 1186–1195. [CrossRef]
31. Bregni, G.; Galli, G.; Gevorgyan, A.; de Braud, F.; Di Cosimo, S. Trastuzumab cardiac toxicity: A problem we put our heart into. *Tumori J.* **2015**, *102*, 1–5. [CrossRef] [PubMed]
32. Mohan, N.; Jiang, J.; Dokmanovic, M.; Wu, W.J. Trastuzumab-mediated cardiotoxicity: Current understanding, challenges, and frontiers. *Antib. Ther.* **2018**, *1*, 13–17. [CrossRef]
33. Ewer, M.S.; Ewer, S.M. Cardiotoxicity of anticancer treatments. *Nat. Rev. Cardiol.* **2015**, *12*, 547–558, Erratum in *Nat. Rev. Cardiol.* **2015**, *12*, 620. [CrossRef] [PubMed]
34. Riccio, G.; Coppola, C.; Piscopo, G.; Capasso, I.; Maurea, C.; Esposito, E.; De Lorenzo, C.; Maurea, N. Trastuzumab and target-therapy side effects: Is still valid to differentiate anthracycline Type I from Type II cardiomyopathies? *Hum. Vaccines Immunother.* **2016**, *12*, 1124–1131. [CrossRef]
35. ElZarrad, M.K.; Mukhopadhyay, P.; Mohan, N.; Hao, E.; Dokmanovic, M.; Hirsch, D.S.; Shen, Y.; Pacher, P.; Wu, W.J.; Peng, T. Trastuzumab alters the expression of genes essential for cardiac function and induces ultrastructural changes of cardiomyocytes in mice. *PLoS ONE* **2013**, *8*, e79543, Erratum in *PLoS ONE* **2014**, *9*. [CrossRef]
36. Mohan, N.; Shen, Y.; Endo, Y.; ElZarrad, M.K.; Wu, W.J. Trastuzumab, but Not Pertuzumab, Dysregulates HER2 Signaling to Mediate Inhibition of Autophagy and Increase in Reactive Oxygen Species Production in Human Cardiomyocytes. *Mol. Cancer Ther.* **2016**, *15*, 1321–1331. [CrossRef]
37. Yu, A.F.; Singh, J.C.; Wang, R.; Liu, J.E.; Eaton, A.; Oeffinger, K.C.; Steingart, R.M.; Hudis, C.A.; Dang, C.T. Cardiac Safety of Dual Anti-HER2 Therapy in the Neoadjuvant Setting for Treatment of HER2-Positive Breast Cancer. *Oncologist* **2017**, *22*, 642–647. [CrossRef]
38. Pohl, C.; Dikic, I. Cellular quality control by the ubiquitin-proteasome system and autophagy. *Science* **2019**, *366*, 818–822. [CrossRef]
39. Manasanch, E.E.; Orlowski, R.Z. Proteasome inhibitors in cancer therapy. *Nat. Rev. Clin. Oncol.* **2017**, *14*, 417–433. [CrossRef]
40. Sundaravel, S.H.; I Marar, R.; A Abbasi, M.; Baljevic, M.; Stone, J.R. Bortezomib-Induced Reversible Cardiomyopathy: Recovered With Guideline-Directed Medical Therapy. *Cureus* **2021**, *13*, e20295. [CrossRef] [PubMed]

41. Papanagnou, E.-D.; Terpos, E.; Kastritis, E.; Papassideri, I.S.; Tsitsilonis, O.E.; Dimopoulos, M.A.; Trougakos, I.P. Molecular responses to therapeutic proteasome inhibitors in multiple myeloma patients are donor-, cell type- and drug-dependent. *Oncotarget* **2018**, *9*, 17797–17809. [CrossRef]
42. Efentakis, P.; Kremastiotis, G.; Varela, A.; Nikolaou, P.-E.; Papanagnou, E.-D.; Davos, C.H.; Tsoumani, M.; Agrogiannis, G.; Konstantinidou, A.; Kastritis, E.; et al. Molecular mechanisms of carfilzomib-induced cardiotoxicity in mice and the emerging cardioprotective role of metformin. *Blood* **2019**, *133*, 710–723. [CrossRef]
43. Lettino, M.; Mascherbauer, J.; Nordaby, M.; Ziegler, A.; Collet, J.P.; Derumeaux, G.; Hohnloser, S.H.; Leclercq, C.; E O'NEill, D.; Visseren, F.; et al. Cardiovascular disease in the elderly: Proceedings of the European Society of Cardiology-Cardiovascular Round Table. *Eur. J. Prev. Cardiol.* **2022**, *29*, 1412–1424. [CrossRef]
44. Bringhen, S.; Milan, A.; D'AGostino, M.; Ferri, C.; Wäsch, R.; Gay, F.; Larocca, A.; Offidani, M.; Zweegman, S.; Terpos, E.; et al. Prevention, monitoring and treatment of cardiovascular adverse events in myeloma patients receiving carfilzomib A consensus paper by the European Myeloma Network and the Italian Society of Arterial Hypertension. *J. Intern. Med.* **2019**, *286*, 63–74. [CrossRef]
45. Waxman, A.J.; Clasen, S.; Hwang, W.T.; Garfall, A.; Vogl, D.T.; Carver, J.; O'Quinn, R.; Cohen, A.D.; Stadtmauer, E.A.; Ky, B.; et al. Carfilzomib-Associated Cardiovascular Adverse Events: A Systematic Review and Meta-analysis. *JAMA Oncol.* **2018**, *4*, e174519. [CrossRef] [PubMed]
46. Das, A.; Dasgupta, S.; Gong, Y.; Shah, U.A.; Fradley, M.G.; Cheng, R.K.; Roy, B.; Guha, A. Cardiotoxicity as an adverse effect of immunomodulatory drugs and proteasome inhibitors in multiple myeloma: A network meta-analysis of randomized clinical trials. *Hematol. Oncol.* **2022**, *40*, 233–242. [CrossRef]
47. Georgiopoulos, G.; Makris, N.; Laina, A.; Theodorakakou, F.; Briasoulis, A.; Trougakos, I.P.; Dimopoulos, M.-A.; Kastritis, E.; Stamatelopoulos, K. Cardiovascular Toxicity of Proteasome Inhibitors: Underlying Mechanisms and Management Strategies: *JACC CardioOncology* State-of-the-Art Review. *JACC CardioOncol.* **2023**, *5*, 1. [CrossRef]
48. Wang, H.; Wang, Y.; Li, J.; He, Z.; Boswell, S.A.; Chung, M.; You, F.; Han, S. Three tyrosine kinase inhibitors cause cardiotoxicity by inducing endoplasmic reticulum stress and inflammation in cardiomyocytes. *BMC Med.* **2023**, *21*, 147. [CrossRef]
49. Wang, H.; Sheehan, R.P.; Palmer, A.C.; Everley, R.A.; Boswell, S.A.; Ron-Harel, N.; Ringel, A.E.; Holton, K.M.; Jacobson, C.A.; Erickson, A.R.; et al. Adaptation of Human iPSC-Derived Cardiomyocytes to Tyrosine Kinase Inhibitors Reduces Acute Cardiotoxicity via Metabolic Reprogramming. *Cell Syst.* **2019**, *8*, 412–426.e7. [CrossRef] [PubMed]
50. Richards, C.J.; Je, Y.; Schutz, F.A.; Heng, D.Y.; Dallabrida, S.M.; Moslehi, J.J.; Choueiri, T.K. Incidence and risk of congestive heart failure in patients with renal and nonrenal cell carcinoma treated with sunitinib. *J. Clin. Oncol.* **2011**, *29*, 3450–3456. [CrossRef] [PubMed]
51. Glen, C.; Adam, S.; McDowell, K.; Waterston, A.; Tan, Y.Y.; Petrie, M.C.; Coats, C.J.; Lang, N.N. Cardiotoxicity of BRAF/MEK Inhibitors: A Longitudinal Study Incorporating Contemporary Definitions and Risk Scores. *JACC CardioOncol.* **2023**, *5*, 628–637. [CrossRef]
52. Banks, M.; Crowell, K.; Proctor, A.; Jensen, B.C. Cardiovascular Effects of the MEK Inhibitor, Trametinib: A Case Report, Literature Review, and Consideration of Mechanism. *Cardiovasc. Toxicol.* **2017**, *17*, 487–493. [CrossRef]
53. Agostinetto, E.; Eiger, D.; Lambertini, M.; Ceppi, M.; Bruzzone, M.; Pondé, N.; Plummer, C.; Awada, A.H.; Santoro, A.; Piccart-Gebhart, M.; et al. Cardiotoxicity of immune checkpoint inhibitors: A systematic review and meta-analysis of randomised clinical trials. *Eur. J. Cancer* **2021**, *148*, 76–91. [CrossRef] [PubMed]
54. Haanen, J.B.; Robert, C. Immune Checkpoint Inhibitors. *Prog. Tumor Res.* **2015**, *42*, 55–66. [CrossRef] [PubMed]
55. Rubio-Infante, N.; Ramírez-Flores, Y.A.; Castillo, E.C.; Lozano, O.; García-Rivas, G.; Torre-Amione, G. A Systematic Review of the Mechanisms Involved in Immune Checkpoint Inhibitors Cardiotoxicity and Challenges to Improve Clinical Safety. *Front. Cell Dev. Biol.* **2022**, *10*, 851032. [CrossRef] [PubMed]
56. Khan, S.; Gerber, D.E. Autoimmunity, checkpoint inhibitor therapy and immune-related adverse events: A review. *Semin. Cancer Biol.* **2020**, *64*, 93–101. [CrossRef]
57. Grabie, N.; Lichtman, A.H.; Padera, R. T cell checkpoint regulators in the heart. *Cardiovasc. Res.* **2019**, *115*, 869–877. [CrossRef]
58. Hanna, A.; Frangogiannis, N.G. Inflammatory Cytokines and Chemokines as Therapeutic Targets in Heart Failure. *Cardiovasc. Drugs Ther.* **2020**, *34*, 849–863. [CrossRef]
59. Lyon, A.R.; Dent, S.; Stanway, S.; Earl, H.; Brezden-Masley, C.; Cohen-Solal, A.; Tocchetti, C.G.; Moslehi, J.J.; Groarke, J.D.; Bergler-Klein, J.; et al. Baseline cardiovascular risk assessment in cancer patients scheduled to receive cardiotoxic cancer therapies: A position statement and new risk assessment tools from the Cardio-Oncology Study Group of the Heart Failure Association of the European Society of Cardiology in collaboration with the International Cardio-Oncology Society. *Eur. J. Heart Fail.* **2020**, *22*, 1945–1960. [CrossRef]
60. Michel, L.; Mincu, R.I.; Mahabadi, A.A.; Settelmeier, S.; Al-Rashid, F.; Rassaf, T.; Totzeck, M. Troponins and brain natriuretic peptides for the prediction of cardiotoxicity in cancer patients: A meta-analysis. *Eur. J. Heart Fail.* **2020**, *22*, 350–361. [CrossRef]

61. Porter, T.R.; Mulvagh, S.L.; Abdelmoneim, S.S.; Becher, H.; Belcik, J.T.; Bierig, M.; Choy, J.; Gaibazzi, N.; Gillam, L.D.; Janardhanan, R.; et al. Clinical Applications of Ultrasonic Enhancing Agents in Echocardiography: 2018 American Society of Echocardiography Guidelines Update. *J. Am. Soc. Echocardiogr.* **2018**, *31*, 241–274. [CrossRef]
62. Mousavi, N.; Tan, T.C.; Ali, M.; Halpern, E.F.; Wang, L.; Scherrer-Crosbie, M. Echocardiographic parameters of left ventricular size and function as predictors of symptomatic heart failure in patients with a left ventricular ejection fraction of 50-59% treated with anthracyclines. *Eur. Heart J. Cardiovasc. Imaging.* **2015**, *16*, 977–984. [CrossRef]
63. Ali, M.T.; Yucel, E.; Bouras, S.; Wang, L.; Fei, H.-W.; Halpern, E.F.; Scherrer-Crosbie, M. Myocardial Strain Is Associated with Adverse Clinical Cardiac Events in Patients Treated with Anthracyclines. *J. Am. Soc. Echocardiogr.* **2016**, *29*, 522–527.e3. [CrossRef]
64. Manganaro, R.; Marchetta, S.; Dulgheru, R.; Ilardi, F.; Sugimoto, T.; Robinet, S.; Cimino, S.; Go, Y.Y.; Bernard, A.; Kacharava, G.; et al. Echocardiographic reference ranges for normal non-invasive myocardial work indices: Results from the EACVI NORRE study. *Eur. Heart J. Cardiovasc. Imaging* **2019**, *20*, 582–590. [CrossRef]
65. Bellenger, N.G.; Burgess, M.I.; Ray, S.G.; Lahiri, A.; Coats, A.J.; Cleland, J.G.; Pennell, D.J. Comparison of left ventricular ejection fraction and volumes in heart failure by echocardiography, radionuclide ventriculography and cardiovascular magnetic resonance; are they interchangeable? *Eur. Heart J.* **2000**, *21*, 1387–1396. [CrossRef] [PubMed]
66. Romano, S.; Judd, R.M.; Kim, R.J.; Kim, H.W.; Klem, I.; Heitner, J.; Shah, D.J.; Jue, J.; White, B.E.; Shenoy, C.; et al. Association of Feature-Tracking Cardiac Magnetic Resonance Imaging Left Ventricular Global Longitudinal Strain With All-Cause Mortality in Patients With Reduced Left Ventricular Ejection Fraction. *Circulation* **2017**, *135*, 2313–2315. [CrossRef] [PubMed]
67. Thavendiranathan, P.; Shalmon, T.; Fan, C.-P.S.; Houbois, C.; Amir, E.; Thevakumaran, Y.; Somerset, E.; Malowany, J.M.; Urzua-Fresno, C.; Yip, P.; et al. Comprehensive Cardiovascular Magnetic Resonance Tissue Characterization and Cardiotoxicity in Women With Breast Cancer. *JAMA Cardiol.* **2023**, *8*, 524–534. [CrossRef]
68. Farhad, H.; Staziaki, P.V.; Addison, D.; Coelho-Filho, O.R.; Shah, R.V.; Mitchell, R.N.; Szilveszter, B.; Abbasi, S.A.; Kwong, R.Y.; Scherrer-Crosbie, M.; et al. Characterization of the Changes in Cardiac Structure and Function in Mice Treated With Anthracyclines Using Serial Cardiac Magnetic Resonance Imaging. *Circ. Cardiovasc. Imaging* **2016**, *9*, 12. [CrossRef]
69. Maffei, E.; Messalli, G.; Martini, C.; Nieman, K.; Catalano, O.; Rossi, A.; Seitun, S.; Guaricci, A.I.; Tedeschi, C.; Mollet, N.R.; et al. Left and right ventricle assessment with Cardiac CT: Validation study vs. Cardiac MR. *Eur. Radiol.* **2012**, *22*, 1041–1049. [CrossRef]
70. Sharma, A.; Einstein, A.J.; Vallakati, A.; Arbab-Zadeh, A.; Mukherjee, D.; Lichstein, E. Meta-analysis of global left ventricular function comparing multidetector computed tomography with cardiac magnetic resonance imaging. *Am. J. Cardiol.* **2014**, *113*, 731–738. [CrossRef] [PubMed]
71. Knuuti, J.; Wijns, W.; Saraste, A.; Capodanno, D.; Barbato, E.; Funck-Brentano, C.; Prescott, E.; Storey, R.F.; Deaton, C.; Cuisset, T.; et al. 2019 ESC Guidelines for the diagnosis and management of chronic coronary syndromes. *Eur. Heart J.* **2020**, *41*, 407–477, Erratum in *Eur. Heart J.* **2020**, *41*, 4242. [CrossRef]
72. Sueta, D.; Kidoh, M.; Oda, S.; Egashira, K.; Yamamoto, E.; Kaikita, K.; Matsushita, K.; Yamamoto, Y.; Hirai, T.; Tsujita, K. Usefulness of Cardiac Computed Tomography in the Diagnosis of Anti-Cancer Therapy-Related Cardiac Dysfunction-Consistency With Magnetic Resonance Imaging. *Circ. J.* **2021**, *85*, 393–396. [CrossRef] [PubMed]
73. El-Aziz, M.A.A.; Othman, A.I.; Amer, M.; El-Missiry, M.A. Potential protective role of angiotensin-converting enzyme inhibitors captopril and enalapril against adriamycin-induced acute cardiac and hepatic toxicity in rats. *J. Appl. Toxicol.* **2001**, *21*, 469–473. [CrossRef] [PubMed]
74. Vaynblat, M.; Shah, H.R.; Bhaskaran, D.; Ramdev, G.; Davis, W.J., 3rd; Cunningham, J.N., Jr.; Chiavarelli, M. Simultaneous angiotensin converting enzyme inhibition moderates ventricular dysfunction caused by doxorubicin. *Eur. J. Heart Fail.* **2002**, *4*, 583–586. [CrossRef]
75. Cernecka, H.; Ochodnicka-Mackovicova, K.; Kucerova, D.; Kmecova, J.; Nemcekova, V.; Doka, G.; Kyselovic, J.; Krenek, P.; Ochodnicky, P.; Klimas, J. Enalaprilat increases PPARβ/δ expression, without influence on PPARα and PPARγ, and modulate cardiac function in sub-acute model of daunorubicin-induced cardiomyopathy. *Eur. J. Pharmacol.* **2013**, *714*, 472–477. [CrossRef]
76. Boucek, R.J., Jr.; Steele, A.; Miracle, A.; Atkinson, J. Effects of angiotensin-converting enzyme inhibitor on delayed-onset doxorubicin-induced cardiotoxicity. *Cardiovasc. Toxicol.* **2003**, *3*, 319–330. [CrossRef]
77. Blaes, A.H.; Gaillard, P.; Peterson, B.A.; Yee, D.; Virnig, B. Angiotensin converting enzyme inhibitors may be protective against cardiac complications following anthracycline chemotherapy. *Breast Cancer Res. Treat.* **2010**, *122*, 585–590. [CrossRef]
78. Lee, S.; Alsamarrai, A.; Xiao, A.; Wang, T.K.M. Prevention of anthracycline and trastuzumab-induced decline in left ventricular ejection fraction with angiotensin-converting enzyme inhibitors or angiotensin receptor blocker: A narrative systematic review of randomised controlled trials. *Intern. Med. J.* **2024**, *54*, 1254–1263. [CrossRef]
79. Cardinale, D.; Colombo, A.; Sandri, M.T.; Lamantia, G.; Colombo, N.; Civelli, M.; Martinelli, G.; Veglia, F.; Fiorentini, C.; Cipolla, C.M.; et al. Prevention of high-dose chemotherapy-induced cardiotoxicity in high-risk patients by angiotensin-converting enzyme inhibition. *Circulation* **2006**, *114*, 2474–2481. [CrossRef]

80. Oliveira, P.J.; Bjork, J.A.; Santos, M.S.; Leino, R.L.; Froberg, M.; Moreno, A.J.; Wallace, K.B. Carvedilol-mediated antioxidant protection against doxorubicin-induced cardiac mitochondrial toxicity. *Toxicol. Appl. Pharmacol.* **2004**, *2*, 159–168. [CrossRef] [PubMed]
81. Santos, D.L.; Moreno, A.J.; Leino, R.L.; Froberg, M.K.; Wallace, K.B. Carvedilol protects against doxorubicin-induced mitochondrial cardiomyopathy. *Toxicol. Appl. Pharmacol.* **2002**, *185*, 218–227. [CrossRef]
82. Carrasco, R.; Ramirez, M.C.; Nes, K.; Schuster, A.; Aguayo, R.; Morales, M.; Ramos, C.; Hasson, D.; Sotomayor, C.G.; Henriquez, P.; et al. Prevention of doxorubicin-induced Cardiotoxicity by pharmacological non-hypoxic myocardial preconditioning based on Docosahexaenoic Acid (DHA) and carvedilol direct antioxidant effects: Study protocol for a pilot, randomized, double-blind, controlled trial (CarDHA trial). *Trials* **2020**, *21*, 137. [CrossRef] [PubMed]
83. Wei, Y.; Whaley-Connell, A.T.; Habibi, J.; Rehmer, J.; Rehmer, N.; Patel, K.; Hayden, M.; DeMarco, V.; Ferrario, C.M.; Ibdah, J.A.; et al. Mineralocorticoid receptor antagonism attenuates vascular apoptosis and injury via rescuing protein kinase B activation. *Hypertension* **2009**, *53*, 158–165. [CrossRef] [PubMed]
84. Varghese, S.S.; Eekhoudt, C.R.; Jassal, D.S. Mechanisms of anthracycline-mediated cardiotoxicity and preventative strategies in women with breast cancer. *Mol. Cell Biochem.* **2021**, *476*, 3099–3109. [CrossRef] [PubMed]
85. Gabizon, A.; Shmeeda, H.; Barenholz, Y. Pharmacokinetics of pegylated liposomal Doxorubicin: Review of animal and human studies. *Clin. Pharmacokinet.* **2003**, *42*, 419–436. [CrossRef] [PubMed]
86. Acar, Z.; Kale, A.; Turgut, M.; Demircan, S.; Durna, K.; Demir, S.; Meriç, M.; Ağaç, M.T. Efficiency of atorvastatin in the protection of anthracycline-induced cardiomyopathy. *J. Am. Coll. Cardiol.* **2011**, *58*, 988–989. [CrossRef] [PubMed]
87. Riad, A.; Bien, S.; Westermann, D.; Becher, P.M.; Loya, K.; Landmesser, U.; Kroemer, H.K.; Schultheiss, H.P.; Tschöpe, C. Pretreatment with statin attenuates the cardiotoxicity of Doxorubicin in mice. *Cancer Res.* **2009**, *69*, 695–699. [CrossRef]
88. Abdel-Qadir, H.; Bobrowski, D.; Zhou, L.; Austin, P.C.; Calvillo-Argüelles, O.; Amir, E.; Lee, D.S.; Thavendiranathan, P. Statin Exposure and Risk of Heart Failure After Anthracycline- or Trastuzumab-Based Chemotherapy for Early Breast Cancer: A Propensity Score–Matched Cohort Study. *J. Am. Heart Assoc.* **2021**, *10*, e018393. [CrossRef]
89. Shahid, I.; Yamani, N.; Ali, A.; Kumar, P.; Figueredo, V.; Unzek, S.; Mookadam, F. Meta-analysis Evaluating the Use of Statins to attenuate Cardiotoxicity in Cancer Patients receiving Anthracyclines and Trastuzumab-based Chemotherapy. *Am. J. Cardiol.* **2021**, *156*, 142–145. [CrossRef]
90. Sobiborowicz-Sadowska, A.M.; Kamińska, K.; Cudnoch-Jędrzejewska, A. Neprilysin Inhibition in the Prevention of Anthracycline-Induced Cardiotoxicity. *Cancers* **2023**, *15*, 312. [CrossRef]
91. Hu, F.; Yan, S.; Lin, L.; Qiu, X.; Lin, X.; Wang, W. Sacubitril/valsartan attenuated myocardial inflammation, fibrosis, apoptosis and promoted autophagy in doxorubicin-induced cardiotoxicity mice via regulating the AMPKα-mTORC1 signaling pathway. *Mol Cell Biochem.* **2025**, *480*, 1891–1908, Erratum in *Mol Cell Biochem.* **2025**, *480*, 3217. [CrossRef] [PubMed]
92. Mecinaj, A.; Vinje-Jakobsen, V.; Ngo, D.T.M.; Sverdlov, A.L.; Myhre, P.L. The SARAH trial: More evidence on the role of neurohormonal blockers in prevention of anthracycline-induced cardiotoxicity. *Heart Fail. Rev.* **2025**, *30*, 627–631. [CrossRef]
93. Dabour, M.S.; George, M.Y.; Daniel, M.R.; Blaes, A.H.; Zordoky, B.N. The Cardioprotective and Anticancer Effects of SGLT2 Inhibitors: *JACC: CardioOncology* State-of-the-Art Review. *JACC CardioOncol.* **2024**, *6*, 159–182. [CrossRef] [PubMed]
94. Abdel-Qadir, H.; Carrasco, R.; Austin, P.C.; Chen, Y.; Zhou, L.; Fang, J.; Su, H.M.H.; Lega, I.C.; Kaul, P.; Neilan, T.G.; et al. The Association of Sodium-Glucose Cotransporter 2 Inhibitors With Cardiovascular Outcomes in Anthracycline-Treated Patients With Cancer. *JACC CardioOncol.* **2023**, *5*, 318–328. [CrossRef]
95. Daniele, A.J.; Gregorietti, V.; Costa, D.; López-Fernández, T. Use of EMPAgliflozin in the prevention of CARDiotoxicity: The EMPACARD-PILOT trial. *Cardiooncology* **2024**, *10*, 58. [CrossRef]
96. Quagliariello, V.; Berretta, M.; Bisceglia, I.; Giacobbe, I.; Iovine, M.; Giordano, V.; Arianna, R.; Barbato, M.; Izzo, F.; Maurea, C.; et al. The sGCa Vericiguat Exhibit Cardioprotective and Anti-Sarcopenic Effects through NLRP-3 Pathways: Potential Benefits for Anthracycline-Treated Cancer Patients. *Cancers* **2024**, *16*, 1487. [CrossRef]
97. Mir, A.; Badi, Y.; Bugazia, S.; Nourelden, A.Z.; Fathallah, A.H.; Ragab, K.M.; Alsillak, M.; Elsayed, S.M.; Hagrass, A.I.; Bawek, S.; et al. Efficacy and safety of cardioprotective drugs in chemotherapy-induced cardiotoxicity: An updated systematic review & network meta-analysis. *Cardiooncology* **2023**, *9*, 10. [CrossRef] [PubMed]
98. Fonoudi, H.; Jouni, M.; Cejas, R.B.; Magdy, T.; Blancard, M.; Ge, N.; Shah, D.A.; Lyra-Leite, D.M.; Neupane, A.; Gharib, M.; et al. Functional Validation of Doxorubicin-Induced Cardiotoxicity-Related Genes. *JACC CardioOncol.* **2024**, *6*, 38–50. [CrossRef]
99. Muckiene, G.; Vaitiekus, D.; Zaliaduonyte, D.; Bartnykaite, A.; Plisiene, J.; Zabiela, V.; Juozaityte, E.; Jurkevicius, R. The Impact of Polymorphisms in ATP-Binding Cassette Transporter Genes on Anthracycline-Induced Early Cardiotoxicity in Patients with Breast Cancer. *J. Cardiovasc. Dev. Dis.* **2023**, *10*, 232. [CrossRef]
100. Visscher, H.; Ross, C.J.; Rassekh, S.R.; Sandor, G.S.; Caron, H.N.; Van Dalen, E.C.; Kremer, L.C.; Van Der Pal, H.J.; Rogers, P.C.; Rieder, M.J.; et al. Validation of variants in SLC28A3 and UGT1A6 as genetic markers predictive of anthracycline-induced cardiotoxicity in children. *Pediatr. Blood Cancer.* **2013**, *60*, 1375–1381. [CrossRef]

101. Zemmour, H.; Planer, D.; Magenheim, J.; Moss, J.; Neiman, D.; Gilon, D.; Korach, A.; Glaser, B.; Shemer, R.; Landesberg, G.; et al. Non-invasive detection of human cardiomyocyte death using methylation patterns of circulating DNA. *Nat. Commun.* **2018**, *9*, 1443. [CrossRef] [PubMed]
102. Gu, J.; Hu, W.; Song, Z.-P.; Chen, Y.-G.; Zhang, D.-D.; Wang, C.-Q. Resveratrol-induced autophagy promotes survival and attenuates doxorubicin-induced cardiotoxicity. *Int. Immunopharmacol.* **2016**, *32*, 1–7. [CrossRef] [PubMed]
103. Sun, X.; Du, J.; Meng, H.; Liu, F.; Yang, N.; Deng, S.; Wan, H.; Ye, D.; Song, E.; Zeng, H. Targeting autophagy with SAR405 alleviates doxorubicin-induced cardiotoxicity. *Cell Biol. Toxicol.* **2023**, *39*, 3255–3267. [CrossRef] [PubMed]

BioChem

MDPI

Review

Therapeutic Potential of Essential Oils and Their Bioactive Compounds Against Colon Cancer: Focus on Colon-Specific Micro- and Nanocarriers

Yana Gvozdeva [1,2,*] and Petya Georgieva [1]

1 Department of Pharmaceutical Technology and Biopharmacy, Faculty of Pharmacy, Medical University of Plovdiv, 4002 Plovdiv, Bulgaria; petya.georgieva@mu-plovdiv.bg
2 Research Institute at Medical University of Plovdiv (RIMU), 4002 Plovdiv, Bulgaria
* Correspondence: yana.gvozdeva@mu-plovdiv.bg

Abstract

Colon cancer ranks among the most prevalent and lethal cancers worldwide. Lifestyle and dietary factors—such as high consumption of processed foods, red meat, and alcohol, coupled with sedentary behavior—are key contributors to its development. Despite the availability of standard treatments like surgery, chemotherapy, and radiotherapy, colon cancer remains a significant cause of cancer-related deaths. These conventional approaches are often limited by severe side effects, toxicity, recurrence, and the emergence of drug resistance, highlighting the urgent need for alternative therapeutic strategies. Essential oils are a potential cancer-treatment candidate owing to their diverse composition and favorable safety profile. Numerous studies have revealed essential oils' promising cytotoxic, antioxidant, and anti-inflammatory effects, supporting their potential role in cancer prevention and treatment. Nevertheless, applying volatile oils to the colon faces several limitations, mainly due to their low bioavailability. Furthermore, conditions within the gastrointestinal tract also contribute to the reduced therapeutic efficacy of essential oils. Novel and promising strategies have been developed to overcome the limitations associated with the application of essential oils. The utilization of targeted drug delivery systems has improved the stability of essential oils and enhanced their therapeutic potential in colon cancer treatment. Moreover, even though essential oils cannot replace conventional chemotherapy, they can mitigate some of its adverse effects and improve the efficacy of associated chemotherapy drugs. This review explores the potential of essential oils and their bioactive compounds in colon cancer therapy and highlights current advancements in micro- and nanoencapsulation techniques for their targeted delivery to the colon.

Keywords: anticancer essential oils; bioactive compounds; colon cancer; colon-targeted drug delivery; micro- and nanoapproaches

1. Introduction

Colon cancer (CC) ranks as the third most common cancer globally, following breast and lung cancers, and holds the second-highest mortality rate among all malignancies [1]. By 2035, the number of deaths from rectal cancer is projected to rise by 60%, while fatalities from CC are expected to increase by 71.5% [2]. Although advancements in diagnostics, treatment options, and screening programs have been made, the rates of illness and death from CC remain significant [3]. There is a growing consensus that the most effective strategy for combating cancer lies in its prevention. One promising approach involves the long-term use of oral agents, a method known as chemoprevention. This strategy entails the use of pharmaceutical, nutritional, or biological substances to prevent, delay, or even reverse the development of cancer. CC, in particular, is considered a strong candidate for chemoprevention due to its extended premalignant phase, which offers a valuable window for early intervention [4]. Natural compounds derived from plants have been highlighted as potential chemopreventive agents, offering cancer-inhibiting properties with reduced toxicity compared to conventional synthetic drugs. Moreover, traditional cancer treatments often come with a range of adverse side effects for patients. Therefore, there is a pressing need to explore more effective and less harmful therapeutic approaches [2,5].

The use of natural products in treating various diseases has long been a foundational element of modern medicine. In the field of cancer therapy, natural compounds continue to play a critical role, with many therapeutic agents either derived directly from natural sources or inspired by them [6]. Essential oils (EOs), which are complex mixtures containing up to 300 distinct bioactive compounds, can be extracted from a variety of medicinal or aromatic plants [7]. Although research into their anticancer potential is relatively recent, EOs have already been shown to influence several molecular pathways, including the induction of apoptosis via both intrinsic and extrinsic mechanisms. While EOs are not a replacement for conventional chemotherapy, they offer notable benefits, such as alleviating some of chemotherapy's side effects and enhancing the overall efficacy of cancer drugs, ultimately contributing to an improved quality of life for patients [2].

EOs have been widely shown to exhibit remarkable activity against cancer cells. Numerous EOs have been investigated across different experimental models in the search for novel therapies for CC, yielding highly encouraging results. These oils demonstrate a broad spectrum of bioactive properties, including cytotoxic, antiproliferative, and antimetastatic effects, which they exert through diverse mechanisms of action [4]. For example, linalool—a naturally occurring terpene alcohol found in various fruits, herbs, and EOs—is known for its antibacterial, anti-inflammatory, and antioxidant properties. Beyond these, linalool has demonstrated anticancer potential against several types of cancer, including prostate, colon, leukemia, and cervical cancers. Its anticancer activity is believed to be mediated through mechanisms such as the induction of apoptosis, generation of oxidative stress, cell cycle arrest, and modulation of the immune response [8].

Despite their potent anticancer properties, the clinical application of EOs remains limited due to several inherent limitations, including poor bioavailability, low solubility, chemical instability, high volatility, and sensitivity to environmental factors such as light and humidity [9]. In recent years, advancements in micro- and nanotechnology have opened new avenues for addressing these limitations. Innovative methods for encapsulating EOs into micro- and nanostructures are being developed, with ongoing efforts to identify suitable polymers and delivery carriers [10].

Micro- and nano-scale drug delivery systems have shown significant promise in enhancing the therapeutic potential of EOs against various tumor types. These delivery platforms offer numerous advantages, such as controlled and sustained release, increased permeability to tumor tissues, improved bioavailability and solubility, and ultimately enhanced anticancer efficacy [10,11].

In this context, the present review explores the chemical composition of EOs, their potential role in the prevention and treatment of CC, and the benefits and limitations of micro- and nanotechnology-based delivery systems. It also discusses the classification and properties of commonly used micro- and nanomaterials and highlights emerging colon-targeted carriers. Most importantly, the review emphasizes recent advances in micro- and nanosystems that have demonstrated promising capabilities in enhancing the anticancer activity of EOs across various cancer cell lines and tumor models [9–12].

2. Colon Cancer and Conventional Therapies

CC ranks as the second most frequently diagnosed cancer among women and the third among men. It represents about 10% of all annual cancer cases and related deaths. Global death rates from CC have increased, largely due to modern dietary patterns and lifestyles characterized by high meat and alcohol consumption and insufficient physical activity [13]. Several factors contribute to the development of CC. Research has shown that individuals face a higher risk of CC if they or their family members have a history of cancer, colon polyps, inflammatory bowel diseases, diabetes mellitus, or if they have undergone a cholecystectomy. Inflammatory bowel disease (IBD) is considered the third most significant risk factor for developing CC. IBD encompasses chronic, incurable conditions—primarily Crohn's disease and ulcerative colitis—that disrupt the immune function of the gastrointestinal tract (GIT), resulting in persistent, uncontrolled inflammation [3].

CC typically begins as small, benign growths known as polyps that develop in the colon lining [14]. Over time, these can turn into malignant tumors capable of spreading to other parts of the body. The progression of CC is divided into four main stages. In the early stages (0 and 1), the abnormal cells are confined to the inner layers of the colon. By stage 2, the cancer penetrates the outer layers of the colon wall but has not yet spread to the lymph nodes; at this stage, these growths are often referred to as "adenomas." In stage 3, the cancer reaches the lymph nodes and becomes known as "carcinoma." Finally, in stage 4, the disease metastasizes and spreads to distant organs [15].

In the early stages, CC often progresses without obvious symptoms. Still, some individuals may experience subtle signs such as slight bleeding, tiredness, loss of appetite, anemia, unexplained weight loss, chronic constipation, or alternating constipation and diarrhea [2]. The mainstay of treatment is surgical removal of the tumor, which may be complemented by chemotherapy or radiotherapy—either separately or in combination—as part of adjuvant or neoadjuvant treatment plans. In recent years, targeted therapies and immunotherapies have become valuable options for managing more advanced, aggressive, or metastatic CC cases [16].

Surgical intervention remains one of the most common treatment methods for patients with CC. However, a major drawback of surgical resection is the potential removal of healthy portions of the colon along with the tumor. Another treatment option is radiation therapy, which can lead to side effects such as fatigue, skin irritation, gastrointestinal discomfort, and, in some cases, bloody stools and infertility in both men and women [17]. Chemotherapy using synthetic anticancer drugs is also widely employed. These drugs aim to kill cancer cells by inhibiting their proliferation or halting cell division. While effective, they often cause side effects such as nausea, diarrhea, neuropathy, and mouth sores. Targeted therapies represent another strategy, focusing on the genes responsible for tumor growth [18]. These treatments work by blocking or deactivating cancer-causing genes, selectively inhibiting cancer cell growth and division while sparing healthy cells. Nonetheless, some patients experience rashes on the face and upper body [19]. Additionally, cancer immunotherapy has shown promise in improving survival rates among patients with advanced or metastatic cancers. However, its effectiveness in CC remains limited, with only about 12% of patients responding to this approach [20]. The effectiveness of immunotherapy varies significantly between patients, making it challenging to predict treatment outcomes. Additionally, immunotherapy can trigger immune-related adverse events that may affect vital organs. Identifying risk factors, such as individual patient characteristics, concurrent medications, and the specific type of cancer [21]. Another major limitation is the high cost of newer immunotherapy treatments, which makes them inaccessible to many patients. Given this context, there is a growing need to explore alternative strategies or combination therapies that can offer more effective outcomes in the treatment of CC [19].

Current cancer research is increasingly focused on developing novel therapeutic strategies and more effective chemotherapeutic agents. Notably, around 60% of cancer treatment drugs are derived from natural products [22]. Today, aromatic plants are recognized as valuable sources of therapeutic compounds, and it is anticipated that several new pharmaceuticals will be developed from these plants in the coming decade [23]. Medicinal plants in particular are a rich and reliable source of bioactive compounds (BACs), offering a sustainable and green approach to CC treatment. Compounds such as terpenoids, saponins, volatile oils, flavonoids, phenolics, quinones, and alkaloids have demonstrated strong cytotoxic effects against CC cells, often with fewer side effects and lower risk compared to conventional treatments. Plant-derived compounds are also known to support CC management by slowing tumor growth, alleviating the side effects of chemotherapy and radiation, and targeting cancer at the molecular level [24]. When used alongside conventional therapies, natural compounds can enhance treatment effectiveness by increasing the sensitivity of cancer cells to standard treatments, allowing for lower drug dosages, minimizing resistance, and reducing adverse side effects. This integrative approach holds significant promise as a valuable therapeutic strategy for managing the progression and metastasis of CC [1]. Although no dietary supplement can completely treat, cure, or prevent cancer, some therapies with BACs may help reduce the risk of developing it, relieve symptoms, and improve quality of life. Nature has long served as a rich reservoir of molecules for drug discovery and innovation. Over half of all anticancer drugs approved globally between 1940 and 2010 are either natural products or derived from them. Within the wide array of natural compounds, phytochemicals—especially EOs—have attracted significant attention due to their diverse and potent biological activities [25].

3. Essential Oils and Their Bioactive Compounds' Potential for Colon Cancer Treatment

EOs are volatile blends of numerous compounds produced by aromatic plants as secondary metabolites, notable for their distinctive and potent scents. These oils typically contain a mixture of 20 to 60 different essential oil (EO) constituents, with terpenes being the most abundant. The components of EOs are categorized according to their chemical structures in terpene hydrocarbons, oxygenated terpenes, and other oxygenated compounds like phenols, esters, ketones, lactones, coumarins, monoterpene alcohols, and sesquiterpene alcohols [26]. EOs are often regarded as more potent than their constituents due to their synergistic interactions. The concentration of individual components in EOs can be affected by various microclimatic and environmental factors, including plant species, seasonal timing, geographic location, climate conditions, soil type, leaf maturity, and the method of extraction. Among these, seasonal changes are particularly significant, often playing a major role in altering the chemical composition of plant-derived EOs [27]. EOs can be extracted from plant materials using a variety of techniques, such as hydro-distillation, steam distillation, cold pressing, solvent extraction, as well as ultrasound and microwave-assisted methods. A distinctive characteristic of EOs is their vast chemical diversity and wide range of botanical sources. Globally, over 17,000 plant species are known to produce EOs, yet out of the more than 3000 identified EOs, only about 10% are produced on a commercial scale [9].

For centuries, EOs have been used in the treatment of various ailments, and their therapeutic use continues to grow in both whole-oil and isolated-compound forms. Recently, researchers have increasingly explored the anticancer properties of EOs, recognizing that their mechanisms of action differ from those of conventional cytotoxic drugs [28].

The complex and variable composition of EOs, along with the diverse nature of cancer types, makes it difficult to pinpoint a single, specific mechanism of action. Early studies primarily investigated the antioxidant and anti-inflammatory effects of EOs, proposing their potential role in cancer treatment. Notably, cancer cells typically exhibit moderately elevated levels of reactive oxygen species (ROS) compared to healthy cells. Additionally, chronic inflammation is recognized as a key feature in the development and progression of cancer [25]. EOs exhibit diverse bioactive properties, including cytotoxic, antiproliferative, and antimetastatic effects on cancer cells through multiple mechanisms of action. Numerous studies have shown the advantages of EOs and their ability to fight cancer cells via target systems (Figure 1). In the context of CC, the effects of EOs have been investigated through in vitro studies using various human CC cell lines, including HT-29, Caco-2, LoVo, Colo-205, SW480, and HCT-116. Among these, HCT-116 cells have been the most extensively studied, as they serve as a model for exploring the molecular mechanisms underlying tumor metastasis [4,29,30].

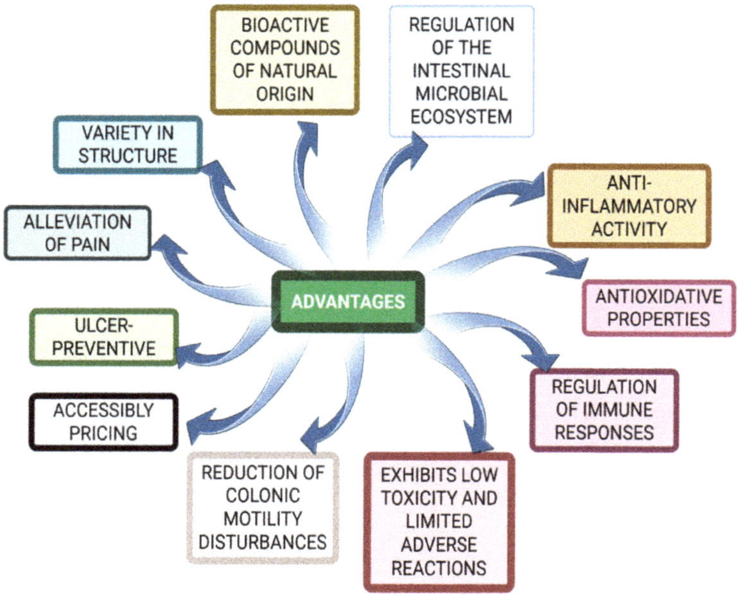

Figure 1. Opportunities and advantages of essential oils and their targeted therapeutic applications. Created at https://BioRender.com (accessed on 7 June 2025).

EOs can trigger programmed cell death in cancer cells through mechanisms such as apoptosis, necrosis, cell cycle arrest, and the disruption of key cellular organelles. These effects are typically accompanied by some critical events leading to cell death, such as increased membrane fluidity and changes in the pH gradient. The three primary classes of EO constituents—phenols, aldehydes, and alcohols—are believed to play central roles in these processes. For example, plant-derived isoprenoids have shown potential in reducing tumor cell size in cancer patients [31].

Oxidative stress is a defining feature of the tumor microenvironment, where the inhibition of the electron transport chain leads to the accumulation of ROS. This buildup can cause extensive damage to cellular components, including proteins, lipids, and nucleic acids such as deoxyribonucleic acid (DNA) [10]. Due to their already elevated baseline levels of oxidative stress, cancer cells are more vulnerable than normal cells to further increases in ROS. Consequently, they are highly susceptible to agents that promote ROS generation [32]. According to the literature, natural compounds found in EOs could be potent inducers of ROS, capable of disrupting critical cellular signaling pathways. Furthermore, EOs can stimulate the upregulation of both non-enzymatic antioxidants (like glutathione) and enzymatic antioxidants (such as superoxide dismutase, glutathione peroxidase, and catalase), which are essential for mitigating oxidative stress during tumor progression [10].

The antimutagenic properties of EOs contribute significantly to cancer prevention by interfering with various stages of tumor initiation. These mechanisms include enhancing enzymatic detoxification of mutagens, blocking their entry into cells, inhibiting the metabolic activation of pro-mutagens, directly neutralizing mutagens, and scavenging free radicals [10]. Thymol, geraniol, and fennel EOs have shown notable antimutagenic effects, largely attributed to their strong antioxidant properties [4].

Numerous studies have shown that inflammation contributes significantly to tumor development. This is why preventing or reducing inflammation can halt the progression of precancerous conditions into malignant tumors. As a result, identifying new and effective strategies to counter inflammation-related carcinogenesis has become a central focus in cancer research [33]. The dual antioxidant and anti-inflammatory properties of EOs compounds are particularly relevant in CC, where chronic inflammation and oxidative stress are known to drive the neoplastic transformation and progression of CC cells [34].

Since EOs contain a variety of biologically active components with different effects, one type of oil can exhibit antifungal, antibacterial, antiseptic, cytotoxicity, anti-inflammatory, antimutagenic, antioxidant, anticancer, antiradical, and chemopreventive effects. Examples include the EO of *Stachys germanica* L. (Lamiaceae) [35,36], *Tanacetum parthenium* L. (Asteraceae) [37], *Salvia miltiorrhiza* and *Salvia verticillata* L. (Lamiaceae) [38], *Carum carvi* L. (Apiaceae) [39], *Cymbopogon flexuosus* (Poaceae), *Laurus nobilis* (Lauraceae), *Melissa officinalis* (Lamiaceae), *Aristolochia mollissima* (Aristolochiaceae), *Cyperus rotundus* (Cyperaceae), *Lindera strychnifolia* (Lauraceae), *Salicornia europaea* L. and *Suaeda maritima* L. (Chenopodiaceae) [40], *Melaleuca alternifolia* (Myrtaceae), *Ocimum basilicum* (Lamiaceae), *Nigella sativa* (Ranunculaceae) [41], *Boswellia sacra* (Burseraceae), *Curcuma aromatica; Curcuma longa; Curcuma wenyujin* (Zingiberaceae), *Zanthoxylum rhoifolium* (Rutaceae), *Rosmarinus officinalis* L. (Lamiaceae) [42], *Juniperus excelsa* M. Bieb and *Juniperus sabina* L. (Cupressaceae) [43], etc., and their main constituents. Several studies have shown the antitumor activity of certain biologically active components in Eos, such as 1,8-Cineole, allicin, Allyl Isothiocyanate, β-Bisabolene, camphene, Cinnamaldehyde, Iso-egomaketone, elemene, eugenol, farnesol, Furanodiene, geraniol, Hinokitiol, d-limonene, Linalool, Myrtenal, and Terpinen-4-Ol [41]. Therefore, EOs and their components hold potential as therapeutic agents and may be considered for use alongside conventional treatments.

Various EOs and their components have been evaluated in vitro and/or in vivo using the above-mentioned CC models, demonstrating notable antitumor activity.

Fan et al. (2015) tested the effect of carvacrol on CC cell lines such as HCT-116 and LoVo. It was found that carvacrol reduces cell proliferation and induces cell cycle arrest, alongside decreasing cell invasion and migration capabilities [29].

EO derived from blood oranges (*Citrus sinensis*) has demonstrated pro-apoptotic and anti-angiogenic effects in CC cells [44]. Similarly, the volatile oil extracted from *Artemisia campestris* L. exhibited notable antitumor activity against HT-29 CC cells, supporting its potential for further investigation as a candidate for chemoprevention and therapeutic development [4].

The study of Cianfaglione et al. (2017) evaluated the anticancer properties of *Eryngium campestre* L. and *Eryngium amethystinum* L., revealing IC_{50} values ranging from 1.5 to

2.99 µg/mL for *E. campestre* and 1.65 to 5.32 µg/mL for *E. amethystinum*. These values were found to be comparable to those of the standard chemotherapeutic agent cisplatin [45].

Qi et al. (2018) found that geraniol exhibited strong cytotoxicity against the Colo-205 cell line [30], but did not induce apoptosis in Caco-2 cells. In the latter, it displayed a cytostatic effect by halting the cell cycle in the S phase, highlighting how the same compound can elicit varying responses depending on the CC cell model. Additionally, geraniol has shown anticancer potential by modulating polyamine metabolism—an emerging target in cancer therapy [46].

Chauhan et al. (2018) studied the cytotoxic effects of thymol on HCT-116 cells by promoting ROS generation, causing DNA damage, and triggering cell death via mitochondrial pathways [47]. Interestingly, at low concentrations, thymol, along with geraniol, nerolidol, and methyleugenol, exerted genoprotective effects in HT-29 cells by mitigating oxidative stress and DNA methylation damage [48]. These findings underscore the critical importance of effective dosing and bioavailability in the colon, as these compounds undergo intestinal absorption that may influence their local concentration and therapeutic potential.

Another study of Larbi et al. (2019) showed that EO from *Brocchia cinerea* (Delile) Vis. demonstrated a more selective anticancer effect in two human cancer cell lines—HCT-116 (CC) and HePG2 (liver cancer). Notably, it exhibited a significant inhibitory effect, reducing CC cell growth by 66.9% [49].

Cuminum cyminum L., a widely used spice from the Apiaceae family, is rich in diverse BACs, including alkaloids, flavonoids, and terpenoids. Among these, cinnamaldehyde stands out as the primary active constituent, contributing significantly to the plant's pharmacological and clinical potential [50]. In animal studies, cumin supplementation demonstrated a protective effect against chemically induced CC in rats. Cinnamaldehyde, the main component of its EO, has been associated with notable antimicrobial, anticancer, and antimutagenic activities. Gas–mass spectrometry analysis of cumin seed extracts revealed the presence of several volatile constituents, including α-pinene, β-pinene, cuminaldehyde, and p-cymene [51].

The fruit rinds of *Garcinia indica* and *Garcinia cambogia* are natural sources of garcinol, a polyisoprenylated benzophenone with demonstrated anticancer properties. Extensive research involving cancer cell lines and animal models has highlighted garcinol's potential in cancer therapy. It exerts strong anticancer effects by inducing both apoptosis and autophagy, and by enhancing the sensitivity of gastric cancer cells to chemotherapeutic agents. In CC, garcinol functions as a powerful suppressor of tumor progression by modulating key molecular pathways. It downregulates critical oncogenes and signaling molecules, thereby reducing tumor cell aggressiveness and invasiveness [52].

Eucalyptus camaldulensis, a fast-growing tree species from the Myrtaceae family, produces EO composed primarily of cineole, which accounts for about 60–80% of its content. In the study by Taheri et al. (2020), the EO of *E. camaldulensis* demonstrated notable cytotoxic effects against the CC cell line CaCo-2 in a manner dependent on both concentration and exposure time. The anticancer properties of this EO are likely linked to its ability to reduce free radical levels and promote apoptosis in these cancer cells [53].

Eugenia caryophyllata (clove), a member of the Myrtaceae family, has been used for various therapeutic applications, including antimicrobial, antioxidant, and anti-inflammatory purposes. Scientific studies have shown that clove contains 15–20% EO rich in phenolic compounds, which contribute to its diverse biological activities such as antibacterial, antioxidant, and antifungal effects. Clove EO has demonstrated antiproliferative activity against multiple cancer cell lines, including HTh-7 thyroid cancer, HT-29 and Caco-2 CC, HepG2 liver cancer, and MCF-7 breast cancer cells. Notably, the EO exhibited stronger anticancer effects than its primary constituent, eugenol, suggesting a possible synergistic interaction with other minor components [4].

Carnosic acid (CA), a phenolic diterpenoid abundantly found in plants of the Lamiaceae family, has gained significant attention for its strong anticancer potential and low toxicity profile. Extensive in vitro and in vivo studies have identified a wide range of molecular targets influenced by CA, including enzymes, kinases, apoptotic regulators, transcription factors, growth factors, oncoproteins, tumor suppressor genes, and various receptors. These targets are associated with key processes such as cell proliferation, survival, angiogenesis, migration, and invasion, positioning CA as a promising agent for cancer prevention and therapy [54].

Barni et al. (2012) demonstrated that CA, derived from *Rosmarinus officinalis*, inhibited the proliferation of Caco-2 CC cells by inducing apoptosis and suppressing cell adhesion, migration, and proteolytic enzyme activity. These effects were likely mediated through the downregulation of COX-2 mRNA expression. The study suggested that CA may influence multiple signaling pathways involved in cancer cell growth and death, supporting its potential use as a chemopreventive or chemotherapeutic agent in CC [55].

Petrocelli et al. (2021) demonstrated that eugenol, an allylbenzene, and cinnamaldehyde, a phenylpropanoid, exhibit specific antitumor effects against CC cells within defined dosage and exposure timeframes. In their study, treatment with cinnamaldehyde and eugenol for 72 h induced apoptosis, necrosis, and cell cycle arrest in Caco-2 and SW-620 CC cell lines, while showing no cytotoxic effects on the normal colonic cell line NCM-460. These findings suggest that, when combined with targeted delivery to the colon, both compounds hold promise for CC prevention or therapy [34].

Research has indicated that plant-derived compounds tend to be less effective when used as isolated molecules. The concept of drug synergism—enhancing therapeutic impact through combination therapies—has gained significant interest. Co-administering conventional drugs with plant-derived BACs can enhance their cytotoxic effects against cancer cells or cell lines, improving overall treatment efficacy [24].

Drug resistance to existing therapies is the main reason for patient mortality. Overcoming this resistance is crucial to improving survival. Combining natural compounds with targeted therapies has shown promise in enhancing clinical efficacy while minimizing side effects. Bioactive natural compounds from EOs, for example, exhibit anticancer effects both in vitro and in vivo through various mechanisms [56]. When used alongside conventional treatments, these compounds can enhance therapeutic sensitivity, allow for lower drug dosages, reduce resistance, and mitigate adverse effects. This approach holds significant potential as a therapeutic strategy for managing the progression and metastasis of CC [1]. Certain specific EO components have been thoroughly investigated in vitro and in animal models, demonstrating notable anticancer properties when used in conjunction with chemotherapy drugs.

β-Elemene, a bioactive compound derived from plants of the *Curcuma genus* (Zingiberaceae family), has demonstrated significant anticancer properties. It is a sesquiterpene compound derived from turmeric that modulates the expression of several key molecules that play critical roles in tumor angiogenesis and metastasis [57]. According to Wang et al. (2022), β-elemene inhibited the proliferation of CC cells by inducing cell cycle arrest at the G2/M phase. It also triggered hallmark features of apoptosis, including nuclear chromatin condensation, externalization of phosphatidylserine on the cell membrane, and reduced mitochondrial membrane potential. In addition to its pro-apoptotic effects, β-elemene elevated intracellular ROS levels, suggesting a role in regulating metabolic stress pathways. In in vivo studies, β-elemene treatment led to a reduction in tumor volume and induced both apoptosis and autophagy in xenograft models using nude mice [58].

The study of Chen et al. (2020) comprehensively evaluated the role of β-elemene in inducing ferroptosis and enhancing the sensitivity of Kirsten rat sarcoma (KRAS)-mutant CC cells to cetuximab. In vitro experiments demonstrated that the combination of β-elemene and cetuximab triggered ferroptotic cell death in KRAS-mutant CC cell lines HCT116 and LoVo. In vivo results further confirmed that this combination inhibited tumor growth by promoting ferroptosis and reduced cancer cell migration through modulation of epithelial–mesenchymal transition. These findings highlight β-elemene as a potential ferroptosis inducer and suggest that its combined use with cetuximab may offer an effective therapeutic approach for CC patients harboring KRAS mutations [59].

Although EOs and their BACs exhibit remarkable biological properties, several challenges (Figure 2) hinder their widespread use in clinical settings and therapies.

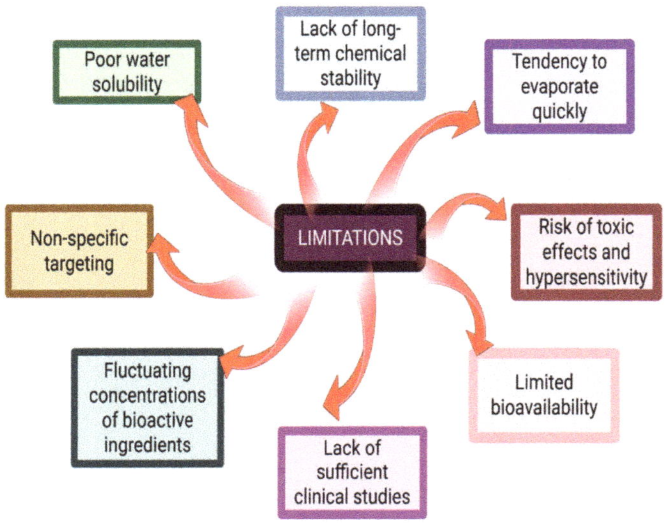

Figure 2. Challenges and limitations in the application of essential oils in targeted therapies. Created at https://BioRender.com (accessed on 13 June 2025).

The direct application of EOs in their free form is often limited by several critical factors. These include their low stability and vulnerability to degradation through volatilization and/or oxidation when exposed to environmental stressors such as oxygen, light, and temperature. EOs also exhibit poor water solubility and may interact unfavorably with food components or pharmaceutical compounds [60]. Oxygen exposure in particular has been shown to cause significant physicochemical and compositional changes in EOs, compromising their stability, quality, and functional properties. Temperature similarly affects EO stability, as higher temperatures tend to accelerate chemical reactions and degradation processes [61]. Light exposure can further exacerbate the instability of EOs, negatively influencing both their integrity and functional efficacy [62]. Another key limitation in EO application arises from its potential interactions with food constituents such as proteins, lipids, and other bioactive molecules. These interactions can diminish their effectiveness and hinder their practical use [63]. To address these challenges, innovative delivery systems—such as microencapsulation and nanoencapsulation—are being developed to enhance the stability, bioavailability, and overall effectiveness of EOs [60].

4. Encapsulation Strategies for Colon-Targeted Delivery of EOs and Their Bioactive Compounds: Advantages, Limitations, and Progress by Incorporating Essential Oils into Specific Colon-Targeted Drug Delivery Systems

Free EOs are typically administered orally by diluting them in milk, olive oil, or other vegetable oils. However, encapsulated EOs are more commonly used for oral intake, particularly in food supplements and functional foods [64]. It is believed that the biological effects of microencapsulated EOs differ from those of conventional EOs, owing to the influence of microencapsulation technology. Encapsulation of EOs or their isolated bioactive constituents allows for targeted and controlled release, protects the oils from degradation and loss, and helps to mask their often unpleasant taste and odor during consumption [65].

Encapsulation of EOs involves enclosing the oils within a uniform or composite matrix—often referred to as a shell, wall, or carrier material—at either the micro or nano size [60]. While the terms "nano" and "micro" generally refer to sizes ranging from 1 to 1000 nm and 1 to 1000 μm, respectively, in practice, nanoencapsulation typically produces capsules between 1 nm and several hundred nanometers in diameter, whereas microencapsulation results in capsules ranging from 1 μm to several hundred micrometers [66]. Encapsulating EOs has emerged as a preferred strategy to enhance their medical efficacy. The bioavailability of both the encapsulating matrix and the EO components is complex, requiring careful consideration of the stability of micro- or nanoencapsulated carriers within the GIT, as well as their potential to improve systemic absorption and therapeutic efficacy at the site of action [64,65].

This encapsulation process shields EOs from harmful environmental conditions such as oxidation, evaporation, and degradation. Thanks to their reduced size and increased surface-area-to-volume ratio, microparticles (MPs) and nanoparticles (NPs) enhance the bioavailability of EOs and enable more efficient diffusion to target sites—for example, targeted delivery to the colon. Additionally, encapsulation requires smaller quantities of EOs, which helps to minimize potential toxicity and lower production costs [60,67]. Encapsulation serves to limit the interaction between the core substance and its surrounding environment, thereby reducing the evaporation or loss of core compounds. It also enhances the handling and stability of the encapsulated material, enables controlled release, minimizes unwanted odors and off-flavors, and promotes uniform dispersion of the active ingredient within food products [68]. One of the major challenges associated with plant-derived compounds, particularly EOs, is their limited bioavailability. EOs generally exhibit poor solubility in systemic fluids, which can cast doubt on their therapeutic potential. Moreover, many EOs are unstable in physiological conditions, hindering their ability to reach cancer cells at effective concentrations [24].

EO encapsulation requires the use of a wall or carrier material. A wide variety of substances can serve this purpose, including gums (sodium alginate), carbohydrates (starches, cellulose derivatives), chitosan, lipids (including stearic acid, waxes), polymethylmetacrylates, and proteins (albumin, casein), among others. Several techniques are employed for encapsulation, including spray drying (atomization), coacervation, nano-/microemulsions, nanoprecipitation, high pressure, homogenization, coatings, freeze-drying, in situ polymerization, supercritical fluid, and others [69].

The encapsulation approach enables controlled and sustained release of EOs over time, thereby improving their therapeutic performance. The primary objective is to develop advanced delivery systems that not only regulate the release of EOs but also enhance their physical stability, minimize off-target effects, reduce volatility, improve therapeutic activity, lower toxicity, and ultimately increase patient compliance and convenience [70]. Sustained-release formulations can help maintain EO concentrations within the therapeutic window for extended periods, thereby boosting treatment effectiveness [71]. Encapsulated EOs can be released through various mechanisms, such as mechanical stress, temperature fluctuations, diffusion from MPs and NPs, and stimuli-responsive triggers—including pH changes, nanocarrier biodegradation, or dissolution [9].

One potential challenge in using these compounds to prevent CC is ensuring their selective release into the colon at appropriate concentrations. This is complicated by the fact that, physiologically, they are quickly absorbed in the small intestine after ingestion and tend to have short half-lives in the bloodstream due to rapid liver metabolism [34]. Targeted delivery to the ileo-colon region is a highly effective strategy for the localized treatment of various bowel disorders, including ulcerative colitis, Crohn's disease, and CC. Moreover, delivering sensitive compounds—such as proteins, peptides, and EOs—specifically to this region can significantly enhance their systemic bioavailability [72]. To achieve this, colon-specific drug delivery systems (CDDS) must be designed to protect EOs from degradation in the acidic environment of the stomach and the enzymatic activity of the small intestine, ensuring that release and absorption occur primarily upon reaching the colon [73]. Targeted drug delivery to the colon enhances drug concentration at the disease site, allowing for lower doses and consequently reducing side effects [74].

The GIT's complex physiology—including variable pH levels, fluid volumes, and transit times—can significantly impact the performance of such systems. These factors vary both between and within individuals, and are further influenced by the presence of food and digestive enzymes, adding to the complexity [73]. Numerous studies have demonstrated that the colon is an effective site for the absorption of drugs and biologically active compounds, largely due to its lower enzymatic activity, which reduces the risk of degradation. Designing CDDS involves several challenges and limitations, as the oral dosage must go through the entire GIT before reaching its intended site of action. As a result, the primary objective of CDDS is to protect active substances from the harsh acidic environment of the stomach and enzymatic hydrolysis in the duodenum and jejunum [75]. Traditional oral dosage forms are largely ineffective for delivering substantial amounts of BACs to the colon, as these compounds are often absorbed or degraded in the upper segments of the GIT. Oral CDDS systems are designed to minimize or prevent drug release in the upper GIT, ensuring targeted delivery to the colorectal region for either local or systemic therapeutic effects. To achieve effective colon targeting, it is essential to consider the physiological conditions that influence drug delivery in this region. To overcome these barriers, several colon-targeting strategies have been developed, including pH-sensitive systems, time-dependent release mechanisms, microbiota-triggered delivery systems, and osmotically controlled CDDS [72,76,77], which will be discussed in greater detail in the following paragraphs. Some of them have been developed that exploit physiological conditions of the gastrointestinal tract, specifically the elevated luminal pH of the ileum and the enzymatic activity of the colonic microbiota. These microbial enzymes include pectinase, amylase, dextranase, glycosidase, and azoreductase [72]. The dynamic physiological conditions of the gastrointestinal tract, coupled with pathophysiological alterations at disease sites, pose significant challenges for the rational design of colon-targeted formulations. Such variability often compromises site-specific delivery, resulting in suboptimal in vivo performance. For instance, fluctuations in luminal pH, influenced by both intrinsic and extrinsic factors, can markedly reduce the reliability of pH-dependent systems, leading to premature drug release in the upper gastrointestinal tract or incomplete release at the target site [78]. Each approach for colon-targeted delivery offers unique benefits and limitations, but combining two or more strategies is a promising way to overcome individual drawbacks and enhance the overall efficiency of colon-targeted drug delivery [79].

What makes it possible to encapsulate and deliver EOs and their BACs directly to the colon? This is primarily enabled through the use of specialized polymers. Which polymers are specific for targeting the colon? The most widely employed are pH-sensitive polymers, followed by biodegradable polymers such as azopolymers, which remain stable throughout the gastric and small intestinal environments but undergo enzymatic cleavage in the colon through azoreduction mediated by azoreductase enzymes produced by the resident colonic microflora (e.g., azobacteria) [80].

Among pH-sensitive excipients, methacrylic acid copolymers (Eudragit® series) are the most extensively utilized. Their dissolution profiles are pH-dependent: Eudragit® L100 dissolves at pH 6.0, Eudragit® S100 at pH 7.0, Eudragit® L30D at pH 5.6, Eudragit® FS30D at pH 6.8, and Eudragit® L100-55 at pH 5.5 [81]. These polymers can be applied either individually or in combination to develop CDDS. Additional enteric pH-responsive polymers include polyvinyl acetate phthalate, hydroxypropyl methylcellulose phthalate, cellulose acetate phthalate, and cellulose acetate trimellitate [82].

Another class of carriers comprises polysaccharide-based matrices, which resist degradation in the upper gastrointestinal tract but undergo enzymatic hydrolysis upon exposure to colonic microbiota. Representative polysaccharides for CDDS include amylose, guar gum, pectin, chitosan, inulin, cyclodextrins, chondroitin sulfate, dextrans, and locust bean gum [83].

Time-dependent systems provide an alternative targeting strategy by delaying drug release until the formulation reaches the colon [84]. While gastric emptying is highly variable, small intestinal transit is comparatively consistent, enabling the design of chronotherapeutic systems that resist gastric acidity and release the drug following a predetermined lag phase corresponding to orocecal transit [85].

In addition to conventional polymeric approaches, several patented delivery technologies—such as Phloral® Technology, CODES® Technology, and Opticore® Technology—have been developed [86]. These systems integrate polymers with complementary mechanisms of action to enhance robustness and precision in colonic targeting.

4.1. Microencapsulation Approaches

Microencapsulation is described as the process of coating drugs or BACs within a homogeneous or heterogeneous matrix to form small capsules. This encapsulation protects the core material and can facilitate its controlled release [87].

This technique was initially developed to shield active ingredients from environmental and external factors. Microencapsulaton approach holds significant importance in the pharmaceutical field for several key reasons: it protects bioactive ingredients from environmental factors such as heat, humidity, air, and light; it helps preserve their bioactivity; it improves ease of handling; it enables controlled release under specific conditions; it masks unpleasant tastes and odors; and it allows for precise dosing, especially when working with small quantities of active compounds [88]. In oil microencapsulation, emulsification is a crucial step in which BACs are typically dispersed within an aqueous solution. The process typically begins with emulsifying or dispersing the oils in an aqueous solution containing a wall material that also serves as an emulsifier. This forms an emulsion, which is the basis for obtaining the final microstructures. Nevertheless, the primary purpose of encapsulating an active ingredient is to achieve controlled release, as already said [89]. Over time, however, the ability to achieve controlled release has become a key focus in the advancement of encapsulation technologies. These techniques are generally categorized into chemical, physicochemical, or mechanical methods. Some of the most widely used approaches for protecting lipid-based compounds include coacervation, spray drying, liposome formation, fluid bed spray coating, and emulsification followed by solvent evaporation or extraction [88]. Microencapsulated structures can be classified into two types: microcapsules or microspheres. This division refers to polymers and methods used in the process of microencapsulation. Common wall materials used in microencapsulation include carbohydrates (such as starches, cellulose, and chitosan), gums (like sodium alginate and carrageenan), lipids (including waxes, oils, and fats), and proteins (such as casein and peptides) [87].

Microspheres can be obtained in a range of sizes, from micrometers to millimeters. Their small size provides a larger surface area, giving them an advantage in drug delivery [90]. Most micro-scale carriers designed for colon-targeted therapy fall within a size range of 10–300 μm. Their enhanced therapeutic effectiveness is largely due to reduced EOs absorption in the small intestine and slower transit through the colon, aided by the streaming effect [91]. As already mentioned, by integrating various controlled release strategies—such as polymeric or swellable matrices, enzymatic degradation of the carrier, and pH-responsive coatings—these systems can enhance colon-specific EOs delivery. However, this complexity also brings a heightened risk of incomplete EOs release, particularly due to individual variations in gastrointestinal physiology [92].

4.2. Nanoencapsulation Approaches

Nanoencapsulation technology presents both a significant challenge and a promising opportunity for the delivery of EOs. Due to their subcellular dimensions, nanocarriers can enhance the bioactivity of EOs by enabling deeper tissue penetration and facilitating cellular uptake. Additionally, they offer the ability to control and modulate the release of active compounds at the targeted site [93]. The size and shape of NPs play a crucial role in the effective delivery and penetration of the bioactive compound. NPs can be engineered in various forms, such as spheres, nanotubes, nanorods, nanoflowers, nanostars, and nanoprisms, to enhance therapeutic outcomes. Among these, spherical NPs have demonstrated superior penetration into cancer cells. Research shows that NPs ranging from 10 to 400 nm can effectively pass through cancer cell membranes, while larger NPs—over 500 nm—can still enter cells via endocytosis [19]. Various EO delivery strategies have been used for cancer therapy, including the use of polymeric nanocarriers (dendrimers, micelles, and nanofibers), metallic nanocarriers (magnetic and gold NPs), and carbon-based nanocarriers (multi-wall carbon nanotubes and graphene), as well as lipid-based systems such as liposomes, solid lipid NPs (SLNs), and nanoemulsions (NEs) [10]. Among these, NEs are the most straightforward and cost-effective to formulate and handle [94]. NEs are created by mixing two immiscible liquids to form a heterogeneous system in which the dispersed droplets are smaller than 1000 nm [9]. NEs offer numerous benefits according to their unique properties, such as a high surface area and small droplet size, including enhanced solubility and bioavailability, improved stability, increased dispersibility of hydrophobic drugs, and better sustainability [95]. Unlike traditional emulsions, which typically appear milky, NEs are generally translucent [96]. Some studies have shown that liposomes have become increasingly prominent as carriers for antitumor drugs or BACs. Despite their potential, only a limited number of liposomal drug formulations have successfully reached the market, highlighting the significant challenges in translating liposome technology from laboratory research to clinical practice [97]. One of the most critical issues is the stability of liposomes, which directly impacts drug loading, permeability, and release rates during formulation, storage, and metabolism [98]. Their inherent physical and chemical instability poses a major barrier to clinical application. The liposomal membrane, composed of dynamic phospholipid bilayers, is prone to constant transmembrane movement, which can cause particle aggregation, sedimentation, and ultimately lead to structural instability.

Chemically, liposomes are vulnerable to degradation through phospholipid oxidation and hydrolysis, two primary factors contributing to their instability [99]. Additionally, finding suitable sterilization methods for liposomal formulations remains a significant technical challenge in their development for medical use. SLNs were introduced in the 1990s as a novel system for encapsulating lipophilic drugs, offering an alternative to traditional lipid-based carriers like emulsions and liposomes [25]. SLNs provide several key advantages, including excellent physical stability, high drug loading capacity, controlled drug release, and the potential for surface modification to enable both passive and active targeting of cells [100]. Being biocompatible, non-toxic, and suitable for large-scale production makes them especially attractive for pharmaceutical applications. Furthermore, incorporating liquid lipids—such as EOs—into the solid lipid matrix leads to the formation of nanostructured lipid carriers [101]. This modification enhances the system's stability, drug loading efficiency, and encapsulation capacity while minimizing drug expulsion during storage [25].

In healthy tissues, NPs are generally unable to enter or accumulate within cells because of the intact cell membrane, tight cell junctions, and efficient lymphatic drainage [102]. However, tumor cells disrupt the normal physiological environment as they grow and invade the surrounding tissue. This disruption leads to several abnormal changes, including damage to cell membranes, development of hypoxic (low oxygen) regions, initiation of new blood vessel growth (angiogenesis), increased tissue permeability, and impaired lymphatic drainage [103]. These changes create a unique tumor microenvironment that NPs can exploit. By taking advantage of the enhanced permeability and retention (EPR) effect in tumors, NPs can more effectively and passively deliver their therapeutic cargo to cancer cells [10]. Nanoparticle-based drug formulations typically require lower doses than conventional therapies. Additionally, NPs tend to be taken up by macrophages at sites of inflammation in the colon, which helps retain the drug at the target site for an extended period [79].

Nano-scaled drug delivery systems face several challenges in reaching their target site. However, one of the key advantages of NPs as drug delivery systems is their ability to release drugs in a controlled, pH-dependent manner. This feature helps maintain drug stability as they pass through varying pH environments in the GIT [104]. Additionally, encapsulating active compounds within nanocarriers protects them from enzymatic degradation by intestinal microbiota, which could otherwise bind to and break down bioactive molecules, reducing their pharmacological effectiveness [105]. Their surface properties can be customized—such as by adjusting their charge—to target specific conditions like colonic diseases. For instance, polycationic NPs can bind to inflamed intestinal tissues, while polyanionic NPs can interact with positively charged molecules such as cytokines found in CC [79]. It has been shown that negatively charged NPs with a diameter of 200 nm or smaller can more effectively cross the intestinal mucosal barrier [106].

4.3. Comparison Between Micro- and Nano-Scaled Colon-Targeted Delivery Systems

Microparticles (MPs) in particular tend to exhibit mucoadhesive properties, allowing them to adhere to the intestinal mucosal surface, though their absorption across the epithelial barrier remains limited. Studies have reported the accumulation of MPs in the rectal mucosa of patients, whereas NPs were detected only in minimal amounts at that site. In contrast, NPs can penetrate the systemic circulation, enabling them to exert systemic effects [73]. Compared to MPs, NPs offer a significantly larger surface area and greater flexibility for surface modification, which enhances their ability to deliver antitumor drugs to targeted sites while reducing the risk of opsonization. Moreover, NPs improve drug diffusion and accumulation within tumor cells by leveraging the EPR effect [75]. Most nanomedicines typically exhibit a low drug-loading capacity (less than 10%), which poses challenges for clinical translation. Consequently, there is growing interest in nanomedicines with drug loadings exceeding 10% [9]. One of the key features of NPs is their biocompatibility and high bioavailability, making them appropriate for drug delivery in cancer therapy. A major challenge in current cancer treatment is efficiently delivering drugs into cancerous cells—a task that is difficult for larger molecules. To penetrate the cancer cells, the carrier structure must be small, and NPs offer an advantage due to their nano-scale size, which allows easier cellular entry [19]. However, smaller particles may penetrate more deeply into the tissue, passing beyond the epithelium into the submucosa, where they tend to remain for longer periods. Previous studies have demonstrated that accumulated NPs in the colon can remain in cancerous regions for several days. A sustained release over this period appears beneficial for establishing a reservoir of EOs at the target site [91].

Ultimately, both micro- and nanoencapsulation techniques can significantly improve the physicochemical properties and stability of EOs by enhancing their water dispersibility, reducing volatility, and protecting them from environmental degradation [107]. Furthermore, nano- and micro-particulate carriers offer considerable promise for improving both colonic targeting and mucosal uptake. Therefore, this review focuses precisely on them as an alternative strategy for delivering EOs and their BACs to the colon. Consequently, formulation strategies incorporating particle size reduction are increasingly recognized as advantageous for enhancing the therapeutic efficacy of CDDS [108]. Figure 3 represents a comparison between micro- and nano-scaled carriers.

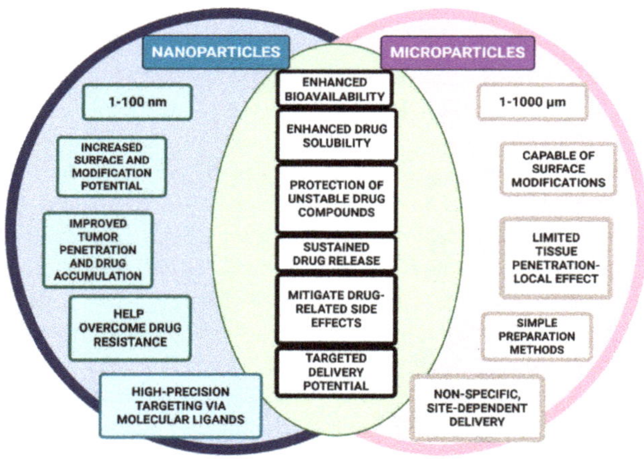

Figure 3. Nanoparticles vs. microparticles: a comparative analysis of properties, applications, and behaviors. Created at https://BioRender.com (accessed on 13 June 2025).

Developing EOs micro- and nano-scale delivery systems for colon targeting, using various strategies, presents a promising approach for colon-specific therapy by enabling controlled EOs release and enhanced therapeutic outcomes. However, the release of EOs should not be considered the ultimate goal of these delivery systems. Since the colon naturally tends to expel metabolites rather than absorb them, further research is needed, focusing especially on the uptake of specific BACs within the colonic environment [90].

5. Advanced Micro- and Nano-Delivery Systems for Colon-Specific Targeting of Essential Oils and Their Bioactive Compounds

As already noted, micro- and nanotechnology-based strategies have great potential for delivering EOs or their BACs directly to the colon, maximizing their therapeutic benefits while minimizing systemic side effects. These advanced approaches exploit the unique characteristics of polymers and the preparation method to encapsulate EOs, protect them from degradation, and provide targeted release in the colon with the potential to treat or prevent CC. Many studies have demonstrated the incorporation of EOs or their bioactive components into micro- and nanostructures, some of which are presented in Figure 4 and described below.

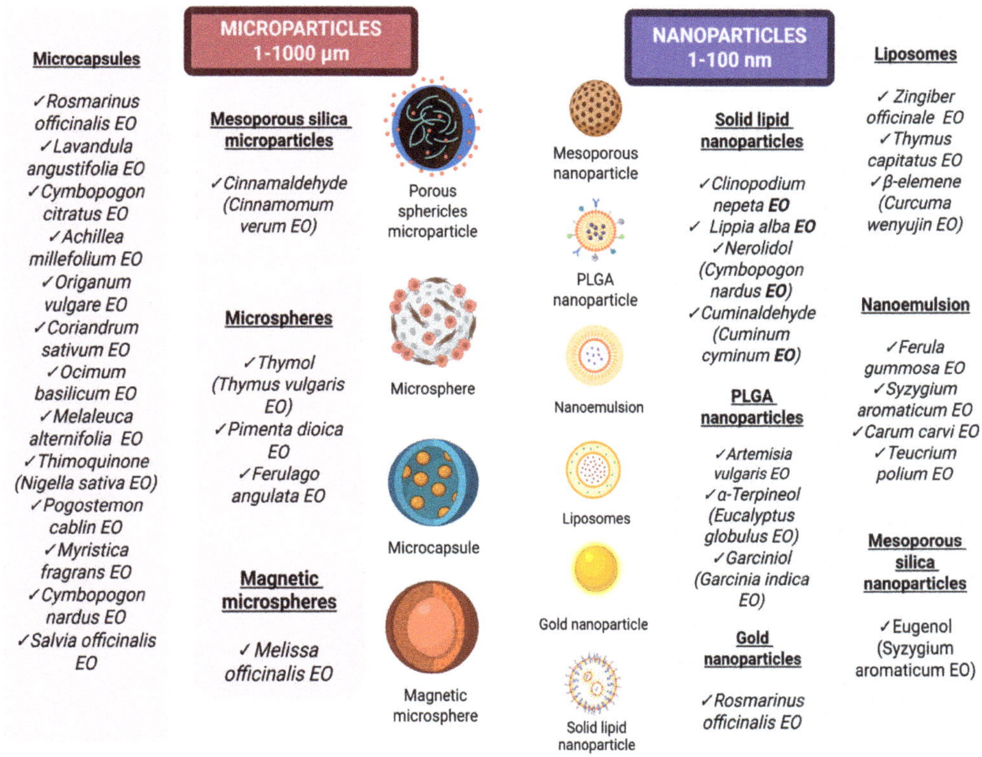

Figure 4. Colon-targeted delivery of essential oils and their bioactive compounds via micro- and nanoformulations. Created at https://BioRender.com (accessed on 14 June 2025).

Zhai et al. (2022) developed polyethylene glycol (PEG)ylated β-elemene (from Curcuma, Zingiberaceae) liposomes and evaluated them for antitumor activity [109]. Rodenak-Kladniew et al. (2023) explored the encapsulation of EOs derived from *Clinopodium nepeta* and *Lippia alba* into SLNs. Their study aimed to improve the anticancer efficacy of these EOs by incorporating them into biocompatible lipid-based carriers [25]. Waad A. Al-Otaibi (2022) developed an NE containing EO derived from *Teucrium polium* [32]. Parvez et al. (2022) prepared nerolidol-loaded SLNs by the emulsion solvent evaporation method [110]. Poyatos-Racionero, E et al. (2021) prepared lactose-gated mesoporous silica MPs loaded with cinnamaldehyde [111]. Josef Jampilekk et al. (2022) formulated cuminaldehyde-loaded SLNs [95]. Rassu et al. (2014) developed alginate and soy protein isolate-based microspheres with thymol [112]. Berraaouan et al. (2023) formulated calcium alginate and calcium alginate/montmorillonite hybrid microcapsules with *Romarinus officinalis* EO [113]. Turasan et al. (2015) used the freeze-drying encapsulation technique for microencapsulation of rosemary EO with two coating materials—maltodextrin and whey protein—in different coating ratios [114]. Alshahrani et al. (2022) synthesized gold NPs loaded with *Rosmarinus officinalis* EO [115]. Szentmiklósi et al. (2024) encapsulated *Lavandula officinalis* EO into gum arabic and gelatin type A microcapsules [116]. Alencar et al. (2022) encapsulated *Cymbopogon citratus* EO into microcapsules [117]. Szentmiklósi et al. (2024) loaded *Achillea millefolium* EO into microcapsules [118]. Dima et al. (2014) synthesized microspheres with *Pimenta dioica* EO [119]. Wang et al. (2024) formulated gelatin–sodium alginate-based microcapsules loaded with *Origanum vulgare* EO [120]. Alirezaei et al. (2022) encapsulated *Artemisia vulgaris* L. EO into Poly(lactic-co-glycolic acid)-based NPs modified with chitosan-folic acid [121]. Keshavarz et al. (2024) investigated the apoptotic and antiproliferative effects of thymol-loaded nanoliposomes [122]. Nosrat et al. (2022) encapsulated *Ferula gummosa* EO into an NE [123]. Abadi et al. (2022) prepared an NE system containing the EOs of clove (*Syzygium aromaticum* L.) [124]. Khatamian et al. (2019) developed an NE containing carvi EO and evaluated its anticancer effects on HT-29 human CC cells, focusing on its ability to induce apoptosis. Their findings showed that the nanoformulations significantly reduced the viability of HT-29 cells, suggesting selective toxicity. Moreover, treatment with the NEs led to a marked upregulation of caspase-3 expression, indicating strong pro-apoptotic activity in the cancer cells. Khatamian et al. proved that an NE of carvi oil enhanced its anticancer effect against human CC cells [125].

The research conducted by Wijewantha et al. (2023) focuses on the preparation of enzyme-responsive NPs encapsulating eugenol (eNPs–EUG), aiming to develop novel antimetastatic therapies. Eugenol (2-methoxy-4-(2-propenyl)phenol), a natural phenolic compound primarily found in clove oil (*Syzygium aromaticum* L., Myrtaceae), has long been used in traditional Asian medicine for its antimicrobial, antioxidant, anti-inflammatory, antiseptic, and analgesic properties. More recently, it has gained attention for its pro-apoptotic and antiproliferative effects against various cancers, including CC. However, eugenol's clinical utility is hindered by its rapid metabolism, quick excretion, and toxic effects on normal cells at high doses. To overcome these limitations, Wijewantha and colleagues engineered "smart" eNPs–EUG, designed to deliver high concentrations of the drug directly to tumor cells while minimizing exposure to healthy tissues. Their findings show that eugenol triggers apoptosis in CC cells across different tumor grades in a dose- and time-dependent fashion. Moreover, it effectively suppresses cancer cell migration, invasion, and the population of CC stem cells—key drivers of metastasis and chemoresistance. These engineered NPs demonstrated selective uptake by cancer cells with rapid internalization, while showing minimal interaction with healthy colon epithelial cells. Overall, eNPs–EUG significantly enhances the therapeutic impact of eugenol [126].

6. Advanced Drug Delivery Systems' Potential to Improve the Outcomes of CC Treatment

6.1. Cytotoxicity Evaluation of Novel EO's Drug Delivery Strategies in CC Cell Lines

Salihu et al. (2025) investigated the cytotoxic activity of *Thymus capitatus* and *Origanum vulgare* EOs against MCF7, DU145, and HT-29 cancer cell lines, both in their free form and after encapsulation in NEs produced via high-pressure homogenization. The researchers successfully developed stable NEs characterized by high encapsulation efficiency, minimal phase separation, and small droplet sizes. For the HT-29 CC cells, treatment with *T. capitatus* NE and *O. vulgare* NE reduced cell viability to approximately 28% and 26%, respectively, with highly significant effects ($p < 0.001$). Overall, encapsulated oils demonstrated superior cytotoxicity compared to their unencapsulated form. Notably, *T. capitatus* NE showed stronger anticancer activity than *O. vulgare* NE, highlighting its greater potential as a cytotoxic agent [127].

Saffari et al. (2023) explored the anticancer potential of an NE formulated with *Rosa damascena* EO NE. The prepared NE exhibited IC_{50} values of 3.31 μg/mL against HCT116 colon cancer cells and 4.6 μg/mL against HeLa cervical cancer cells, indicating greater sensitivity of CC cells compared to HeLa cells [128].

Almnhavy (2020) investigated the development of poly(lactic acid)-based NPs loaded with *Trachyspermum ammi* seed EO. The NPs were prepared using evaporation and ultrasound-assisted emulsification techniques. Experimental findings revealed that prepared NPs induced apoptosis in HT-29 colon cancer cells, as evidenced by the upregulation of pro-apoptotic genes (Caspase-9 and BAX), downregulation of the anti-apoptotic gene (BCL-2), and flow cytometry analysis showing increased apoptosis with higher doses. Furthermore, the prepared NPs demonstrated a significant dose- and time-dependent cytotoxic effect on HT-29 cells. The anticancer effects of the prepared NPs were attributed to three main mechanisms: reduction in oxidative stress, induction of apoptosis, and inhibition of angiogenesis [129]. This study proved the advantages of the presented EO-loaded nanocarrier as a successful targeted agent against CC cell lines.

In the study by Moghimipour et al. (2023), thymol was successfully encapsulated into solid lipid nanoparticles (SLNs). The resulting SLNs have a mean particle size of 145 nm and an encapsulation efficiency % of 63%. In vitro release experiments demonstrated a sustained release profile of thymol from the SLNs. Moreover, thymol-loaded SLNs (Th-SLNs) exhibited significantly higher cytotoxicity in the HT29 CC cell line compared with free thymol and blank SLNs ($p < 0.05$). The IC_{50} values obtained for free thymol, SLNs, and Th-SLNs were 39.22 ± 0.9, 94.87 ± 1.1, and 7.88 ± 0.7 μM, respectively. The findings suggest that Th-SLNs are more effectively internalized by cancer cells than free thymol, likely due to differences in cellular uptake mechanisms. The enhanced efficiency of Th-SLNs may be attributed to the active endocytosis of SLNs, which facilitates greater intracellular delivery of thymol [130].

Overall, these results highlight the potential of SLNs as a promising nanocarrier system to enhance the therapeutic efficacy of thymol and other essential oil constituents in CC treatment.

6.2. Effect of Advanced Combinations of EOs and Cytotoxic Agents in the Treatment of CC

Advanced drug delivery systems offer significant promise in cancer therapy, as they can incorporate multiple anticancer agents with distinct mechanisms of action.

Similarly, Alkhatib et al. (2021) investigated the incorporation of epirubicin (EPI) into an NE formulated with algae and cinnamon oils via ultrasonication. The apoptotic potential of this formulation was evaluated in HCT116 human CC cells. Findings demonstrated that NE treatment markedly increased intracellular ROS generation and induced higher levels of late apoptosis compared to cells treated with EPI alone. Moreover, the prepared NE suppressed the invasive capacity of HCT116 cells to approximately 33%, in contrast to the 56% reduction observed with EPI treatment. Based on the antiproliferation assay, the IC_{50} value of EPI was reported to be 8.6-fold lower than that of the prepared NE. Collectively, these results indicate that the nanoformulation enhanced the antitumor efficacy of EPI, highlighting its potential as a novel nanotherapeutic strategy [131].

As already mentioned, Al-Otaibi et al. (2022) developed and optimized an NE of *Teucrium polium* L. EO and evaluated its ability to enhance the sensitivity of CC cells (HCT116 wild-type and HT-29 mutant-type) to Oxaliplatin. The combined treatment with NE of Oxaliplatin and *T. polium* demonstrated a synergistic effect, with combination index values of 0.94 for HCT116 and 0.88 for HT-29 cells. Microscopy and flow cytometry analyses confirmed that co-treatment induced a higher proportion of apoptotic cells compared to monotherapy. Mechanistically, the combination promoted apoptosis in both wild-type and mutant p53 CC cells through ROS-mediated mitochondrial pathways. Overall, these results indicate that NEs of Oxaliplatin and *T. polium* may represent a promising therapeutic approach for CC management [32].

The enhanced activity of NE strategies is likely attributed to their nano-scale size, which facilitates greater cellular penetration and promotes improved accumulation at the target site [131]. The findings suggest that encapsulating EOs in NEs not only enhances their biological efficacy but also improves stability, supporting their potential application in therapeutic strategies [127]. In addition, novel drug delivery systems can be designed with imaging functions, enabling non-invasive tracking of drug biodistribution, tumor localization, and therapeutic response. Such innovations hold promise for advancing personalized medicine and optimizing treatment strategies in CC [132].

7. Future Directions

EOs have long been utilized in medicine for their anti-inflammatory, antioxidant, and anticancer properties. Their BACs represent a valuable source of anticancer agents and plant-based therapeutics. The current technological focus is on developing intelligent EOs microcapsules capable of multiple reactions and containing distinct compartments that can simultaneously respond to diverse external stimuli (for example, pH- or time-responsive stimuli). Moreover, enhancing the retention of EOs and their BACs within microcapsules—particularly under environmentally triggered conditions—remains a critical challenge. Future research should explore innovative approaches to achieve sustained release and improved retention of EOs and their BACs within these systems [133]. With continued research, EOs hold promise for incorporation into pharmaceutical formulations, and several are already in various stages of clinical trials. Preliminary studies suggest that whole EOs often exhibit greater effectiveness than their isolated components. However, their full therapeutic potential remains underexplored, largely due to challenges such as the lack of targeted delivery mechanisms. As research into the anticancer applications of EOs is still in its early phases, further investigation—including well-designed clinical trials and the development of advanced, organ-specific delivery systems—is essential to enhance their efficacy in cancer treatment [41].

8. Conclusions

EOs contain a wide array of diversified BACs, including terpenoids, monoterpenes, sesquiterpenes, and other aromatic constituents, and have shown great promise for anticancer therapeutic applications. However, their clinical potential is hindered by challenges such as low stability, poor bioavailability, high volatility, and lack of targeted delivery. This review provides a comparison of micro- and nanocarriers loaded with EOs or their BACs, highlighting their potential as a promising strategy for CC prevention and treatment. However, preclinical research on EOs as anticancer agents remains limited, and many EOs still require thorough safety and toxicity evaluations before advancing to clinical trials. As previously noted, numerous EO-based micro- and nanoformulations have shown promising anticancer activity in vitro; however, substantial progress is still needed before these traditional compounds from folk medicine can be successfully registered and utilized as approved anticancer drugs.

Author Contributions: Conceptualization, Y.G. and P.G.; methodology, Y.G.; investigation, Y.G.; resources, Y.G. and P.G.; writing—original draft preparation, Y.G.; writing—review and editing, Y.G.; visualization, P.G.; supervision, Y.G. and P.G. All authors have read and agreed to the published version of the manuscript.

Funding: This research received no external funding.

Acknowledgments: This research is supported by the Bulgarian Ministry of Education and Science under the National Program "Young Scientists and Postdoctoral Students-2."

Conflicts of Interest: The authors declare no conflicts of interest.

Abbreviations

The following abbreviations are used in this manuscript:

CC	Colon cancer
EO	Essential oil
EOs	Essential oils
BAC	Bioactive compound
GIT	Gastrointestinal tract
CDDS	Colon-specific drug delivery systems
IBD	Inflammatory bowel disease
ROS	Reactive oxygen species
DNA	Deoxyribonucleic acid
CA	Carnosic acid
KRAS	Kirsten rat sarcoma
MPs	Microparticles
NPs	Nanoparticles
SLNs	Solid lipid nanoparticles
NEs	Nanoemulsions
EPI	Epirubicin
EPR	Enhanced permeability and retention effect
PEG	Polyethylene glycol
eNPs–EUG	Enzyme-responsive nanoparticles encapsulating eugenol

References

1. Liang, Z.; Xie, H.; Shen, W.; Shao, L.; Zeng, L.; Huang, X.; Zhu, Q.; Zhai, X.; Li, K.; Qiu, Z.; et al. The Synergism of Natural Compounds and Conventional Therapeutics against Colorectal Cancer Progression and Metastasis. *Front. Biosci.-Landmark* **2022**, *27*, 263. [CrossRef] [PubMed]
2. Douaiher, J.; Ravipati, A.; Grams, B.; Chowdhury, S.; Alatise, O.; Are, C. Colorectal Cancer—Global Burden, Trends, and Geographical Variations. *J. Surg. Oncol.* **2017**, *115*, 619–630. [CrossRef]
3. Sawicki, T.; Ruszkowska, M.; Danielewicz, A.; Niedźwiedzka, E.; Arłukowicz, T.; Przybyłowicz, K.E. A Review of Colorectal Cancer in Terms of Epidemiology, Risk Factors, Development, Symptoms and Diagnosis. *Cancers* **2021**, *13*, 2025. [CrossRef]
4. Garzoli, S.; Alarcón-Zapata, P.; Seitimova, G.; Alarcón-Zapata, B.; Martorell, M.; Sharopov, F.; Fokou, P.V.T.; Dize, D.; Yamthe, L.R.T.; Les, F.; et al. Natural Essential Oils as a New Therapeutic Tool in Colorectal Cancer. *Cancer Cell Int.* **2022**, *22*, 407. [CrossRef]
5. Ranjbarnejad, T.; Saidijam, M.; Moradkhani, S.; Najafi, R. Methanolic Extract of Boswellia Serrata Exhibits Anti-Cancer Activities by Targeting Microsomal Prostaglandin E Synthase-1 in Human Colon Cancer Cells. *Prostaglandins Other Lipid Mediat.* **2017**, *131*, 1–8. [CrossRef]
6. Atanasov, A.G.; Zotchev, S.B.; Dirsch, V.M.; the International Natural Product Sciences Taskforce; Orhan, I.E.; Banach, M.; Rollinger, J.M.; Barreca, D.; Weckwerth, W.; Bauer, R.; et al. Natural Products in Drug Discovery: Advances and Opportunities. *Nat. Rev. Drug Discov.* **2021**, *20*, 200–216. [CrossRef]
7. Chunarkar-Patil, P.; Kaleem, M.; Mishra, R.; Ray, S.; Ahmad, A.; Verma, D.; Bhayye, S.; Dubey, R.; Singh, H.; Kumar, S. Anticancer Drug Discovery Based on Natural Products: From Computational Approaches to Clinical Studies. *Biomedicines* **2024**, *12*, 201. [CrossRef]
8. Xie, L.; Wu, Z.; Liu, Y.; Tang, J.; Lu, C.; Wang, H. α-Linalool from Coriander Root Inhibits the Proliferation and Invasion of a Human Gastric Cancer Cell Line. *Clin. Cancer Investig. J.* **2023**, *12*, 6–14. [CrossRef]
9. AbouAitah, K.; Lojkowski, W. Nanomedicine as an Emerging Technology to Foster Application of Essential Oils to Fight Cancer. *Pharmaceuticals* **2022**, *15*, 793. [CrossRef] [PubMed]
10. Alabrahim, O.A.A.; Lababidi, J.M.; Fritzsche, W.; Azzazy, H.M.E.-S. Beyond Aromatherapy: Can Essential Oil Loaded Nanocarriers Revolutionize Cancer Treatment? *Nanoscale Adv.* **2024**, *6*, 5511–5562. [CrossRef]

11. Awais Khan, R.M.; Akhtar, S.; J Edagwa, B.; Shahnaz, G.; Rehman, S.U.; Rahdar, A.; Kharaba, Z. Doxycycline Monohydrate and Azelaic Acid Co-Loaded Nanoemulgel for the Treatment of Facial Rosacea: In Vitro and In Vivo Evaluation. *J. Drug Deliv. Sci. Technol.* **2024**, *98*, 105894. [CrossRef]
12. Salvioni, L.; Rizzuto, M.A.; Bertolini, J.A.; Pandolfi, L.; Colombo, M.; Prosperi, D. Thirty Years of Cancer Nanomedicine: Success, Frustration, and Hope. *Cancers* **2019**, *11*, 1855. [CrossRef]
13. Islam, M.R.; Akash, S.; Rahman, M.M.; Nowrin, F.T.; Akter, T.; Shohag, S.; Rauf, A.; Aljohani, A.S.M.; Simal-Gandara, J. Colon Cancer and Colorectal Cancer: Prevention and Treatment by Potential Natural Products. *Chem. Biol. Interact.* **2022**, *368*, 110170. [CrossRef]
14. Ale-Ahmad, A.; Kazemi, S.; Daraei, A.; Sepidarkish, M.; Moghadamnia, A.A.; Parsian, H. pH-Sensitive Nanoformulation of Acetyl-11-Keto-Beta-Boswellic Acid (AKBA) as a Potential Antiproliferative Agent in Colon Adenocarcinoma (in Vitro and in Vivo). *Cancer Nanotechnol.* **2024**, *15*, 49. [CrossRef]
15. Dey, A.; Mitra, A.; Pathak, S.; Prasad, S.; Zhang, A.S.; Zhang, H.; Sun, X.-F.; Banerjee, A. Recent Advancements, Limitations, and Future Perspectives of the Use of Personalized Medicine in Treatment of Colon Cancer. *Technol. Cancer Res. Treat.* **2023**, *22*, 1–21. [CrossRef]
16. Sakata, S.; Larson, D.W. Targeted Therapy for Colorectal Cancer. *Surg. Oncol. Clin. N. Am.* **2022**, *31*, 255–264. [CrossRef]
17. Zafar, A.; Khatoon, S.; Khan, M.J.; Abu, J.; Naeem, A. Advancements and Limitations in Traditional Anti-Cancer Therapies: A Comprehensive Review of Surgery, Chemotherapy, Radiation Therapy, and Hormonal Therapy. *Discov. Oncol.* **2025**, *16*, 607. [CrossRef]
18. Anand, U.; Dey, A.; Chandel, A.K.S.; Sanyal, R.; Mishra, A.; Pandey, D.K.; De Falco, V.; Upadhyay, A.; Kandimalla, R.; Chaudhary, A.; et al. Cancer Chemotherapy and beyond: Current Status, Drug Candidates, Associated Risks and Progress in Targeted Therapeutics. *Genes Dis.* **2023**, *10*, 1367–1401. [CrossRef]
19. Khan, F.A.; Albalawi, R.; Pottoo, F.H. Trends in Targeted Delivery of Nanomaterials in Colon Cancer Diagnosis and Treatment. *Med. Res. Rev.* **2022**, *42*, 227–258. [CrossRef] [PubMed]
20. Tolba, M.F. Revolutionizing the Landscape of Colorectal Cancer Treatment: The Potential Role of Immune Checkpoint Inhibitors. *Int. J. Cancer* **2020**, *147*, 2996–3006. [CrossRef] [PubMed]
21. Kaviyarasan, V.; Das, A.; Deka, D.; Saha, B.; Banerjee, A.; Sharma, N.R.; Duttaroy, A.K.; Pathak, S. Advancements in Immunotherapy for Colorectal Cancer Treatment: A Comprehensive Review of Strategies, Challenges, and Future Prospective. *Int. J. Color. Dis.* **2024**, *40*, 1. [CrossRef]
22. Baran, M.; Bitgen, N.; Onder, G.O.; Gurbuz, P.; Yay, A. Cell Proliferation, Apoptosis and Inflammation Response of Melissa Officinalis and Thymus Vulgaris in SW480 Colon Cancer Cells. *S. Afr. J. Bot.* **2023**, *162*, 282–295. [CrossRef]
23. Ivanova, S.; Staynova, R.; Koleva, N.; Ivanov, K.; Grekova-Kafalova, D. Public Perception and Usage Trends of Essential Oils: Findings from a Nationwide Survey. *Cosmetics* **2025**, *12*, 53. [CrossRef]
24. Esmeeta, A.; Adhikary, S.; Dharshnaa, V.; Swarnamughi, P.; Ummul Maqsummiya, Z.; Banerjee, A.; Pathak, S.; Duttaroy, A.K. Plant-Derived Bioactive Compounds in Colon Cancer Treatment: An Updated Review. *Biomed. Pharmacother.* **2022**, *153*, 113384. [CrossRef] [PubMed]
25. Rodenak-Kladniew, B.; Castro, M.A.; Gambaro, R.C.; Girotti, J.; Cisneros, J.S.; Viña, S.; Padula, G.; Crespo, R.; Castro, G.R.; Gehring, S.; et al. Cytotoxic Screening and Enhanced Anticancer Activity of *Lippia alba* and *Clinopodium nepeta* Essential Oils-Loaded Biocompatible Lipid Nanoparticles against Lung and Colon Cancer Cells. *Pharmaceutics* **2023**, *15*, 2045. [CrossRef] [PubMed]
26. Gautam, N.; Mantha, A.K.; Mittal, S. Essential Oils and Their Constituents as Anticancer Agents: A Mechanistic View. *BioMed Res. Int.* **2014**, *2014*, 154106. [CrossRef]
27. Chan, W.-K.; Tan, L.; Chan, K.-G.; Lee, L.-H.; Goh, B.-H. Nerolidol: A Sesquiterpene Alcohol with Multi-Faceted Pharmacological and Biological Activities. *Molecules* **2016**, *21*, 529. [CrossRef] [PubMed]
28. Luca, T.; Napoli, E.; Privitera, G.; Musso, N.; Ruberto, G.; Castorina, S. Antiproliferative Effect and Cell Cycle Alterations Induced by *Salvia officinalis* Essential Oil and Its Three Main Components in Human Colon Cancer Cell Lines. *Chem. Biodivers.* **2020**, *17*, e2000309. [CrossRef]
29. Fan, K.; Li, X.; Cao, Y.; Qi, H.; Li, L.; Zhang, Q.; Sun, H. Carvacrol Inhibits Proliferation and Induces Apoptosis in Human Colon Cancer Cells. *Anti-Cancer Drugs* **2015**, *26*, 813–823. [CrossRef]
30. Qi, F.; Yan, Q.; Zheng, Z.; Liu, J.; Chen, Y.; Zhang, G. Geraniol and Geranyl Acetate Induce Potent Anticancer Effects in Colon Cancer Colo-205 Cells by Inducing Apoptosis, DNA Damage and Cell Cycle Arrest. *J. BUON* **2018**, *23*, 346–352.
31. Sharma, M.; Grewal, K.; Jandrotia, R.; Batish, D.R.; Singh, H.P.; Kohli, R.K. Essential Oils as Anticancer Agents: Potential Role in Malignancies, Drug Delivery Mechanisms, and Immune System Enhancement. *Biomed. Pharmacother.* **2022**, *146*, 112514. [CrossRef] [PubMed]
32. Al-Otaibi, W.A.; AlMotwaa, S.M. Oxaliplatin-Loaded Nanoemulsion Containing *Teucrium polium* L. Essential Oil Induces Apoptosis in Colon Cancer Cell Lines through ROS-Mediated Pathway. *Drug Deliv.* **2022**, *29*, 2190–2205. [CrossRef]

33. Drif, A.I.; Yücer, R.; Damiescu, R.; Ali, N.T.; Abu Hagar, T.H.; Avula, B.; Khan, I.A.; Efferth, T. Anti-Inflammatory and Cancer-Preventive Potential of Chamomile (*Matricaria chamomilla* L.): A Comprehensive In Silico and In Vitro Study. *Biomedicines* **2024**, *12*, 1484. [CrossRef] [PubMed]
34. Petrocelli, G.; Farabegoli, F.; Valerii, M.C.; Giovannini, C.; Sardo, A.; Spisni, E. Molecules Present in Plant Essential Oils for Prevention and Treatment of Colorectal Cancer (CRC). *Molecules* **2021**, *26*, 885. [CrossRef]
35. Ivanova, S.; Pashova, S.; Dyankov, S.; Georgieva, Y.; Ivanov, K.; Benbassat, N.; Koleva, N.; Bozhkova, M.; Karcheva-Bahchevanska, D. Chemical Composition and Future Perspectives of Essential Oil Obtained from a Wild Population of *Stachys germanica* L. Distributed in the Balkan Mountains in Bulgaria. *Int. J. Anal. Chem.* **2023**, *2023*, 4275213. [CrossRef]
36. Pashova, S.; Karcheva-Bahchevanska, D.; Ivanov, K.; Ivanova, S. Genus Stachys—Phytochemistry, Traditional Medicinal Uses, and Future Perspectives. *Molecules* **2024**, *29*, 5345. [CrossRef] [PubMed]
37. Lechkova, B.; Karcheva-Bahchevanska, D.; Ivanov, K.; Todorova, V.; Benbassat, N.; Penkova, N.; Atanassova, P.; Peychev, L.; Hrischev, P.; Peychev, Z.; et al. A Study of the Chemical Composition, Acute and Subacute Toxicity of Bulgarian Tanacetum Parthenium Essential Oil. *Molecules* **2023**, *28*, 4906. [CrossRef]
38. Ivanova, S.; Dzhakova, Z.; Staynova, R.; Ivanov, K. *Salvia verticillata* (L.)—Biological Activity, Chemical Profile, and Future Perspectives. *Pharmaceuticals* **2024**, *17*, 859. [CrossRef]
39. Aly, A.; Maraei, R.; Rezk, A.; Diab, A. Phytochemical Constitutes and Biological Activities of Essential Oil Extracted from Irradiated Caraway Seeds (*Carum carvi* L.). *Int. J. Radiat. Biol.* **2023**, *99*, 318–328. [CrossRef]
40. Dzhoglova, V.; Ivanov, K.; Karcheva-Bahchevanska, D.; Koleva, N.; Bozhkova, M.; Benbassat, N.; Ivanova, S. *Salicornia europaea* L. and *Suaeda maritima* (L.) Dumort: Bioactive Compounds and Future Perspectives. *Biotechnol. Biotechnol. Equip.* **2024**, *38*, 2326291. [CrossRef]
41. Angelini, P.; Tirillini, B.; Akhtar, M.S.; Dimitriu, L.; Bricchi, E.; Bertuzzi, G.; Venanzoni, R. Essential Oil with Anticancer Activity: An Overview. In *Anticancer Plants: Natural Products and Biotechnological Implements*; Akhtar, M.S., Swamy, M.K., Eds.; Springer: Singapore, 2018; pp. 207–231, ISBN 978-981-10-8063-0.
42. Ivanova, S.; Nalbantova, V.; Benbassat, N.; Dzhoglova, V.; Dzhakova, Z.; Koleva, N.; Vasilev, V.; Grekova-Kafalova, D.; Ivanov, K. Comparison between the Chemical Composition of Essential Oils Isolated from Biocultivated Salvia Rosmarinus Spenn. (*Rosmarinus officinalis* L.) and Some Commercial Products. *Pharmacia* **2025**, *72*, 1–13. [CrossRef]
43. Zheljazkov, V.D.; Cantrell, C.L.; Semerdjieva, I.; Radoukova, T.; Stoyanova, A.; Maneva, V.; Kačániová, M.; Astatkie, T.; Borisova, D.; Dincheva, I.; et al. Essential Oil Composition and Bioactivity of Two Juniper Species from Bulgaria and Slovakia. *Molecules* **2021**, *26*, 3659. [CrossRef]
44. Chidambara Murthy, K.N.; Jayaprakasha, G.K.; Patil, B.S. D-Limonene Rich Volatile Oil from Blood Oranges Inhibits Angiogenesis, Metastasis and Cell Death in Human Colon Cancer Cells. *Life Sci.* **2012**, *91*, 429–439. [CrossRef]
45. Cianfaglione, K.; Blomme, E.E.; Quassinti, L.; Bramucci, M.; Lupidi, G.; Dall'Acqua, S.; Maggi, F. Cytotoxic Essential Oils from *Eryngium campestre* and *Eryngium amethystinum* (Apiaceae) Growing in Central Italy. *Chem. Biodivers.* **2017**, *14*, e1700096. [CrossRef]
46. Soda, K. Polyamine Metabolism and Gene Methylation in Conjunction with One-Carbon Metabolism. *Int. J. Mol. Sci.* **2018**, *19*, 3106. [CrossRef] [PubMed]
47. Chauhan, A.K.; Bahuguna, A.; Paul, S.; Kang, S.C. Thymol Elicits HCT-116 Colorectal Carcinoma Cell Death Through Induction of Oxidative Stress. *Anti-Cancer Agents Med. Chem.* **2018**, *17*, 1942–1950. [CrossRef]
48. Thapa, D.; Richardson, A.J.; Zweifel, B.; Wallace, R.J.; Gratz, S.W. Genoprotective Effects of Essential Oil Compounds Against Oxidative and Methylated DNA Damage in Human Colon Cancer Cells. *J. Food Sci.* **2019**, *84*, 1979–1985. [CrossRef] [PubMed]
49. Larbi, B.A.M.; Naima, B.; Elsharkawy, E.R.; Neghmouche, N.S. Phytochemical Characterization, In-Vitro Cytotoxic and Antibacterial Activity of *Cotula cinerea* (Delile) Vis Essential Oil. *J. Nat. Remedies* **2018**, *18*, 107–112. [CrossRef]
50. Singh, N.; Yadav, S.S.; Kumar, S.; Narashiman, B. A Review on Traditional Uses, Phytochemistry, Pharmacology, and Clinical Research of Dietary Spice *Cuminum cyminum* L. *Phytother. Res.* **2021**, *35*, 5007–5030. [CrossRef]
51. Sunohara, Y.; Nakano, K.; Matsuyama, S.; Oka, T.; Matsumoto, H. Cuminaldehyde, a Cumin Seed Volatile Component, Induces Growth Inhibition, Overproduction of Reactive Oxygen Species and Cell Cycle Arrest in Onion Roots. *Sci. Hortic.* **2021**, *289*, 110493. [CrossRef]
52. Patwa, N.; Chauhan, R.; Chauhan, A.; Kumar, M.; Ramniwas, S.; Mathkor, D.M.; Saini, A.K.; Tuli, H.S.; Haque, S.; Slama, P. Garcinol in Gastrointestinal Cancer Prevention: Recent Advances and Future Prospects. *J. Cancer Res. Clin. Oncol.* **2024**, *150*, 370. [CrossRef]
53. Taheri, E.; Ghorbani, S.; Safi, M.; Seyyed Sani, N.; Firouzi Amoodizaj, F.; Heidari, M.; Chavoshi, R.; Hajazimian, S.; Isazadeh, A.; Heidari, M. Inhibition of Colorectal Cancer Cell Line CaCo-2 by Essential Oil of Eucalyptus Camaldulensis Through Induction of Apoptosis. *Acta Med. Iran.* **2020**. [CrossRef]

54. Chen, X.; Wei, C.; Zhao, J.; Zhou, D.; Wang, Y.; Zhang, S.; Zuo, H.; Dong, J.; Zhao, Z.; Hao, M.; et al. Carnosic Acid: An Effective Phenolic Diterpenoid for Prevention and Management of Cancers via Targeting Multiple Signaling Pathways. *Pharmacol. Res.* **2024**, *206*, 107288. [CrossRef] [PubMed]
55. Barni, M.V.; Carlini, M.J.; Cafferata, E.G.; Puricelli, L.; Moreno, S. Carnosic Acid Inhibits the Proliferation and Migration Capacity of Human Colorectal Cancer Cells. *Oncol. Rep.* **2012**, *27*, 1041–1048. [CrossRef] [PubMed]
56. Huang, D.; Sun, W.; Zhou, Y.; Li, P.; Chen, F.; Chen, H.; Xia, D.; Xu, E.; Lai, M.; Wu, Y.; et al. Mutations of Key Driver Genes in Colorectal Cancer Progression and Metastasis. *Cancer Metastasis Rev.* **2018**, *37*, 173–187. [CrossRef]
57. Zhao, P.; Qiu, J.; Pan, C.; Tang, Y.; Chen, M.; Song, H.; Yang, J.; Hao, X. Potential Roles and Molecular Mechanisms of Bioactive Ingredients in Curcumae Rhizoma against Breast Cancer. *Phytomedicine* **2023**, *114*, 154810. [CrossRef]
58. Wang, G.-Y.; Zhang, L.; Geng, Y.-D.; Wang, B.; Feng, X.-J.; Chen, Z.-L.; Wei, W.; Jiang, L. β-Elemene Induces Apoptosis and Autophagy in Colorectal Cancer Cells through Regulating the ROS/AMPK/mTOR Pathway. *Chin. J. Nat. Med.* **2022**, *20*, 9–21. [CrossRef]
59. Chen, P.; Li, X.; Zhang, R.; Liu, S.; Xiang, Y.; Zhang, M.; Chen, X.; Pan, T.; Yan, L.; Feng, J.; et al. Combinative Treatment of β-Elemene and Cetuximab Is Sensitive to KRAS Mutant Colorectal Cancer Cells by Inducing Ferroptosis and Inhibiting Epithelial-Mesenchymal Transformation. *Theranostics* **2020**, *10*, 5107–5119. [CrossRef]
60. Yammine, J.; Chihib, N.-E.; Gharsallaoui, A.; Ismail, A.; Karam, L. Advances in Essential Oils Encapsulation: Development, Characterization and Release Mechanisms. *Polym. Bull.* **2024**, *81*, 3837–3882. [CrossRef]
61. Turek, C.; Stintzing, F.C. Stability of Essential Oils: A Review. *Compr. Rev. Food Sci. Food Saf.* **2013**, *12*, 40–53. [CrossRef]
62. Turek, C.; Stintzing, F.C. Impact of Different Storage Conditions on the Quality of Selected Essential Oils. *Food Res. Int.* **2012**, *46*, 341–353. [CrossRef]
63. Niu, F.; Pan, W.; Su, Y.; Yang, Y. Physical and Antimicrobial Properties of Thyme Oil Emulsions Stabilized by Ovalbumin and Gum Arabic. *Food Chem.* **2016**, *212*, 138–145. [CrossRef]
64. Dajic Stevanovic, Z.; Sieniawska, E.; Glowniak, K.; Obradovic, N.; Pajic-Lijakovic, I. Natural Macromolecules as Carriers for Essential Oils: From Extraction to Biomedical Application. *Front. Bioeng. Biotechnol.* **2020**, *8*, 563. [CrossRef] [PubMed]
65. Stevanović, Z.D.; Bošnjak-Neumüller, J.; Pajić-Lijaković, I.; Raj, J.; Vasiljević, M. Essential Oils as Feed Additives—Future Perspectives. *Molecules* **2018**, *23*, 1717. [CrossRef]
66. Alu'datt, M.H.; Alrosan, M.; Gammoh, S.; Tranchant, C.C.; Alhamad, M.N.; Rababah, T.; Zghoul, R.; Alzoubi, H.; Ghatasheh, S.; Ghozlan, K.; et al. Encapsulation-Based Technologies for Bioactive Compounds and Their Application in the Food Industry: A Roadmap for Food-Derived Functional and Health-Promoting Ingredients. *Food Biosci.* **2022**, *50*, 101971. [CrossRef]
67. Movahedi, F.; Nirmal, N.; Wang, P.; Jin, H.; Grøndahl, L.; Li, L. Recent Advances in Essential Oils and Their Nanoformulations for Poultry Feed. *J. Anim. Sci. Biotechnol.* **2024**, *15*, 110. [CrossRef]
68. Avila Gandra, E.; Radünz, M.; Helbig, E.; Dellinghausen Borges, C.; Kuka Valente Gandra, T. A Mini-Review on Encapsulation of Essential Oils. *J. Anal. Pharm. Res.* **2018**, *7*, 00205. [CrossRef]
69. Zhan, L.; Lin, Z.; Li, W.; Qin, Y.; Sun, Q.; Ji, N.; Xie, F. The Construction of Sodium Alginate/Carboxymethyl Chitosan Microcapsules as the Physical Barrier to Reduce Corn Starch Digestion. *Foods* **2024**, *13*, 1355. [CrossRef]
70. Yousefi, S.; Weisany, W.; Hosseini, S.E.; Ghasemlou, M. Mechanisms of Nanoencapsulation to Boost the Antimicrobial Efficacy of Essential Oils: A Review. *Food Hydrocoll.* **2024**, *150*, 109655. [CrossRef]
71. Fakhariha, M.; Rafati, A.A.; Garmakhany, A.D.; Asl, A.Z. Nanoencapsulation Enhances Stability, Release Behavior, and Antimicrobial Properties of Sage and Thyme Essential Oils. *Sci. Rep.* **2025**, *15*, 18373. [CrossRef]
72. Lee, S.H.; Bajracharya, R.; Min, J.Y.; Han, J.-W.; Park, B.J.; Han, H.-K. Strategic Approaches for Colon Targeted Drug Delivery: An Overview of Recent Advancements. *Pharmaceutics* **2020**, *12*, 68. [CrossRef] [PubMed]
73. Curcio, C.; Greco, A.S.; Rizzo, S.; Saitta, L.; Musumeci, T.; Ruozi, B.; Pignatello, R. Development, Optimization and Characterization of Eudraguard®-Based Microparticles for Colon Delivery. *Pharmaceuticals* **2020**, *13*, 131. [CrossRef]
74. Gulbake, A. Insight to Drug Delivery Aspects for Colorectal Cancer. *World J. Gastroenterol.* **2016**, *22*, 582. [CrossRef]
75. Ying, K.; Bai, B.; Gao, X.; Xu, Y.; Wang, H.; Xie, B. Orally Administrable Therapeutic Nanoparticles for the Treatment of Colorectal Cancer. *Front. Bioeng. Biotechnol.* **2021**, *9*, 670124. [CrossRef]
76. Iswandana, R.; Putri, K.S.S.; Larasati, S.A.; Gunawan, M.; Putri, F.A. Delivery of Potential Drugs to The Colon: Challenges and Strategies. *Indones. J. Pharm.* **2022**, *33*, 307–332. [CrossRef]
77. Rifisha Basumatary, M.Z.A. Development of the Formulation and Assessment of Colon Targeted Drug Delivery or Oral Administered Prednisolone. *Int. J. Pharm. Sci.* **2025**, *3*, 9172–9213. Available online: https://zenodo.org/records/15225136 (accessed on 24 June 2025).
78. Mohajeri, S.; Moayedi, S.; Mohajeri, S.; Yadegar, A.; Haririan, I. Targeting Pathophysiological Changes Using Biomaterials-Based Drug Delivery Systems: A Key to Managing Inflammatory Bowel Disease. *Front. Pharmacol.* **2022**, *13*, 1045575. [CrossRef] [PubMed]

79. Choukaife, H.; Seyam, S.; Alallam, B.; Doolaanea, A.A.; Alfatama, M. Current Advances in Chitosan Nanoparticles Based Oral Drug Delivery for Colorectal Cancer Treatment. *Int. J. Nanomed.* **2022**, *17*, 3933–3966. [CrossRef] [PubMed]
80. Patel, M.; Shah, T.; Amin, A. Therapeutic Opportunities in Colon-Specific Drug-Delivery Systems. *Crit. Rev. Ther. Drug Carrier Syst.* **2007**, *24*, 147–202. [CrossRef]
81. Dos Santos, J.; Da Silva, G.S.; Velho, M.C.; Beck, R.C.R. Eudragit®: A Versatile Family of Polymers for Hot Melt Extrusion and 3D Printing Processes in Pharmaceutics. *Pharmaceutics* **2021**, *13*, 1424. [CrossRef]
82. Fang, Y.; Wang, G.; Zhang, R.; Liu, Z.; Liu, Z.; Wu, X.; Cao, D. Eudragit L/HPMCAS Blend Enteric-Coated Lansoprazole Pellets: Enhanced Drug Stability and Oral Bioavailability. *AAPS PharmSciTech* **2014**, *15*, 513–521. [CrossRef]
83. Ibrahim, I.M. Advances in Polysaccharide-Based Oral Colon-Targeted Delivery Systems: The Journey So Far and the Road Ahead. *Cureus* **2023**, *15*, e33636. [CrossRef]
84. Gazzaniga, A.; Moutaharrik, S.; Filippin, I.; Foppoli, A.; Palugan, L.; Maroni, A.; Cerea, M. Time-Based Formulation Strategies for Colon Drug Delivery. *Pharmaceutics* **2022**, *14*, 2762. [CrossRef] [PubMed]
85. Amidon, S.; Brown, J.E.; Dave, V.S. Colon-Targeted Oral Drug Delivery Systems: Design Trends and Approaches. *AAPS PharmSciTech* **2015**, *16*, 731–741. [CrossRef]
86. Yeung, C.; McCoubrey, L.E.; Basit, A.W. Advances in Colon-Targeted Drug Technologies. *Curr. Opin. Gastroenterol.* **2025**, *41*, 9–15. [CrossRef] [PubMed]
87. Veiga, R.D.S.D.; Aparecida Da Silva-Buzanello, R.; Corso, M.P.; Canan, C. Essential Oils Microencapsulated Obtained by Spray Drying: A Review. *J. Essent. Oil Res.* **2019**, *31*, 457–473. [CrossRef]
88. Sánchez-Osorno, D.M.; López-Jaramillo, M.C.; Caicedo Paz, A.V.; Villa, A.L.; Peresin, M.S.; Martínez-Galán, J.P. Recent Advances in the Microencapsulation of Essential Oils, Lipids, and Compound Lipids through Spray Drying: A Review. *Pharmaceutics* **2023**, *15*, 1490. [CrossRef]
89. Sousa, V.I.; Parente, J.F.; Marques, J.F.; Forte, M.A.; Tavares, C.J. Microencapsulation of Essential Oils: A Review. *Polymers* **2022**, *14*, 1730. [CrossRef]
90. Shendge, R.S.; Zalte, T.S.; Khade, S.B. Polymeric Microspheres Redefining the Landscape of Colon-Targeted Delivery: A Contemporary Update. *Eur. J. Med. Chem. Rep.* **2024**, *11*, 100156. [CrossRef]
91. Collnot, E.-M.; Ali, H.; Lehr, C.-M. Nano- and Microparticulate Drug Carriers for Targeting of the Inflamed Intestinal Mucosa. *J. Control. Release* **2012**, *161*, 235–246. [CrossRef]
92. Dubey, R.; Dubey, R.; Omrey, P.; Vyas, S.P.; Jain, S.K. Development and Characterization of Colon Specific Drug Delivery System Bearing 5-ASA and Camylofine Dihydrochloride for the Treatment of Ulcerative Colitis. *J. Drug Target.* **2010**, *18*, 589–601. [CrossRef]
93. Perlatti, B.; Souza Bergo, P.L.D.; Fernandes Da Silva, M.F.D.G.; Batista, J.; Rossi, M. Polymeric Nanoparticle-Based Insecticides: A Controlled Release Purpose for Agrochemicals. In *Insecticides—Development of Safer and More Effective Technologies*; Trdan, S., Ed.; InTech: Houston, TX, USA, 2013; ISBN 978-953-51-0958-7.
94. Pavoni, L.; Perinelli, D.R.; Bonacucina, G.; Cespi, M.; Palmieri, G.F. An Overview of Micro- and Nanoemulsions as Vehicles for Essential Oils: Formulation, Preparation and Stability. *Nanomaterials* **2020**, *10*, 135. [CrossRef] [PubMed]
95. Jampilek, J.; Kralova, K. Anticancer Applications of Essential Oils Formulated into Lipid-Based Delivery Nanosystems. *Pharmaceutics* **2022**, *14*, 2681. [CrossRef]
96. De Oliveira Filho, J.G.; Miranda, M.; Ferreira, M.D.; Plotto, A. Nanoemulsions as Edible Coatings: A Potential Strategy for Fresh Fruits and Vegetables Preservation. *Foods* **2021**, *10*, 2438. [CrossRef]
97. Rommasi, F.; Esfandiari, N. Liposomal Nanomedicine: Applications for Drug Delivery in Cancer Therapy. *Nanoscale Res. Lett.* **2021**, *16*, 95. [CrossRef] [PubMed]
98. Wang, S.; Chen, Y.; Guo, J.; Huang, Q. Liposomes for Tumor Targeted Therapy: A Review. *Int. J. Mol. Sci.* **2023**, *24*, 2643. [CrossRef]
99. Gatto, M.S.; Johnson, M.P.; Najahi-Missaoui, W. Targeted Liposomal Drug Delivery: Overview of the Current Applications and Challenges. *Life* **2024**, *14*, 672. [CrossRef]
100. Lüdtke, F.L.; Silva, T.J.; Da Silva, M.G.; Hashimoto, J.C.; Ribeiro, A.P.B. Lipid Nanoparticles: Formulation, Production Methods and Characterization Protocols. *Foods* **2025**, *14*, 973. [CrossRef]
101. Queiroz, M.D.C.V.; Muehlmann, L.A. Characteristics and Preparation of Solid Lipid Nanoparticles and Nanostructured Lipid Carriers. *J. Nanotheranostics* **2024**, *5*, 188–211. [CrossRef]
102. Barua, S.; Mitragotri, S. Challenges Associated with Penetration of Nanoparticles across Cell and Tissue Barriers: A Review of Current Status and Future Prospects. *Nano Today* **2014**, *9*, 223–243. [CrossRef] [PubMed]
103. Prieložná, J.; Mikušová, V.; Mikuš, P. Advances in the Delivery of Anticancer Drugs by Nanoparticles and Chitosan-Based Nanoparticles. *Int. J. Pharm. X* **2024**, *8*, 100281. [CrossRef]
104. Enin, H.A.A.; Alquthami, A.F.; Alwagdani, A.M.; Yousef, L.M.; Albuqami, M.S.; Alharthi, M.A.; Alsaab, H.O. Utilizing TPGS for Optimizing Quercetin Nanoemulsion for Colon Cancer Cells Inhibition. *Colloids Interfaces* **2022**, *6*, 49. [CrossRef]

105. López-Cabanillas Lomelí, M.; Tijerina-Sáenz, A.; García-Hernández, D.G.; Hernández-Salazar, M.; Salas García, R.; González-Llerena, J.L.; Verde-Star, M.J.; Cordero-Díaz, A.; Heya, M.S. Colon Cancer: Overview on Improved Therapeutic Potential of Plant-Based Compounds Using Nanotechnology. *Sci. Pharm.* **2024**, *93*, 1. [CrossRef]
106. Aboulthana, W.M.; Ibrahim, N.E.-S.; Osman, N.M.; Seif, M.M.; Hassan, A.K.; Youssef, A.M.; El-Feky, A.M.; Madboli, A.A. Evaluation of the Biological Efficiency of Silver Nanoparticles Biosynthesized Using *Croton tiglium* L. Seeds Extract against Azoxymethane Induced Colon Cancer in Rats. *Asian Pac. J. Cancer Prev.* **2020**, *21*, 1369–1389. [CrossRef]
107. Pavoni, L.; Pavela, R.; Cespi, M.; Bonacucina, G.; Maggi, F.; Zeni, V.; Canale, A.; Lucchi, A.; Bruschi, F.; Benelli, G. Green Micro- and Nanoemulsions for Managing Parasites, Vectors and Pests. *Nanomaterials* **2019**, *9*, 1285. [CrossRef] [PubMed]
108. McCoubrey, L.E.; Favaron, A.; Awad, A.; Orlu, M.; Gaisford, S.; Basit, A.W. Colonic Drug Delivery: Formulating the next Generation of Colon-Targeted Therapeutics. *J. Control. Release* **2023**, *353*, 1107–1126. [CrossRef]
109. Zhai, B.; Wu, Q.; Wang, W.; Zhang, M.; Han, X.; Li, Q.; Chen, P.; Chen, X.; Huang, X.; Li, G.; et al. Preparation, Characterization, Pharmacokinetics and Anticancer Effects of PEGylated β-Elemene Liposomes. *Cancer Biol. Med.* **2020**, *17*, 60–75. [CrossRef]
110. Parvez, S.; Karole, A.; Mudavath, S.L. Fabrication, Physicochemical Characterization and In Vitro Anticancer Activity of Nerolidol Encapsulated Solid Lipid Nanoparticles in Human Colorectal Cell Line. *Colloids Surf. B Biointerfaces* **2022**, *215*, 112520. [CrossRef] [PubMed]
111. Poyatos-Racionero, E.; González-Álvarez, I.; Sánchez-Moreno, P.; Sitia, L.; Gatto, F.; Pompa, P.P.; Aznar, E.; González-Álvarez, M.; Martínez-Máñez, R.; Marcos, M.D.; et al. Lactose-Gated Mesoporous Silica Particles for Intestinal Controlled Delivery of Essential Oil Components: An In Vitro and In Vivo Study. *Pharmaceutics* **2021**, *13*, 982. [CrossRef] [PubMed]
112. Rassu, G.; Nieddu, M.; Bosi, P.; Trevisi, P.; Colombo, M.; Priori, D.; Manconi, P.; Giunchedi, P.; Gavini, E.; Boatto, G. Encapsulation and Modified-Release of Thymol from Oral Microparticles as Adjuvant or Substitute to Current Medications. *Phytomedicine* **2014**, *21*, 1627–1632. [CrossRef]
113. Berraaouan, D.; Essifi, K.; Addi, M.; Hano, C.; Fauconnier, M.-L.; Tahani, A. Hybrid Microcapsules for Encapsulation and Controlled Release of Rosemary Essential Oil. *Polymers* **2023**, *15*, 823. [CrossRef]
114. Turasan, H.; Sahin, S.; Sumnu, G. Encapsulation of Rosemary Essential Oil. *LWT Food Sci. Technol.* **2015**, *64*, 112–119. [CrossRef]
115. Alshahrani, M.Y.; Ibrahim, E.H. Gold Nanoparticles (AuNPs) and Rosmarinus Officinalis Extract and Their Potentials to Prompt Apoptosis and Arrest Cell Cycle in HT-29 Colon Cancer Cells. *J. King Saud Univ. Sci.* **2022**, *34*, 102304. [CrossRef]
116. Székely-Szentmiklósi, I.; Rédai, E.M.; Szabó, Z.-I.; Kovács, B.; Albert, C.; Gergely, A.-L.; Székely-Szentmiklósi, B.; Sipos, E. Microencapsulation by Complex Coacervation of Lavender Oil Obtained by Steam Distillation at Semi-Industrial Scale. *Foods* **2024**, *13*, 2935. [CrossRef]
117. De Oliveira Alencar, D.D.; De Souza, E.L.; Da Cruz Almeida, E.T.; Da Silva, A.L.; Oliveira, H.M.L.; Cavalcanti, M.T. Microencapsulation of Cymbopogon Citratus D.C. Stapf Essential Oil with Spray Drying: Development, Characterization, and Antioxidant and Antibacterial Activities. *Foods* **2022**, *11*, 1111. [CrossRef] [PubMed]
118. Székely-Szentmiklósi, I.; Rédai, E.M.; Kovács, B.; Gergely, A.-L.; Albert, C.; Szabó, Z.-I.; Székely-Szentmiklósi, B.; Sipos, E. Investigation of Yarrow Essential Oil Composition and Microencapsulation by Complex Coacervation Technology. *Appl. Sci.* **2024**, *14*, 7867. [CrossRef]
119. Dima, C.; Cotârlet, M.; Alexe, P.; Dima, S. Microencapsulation of Essential Oil of Pimento [*Pimenta dioica* (L) Merr.] by Chitosan/k-Carrageenan Complex Coacervation Method. *Innov. Food Sci. Emerg. Technol.* **2014**, *22*, 203–211. [CrossRef]
120. Wang, K.; Wang, Y.; Cheng, M.; Wang, Y.; Zhao, P.; Xi, X.; Lu, J.; Wang, X.; Han, X.; Wang, J. Preparation and Characterization of Active Films Based on Oregano Essential Oil Microcapsules/Soybean Protein Isolate/Sodium Carboxymethyl Cellulose. *Int. J. Biol. Macromol.* **2024**, *258*, 128985. [CrossRef]
121. Alirezaei, M.; Ghobeh, M.; Es-haghi, A. Poly(Lactic-Co-Glycolic Acid)(PLGA)-Based Nanoparticles Modified with Chitosan-Folic Acid to Delivery of *Artemisia vulgaris* L. Essential Oil to HT-29 Cancer Cells. *Process Biochem.* **2022**, *121*, 207–215. [CrossRef]
122. Keshavarz, F.; Soltanshahi, M.; Khosravani, F.; Bakhshiyan, F.; Ghanbari, A.; Hassanzadeh, S.; Amirpour, M.; Ghalamfarsa, G. Thymol-Loaded Liposomes Effectively Induced Apoptosis and Decreased EGFR Expression in Colorectal Cancer Cells. *Naunyn-Schmiedeberg's Arch. Pharmacol.* **2024**, *397*, 5157–5165. [CrossRef]
123. Nosrat, T.; Tabrizi, M.H.; Etminan, A.; Irani, M.; Zarei, B.; Rahmati, A. In Vitro and In Vivo Anticancer Activity of Ferula Gummosa Essential Oil Nanoemulsions (FGEO-NE) for the Colon Cancer Treatment. *J. Polym. Environ.* **2022**, *30*, 4166–4177. [CrossRef]
124. Abadi, A.V.M.; Karimi, E.; Oskoueian, E.; Mohammad, G.R.K.S.; Shafaei, N. Synthesis, Characterization and Cytoxicity Evaluation of *Syzygium aromaticum* L. Bud (Clove) Essential Oil Nanoemulsion. *Res. Sq.* **2021**. Available online: https://www.researchsquare.com/article/rs-952488/v1 (accessed on 24 June 2025).
125. Khatamian, N.; Homayouni Tabrizi, M.; Ardalan, P.; Yadamani, S.; Darchini Maragheh, A. Synthesis of *Carum carvi* Essential Oil Nanoemulsion, the Cytotoxic Effect, and Expression of Caspase 3 Gene. *J. Food Biochem.* **2019**, *43*, e12956. [CrossRef] [PubMed]
126. Wijewantha, N.; Sane, S.; Eikanger, M.; Antony, R.M.; Potts, R.A.; Lang, L.; Rezvani, K.; Sereda, G. Enhancing Anti-Tumorigenic Efficacy of Eugenol in Human Colon Cancer Cells Using Enzyme-Responsive Nanoparticles. *Cancers* **2023**, *15*, 1145. [CrossRef]

127. Basholli Salihu, M.; Haloci, E.; Kryeziu, T.; Ahmeti, J.; Zogjani, B.; Bağcı, U.; Lupci, V.; Jakupi, X.; Zimmer, A.; Shala, A. Nanoemulsion of Thymus Capitatus and Origanum Vulgare Essential Oil: Stability, Antimicrobial and Cytotoxic Properties. *J. Res. Pharm.* **2025**, *29*, 852–870. [CrossRef]
128. Saffari, I.; Motallebi Moghanjoghi, A.; Sharafati Chaleshtori, R.; Ataee, M.; Khaledi, A. Nanoemulsification of Rose (*Rosa damascena*) Essential Oil: Characterization, Anti-Salmonella, In Vitro Cytotoxicity to Cancer Cells, and Advantages in Sheep Meat Application. *J. Food Qual.* **2023**, *2023*, 1–15. [CrossRef]
129. Almnhawy, M.; Jebur, M.; Alhajamee, M.; Marai, K.; Tabrizi, M.H. PLGA-Based Nano-Encapsulation of *Trachyspermum ammi* Seed Essential Oil (TSEO-PNP) as a Safe, Natural, Efficient, Anticancer Compound in Human HT-29 Colon Cancer Cell Line. *Nutr. Cancer* **2021**, *73*, 2808–2820. [CrossRef]
130. Moghimipour, E.; Abedini, M.; Handali, S. Targeted Solid Lipid Nanoparticles Formulation for Colon Cancer Treatment and Cytotoxicity Assessment Using HT29 Cell Line. *Jundishapur J. Nat. Pharm. Prod.* **2023**, *18*, e135987. [CrossRef]
131. Alkhatib, M.H.; Aljadani, M.A.; Mahassni, S.H. Carrying Epirubicin on Nanoemulsion Containing Algae and Cinnamon Oils Augments Its Apoptotic and Anti-Invasion Effects on Human Colon Cancer Cells. *Am. J. Transl. Res.* **2020**, *12*, 2463–2472. [PubMed]
132. Jain, A.; Bhattacharya, S. Recent Advances in Nanomedicine Preparative Methods and Their Therapeutic Potential for Colorectal Cancer: A Critical Review. *Front. Oncol.* **2023**, *13*, 1211603. [CrossRef]
133. Zhao, Y.; Wang, Y.; Zhang, Z.; Li, H. Advances in Controllable Release Essential Oil Microcapsules and Their Promising Applications. *Molecules* **2023**, *28*, 4979. [CrossRef]

BioChem MDPI

Opinion

An Opinion on the Supplementation of Folic Acid 1 mg + Iron (Ferrous Sulfate) 90 mg in the Prevention and Treatment of Anemia

João Gomes [1], Joana Brandão Silva [2,3], César Vinícius José [2] and Hugo Ribeiro [2,4,5,6,*]

1 Service of Hematology, Local Health Unit of Coimbra, 3004-561 Coimbra, Portugal
2 Faculty of Medicine, University of Coimbra, 3000-548 Coimbra, Portugal
3 Abel Salazar School of Medicine and Biomedical Sciences, University of Porto, 4050-313 Porto, Portugal
4 Community Palliative Care Support Team Gaia, 4400-129 Vila Nova de Gaia, Portugal
5 Centre for Innovative Biomedicine and Biotechnology, 3000-548 Coimbra, Portugal
6 Faculty of Medicine, University of Porto, 4200-319 Porto, Portugal
* Correspondence: hribeiroff@gmail.com

Abstract

Introduction: Anemia, characterized by a reduction in hemoglobin concentration, is a widespread health concern globally, impacting individuals across various demographics. Iron deficiency, often compounded by inadequate folic acid levels, is a primary driver. This review aims to consolidate current evidence and offer a practical recommendation regarding the role of folic acid 1 mg + iron (ferrous sulfate) 90 mg supplementation in both preventing and treating anemia. **Objective:** We aimed to provide a comprehensive review and recommendation regarding the use of folic acid 1 mg + iron (ferrous sulfate) 90 mg supplementation in the prevention and treatment of anemia in adults, based on current evidence and clinical experience. **Methods:** A thorough literature review was conducted, encompassing studies, guidelines, and meta-analyses related to iron deficiency, anemia, and folic acid supplementation. This review incorporated data from sources such as the World Health Organization (WHO), the European Hematology Association (EHA), and Cochrane Database. Clinical experience of the authors was also taken into account. **Results:** Anemia, a prevalent hematological condition, affects a significant portion of the global population. The risk factors for iron deficiency and iron deficiency anemia include age, menstruation, pregnancy, dietary restrictions, chronic diseases, and inflammatory conditions. Accurate diagnosis of anemia involves reticulocyte count, morphological classification, and identification of the underlying etiology. Oral iron salts, particularly ferrous sulfate, are the first-line treatment for uncomplicated iron deficiency anemia, with lower doses or alternate-day dosing improving tolerability. Adequate folic acid availability is crucial for erythropoiesis, and supplementation is safe and enhances treatment response, especially in mixed deficiency anemia. A fixed-dose combination of folic acid 1 mg + iron (ferrous sulfate) 90 mg is effective and well-tolerated for the treatment of iron deficiency anemia, mixed nutritional anemia, and iron deficiency without anemia in adults. **Conclusions:** Based on extensive scientific evidence and clinical experience, the combination of folic acid 1 mg + iron (ferrous sulfate) 90 mg is a valuable therapeutic option for the prevention and treatment of anemia. This combination should be indicated for iron and folic acid deficiency during pregnancy, lactation, and the postpartum period and for the prophylaxis and treatment of anemia during pregnancy and in adults in general. This approach enables correction of folate deficiencies, optimizing treatment response and ensuring sufficient folic acid levels, particularly in cases of incomplete adherence or missed doses, and is critical during pregnancy to minimize the risk of neural tube defects.

Keywords: anemia; iron deficiency; folic acid; supplementation; ferrous sulfate

1. Introduction

Anemia, characterized by a reduction in hemoglobin concentration, poses a significant global health challenge [1]. Its origins lie in a variety of pathophysiological mechanisms, including iron deficiency, impaired erythropoiesis, chronic inflammation, and genetic abnormalities [2–5]. Understanding these underlying mechanisms is crucial for tailoring effective treatment strategies. This article delves into the intricate pathophysiology of anemia, exploring the interplay of factors such as iron metabolism, folic acid and vitamin B12 deficiencies, erythropoietin regulation, and the impact of inflammatory processes on red blood cell production and survival.

A comprehensive review of the current evidence base regarding the most effective treatments for anemia is presented, considering the role of oral iron supplementation, folic acid and vitamin B12 replacement, erythropoiesis-stimulating agents, and emerging therapeutic approaches, consolidating current evidence from studies, guidelines, and meta-analyses and incorporating data from organizations such as the World Health Organization (WHO), European Hematology Association (EHA), and Cochrane Database. These interventions are examined in light of specific anemia subtypes and patient populations, with a focus on optimizing treatment outcomes and minimizing adverse effects.

This review integrates the clinical experience of the authors, who have over a decade of expertise in hematology and primary care. Their insights, gained from using a fixed-dose combination of folic acid 1 mg + iron (ferrous sulfate) 90 mg, offer a practical perspective on the real-world application of evidence-based strategies for the prevention and treatment of anemia in diverse clinical settings.

2. Erythropoiesis

Erythropoiesis is the process of producing and maturing red blood cells occurring in the bone marrow, followed by their release into the bloodstream [6]. This process involves several growth factors such as interleukin-3 (IL-3), erythropoietin (EPO), and substrates like iron, vitamin B12, folic acid, and heme [7,8], as shown in Figure 1.

The well-established model indicates that erythropoiesis starts with hematopoietic stem cells in the bone marrow differentiating into myeloid progenitors. They then develop into erythroid progenitors, specifically burst-forming unit-erythroid (BFU-E) and colony-forming unit-erythroid (CFU-E). CFU-E cells further mature through several stages—proerythroblast, basophilic, polychromatic, and orthochromatic erythroblast—where they produce hemoglobin and prepare for enucleation. During this process, the orthochromatic erythroblast ejects its nucleus, forming a reticulocyte. This reticulocyte then matures into an erythrocyte in approximately one week [9].

Under normal physiological conditions, bone marrow can produce approximately 200 billion red blood cells per day, with each pronormoblast developing into 16 erythrocytes [10], as illustrated in Figure 2. This process can be divided into three distinct phases, cellular differentiation, proliferation, and maturation—each regulated by specific mediators and regulatory factors [11–13].

During the maturation process, the cell decreases in size and volume, followed by the disappearance of the nucleus and a proportional increase in cytoplasmic volume and pigmentation (shifting from bluish to reddish), thereby becoming more acidophilic [6].

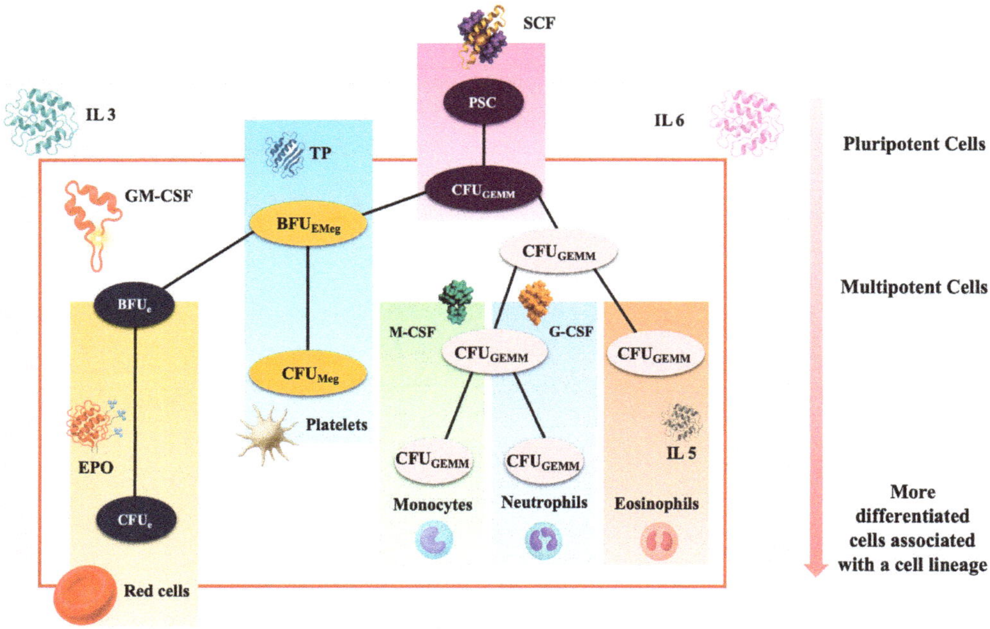

Figure 1. Hematopoiesis and growth factors SCF: stem cell factor; CSF: colony-stimulating factor; GM: granulocyte–macrophage; M: monocyte; G: granulocyte; EPO: erythropoietin; IL: interleukin; TP: thrombopoiesis; Meg: *Megakaryocyte*; BFU_E: burst forming unit-erythroid; CFU: colony forming unit; CFU_E: colony forming unit-erythroid; GEMM: granulocyte, erythrocyte, monocyte, megakaryocyte.

Iron is a fundamental substrate for erythropoiesis. It exists in inorganic form (as ferric [Fe^{3+}] and ferrous [Fe^{2+}] ions) and in organic form (bound to hemoglobin and myoglobin, and found in sources such as meat, eggs, and certain vegetables) [14,15].

Most dietary iron exists as [Fe^{3+}], which needs to be reduced to [Fe^{2+}] before it can be absorbed through the brush border membrane via divalent metal ion transporter 1 (DMT1). This reduction is probably facilitated by the brush border duodenal cytochrome b reductase (DcytB), although other reductases might also play a role. Organic iron is absorbed through the intestinal epithelium by the heme carrier protein 1 (HCP1), where it is degraded by heme oxygenase [14,15].

Hephaestin oxidase converts iron into its ferric form [Fe^{3+}], and transferrin binds two Fe^{3+} ions, transporting them to transferrin receptors in hepatocytes, where the iron is stored in ferritin subunits and can be mobilized as needed [16,17].

Folic acid is also crucial for erythropoiesis. It is found in green leafy vegetables, other vegetables, and fruits. In the enterocyte, it is converted into dihydrofolate (DHF), then into tetrahydrofolate (THF), and finally into methylfolate—the most bioactive form—which is then distributed to body tissues [18,19].

One of the main regulators of erythropoiesis is oxygen. When the partial pressure of O_2 (pO_2) decreases, adult kidneys, particularly the interstitial peritubular cells (IPCs), activate EPO (erythropoietin) production. EPO is secreted into the plasma, stimulating bone marrow erythroid progenitors through the EPO receptor to mature into red blood cells (RBCs) [7,20,21]. The interaction of EPO with the EPO receptor in proerythroid cells triggers proliferation, enhances cell survival, and promotes erythroid-specific differentiation.

Folic acid, vitamin B12, and vitamin B3 are involved in nucleic acid synthesis (during the proliferation phase) [6,8,13,18]. In the maturation phase, substrates such as iron, copper, and vitamin B6 are necessary, along with adequate hormonal levels, including androgens, corticosteroids, growth hormone, and thyroid hormones [6–8,13,16].

Erythropoiesis takes approximately 8 days, and erythrocytes have an expected lifespan of 120 days [6].

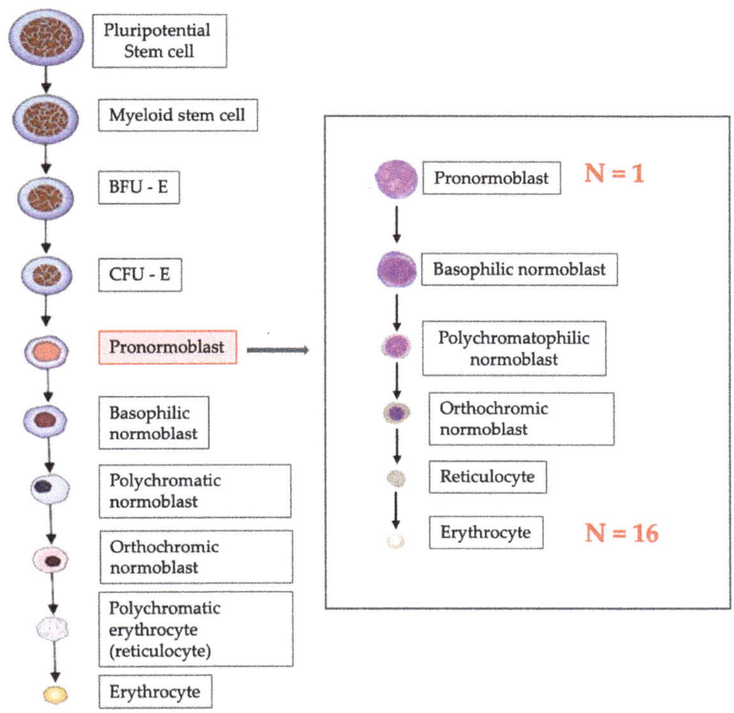

Figure 2. Erythropoiesis—from pluripotent stem cell to erythrocyte. BFU-E: burst forming unit-erythroid; CFU-E: colony forming unit-erythroid.

3. Anemias—Identification and Classification

Anemia is a common hematological condition in adults in clinical practice. While anemia is a global health concern, this review focuses specifically on the epidemiology, diagnostic approaches, and treatment recommendations relevant to the Portuguese population. We will examine Portuguese-specific data and guidelines to provide a contextually relevant perspective on the management of anemia in this region.

The World Health Organization (WHO) estimates that approximately 15% of the Portuguese population is affected by anemia, while the study of prevalence of anemia and iron deficiency in Portugal (EMPIRE) reported a prevalence of 20% [22]. Among individuals aged 80 years and older, the prevalence rate rises to approximately 31.4% [22,23].

Anemia is defined by the WHO as a reduction in hemoglobin concentration in peripheral blood below the normal values for age and sex, less than 12 g/dL in women and 13 g/dL in men [1]. In the elderly population, there is a higher prevalence of predisposing conditions for anemia [24,25], as well as an increased frequency of associated complications, as shown in Table 1.

In the elderly population, anemia represents a significant factor for morbidity, mortality, and frailty. However, in many cases, it is underestimated and undertreated [24].

Table 1. Consequences of anemia in the elderly (adapted from Emmanuel Andrès et al., 2013 [10]).

Anemia in Elderly Patients
Alteration of cognitive function
Iatrogenic injury
Descompasation of underlying disorders
Hospitalization
Increased mortality
Increased debility
Reduction in mobility and bone density; Increased risk of falls and fractures
Reduction of quality of life

The main causes of anemia are iron deficiency (29%), chronic diseases (27.5%), hemorrhagic losses (17.5%), and hemolysis (17.5%), while other causes account for about 9% of cases [24,25].

Table 2 presents the most common causes of anemia in older adults and their respective prevalence rates.

Table 2. Causes and prevalence of anemia in the elderly [22,23].

Cause	Percentage of All Anemia Cases
Deficiency of iron	16.6
Deficiency of folate	6.4
Deficiency of vitamin B12	5.9
Deficiency of folate and vitamin B12	2.0
Deficiency of iron, folate and vitamin B12	3.4
Renal insufficiency	8.2
Inflammation (chronic disease)	19.7
Renal insufficiency and inflammation	4.3

Another study conducted in Japan identified slightly different causes of anemia [24], as shown in Figure 3.

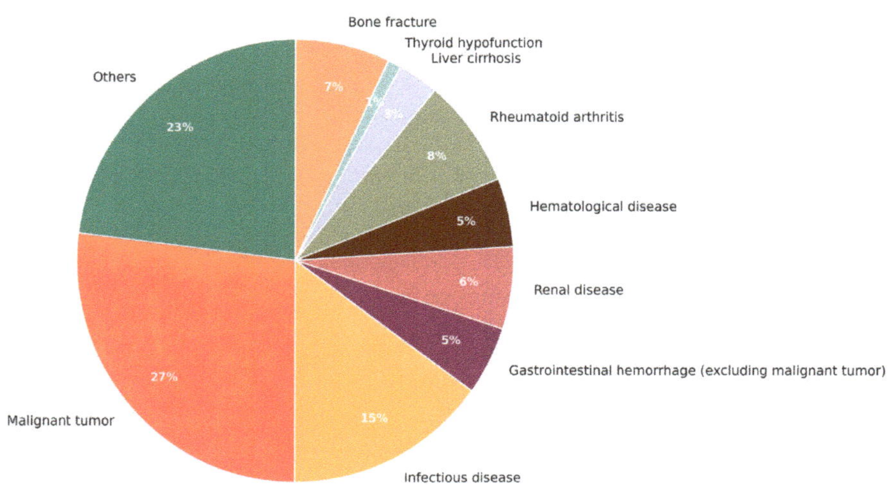

Figure 3. Causes and prevalence of anemia in the elderly according to a Japanese study [26].

For an accurate etiological diagnosis of anemia, classification is essential [1]. One of the most relevant classification systems is based on reticulocyte count, which allows differentiation between non-regenerative and regenerative anemia [2,27], as illustrated in Table 3.

Table 3. Classification of anemias based on regenerative capacity.

Non-Regenerative	Regenerative
Aplastic anaemia	Haemolysis
Pure red cell aplasia	Immune
Myelodysplasic syndrome	Non-immune • Congenital: membrane, SS, thalassaemia, enzymophathies, unstable Hb • Acquired: PNH, drugs, microangiopathy, hyperslpenism
Deficiency states	
Marrow infiltration/fibrosis	
Inflammatory anaemia	
Erytropoietin underproduction	Haemorrhage

Another important classification is the morphological one [3,4,28], which is based on the analysis of mean corpuscular volume (MCV), as illustrated in Table 4.

Table 4. Morphological classification of anemias.

Normocytic (MCV 80-100 fL)	Microcytic (MCV < 80 fL)	Macrocytic (MCV >100 fL)
- Hemorrhagic anemia - Early iron deficiency anemia - Anemia of chronic disease - Anemia associated with bone marrow suppression - Anemia associated with chronic renal insufficiency - Anemia associated with endocrine dysfunction - Autoimmune hemolytic anemia - Anemia associated with hypothyroidism or hypopituitarism - Hereditary spherocytosis - Hemolytic anemia associated with paroxysmal nocturnal hemoglobinuria	- Iron deficiency anemia - Thalassemias - Anemia of chronic disease - Sideroblastic anemia - Anemia associated with copper deficiency - Anemia associated with lead poisoning	- Folic acid deficiency anemia - Anemia associated with vitamin B12 deficiency - Drug-induced hemolytic anemia - Anemia associated with reticulocytosis - Anemia associated with liver disease - Anemia associated with etanol abuse - Anemia associated with acute myelodysplastic syndrome

Laboratory diagnosis includes a complete blood count (including RDW), a peripheral blood smear, iron metabolism evaluation, measurements of vitamin B12 and folic acid, erythropoietin (EPO), thyroid-stimulating hormone (TSH), C-reactive protein (CRP), and creatinine. In some cases, a bone marrow aspirate and biopsy may berequired—for example, in situations of pancytopenia, monoclonal gammopathy, suspected myelodysplastic syndrome, or unexplained progressive anemia [2,4].

Table 5 offers a systematic overview of all the steps needed for accurately diagnosing anemia, from collecting clinical history to laboratory assessment [2–5].

Therefore, the diagnostic approach to anemia involves three key steps: (1) reticulocyte count, which differentiates between a central production problem and increased peripheral destruction; (2) morphological classification of the anemia; and (3) identification of the underlying etiology.

Practically, the reticulocyte count serves as a critical decision point guiding subsequent diagnostic investigations, as depicted in Table 6.

After the morphological classification of anemia, further important steps are needed to identify the underlying cause(s) and to guide a more precise treatment plan, as shown in Figure 4.

Table 5. Diagnostic algorithm for the evaluation of a patient with anemia.

Evaluation		
History	**Physical Examination**	**Laboratory Investigations**
• Time of onset • Severity • Age • Dissociation Curve Hb/O_2 • Positive family history of anemia • Ethnic origin • Acuteness of onset of anemia • Bleeding: stool, urine, lungs, menses • Infections: Parvovírus B19, Hepatitis, HIV • Phlebotomy • Jaundice, dark urine • Petechiae • Symptoms from other organs (CNS, gastrites) • Masses, nodes • Alchohol consumption • Diet:low fruit, no meat • Pica (abdnormal food habits) • Drugs • Previus transfusion	• Pallor, ictereus, petechiae • Temperature • Lymph nodes • Enlarged spleen • Tachycardia, hypotension	• WBC and platelet count (aplastic anaemia) • Reticulocyte count (red cell aplasia) • MCV (iron or vitamin deficiency, thalassaemias) • RBC morphology: ○ double RBC population (transfusion, sideroblastic anaemia, high HbF) ○ RBC agglutination (cold agglutinins) ○ schizocytes (microangiopathy) ○ spherocytes (imune haemolysis) • Bilirubin, LDH, haptoglobin, Coombs test (haemolysis) • Kidney function (EPO deficiency) • Occult blood in the stools • C-reactive protein • Iron studies + vitamins (serum iron, TIBC, ferritinm B12, folates) depending on MCV

Table 6. Causes of anemia according to reticulocyte count.

Reticulocyte Count	
Low	**High**
Hypoproliferative anemia: • Iron deficiency • B12 or folate deficiency • Anemia of chronic inflammation • Cytotoxic chemotherapy • Renal Insufficiency • Myeolophtisic anemia	Increased loss of red blood cells: • Acute hemorrhage • Auto-imune hemolytic anemia • Microangiopatic anemia

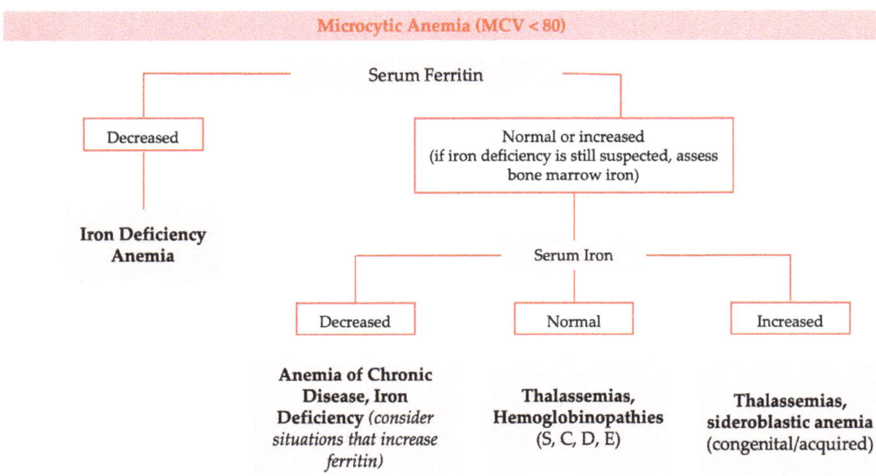

Figure 4. *Cont.*

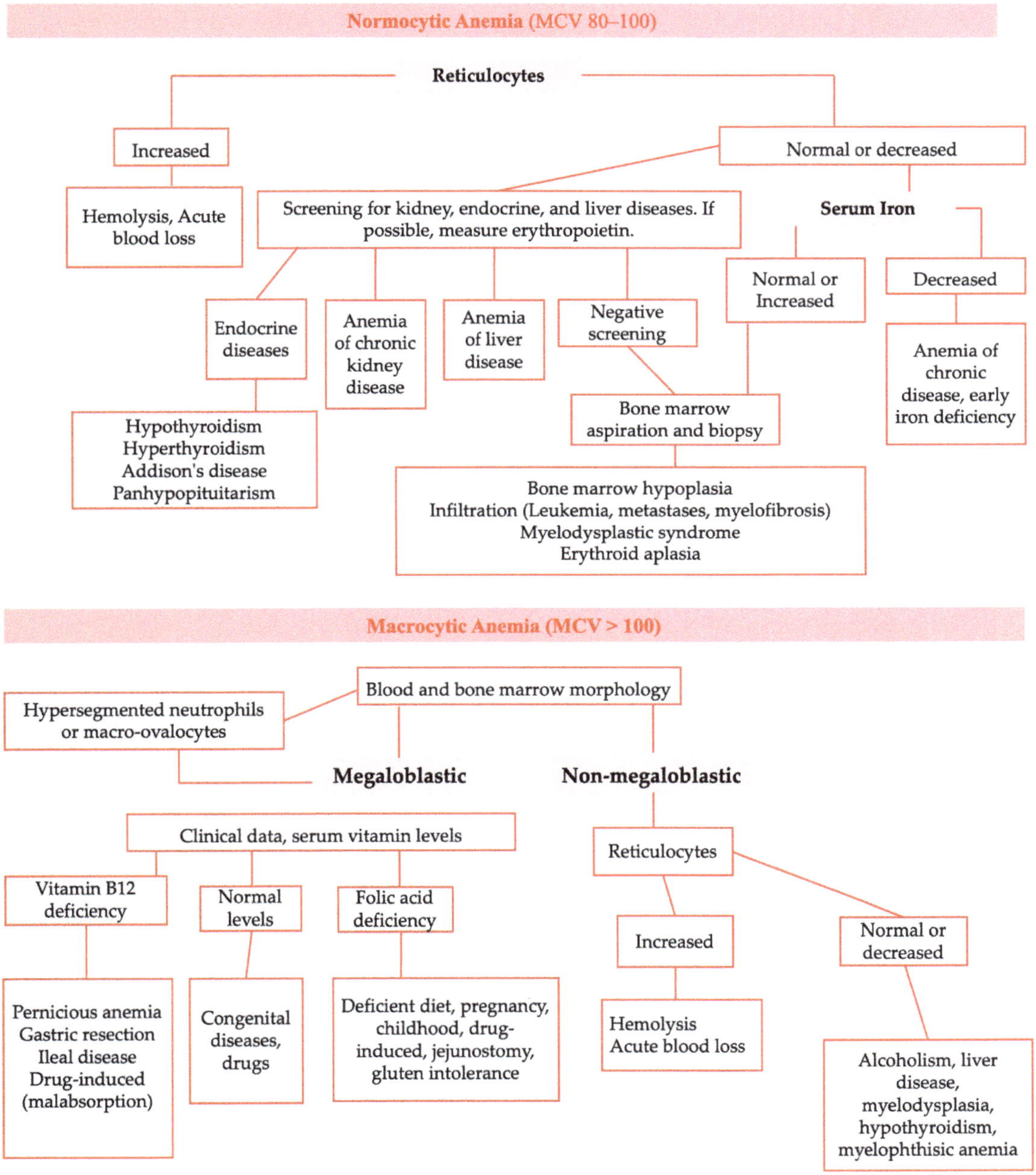

Figure 4. Subsequent laboratory evaluations to be considered after morphological classification of anemia.

4. Prevention of Anemia

The primary risk groups for iron deficiency and iron deficiency anemia include children and adolescents (due to increased needs related to growth), premature infants, menstruating women of reproductive age, pregnant women, older adults, frequent blood donors, and those on strict vegetarian or restrictive diets. People with chronic inflammatory diseases, inflammatory bowel disease, malabsorption syndromes, or chronic hemorrhagic diathesis are also at higher risk for iron deficiency [29].

Pregnant women, older adults, and individuals with inflammatory bowel disease are also at increased risk of mixed nutritional anemias, with concurrent folic acid deficiency [29]. The World Health Organization (WHO) recommends oral iron and folic acid supplementation for the prevention and treatment of iron deficiency anemia in women of reproductive age in countries with high prevalence of this condition [30,31].

A Cochrane review on daily oral iron supplementation during pregnancy—initially published in 2006 and subject to successive updates, the most recent in 2024 [32]—consistently found that such supplementation is associated with a reduction in maternal anemia and iron deficiency at term. Based on 40 clinical trials, the review also concluded that daily oral iron supplementation during pregnancy probably reduces the risk of low birth weight. The included studies used iron supplementation with ferrous sulfate, either alone or combined with folic acid. The authors recommended this strategy as an effective preventive measure for reducing the risk of anemia and iron deficiency during pregnancy [32].

The WHO recommends a daily oral intake of at least 30–60 mg of elemental iron and 400 μg of folic acid during pregnancy to prevent maternal anemia, postpartum sepsis, low birth weight, and preterm delivery, with ferrous salts such as ferrous sulfate being the first-line treatment for this purpose [33]. Although it has some limitations—especially when inflammation is present—ferritin remains the most specific biomarker for evaluating iron stores and aiding in the diagnosis of iron deficiency, even without anemia [29]. Figure 5, adapted from the latest recommendations of the European Hematology Association (EHA) on preventing and treating iron deficiency and iron deficiency anemia [29], summarizes the diagnostic criteria for iron deficiency without anemia, establishing serum ferritin cutoff values to support diagnosis across different age groups and clinical settings.

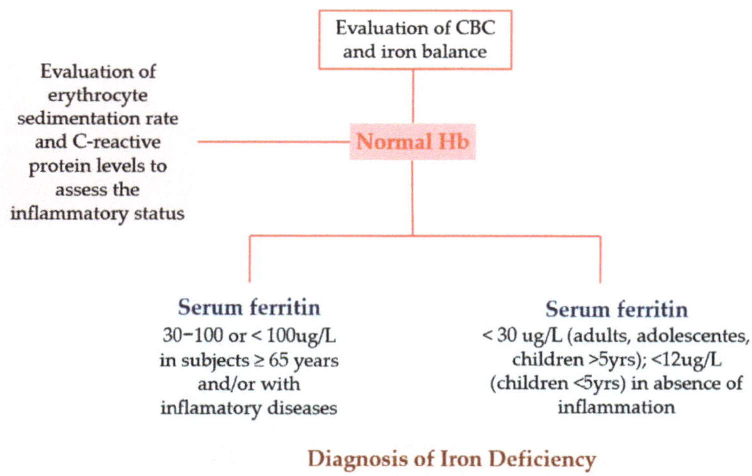

Figure 5. Diagnosis of iron deficiency without anemia.

In cases where iron deficiency is identified in the absence of anemia, the WHO recommends treatment with daily oral iron and folic acid for a duration of 3 months [31], or alternatively, using lower doses such as 40 mg of elemental iron per day or less frequent dosing regimens such as 120 mg of elemental iron every other day [30,34], as a strategy to prevent the development of anemia. The use of lower iron doses or less frequent dosing schedules has been associated with improved tolerability, while remaining effective in preventing anemia in iron-deficient populations [34–36].

The recommendation to treat iron deficiency even in the absence of anemia is broadly accepted for the general population and particularly emphasized during pregnancy and in preoperative settings [37].

Potential causes of iron deficiency should always be investigated and addressed, including heavy menstrual bleeding in women of reproductive age, occult gastrointestinal bleeding, or dietary factors (such as low intake of bioavailable iron-rich foods, and excessive consumption of dairy products, tea, or iced tea) [38].

In addition to preventing anemia, these interventions are also important for minimizing other adverse effects of iron deficiency, such as worsening cognitive decline in older adults [29] or delays in neurodevelopment during childhood [39].

5. Treatment of Anemia

The updated guidelines from the European Hematology Association (EHA) for the treatment and prevention of iron deficiency and iron deficiency anemia, published in July 2024, recommend oral iron salts—particularly ferrous sulfate—as first-line treatment in the vast majority of cases of uncomplicated iron deficiency anemia in adults [29].

Although elemental iron doses of 100–200 mg/day were traditionally recommended, the authors now indicate that lower daily doses or alternate-day dosing regimens of elemental iron are associated with similar or even improved absorption, better gastrointestinal tolerability, and greater adherence, with fewer adverse effects [29].

Figure 6, adapted from these same guidelines, outlines the therapeutic approach to iron deficiency anemia.

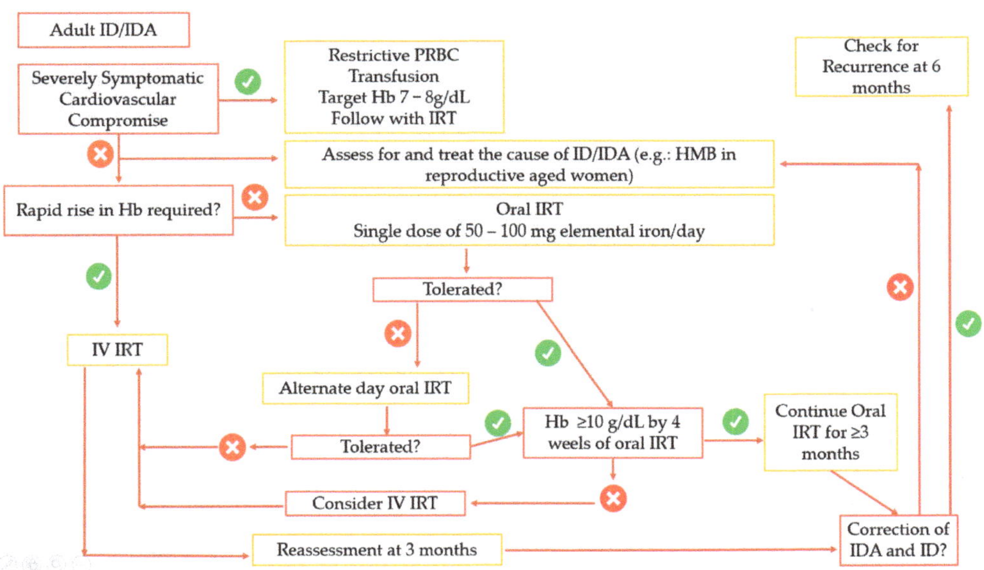

Figure 6. European Hematology Association guidelines on therapeutic approach to iron deficiency anemia [29]. IDA—Iron Deficiency Anemia; ID—Iron Deficiency; IRT—Iron Replacement Therapy; PRBC—Packed Red Blood Cells; Hb—Hemoglobin; HMB—Heavy Menstrual Bleeding; IV—Intravenous.

Although some studies have suggested a marginal increase in absorption with alternate-day dosing regimens, a recent randomized controlled trial found no significant differences in efficacy or tolerability between daily and alternate-day administration of ferrous sulfate [40].

Oral iron treatment should be continued until anemia is corrected and then maintained for at least an additional 3 months to replenish iron stores and minimize the risk of recurrence [29,41]. The management of iron deficiency anemia should always include investigation and, whenever possible, treatment of the underlying causes [41]. Intravenous iron formulations should be reserved for cases requiring a rapid increase in hemoglobin levels or when oral iron is not tolerated or adequately absorbed [29,42]. Intravenous iron is thus considered a complementary strategy to oral therapy—first-line in most cases—with distinct indications, and not a replacement.

Oral ferrous sulfate is the most widely used treatment option globally and may be considered the gold standard first-line therapy for iron deficiency anemia [42]. This is reflected in several international guidelines, including those of the EHA [29], the British Society of Gastroenterology [34], and the American Gastroenterological Association [43]. The main adverse effects of this formulation are gastrointestinal and are comparable to those of other oral iron formulations. When administered as a single daily dose or on alternate days, ferrous sulfate shows a lower incidence of adverse effects that promotes treatment maintenance [42]. Some authors also note its advantages over other formulations due to its lower cost, extensive clinical experience, and the availability of robust data supporting the use of lower doses or intermittent dosing regimens [42,43].

Proper treatment of iron deficiency anemia has been associated with increased hemoglobin levels, reduction in symptoms and comorbidities attributable to anemia, and decreased need for blood component transfusions [44].

Iron replacement therapy in iron deficiency anemia is associated with an erythropoietic response, characterized by an increase in reticulocyte count that can be observed within the first days or weeks after initiating supplementation [45].

This stimulation of erythropoiesis requires sufficient folic acid availability [46]. Since it is safe to supplement orally—even at daily doses above 5 mg [47]—folic acid can be used alongside iron salts without concern. This combination not only promotes erythropoiesis but also improves treatment response in cases of mixed deficiency anemia [29]. In Portugal, the following fixed-dose formulations are authorized:

1. A fixed combination of folic acid 0.35 mg + ferrous sulfate 247.25 mg, approved for the prophylaxis and treatment of iron and folic acid deficiencies during pregnancy, postpartum, and prolonged lactation, or in cases where iron and folic acid were not regularly taken during lactation;
2. A fixed combination of folic acid 0.35 mg + ferrous sulfate 325 mg, approved for the prophylaxis and treatment of anemia during pregnancy, for treating iron deficiency, and for preventing concurrent folic acid deficiency in adults. The safety profile of folic acid is well established, with no significant toxicity reported. Concerns raised in the past regarding potential harmful (e.g., carcinogenic) effects of high folic acid concentrations after supplementation have not been confirmed. A large meta-analysis involving 49,621 participants found no such risk [48]. On the contrary, prenatal folic acid supplementation has shown protective effects not only against neural tube defects but also against various pediatric cancers, reducing the incidence of leukemia, brain tumors, and neuroblastoma [49].

Doses ranging from 400 μg to 5 mg of folic acid per day are considered non-toxic and can be safely used during pregnancy. For pregnant women at higher risk of folic acid deficiency, neural tube defects, or poor treatment adherence, higher doses such as 5 mg/day are recommended [50].

6. Limitations and Future Perspectives

While this review provides a comprehensive overview of the use of folic acid 1 mg + iron (ferrous sulfate) 90 mg in the treatment of anemia, it is important to acknowledge several limitations in the existing evidence base.

Firstly, there is a relative paucity of high-quality, randomized controlled trials specifically evaluating the effectiveness of this fixed-dose combination, particularly in certain subpopulations such as the elderly, pregnant women, and individuals with chronic inflammatory conditions. Many of the recommendations are based on observational studies, expert opinion, and extrapolations from studies of iron or folic acid supplementation alone. Further research is needed to directly compare the efficacy and safety of this combination to alternative treatment strategies, such as higher doses of iron or separate supplementation with iron and folic acid.

Secondly, although this review focuses on the Portuguese context, there is a lack of specific data on the prevalence and management of anemia within this population. Further studies are needed to assess the effectiveness of current treatment guidelines and to identify any unique challenges or opportunities for improving anemia care in Portugal.

Thirdly, the studies included in this review were heterogeneous in terms of patient characteristics, diagnostic criteria, interventions, and outcome measures. This heterogeneity makes it difficult to draw firm conclusions about the overall effectiveness of the folic acid 1 mg + iron (ferrous sulfate) 90 mg combination and limits the generalizability of the findings.

Finally, there is a lack of long-term data on the safety and efficacy of this combination. Future studies should assess the long-term effects on patient-centered outcomes such as quality of life, functional status, and cognitive function, as well as the potential for adverse effects with prolonged use.

Future research should focus on the following: conducting well-designed, randomized controlled trials to directly compare the folic acid 1 mg + iron (ferrous sulfate) 90 mg combination to alternative treatment strategies in diverse patient populations; performing studies to assess the cost-effectiveness of this combination compared to other treatments; investigating the optimal duration of treatment with this combination; identifying biomarkers that can predict response to treatment; developing strategies to improve adherence to treatment; and conducting research to better understand the underlying mechanisms by which iron and folic acid affect erythropoiesis.

Addressing these limitations through rigorous scientific investigation will help to refine our understanding of the role of folic acid and iron supplementation in the treatment of anemia and to optimize clinical practice guidelines for the benefit of patients.

7. Suggestions and Recommendations Based on Clinical Studies/Experiences

While the evidence reviewed supports the use of a fixed-dose combination of folic acid 1 mg + iron (ferrous sulfate) 90 mg for the treatment of iron deficiency anemia, further studies are needed to specifically evaluate the clinical benefits and long-term outcomes associated with this particular dosage regimen, especially in diverse patient populations and in comparison to other available formulations.

The authors have over 10 years of extensive experience in both hospital-based settings (Clinical Hematology) and community-based care (Primary Health Care and Palliative Care), using a fixed oral daily dose of Folic Acid 1 mg + Iron (Ferrous Sulfate) 90 mg for the treatment of iron deficiency anemia, mixed nutritional anemia (iron and folic acid deficiency), and iron deficiency without anemia in adults.

This clinical experience demonstrates good therapeutic efficacy and satisfactory tolerability, allowing for strong adherence to treatment. The most commonly observed adverse effects are gastrointestinal (constipation, diarrhea, and, less frequently, nausea), which are generally manageable through adjustments to the administration schedule—such as taking the supplement with meals rather than on an empty stomach (which improves tolerability), provided the food contains vitamin C (e.g., citrus fruits), in order not to significantly impair iron absorption in the gastrointestinal tract.

The groups with the highest prevalence of iron deficiency anemia in the Portuguese population are women of reproductive age [22] and the elderly [23]. In the authors' clinical practice, switching from formulations containing only oral iron salts to fixed combinations of oral iron with folic acid (with or without concomitant vitamin B12 supplementation, depending on the clinical context) frequently leads to improved anemia response.

Additionally, the authors' extensive clinical experience in the treatment of chronic hemolytic anemias supports the safety of oral folic acid supplementation, both at a dose of 1 mg and at doses up to 5 mg per day, with no observed toxicity or adverse effects attributable to folic acid—even in patients who have been taking daily supplementation for several years or even throughout their entire lives.

Therefore, the use of 1 mg folic acid, being free from toxicity or increased risk, is considered by the authors to be a valuable therapeutic option. It enables correction of folate deficiencies—an essential factor in hematopoiesis—optimizing treatment response and ensuring sufficient folic acid levels even in cases of incomplete adherence or missed doses. This is particularly important during pregnancy, where adequate supplementation helps minimize the risk of neural tube defects. Nevertheless, specific evaluation and recommendations appropriate for the pediatric age group are required.

8. Recommendation

Based on the scientific evidence presented throughout this document, the fact that ferrous sulfate is considered the gold standard in the treatment of iron deficiency anemia, its inclusion as a first-line therapy for both iron deficiency anemia and non-anemic iron deficiency in major and recent international guidelines, as well as the authors' many years of clinical experience in the treatment and prevention of iron deficiency anemia, non-anemic iron deficiency, and mixed nutritional anemias, the authors consider there to be robust evidence supporting the recommendation of Folic Acid 1 mg + Iron (Ferrous Sulfate) 90 mg as a treatment option.

This combination should be indicated not only for the prevention and treatment of iron and folic acid deficiency during pregnancy, lactation, and the postpartum period but also for the prophylaxis and treatment of anemia during pregnancy and in adults in general.

Author Contributions: Conceptualization—J.G. and H.R.; Methodology—J.G. and H.R.; Software—H.R.; Validation—J.B.S., C.V.J. and H.R.; Formal analysis—J.G. and H.R.; Investigation—J.G. and H.R.; Resources—HR; Data Curation—J.G., J.B.S. and H.R.; Writing—Original Draft—J.G. and H.R.; Writing—Review and Editing—J.B.S. and C.V.J.; Visualization—C.V.J. and H.R.; Supervision—H.R.; Project administration—H.R.; Funding acquisition—H.R. All authors have read and agreed to the published version of the manuscript.

Funding: This research received no external funding.

Data Availability Statement: Data can be obtained upon reasonable request.

Conflicts of Interest: J.G. and H.R. declare that they received a grant to conduct a review of current evidence of Folic Acid 1 mg + Iron (Ferrous Sulfate) 90 mg by Bial.

References

1. World Health Organization. *Guideline on Haemoglobin Cutoffs to Define Anaemia in Individuals and Populations;* WHO: Geneva, Switzerland, 2024.
2. Garcia-Casal, M.N.; Dary, O.; Jefferds, M.E.; Pasricha, S. Diagnosing anemia: Challenges at the public health level. *Ann. N. Y. Acad. Sci.* **2023**, *1524*, 37–50. [CrossRef] [PubMed]
3. Cappellini, M.D.; Motta, I. Anemia in Clinical Practice—Does Hemoglobin Change with Aging? *Semin. Hematol.* **2015**, *52*, 261–269. [CrossRef]
4. Kundrapu, S.; Noguez, J. Laboratory Assessment of Anemia. *Adv. Clin. Chem.* **2018**, *83*, 197–225. [CrossRef]
5. De Franceschi, L.; Iolascon, A.; Taher, A.; Cappellini, M.D. Clinical management of iron deficiency anemia in adults: Systemic review on advances in diagnosis and treatment. *Eur. J. Intern. Med.* **2017**, *42*, 16–23. [CrossRef]
6. Hillman, R.S.; Finch, C.A. Erythropoiesis. *N. Engl. J. Med.* **1971**, *285*, 99–101. [CrossRef]
7. Chung, E.Y.; Palmer, S.C.; Saglimbene, V.M.; Craig, J.C.; Tonelli, M.; Strippoli, G.F. Erythropoiesis-stimulating agents for anaemia in adults with chronic kidney disease: A network meta-analysis. *Cochrane Database Syst. Rev.* **2023**, *2023*, CD010590. [CrossRef]
8. Zivot, A.; Lipton, J.M.; Narla, A.; Blanc, L. Erythropoiesis: Insights into pathophysiology and treatments in 2017. *Mol. Med.* **2018**, *24*, 11. [CrossRef]
9. Cha, H.J. Erythropoiesis: Insights from a genomic perspective. *Exp. Mol. Med.* **2024**, *56*, 2099–2104. [CrossRef] [PubMed]
10. Andrès, E.; Serraj, K.; Federici, L.; Vogel, T.; Kaltenbach, G. Anemia in elderly patients: New insight into an old disorder. *Geriatr. Gerontol. Int.* **2012**, *13*, 519–527. [CrossRef]
11. Jagannathan-Bogdan, M.; Zon, L.I. Hematopoiesis. *Development* **2013**, *140*, 2463–2467. [CrossRef] [PubMed]
12. Orkin, S.H.; Zon, L.I. Hematopoiesis: An Evolving Paradigm for Stem Cell Biology. *Cell* **2008**, *132*, 631–644. [CrossRef]
13. Orkin, S.H. Hematopoiesis: How does it happen? *Curr. Opin. Cell Biol.* **1995**, *7*, 870–877. [CrossRef]
14. Anderson, G.J.; Frazer, D.M.; McLaren, G.D. Iron absorption and metabolism. *Curr. Opin. Gastroenterol.* **2009**, *25*, 129–135. [CrossRef]
15. Piskin, E.; Cianciosi, D.; Gulec, S.; Tomas, M.; Capanoglu, E. Iron Absorption: Factors, Limitations, and Improvement Methods. *ACS Omega* **2022**, *7*, 20441–20456. [CrossRef] [PubMed]
16. Conrad, M.E.; Umbreit, J.N. Iron absorption and transport: An update. *Am. J. Hematol.* **2000**, *64*, 287–298. [CrossRef]
17. von Siebenthal, H.K.; Moretti, D.; Zimmermann, M.B.; Stoffel, N.U. Effect of dietary factors and time of day on iron absorption from oral iron supplements in iron deficient women. *Am. J. Hematol.* **2023**, *98*, 1356–1363. [CrossRef]
18. Shulpekova, Y.; Nechaev, V.; Kardasheva, S.; Sedova, A.; Kurbatova, A.; Bueverova, E.; Kopylov, A.; Malsagova, K.; Dlamini, J.C.; Ivashkin, V. The Concept of Folic Acid in Health and Disease. *Molecules* **2021**, *26*, 3731. [CrossRef] [PubMed]
19. Donnelly, J.G. Folic Acid. *Crit. Rev. Clin. Lab. Sci.* **2001**, *38*, 183–223. [CrossRef]
20. Tsiftsoglou, A.S. Erythropoietin (EPO) as a Key Regulator of Erythropoiesis, Bone Remodeling and Endothelial Transdifferentiation of Multipotent Mesenchymal Stem Cells (MSCs): Implications in Regenerative Medicine. *Cells* **2021**, *10*, 2140. [CrossRef]
21. Jelkmann, W. Physiology and Pharmacology of Erythropoietin. *Transfus. Med. Hemother.* **2013**, *40*, 302–309. [CrossRef] [PubMed]
22. Fonseca, C.; Marques, F.; Nunes, A.R.; Belo, A.; Brilhante, D.; Cortez, J. Prevalence of anaemia and iron deficiency in Portugal: The EMPIRE study. *Intern. Med. J.* **2016**, *46*, 470–478. [CrossRef]
23. Robalo Nunes, A.; Fonseca, C.; Marques, F.; Belo, A.; Brilhante, D.; Cortez, J. Prevalence of anemia and iron deficiency in older Portuguese adults: An EMPIRE substudy. *Geriatr. Gerontol. Int.* **2017**, *17*, 1814–1822. [CrossRef]
24. Katsumi, A.; Abe, A.; Tamura, S.; Matsushita, T. Anemia in older adults as a geriatric syndrome: A review. *Geriatr. Gerontol. Int.* **2021**, *21*, 549–554. [CrossRef]
25. Stauder, R.; Thein, S.L. Anemia in the elderly: Clinical implications and new therapeutic concepts. *Haematologica* **2014**, *99*, 1127–1130. [CrossRef]
26. Masatsugu, O.H.T.A. Management of Anemia in the Elderly. *JMAJ* **2009**, *52*, 219–223. [CrossRef]
27. Vohra, R.; Hussain, A.; Dudyala, A.K.; Pahareeya, J.; Khan, W.; Mumtaz, W. Multi-class classification algorithms for the diagnosis of anemia in an outpatient clinical setting. *PLoS ONE* **2022**, *17*, e0269685. [CrossRef]
28. Clinton, O.; Micheal, K.; Namyalo, A.K.; Mary, M.; Mike, M.; Muwanguzi, E.; Okongo, B.; Wagubi, R. Anaemia, Morphological Classification and Its Associated Risk Factors Among Lactating Mothers at Mbarara City Council Health Centre IV, Southwestern Uganda. *J. Blood Med.* **2022**, *13*, 473–481. [CrossRef]
29. Iolascon, A.; Andolfo, I.; Russo, R.; Sanchez, M.; Busti, F.; Swinkels, D.; Martinez, P.A.; Bou-Fakhredin, R.; Muckenthaler, M.U.; Unal, S.; et al. Recommendations for diagnosis, treatment, and prevention of iron deficiency and iron deficiency anemia. *HemaSphere* **2024**, *8*, e108. [CrossRef] [PubMed]
30. World Health Organization. *Guideline: Intermittent Iron and Folic Acid Supplementation in Menstruating Women;* WHO: Geneva, Switzerland, 2011.

31. Roche, M.L.; Samson, K.L.; Green, T.J.; Karakochuk, C.D.; Martinez, H. Perspective: Weekly Iron and Folic Acid Supplementation (WIFAS): A Critical Review and Rationale for Inclusion in the Essential Medicines List to Accelerate Anemia and Neural Tube Defects Reduction. *Adv. Nutr. Int. Rev. J.* **2021**, *12*, 334–342. [CrossRef]
32. Finkelstein, J.L.; Cuthbert, A.; Weeks, J.; Venkatramanan, S.; Larvie, D.Y.; De-Regil, L.M.; Garcia-Casal, M.N.; Cochrane Central Editorial Service. Daily oral iron supplementation during pregnancy. *Cochrane Database Syst. Rev.* **2024**, *2024*, CD004736. [CrossRef] [PubMed]
33. World Health Organization. *WHO Recommendations on Antenatal Care;* WHO: Geneva, Switzerland, 2016.
34. Stoffel, N.U.; von Siebenthal, H.K.; Moretti, D.; Zimmermann, M.B. Oral iron supplementation in iron-deficient women: How much and how often? *Mol. Asp. Med.* **2020**, *75*, 100865. [CrossRef]
35. Simic, S.; Karczewski, M.; Klapdor, S.; Nowak, A.; Schubert, M.; Moretti, D.; Swinkels, D.W.; Beuschlein, F.; Saleh, L.; Suter, P.; et al. Krayenbuehl PA. Effectiveness of low-dose iron treatment in non-anaemic iron-deficient women: A prospective open-label single-arm trial. *Swiss Med. Wkly.* **2023**, *153*, 40079. [CrossRef] [PubMed]
36. Rimon, E.; Kagansky, N.; Kagansky, M.; Mechnick, L.; Mashiah, T.; Namir, M.; Levy, S. Are we giving too much iron? Low-dose iron therapy is effective in octogenarians. *Am. J. Med.* **2005**, *118*, 1142–1147. [CrossRef]
37. Al-Naseem, A.; Sallam, A.; Choudhury, S.; Thachil, J. Iron deficiency without anaemia: A diagnosis that matters. *Clin. Med.* **2021**, *21*, 107–113. [CrossRef] [PubMed]
38. Santoyo-Sánchez, A.; Aponte-Castillo, J.A.; Parra-Peña, R.I.; Ramos-Peñafiel, C.O. Dietary recommendations in deficiency anaemia. *Rev. Med. Hosp. Gen. Mex.* **2015**, *78*, 144–150.
39. Lozoff, B. Iron Deficiency and Child Development. *Food Nutr. Bull.* **2007**, *28* (Suppl. 4), S560–S571. [CrossRef]
40. Pasupathy, E.; Kandasamy, R.; Thomas, K.; Basheer, A. Alternate day vs daily oral iron: A randomized trial. *Sci. Rep.* **2023**, *13*, 1818.
41. Snook, J.; Bhala, N.; Beales, I.L.P.; Cannings, D.; Kightley, C.; Logan, R.P.; Pritchard, D.M.; Sidhu, R.; Surgenor, S.; Thomas, W.; et al. British Society of Gastroenterology guidelines for the management of iron deficiency anaemia in adults. *Gut* **2021**, *70*, 2030–2051. [CrossRef]
42. Pantopoulos, K. Oral iron supplementation: New formulations, old questions. *Haematologica* **2024**, *109*, 2790–2801. [CrossRef]
43. DeLoughery, T.G.; Jackson, C.S.; Ko, C.W.; Rockey, D.C. AGA Update on Iron Deficiency Anemia. *Clin. Gastroenterol. Hepatol.* **2024**, *22*, 1575–1583. [CrossRef]
44. Clevenger, B.; Gurusamy, K.; Klein, A.A.; Murphy, G.J.; Anker, S.D.; Richards, T. Systematic review and meta-analysis of iron therapy in anaemic adults without chronic kidney disease: Updated and abridged Cochrane review. *Eur. J. Heart Fail.* **2016**, *18*, 774–785. [CrossRef] [PubMed]
45. Vollset, S.E.; Clarke, R.; Lewington, S.; Ebbing, M.; Halsey, J.; Lonn, E.; Armitage, J.; Manson, J.E.; Hankey, G.J.; Spence, J.D.; et al. Effects of folic acid supplementation on overall and site-specific cancer incidence during the randomised trials: Meta-analyses of data on 50,000 individuals. *Lancet* **2013**, *381*, 1029–1036. [CrossRef] [PubMed]
46. Koury, M.J.; Ponka, P. The Roles of Folate, Vitamin B12, and Iron. *Annu. Rev. Nutr.* **2004**, *24*, 105–131. [CrossRef] [PubMed]
47. E Butterworth, C.; Tamura, T. Folic acid safety and toxicity: A brief review. *Am. J. Clin. Nutr.* **1989**, *50*, 353–358. [CrossRef]
48. Parodi, E.; Giraudo, M.T.; Davitto, M.; Ansaldi, G.; Mondino, A.; Garbarini, L.; Franzil, A.; Mazzone, R.; Russo, G.; Ramenghi, U. Reticulocyte Parameters. *J. Pediatr. Hematol.* **2012**, *34*, e249–e252. [CrossRef]
49. Barua, S.; Kuizon, S.; Junaid, M.A. Folic acid in pregnancy and disease. *J. Biomed. Sci.* **2014**, *21*, 77. [CrossRef]
50. Bar-Oz, B.; Koren, G.; Nguyen, P.; Kapur, B.M. Folate fortification—Are we there yet? *Reprod. Toxicol.* **2008**, *25*, 408–412. [CrossRef]

BioChem

MDPI

Review

Osteoporosis: Focus on Bone Remodeling and Disease Types

Chiara Castellani [1,2,*], Erica De Martino [1] and Paolo Scapato [1]

1 UOSD Reumatologia, Ospedale San Camillo De Lellis, 02100 Rieti, Italy; e.demartino@asl.rieti.it (E.D.M.); p.scapato@asl.rieti.it (P.S.)
2 Molecular Medicine Department, Sapienza University of Rome, 00185 Roma, Italy
* Correspondence: c.castellani@asl.rieti.it

Abstract

Osteoporosis is a common skeletal disease that leads to increased bone fragility, associated with increased risk of fracture and consequent significant morbidity and mortality, and is a global public health problem. It results from a chronic imbalance in bone remodeling, where bone resorption by osteoclasts exceeds bone formation by osteoblasts. Aging, hormonal changes, comorbidities, and drugs influence the process that leads to osteoporosis. In this review, we delve into the pathogenesis of primary and secondary osteoporosis after a summary of the normal physiology of bone homeostasis. Primary osteoporosis includes postmenopausal osteoporosis, driven largely by estrogen deficiency, and age-related (senile) osteoporosis, associated with reduced bone formation. An insight into male osteoporosis and osteoporosis treatment is also provided. Secondary osteoporosis can derive from underlying conditions, such as endocrine disorders, chronic inflammatory and genetic diseases, or prolonged use of glucocorticoids. Clinically, osteoporosis is often unacknowledged, underlining the importance of early risk assessment and diagnosis. A thorough understanding of the disease, its subtypes, and its underlying pathogenetic mechanisms is essential for early diagnosis and individualized treatment, all targeted to effective fracture prevention.

Keywords: osteoporosis; pathogenesis; secondary osteoporosis

1. Introduction

Osteoporosis (OP) is a systemic skeletal disorder characterized by reduced bone mass and deterioration of bone microarchitecture, resulting in increased fragility and fracture risk [1,2]. Fragility fractures, especially in the proximal femur, vertebrae, and wrist, are major contributors to disability and loss of independence in older adults [2]. The global burden of OP is significant: in 2019, there were approximately 178 million new fracture cases worldwide (95% UI 162–196), a 33.4% increase since 1990 [3]. As the population ages, the incidence of OP is expected to rise further, with substantial health and economic impacts [3].

OP can have various causes: it is classified as primary (post-menopausal and senile) or secondary. Secondary OP results from underlying conditions or medications, including endocrine, genetic, and gastrointestinal diseases [1–3]. This review summarizes the main pathological mechanisms behind this increasingly common disease with diverse causes, after examining bone homeostasis physiology. Therapeutic strategies will also be discussed. Greater understanding of OP can improve treatment options and prevention, emphasizing early diagnosis and personalized care to address this growing health issue.

2. Bone Physiology

Key regulators of calcium-phosphate homeostasis include parathyroid hormone (PTH), calcitonin, fibroblast growth factor 23 (FGF-23), and active vitamin D (calcitriol or 1,25-dihydroxyvitamin D–1,25 (OH)2D) [4]. These hormones control calcium and phosphate metabolism by acting on the kidneys, intestine, and bones [4].

PTH, produced by the parathyroid glands, responds to hypocalcemia [5]. When serum calcium drops, the parathyroid glands secrete more PTH, which is triggered by calcium-sensing receptors (CaSR) detecting reduced serum ionized calcium [4]. PTH elevates serum calcium and phosphate levels by increasing calcium reabsorption from bones and reducing phosphate reabsorption in kidneys, while also stimulating renal synthesis of 1,25(OH)2D [6]. Elevated calcium levels inhibit PTH release through a negative-feedback loop [5]. PTH also boosts 1,25(OH)2D levels by stimulating CYP27B1 (1α-hydroxylase) and downregulating CYP24A1 (24-hydroxylase) [4].

Calcitonin, produced by thyroid parafollicular cells, responds to increased serum calcium by inhibiting osteoclast activity—reducing bone resorption—by promoting the internalization of hydrolyses within cells. In the kidneys, calcitonin decreases calcium reabsorption [7].

1,25(OH)2D promotes calcium and phosphate absorption from the intestine [4]. It is synthesized in the liver from vitamin D3 (produced from 7-dehydrocholesterol via sunlight exposure or obtained through diet and supplements) and then converted to calcitriol in the kidneys. This active form binds to vitamin D receptors in organs such as the parathyroid glands, intestines, kidneys, and bones, increasing serum calcium by absorption and release from bones. PTH and 1,25(OH)2D stimulate bone resorption by activating osteoclasts and promoting RANKL production by osteoblasts [8].

FGF23, secreted by osteocytes and stimulated by PTH, regulates phosphate elimination and influences renal synthesis of 1,25(OH)2D, forming part of the feedback loops that integrate calcium homeostasis [4,9–11].

The primary cells involved in bone health are osteoblasts, osteoclasts, and osteocytes. Osteoblasts originate from bone mesenchymal stem cells, deposit on bone surfaces, and produce collagen to form hydroxyapatite, leading to bone formation. They can become osteocytes by embedding into the matrix or remain as lining cells [12]. Osteocytes, residing in lacunae, regulate mineral metabolism and bone mass; their death causes skeletal fragility. Conditions such as immobilization and glucocorticoid therapy can induce osteocyte necrosis or apoptosis [13]. Osteoclasts resorb bone by secreting acids and proteases, forming trenches on trabeculae where osteoblasts subsequently synthesize new matrix. Excessive osteoclast activity can lead to bone erosion and loss, especially through the RANL/RANKL pathway [14]. Moreover, the mevalonate pathway influences bone physiology by generating intermediates like FPP and GGPP, which activate GTPases Ras and Rho, essential for osteoclast function, and produce isoprenoids that regulate osteoblast differentiation and mineralization [15–17].

Osteocytes secrete RANKL, which binds to its specific receptor (RANK) on osteoclasts, prompting their differentiation and activation [18]. After RANKL and RANK bind, NF-kB is activated through tumor necrosis factor receptor-related factor-6 (TRAF-6), regulating osteoclast maturation and differentiation [19]. Osteoprotegerin (OPG), a decoy receptor produced by osteoblasts, competes with RANKL to bind RANK, thus diminishing osteoclast differentiation [20] (Figure 1A).

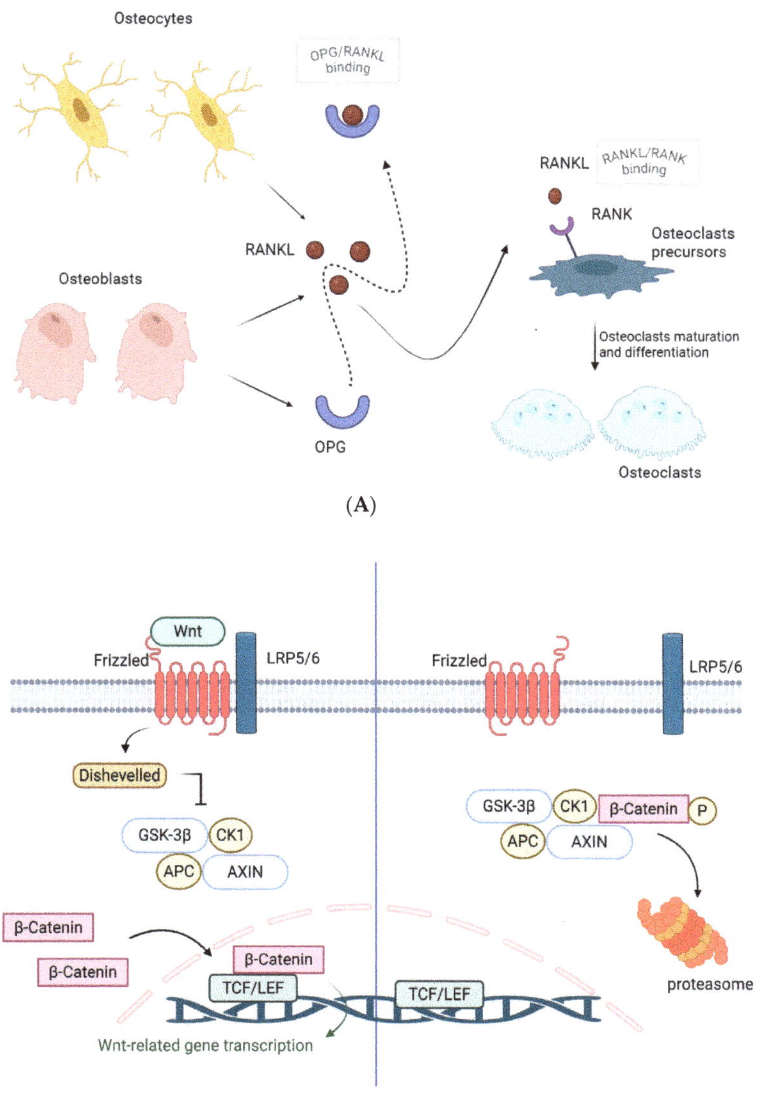

Figure 1. Overview of the main pathways involved in osteoclasts differentiation and bone resorption, and osteoblasts differentiation and proliferation. (**A**) RANKL/RANK/OPG (RANK/L, receptor activator of nuclear factor kb/ligand; OPG, osteoprotegerin), Created in BioRender, https://BioRender.com/ef7fdte (Accessed on 5 September 2025). (**B**) Wnt/βcatenin pathway. Created in BioRender, https://BioRender.com/9vt6eud BioRender.com (Accessed on 5 September 2025). Abbreviations: APC, Adenomatous Polyposis Coli; CK1, casein kinase-1; GSK-3β, Glycogen synthase kinase-3 beta; LRP5/6, lipoprotein receptor-related proteins 5/6; OPG, osteoprotegerin; RANK/L, receptor activator of nuclear factor kb/ligand; TCF/LEF, T cell factor/lymphoid enhancer factor family.

IL-1 can induce TNFα to stimulate osteoblasts to produce GM-CSF and IL-6 and to induce osteoclast precursor differentiation [21,22], enhancing the role of inflammation in bone disruption. TNFα can also bind to TNFR-1 expressed by osteoclast precursors, activate NF-κB, JNK, p38, or ERK, and promote the differentiation of osteoclast precursors into osteoclasts [23].

Wnt protein in osteoblasts binds to LRP5/6 and Fz (frizzled) receptors, located on the osteoblast membrane. The binding stabilizes intracellular β-catenin by inhibiting the β-catenin degradation complex. Then, β-catenin can translocate into the nucleus, interact with a group of genes (TCF/LEF) encoding for transcription factors, and regulate the expression of osterix and Runx2, transcription factors for osteogenesis. When Wnt does not bind to LRP5/6 and Fz, β-catenin is phosphorylated and stays bound to its degradation complex, leading to disruption in the proteasome, thus not prompting gene transcription [24,25] (Figure 1B).

Another important pathway is that of Hedgehog (Hh), which binds to receptors on MSC membranes and activates intracellular signaling molecules, which are translocated into the nucleus to upregulate the expression of Runx2 [26].

Figure 1A,B summarize the main pathways involved in osteoclasts and osteoblasts differentiation.

3. Osteoporosis Definition and Epidemiology

Bone density values are expressed relative to a reference population in standard deviation (SD) units, with the standard being a young, healthy population. The use of SD in relation to this group is called the T-score [2]. Based on these criteria, the WHO established diagnostic thresholds: normal (T-score ≥ 1 SD), low bone mass or osteopenia (T-score between -1.0 and -2.5 SD), and osteoporosis (T-score ≤ -2.5 SD) [2]. A lower T-score indicates a higher fracture risk. WHO defines osteoporosis as affecting roughly 6.3% of men and 21.2% of women over 50 worldwide, equating to about 500 million affected individuals [27]. It can cause fractures, including vertebral and non-vertebral, like hip fractures [3]. Hip fractures carry a mortality rate of 20–24% in the first year post-fracture, with increased mortality risk persisting over five years [28]. They also lead to loss of function and independence [28]. Nearly one-third of vertebral fractures go unnoticed, with up to 29% unrecognized in Europe [28].

4. Primary Osteoporosis

Primary OP is further divided into two groups: type I, which corresponds to postmenopausal OP (50–70 years of age), and type II, age-related OP, often occurring after 70 years of age [29].

Estrogens' actions are mediated by their receptors, ERα and Erβ, also expressed by osteocytes, osteoblasts, BMSCs, and osteoclasts. Estrogens inhibit RANKL and prompt the secretion of OPG and growth hormone; this results in inhibited osteoclast activity. Also, estrogens stimulate an osteogenic differentiation of BMSCs and regulate the number of osteoblasts. For all these actions mediated by estrogens, menopause leads to bone loss and eventually OP [30–33].

While the decline in estrogen levels initiates bone loss in postmenopausal women, deterioration of skeletal integrity continues also after the moment of menopause, thus suggesting the involvement of additional mechanisms, such as senescent osteoblasts, increased oxidative stress, and alterations in the bone remodeling milieu associated with aging. Many mechanisms have been suggested for the pathogenesis of age-related OP, almost all related to loss of cellular function or decrease in number, and finally, to cell senescence.

Cultured human bone MSCs (mesenchymal stem cells) have shown increased expression of markers related to senescence, such as β-galactosidase, suggesting that MSCs could have an impaired proliferation capacity with aging [34,35]. Zhou et al. demonstrated an increase in the number of apoptotic cells in the same MSC cultures, suggesting that increased apoptosis, and thus an age-related decline in the lifespan of cellular precursors, could be implicated in the aging of bone [35]. Contrasting data on the adipocyte differentiation are available, with some authors suggesting that there could be a preferential polarization of MSC in the sense of adipocyte, instead of osteoblast, differentiation [34].

Reactive Oxygen Species (ROS) have also been implicated in the pathogenesis of age-related OP, since ROS damage is a well-known component of tissue degeneration in the whole body [36]. In mice, oxidative stress decreases osteoblast and osteocyte lifespan [37], and ROS are also able to inhibit the Wnt/β-catenin pathway, fundamental for osteoblastogenesis, by interacting with Forkhead box transcription factors (FoxO) transcription factors. ROS promotes FoxO-mediated transcription factors, which can sequester β-catenin. This causes a diminished proliferation and differentiation of osteoblasts and their progenitors, resulting in decreased bone formation in aging [38].

Other data show an age-dependent decrease in the production of collagen in osteoblast cultures, suggesting that these cells could experience a loss of function and insufficient turnover in the aging bone [34]. Also, an insufficient response to growth factors has been demonstrated in an age-related manner, suggesting that aging osteoblastic cells from cultures are less prone to respond to anabolic pathways in the senescent bone [39]. Not only do they respond less, but they also secrete less growth factors and show an enhanced secretion of inflammatory cytokines and matrix-degrading proteases that alter the bone microenvironment [40].

Genetic senescence mechanisms have also been implicated in the pathogenesis of age-related OP. Recent evidence shows the role of telomere length in the pathogenesis of OP and its link to environmental and lifestyle elements: indeed, a network between genetic predisposition, early smoke exposure, and telomere shortening has been shown to constitute a risk for OP onset [41].

Sirtuins, a family of 7 genes (from SIRT1 to 7), are involved in a plethora of biological activities, such as mitochondrial function, inflammation, antioxidative stress, cellular senescence, and OP [42]. As for bone biology, in murine models, sirtuins regulate bone MSCs, osteoblast function, and osteogenic differentiation by interacting with β-catenin and FoxO, decreasing ROS production, and promoting the transcriptional activity of anabolic genes [42].

Data on the quantity and role of bone MSCs remain inconsistent. Autophagy is crucial for stem cell self-renewal and differentiation, but its specific effects on bone MSCs are not entirely clear. Some studies indicate that autophagy declines during osteogenic differentiation, and proper autophagy activation is necessary to facilitate this process. Regarding aging, findings on bone MSCs' number and function are conflicting: some reports suggest an age-related decrease, while others find no change [43,44].

Growing evidence concerning the role of microRNA is available, showing a potential role of the small, single-stranded RNA in the pathogenesis of OP, with up- or downregulated microRNAs that seem to target various molecules implicated in bone physiology [45–48].

5. Male Osteoporosis

OP is often overlooked and undertreated in men because it is more common in women. However, it also occurs frequently in men, with a global prevalence of about 12%, reaching up to 20% in some regions [49–51]. Men experience higher morbidity and mortality from fractures, likely because these happen 5 to 10 years later than in women, often alongside older age and more comorbidities [52,53]. The lower prevalence of male OP can be explained by mechanical and hormonal factors: men generally have higher peak bone mineral density (BMD) and larger bones, due to androgen effects on periosteal bone growth. This creates a more even distribution of mechanical forces across a wider area, providing mechanical benefits. Androgens stimulate periosteal bone apposition, potentially compensating for endocortical thinning and increased cortical porosity, a key difference between aging men and women [54,55]. The structure of trabecular bone also influences male osteoporosis: men tend to have more trabeculae, and aging causes thinning rather than loss of these structures. Conversely, women experience loss of connectivity and perforation in trabeculae with age [54,55]. Androgen levels decline more gradually in men and usually at older ages; however, illness or medications inducing hypogonadism can accelerate this decline, leading to increased bone resorption, rapid bone loss, and higher fracture risk, like estrogen drops during women's menopause [56]. Estrogen also plays a role in reaching peak bone mass, and deficiency is linked to increased bone remodeling and loss in men. Lower levels of dehydroepiandrosterone (DHEA) are associated with reduced BMD, muscle weakness, falls, and fractures in older men [57–59]. Other risk factors include excessive alcohol intake, chronic glucocorticoid use, vitamin D deficiency (due to diet, less sun exposure, and lower skin conversion capacity), declining muscle function causing more falls, and increased osteoclast activity coupled with inhibited osteoblasts driven by inflammatory cytokines in the elderly [60–63].

6. Secondary Osteoporosis

When OP is caused or exacerbated by other disorders or drug exposure, it is referred to as secondary OP. Identifying secondary causes of osteoporosis is critical, as the treatment depends on the underlying condition. A paragraph is also dedicated to the male's OP, which slightly differs from the women's OP. Table 1 presents the main causes of secondary OP.

Table 1. Main causes of secondary OP.

Causes of Secondary Osteoporosis
Hormonal deficiency (Androgen deficiency -> hypogonadism, diseases, medications; estrogen and DHEA deficiency)
Chronic alcohol use
Medications (antiepileptics, SSRIs, PPIs, heparin, glucocorticoids)
Vitamin D deficiency
Sarcopenia and inflammation
Diabetes mellitus (type 1 and type 2)
Hyperthyroidism and hyperparathyroidism
Hypogonadism (including Turner and Klinefelter syndromes)
Obesity and related factors
Gastrointestinal disorders causing malabsorption (e.g., celiac disease, inflammatory bowel disease, bariatric surgery)
Organ transplants and immunosuppressive therapy
Multiple myeloma and other cancers
Rare genetic diseases (osteogenesis imperfecta, Marfan syndrome, Ehlers-Danlos syndrome, homocystinuria)
Rheumatic and neurological diseases (e.g., Parkinson's disease, multiple sclerosis)
Systemic mastocytosis

DHEA, dehydroepiandrosterone; SSRIs, selective serotonin reuptake inhibitors; PPIs, proton pump inhibitors.

6.1. Glucocorticoids

Glucocorticoid (GC)-induced OP is the most common secondary cause of osteoporosis; bone loss and risk of fracture increase in a dose-dependent manner within the first three to six months of assumption [64]. GC prolonged use is associated with an overall higher risk of vertebral (more frequent) and non-vertebral fractures [65]. Direct effects of GC on bone formation are mediated through upregulation of peroxisome proliferator-activated receptor gamma receptor 2 (PPARγ2) [66] and effects on the Wnt/β- catenin signaling pathway [67,68], resulting in the differentiation of precursor cells to adipocytes with respect to osteoblasts, thus leading to a decrease in bone formation. GCs also have direct effects on bone resorption: they increase the production of macrophage colony-stimulating factor (M-CSF) and RANKL and decrease the production of OPG in osteoblasts and osteocytes, increasing the number and activity of osteoclasts [69,70].

Other important factors to consider regarding GC-induced OP are an increased inflammatory state, associated with GC therapy administration, which contributes to bone loss through increased pro-resorptive cytokines [71] and reduced physical activity, increased renal and intestinal losses of calcium, and reduced production of growth hormone, insulin-like growth factor 1 (IGF1) and IGF1 binding protein (IGF-BP) as well as myopathy, with loss of muscle mass and consequent increased risk of fall [72].

6.2. *Endocrine Disorders*

6.2.1. Diabetes Mellitus

Diabetes mellitus (DM) can impact not only the eyes, kidneys, and heart, but also bone density and bone turnover, causing OP, known as diabetic bone disease [73]. In a cohort of patients with Type 2 DM, 35% reported bone loss, of which approximately 20% could satisfy the criteria for OP diagnosis [74].

Available data reports reduced bone density in children with Type 1 DM [75]. In adult cohorts, a decrease in BMD was predominantly observed in the femur, with a marginal diminution in vertebral BMD [76]. The hyperglycemic environment causes reduced bone turnover due to a slowed bone matrix maturation and mineralization rate [77–79].

Patients with either Type 1 or 2 DM have impaired bone microarchitecture, in consideration of a decrease in the number of trabeculae and morphological disorders. Trabecular thickness of the femoral head is significantly lower in patients with Type 2 DM than in non-diabetic patients [80–83].

The pathogenesis of diabetic bone disease involves three main mechanisms: first, insulin receptors on the surface of osteoblasts trigger intracellular signaling pathways that stimulate cell proliferation and the production of osteocalcin and collagen, both of which are crucial for bone tissue development [84]. Insulin enhances the expression of RUNX2, which influences bone metabolism, boosting osteoblast differentiation and maturation of the bone matrix [85]. Furthermore, it promotes IGF-1, known for its pro-anabolic impacts on osteoblasts, which speeds up collagen production and mineralization of the bone matrix [86]. These events are notably diminished in diabetic individuals. The second mechanism involves the build-up of advanced glycation end products (AGEs), which form cross-links with collagen in the bone matrix, resulting in decreased collagen elasticity and heightened bone fragility [87]. Elevated levels of extracellular free sugars adversely affect the activity of both osteoblasts and osteocytes, leading to osteoblast apoptosis [88] and stimulating osteoclast activation [89]. Lastly, the third mechanism is linked to the chronic inflammatory state associated with Type 2 DM, which promotes the lipogenic differentiation of bone marrow MSC and fat accumulation in the marrow, coupled with the release of free fatty acids and a significant number of inflammatory cytokines [90].

6.2.2. Thyroid and Parathyroid

Among endocrine disorders leading to reduced bone formation, thyroid and parathyroid dysfunction play an important role. Both hyperthyroidism and hyperparathyroidism increase bone turnover [91,92]. Primary hyperparathyroidism generally affects individuals aged more than 50 years and has a prevalence of 233 per 100,000 in women and 85 per 100,000 in men in US cohorts [93]. It is characterized by persistent hypercalcemia with an elevated or inappropriately normal PTH serum level, and it is often associated with a single gland adenoma [94] or with syndromes including multiple endocrine neoplasia (MEN) type 1, MEN type 4, MEN2A [95]. Current guidelines recommend surgery in asymptomatic patients who satisfy the following criteria: age less than 50 years, serum calcium level of >1 mg/dL or >0.25 mmol/L above the upper limit of the reference interval for total calcium (>0.12 mmol/L for ionized calcium), BMD T-score ≤ -2.5 at any site, fragility fracture, glomerular filtration rate less than 60 mL/min, nephrocalcinosis, renal calculi, or high stone risk [96].

6.2.3. Hypogonadism

Several conditions associated with sexual hormone alterations are directly related to secondary OP. Estrogen deficiency is the most common, as in primary amenorrhea, premature ovarian insufficiency, or Turner syndrome: the latter is due to total or partial X chromosome monosomy that leads to gonadal dysgenesis. To note, testosterone alterations lead to OP, as in Klinefelter syndrome or androgen insensitivity syndrome, the latter caused by a lack of response to androgens due to mutations of their receptors [97].

6.2.4. Obesity

Although the association of increased fracture risk with low body mass index (BMI) is well recognized, a relationship between fracture risk and increased BMI can be identified. Many factors contribute to an increased fracture risk, such as the coexistence of comorbidities such as DM, vitamin D deficiency, inflammation and oxidative stress associated with adipose tissue, increased fall risk, altered gut microbiome, sarcopenic obesity, reduced physical activity, and obesity-induced hypogonadism in men. On the other hand, protective factors include the effect of adipose tissue–derived estrogens, which may contribute to reduced hip fracture risk [98].

Given the growth of bariatric surgery, malabsorption derived from it is becoming another common cause of secondary OP. Gastroesophageal reflux and changes in gut microbiome are other consequences of bariatric surgery that result in negative skeletal effects [99].

Bariatric surgery includes Roux-en-Y gastric bypass (RYGB) (bypassing duodenum and proximal jejunum), vertical sleeve gastrectomy, adjustable gastric band, and biliopancreatic diversion with duodenal switch (that combines both gastric reduction and extreme intestinal bypass) [100]. Longitudinal studies demonstrate declining BMD, altered cortical and trabecular microarchitecture, reduced bone strength, and increased levels of bone turnover markers in the first 6 months after surgery, that continue for more than 5 years despite stabilization of body weight [101–103].

6.2.5. Miscellaneous

Among other endocrine disorders associated with secondary OP, hemochromatosis, a genetic disease in which there is excessive intestinal iron absorption with subsequent intracellular iron accumulation, that leads to multiple organ dysfunction (hypogonadism, diabetes, anterior pituitary dysfunction) and consequential bone loss, can be mentioned [104]. In acromegaly, an excessive production of Growth Hormone (GH), OP is described, and it is due to the hypogonadism often associated with this disorder [105].

6.3. GI Disease

Conditions that interfere with normal intestinal absorption of calcium, phosphorus, and vitamin D can cause secondary hyperparathyroidism, stimulating osteoclastic bone resorption and resulting in bone loss. Moreover, some gastrointestinal (GI) diseases are characterized by inflammatory states in which inflammatory cytokines contribute to bone microarchitectural deterioration and are treated with GCs, which contribute to bone loss.

Celiac disease is an important cause of secondary OP: gluten ingestion and related proteins cause inflammatory injury to the mucosa of the small intestine, thus resulting in malabsorption. The most common features of celiac disease are diarrhea, steatorrhea in severe cases, and iron deficiency [97].

Age at diagnosis and time when gluten-free diet is initiated are crucial determinants of bone mass in patients with celiac disease: at diagnosis, about 80% of adults have reduced BMD, and 60% of children have growth retardation. After the initiation of a gluten-free diet, BMD often improves, although some patients have persistent osteopenia despite treatment [106]. Peripheral fractures are more frequent, especially radius fractures. Some studies also indicate a greater fracture prevalence in men and individuals with GI symptoms compared with patients with extra-intestinal or subclinical manifestations [107].

26% to 34% of patients with celiac disease reportedly have lumbar spine OP; femoral OP is less common [108,109]. Higher risk of fracture is reported in patients with celiac disease who are not following a gluten-free diet [110].

The presence of inflammatory bowel disease (IBD), i.e., Crohn's disease (CD) or ulcerative colitis (CU), is another crucial cause of secondary OP due to malabsorption and eventual ileal resection in CD. Moreover, GC treatment is often prolonged in IBD patients [97]. The negative impact of IBD on bone health is multifactorial, resulting in lower peak bone mass and increased bone loss in adults [111]. In addition, falls risk may be increased; low muscle mass and sarcopenia were observed in 21% and 12%, respectively, among 137 IBD patients [112].

Pancreatic insufficiency associated with cholestasis and alcohol abuse should be considered in patients with bone loss [106]. Haaber et al. reported decreased bone mass in 62% of patients with chronic pancreatitis. However, among the patients included in their studies, in 79% pancreatic disease was associated with alcohol abuse; to distinguish between the effects of alcohol and pancreatic insufficiency with OP was thus not possible [113]. Children and adults with cystic fibrosis (CF) can develop exocrine pancreatic insufficiency. Notably, decreased BMD and extremely high fracture rates are also frequent complications of CF. However, patients with CF have many additional risk factors for OP, including calcium and vitamin D malabsorption, delayed puberty, reduced sex steroid production, and increased serum concentrations of cytokines secondary to chronic pulmonary infections [114,115].

Hepatic function is also crucial in bone normal structure, as the liver converts vitamin D to 25-OHD, synthesizes vitamin D transport proteins, and promotes intestinal vitamin D and calcium absorption by bile. Primary biliary cirrhosis (PBC) is associated with OP due to malabsorption of calcium, phosphorus, and vitamin D and increased urinary losses of vitamin D conjugates. Moreover, PBC is predominantly in middle-aged women, and its skeletal effects may be added to losses related to postmenopausal estrogen deficiency [106].

6.4. Transplantation

OP is one of the most common long-term complications after organ transplantation. Medications such as loop diuretics or heparin, end-stage renal or hepatic disease, and especially immunosuppressive therapy with GCs, cyclosporine A, or tacrolimus all impact bone health [116]. Since bone loss occurs immediately after transplantation, preventive therapy against OP is recommended for all patients during the early post-transplant period, regardless of BMD. A meta-analysis found that bisphosphonate treatment during the first year after solid organ transplantation reduces fractures by 47% and vertebral fractures by 76% [117].

6.5. Cancer

Osteolytic lesions, osteoporosis (OP), and osteopenia occur in about 80% of people with Multiple Myeloma (MM) [118]. Serious MM complications include pathological fractures leading to hypercalcemia, spinal cord compression, and pain, all increasing morbidity and mortality [119]. Fractures at diagnosis or later are linked to a higher risk of death [120]. Myeloma cells release cytokines and osteoclast-activating factors that promote bone loss [121]. Bisphosphonates effectively prevent lytic lesions, vertebral fractures, and bone pain [119]. Concerns have recently been raised about adverse bone effects from hormonal deprivation therapy used in prostate and breast cancer treatment. Androgen-deprivation therapy (ADT) for prostate cancer causes systemic bone loss, microarchitecture deterioration, and increased fracture risk [122]. The osteoporosis risk varies with the type, duration, and dose of ADT. Bilateral orchidectomy causes more bone loss than pharmacological ADT, which in turn causes less than GnRH ADT [123,124].

Breast cancer can induce bone loss directly through tumor effects or systemically, and via estrogen deprivation from treatments like chemotherapy, ovarian ablation, or endocrine therapy, plus natural menopause [125]. Tamoxifen, an estrogen receptor antagonist, has partial agonist effects on bone, whereas aromatase inhibitors improve disease-free survival but are associated with significant bone loss and higher fracture risk [126]. Aromatase inhibitors block androgen conversion to estrogens, resulting in serum estradiol levels much lower than in healthy postmenopausal women, leading to faster and more severe bone loss compared to natural menopause [127]. All women with breast cancer should follow OP screening and prevention guidelines similar to the general population.

6.6. Other Drugs

The link between OP, fractures, and medications is debated. Data mainly concern: Depot medroxyprogesterone acetate (DMPA), a progestogen-only contraceptive given as an intramuscular injection every 3 months, which can cause hypoestrogenism and glucocorticoid effects [128]; anti-epileptic drugs, which interfere with calcium and vitamin D metabolism, cause OP through hyponatremia, and increase fall risk [129]; selective serotonin reuptake inhibitors (SSRIs), which have opposing effects of serotonin on bone [130]; proton pump inhibitors, which reduce gastric acidity and thus calcium absorption [131]; and heparin, which stimulates bone resorption and suppresses osteoblast function [132].

6.7. Skeletal Development Disorders and Genetic Causes

Among other causes of secondary osteoporosis, we also highlight rare conditions that should be excluded when diagnosing atypical osteoporosis, especially in young patients. These include inherited skeletal development disorders such as osteogenesis imperfecta, Homocystinuria, Marfan syndrome, Ehlers-Danlos syndrome, rheumatic diseases, neurological conditions like Cerebral Palsy, spina bifida, Duchenne dystrophy, multiple sclerosis, Parkinson's disease, and systemic mastocytosis [97,133]. Genetic factors play a significant role in osteoporosis, with many related genes identified through genome-wide association studies (GWAS) [134,135]. The latest classification of genetic skeletal disorders lists 55 conditions associated with skeletal fragility [136]. Osteogenesis imperfecta (OI), a monogenic disorder caused by mutations in extracellular matrix proteins—mainly type I collagen [137,138]—results in bone weakness due to impaired osteoblast and osteoclast activity, defective matrix mineralization, and disrupted calcium and phosphate balance. Several critical genes for early-onset osteoporosis include LRP5, where biallelic mutations cause childhood osteoporosis and blindness; heterozygous loss-of-function variants lead to milder forms [139,140]. Biallelic WNT1 mutations cause severe skeletal fragility, similar to OI type III, while heterozygous variants produce milder, later-onset cases [141,142]. PLS3, on the X chromosome, encodes Plastin3, an actin-binding protein involved in cytoskeleton remodeling, and its mutations can cause severe male osteoporosis due to its role in bone mineralization [143,144]. Recently, other monogenic osteoporosis forms have been identified, such as SGMS2, which encodes sphingomyelin synthase 2 (SMS2). Heterozygous mutations in this gene are linked to childhood fractures, hyperostotic lesions, osteoporosis, and neurological symptoms like transient facial nerve palsy [145,146]. Among genetic syndromes, Marfan syndrome is strongly associated with severe osteoporosis resulting from connective tissue alterations caused by autosomal dominant mutations in the fibrillin-1 (FBN1) gene on chromosome 15q21.1 [147]. Three new FBN1 mutations and ten FBN3 single-nucleotide polymorphisms (SNPs) have been identified [148,149]. The FBN1 gene encodes fibrillin, a glycoprotein that forms microfibrils, essential components of the suspensory ligament of the lens, and a substrate for elastin in the aorta and other tissues. Abnormal fibrillin can impair bone mineralization and alter the distribution of mechanical strain [150]. Bone mineral density (BMD) studies in children with Marfan syndrome report reductions at the femoral neck, and in women, BMD is decreased in the lumbar spine and total hip [151]. Similar BMD decreases are found in men and women at the hip and radius, demonstrating how abnormal fibrillin impacts both bone mineralization and the distribution of mechanical strain [151].

7. Treatment

The treatment of OP must primarily consider non-pharmacological strategies, whose aim is to remove all the modifiable risk factors that increase the risk of OP. These are: alcohol, smoking, low dietary calcium intake and vitamin D deficiency, low body mass index, frequent falls, and insufficient exercise [152]. Secondary OP causes must be investigated, and their treatment is part of the therapeutic strategy of this type of OP [153].

Pharmacological therapies for OP are divided into two main groups: anti-resorptive and anabolic. The first category is composed of bisphosphonates, denosumab and selective estrogen receptor modulators. Teriparatide, abaloparatide and romosozumab are anabolic drugs. To summarize, anti-resorptive preserve already existing bone, while anabolic drugs aim to restore lost bone [154].

During bone resorption, bisphosphonates are taken up by osteoclasts, where they block the mevalonate pathway. The amino bisphosphonates, such as alendronate, risedronate, and zoledronate, inhibit farnesyl pyrophosphate synthase, thus impairing the resorptive activity of osteoclasts and enhancing apoptosis [154] (Figure 2).

The mevalonate pathway is responsible for the formation of prenyl pyrophosphates (farnesyl and geranylgeranyl pyrophosphates), implicated in the activation of small guanosine triphosphate (GTP)-binding proteins (GTPases—Ras, Rho, Rac, Rab). These GTPases are involved in the differentiation of the osteoclasts and hinder the activation of osteoblasts [154].

Instead, denosumab binds RANKL and prevents it from activating RANK. This mechanism impedes the stimulation of RANK by RANKL, which would prompt the differentiation of osteoclast precursor cells into mature osteoclasts. [154]. As stated in previous sections of this review, osteoclasts' reabsorbing activity is increased by the fall in estrogen concentration after the menopause. This offers the rationale for the use of selective estrogen receptor modulators (raloxifene). [154].

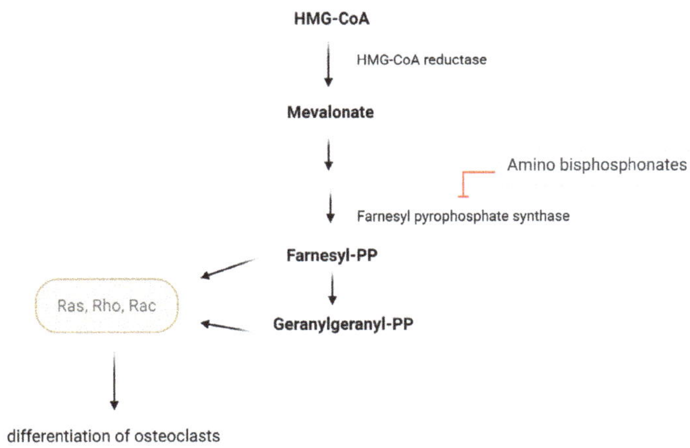

Figure 2. Mevalonate pathway and the action of bisphosphonates. Abbreviations: HMG-CoA, 3-hydroxy-3-methyl-glutaryl-coenzyme A; PP, pyrophosphate. Created in BioRender, https://BioRender.com/n12j0m9 (Accessed on 5 September 2025).

Anabolic drugs can be divided into agonists of parathyroid hormone type 1 receptors (PTH1R) (teriparatide and abaloparatide) and anti-sclerostin antibodies (romosozumab). When the activation of the PTH receptor is continuous, as in primary hyperparathyroidism, when there is an excess of PTH, bone loss is predominant. When PTH receptor activators are administered in pulses, such as teriparatide and abaloparatide, bone gain is achieved [154]. Indeed, when PTH1R is transiently stimulated by PTH or PTHrP (parathyroid hormone-related protein), cAMP-dependent protein kinase A (cAMP/PKA) and phospholipase C-protein kinase C (PKC) pathways are activated. These lead to the transcription of proteins implicated in bone formation, such as OPG [155], the decoy receptor for RANKL [156]. PTHR1 activates Ras and MAPK through the PKC pathway and stimulates bone proliferation in marrow mesenchymal progenitor cells [154]. The PKC pathway is also involved in the estrogen inhibition of PTH-stimulated osteoclasts [138]. As for sclerostin, it binds to the Wnt LRP 5/6 coreceptors, thus inhibiting the canonical Wnt-signaling pathway, which normally prompts differentiation of mesenchymal stem cells to osteoblasts [155]. After the inactivation of the canonical Wnt-signaling, OPG production and bone formation are reduced [156]. Being the decoy receptor of RANKL, in normal conditions, OPG reduces the activity of RANK-RANKL, thus reducing osteoclast differentiation and bone resorption [156]. Romosozumab is an anti-sclerostin antibody: by inhibiting sclerostin, the canonical Wnt-signaling pathway can work by favoring bone formation [156]. Wnt stimulation of the LRP 5/6 and frizzled membrane co-receptor complex allows the migration of β-catenin from the cytoplasm into the nucleus, permitting protein transcription and bone anabolic effects. The same pathways prompt OPG upregulation and are inhibited by sclerostin, which, in turn, is inhibited by PTH [155]. In the FRAME trial of romosozumab, 7180 women were enrolled and randomly assigned to receive either a monthly subcutaneous dose of 210 mg of romosozumab for one year or a placebo. This was followed by denosumab administered every six months for 36 months. The study found a 73% reduction in the risk of new vertebral fractures and a 25% decrease in non-vertebral fractures in the romosozumab group. At 12 months, severe cardiovascular events occurred in 1.1% of patients in the placebo group and 1.2% in the romosozumab group, with these figures rising to 2.2% and 2.3%, respectively, at 24 months. Mortality was 0.5% in the romosozumab group compared to 0.4% in the placebo group [157]. Recent data also indicate that romosozumab induces early increases in bone formation and decreases in bone resorption, as shown by histomorphometric analysis of bone biopsies from the FRAME trial [158]. In Eriksen et al.'s study, higher osteoid and mineralized wall thickness were observed early in the romosozumab group compared to placebo ($p = 0.005$ and $p = 0.004$, respectively), demonstrating the positive effect of the anti-sclerostin antibody on bone mass and fracture risk reduction [158].

8. Conclusions

OP is a complex, multifactorial, often underdiagnosed condition that is a leading cause of morbidity and disability among aging populations, with significant socioeconomic impacts worldwide. Differentiating between primary and secondary OP is not just academically important but essential for accurate diagnosis and effective, targeted treatment. Failing to distinguish these can result in therapeutic errors, underestimating risks, and less effective interventions. Although advancements have been made in understanding the molecular mechanisms—such as specific pathways, senescence roles, and new pharmacological options like anti-sclerostin monoclonal antibodies and anabolic drugs—much remains to be performed. Emphasis is needed on prevention, early diagnosis, personalized therapy, and management of modifiable risk factors. It is crucial to detect OP early, ideally prevent its development, and address risk factors. As shown in this review, its multidisciplinary nature intersects many fields of medicine, making it a common challenge across specialties. Ultimately, managing OP requires a comprehensive, multidisciplinary approach that considers hormonal, genetic, and environmental interactions. Only through personalized, evidence-based strategies can we improve clinical outcomes and lessen the global and individual burdens of this silent, insidious skeletal disease.

Author Contributions: Conceptualization, C.C. and E.D.M.; methodology, E.D.M.; software, C.C.; validation, C.C., P.S. and E.D.M.; formal analysis, C.C., P.S. and E.D.M.; investigation, C.C., P.S. and E.D.M.; resources, C.C. and E.D.M.; data curation, C.C., P.S. and E.D.M.; writing—original draft preparation, C.C. and E.D.M.; writing—review and editing, C.C., P.S. and E.D.M.; visualization, C.C. and E.D.M.; supervision, P.S. All authors have read and agreed to the published version of the manuscript.

Funding: This research received no external funding.

Conflicts of Interest: The authors declare no conflicts of interest.

References

1. Consensus Development Conference. Diagnosis, prophylaxis, and treatment of osteoporosis. *Am. J. Med.* **1993**, *94*, 646–650. [CrossRef] [PubMed]
2. World Health Organization. *Guidelines for Preclinical Evaluation and Clinical Trials in Osteoporosis*; World Health Organization: Geneva, Switzerland, 1998; Available online: https://iris.who.int/handle/10665/42088 (accessed on 20 April 2025).
3. Wu, A.M.; Bisignano, C.; James, S.L.; Abady, G.G.; Abedi, A.; Abu-Gharbieh, E.; Alhassan, R.K.; Alipour, V.; Alipour, J.; Asaad, M.; et al. Global, regional, and national burden of bone fractures in 204 countries and territories, 1990–2019: A systematic analysis from the global burden of disease study 2019. *Lancet Healthy Longev.* **2021**, *2*, e580–e592. [CrossRef]
4. Murray, S.L.; Wolf, M. Calcium and Phosphate Disorders: Core Curriculum 2024. *Am. J. Kidney Dis.* **2024**, *83*, 241–256. [CrossRef]
5. Khan, M.; Jose, A.; Sharma, S. Physiology, Parathyroid Hormone. In *StatPearls*; StatPearls Publishing: Treasure Island, FL, USA, 2025.
6. Leung, E.K.Y. Parathyroid hormone. *Adv. Clin. Chem.* **2021**, *101*, 41–93.
7. Srinivasan, A.; Wong, F.K.; Karponis, D. Calcitonin: A useful old friend. *J. Musculoskelet. Neuronal Interact.* **2020**, *20*, 600–609. [PubMed]
8. Lung, B.E.; Komatsu, D.E.E. Calcitriol. In *StatPearls*; StatPearls Publishing: Treasure Island, FL, USA, 2025.
9. Columbu, C.; Rendina, D.; Gennari, L.; Pugliese, F.; Carnevale, V.; Salcuni, A.S.; Chiodini, I.; Battista, C.; Tabacco, P.; Guarnieri, V.; et al. Phosphate metabolism in primary hyperparathyroidism: A real-life long-term study. *Endocrine* **2025**, *88*, 571–580. [CrossRef]
10. Ho, B.B.; Bergwitz, C. FGF23 signalling and physiology. *J. Mol. Endocrinol.* **2021**, *66*, R23–R32. [CrossRef]

11. Edmonston, D.; Grabner, A.; Wolf, M. FGF23 and klotho at the intersection of kidney and cardiovascular disease. *Nat. Rev. Cardiol.* **2024**, *21*, 11–24. [CrossRef]
12. Blair, H.C.; Larrouture, Q.C.; Li, Y.; Lin, H.; Beer-Stoltz, D.; Liu, L.; Tuan, R.S.; Robinson, L.J.; Schlesinger, P.H.; Nelson, D.J. Osteoblast differentiation and bone matrix formation in vivo and in vitro. *Tissue Eng. Part B Rev.* **2017**, *23*, 268–280. [CrossRef]
13. Nahian, A.; AlEssa, A.M. Histology, Osteocytes. In *StatPearls*; StatPearls Publishing: Treasure Island, FL, USA, 2025.
14. Kitazawa, S.; Haraguchi, R.; Kitazawa, R. Roles of osteoclasts in pathological conditions. *Pathol. Int.* **2025**, *75*, 55–68. [CrossRef] [PubMed]
15. Hasan, W.N.W.; Chin, K.Y.; Jolly, J.J.; Ghafar, N.A.; Soelaiman, I.N. Identifying potential therapeutics for osteoporosis by exploiting the relationship between the mevalonate pathway and bone metabolism. *Endocr. Metab. Immune Disord. Drug Targets* **2018**, *18*, 450–457. [CrossRef]
16. Jung, J.; Park, J.S.; Chun, J.; Al-Nawas, B.; Ziebart, T.; Kwon, Y.D. Geranylgeraniol application in human osteoblasts and osteoclasts for reversal of the effect of bisphosphonates. *Life* **2023**, *13*, 1353. [CrossRef]
17. Xu, W.; Gong, L.; Tang, W.; Lu, G. Nitrogen-containing bisphosphonate induces enhancement of OPG expression and inhibition of RANKL expression via inhibition of farnesyl pyrophosphate synthase to inhibit the osteogenic differentiation and calcification in vascular smooth muscle cells. *BMC Cardiovasc. Disord.* **2024**, *24*, 494. [CrossRef]
18. McDonald, M.M.; Khoo, W.H.; Ng, P.Y.; Xiao, Y.; Zamerli, J.; Thatcher, P.; Kyaw, W.; Pathmanandavel, K.; Grootveld, A.K.; Moran, I.; et al. Osteoclasts recycle via osteomorphs during RANKL-stimulated bone resorption. *Cell* **2021**, *184*, 1330–1347.e13. [CrossRef]
19. Tobeiha, M.; Moghadasian, M.H.; Amin, N.; Jafarnejad, S. RANKL/RANK/OPG pathway: A mechanism involved in exercise-induced bone remodeling. *Biomed. Res. Int.* **2020**, *2020*, 6910312. [CrossRef] [PubMed]
20. Qin, S.; Zhang, Q.; Zhang, L. Effect of OPG gene mutation on protein expression and biological activity in osteoporosis. *Exp. Ther. Med.* **2017**, *14*, 1475–1480. [CrossRef]
21. Wang, T.; He, C. TNF-alpha and IL-6: The link between immune and bone system. *Curr. Drug Targets* **2020**, *21*, 213–227.
22. Amarasekara, D.S.; Yun, H.; Kim, S.; Lee, N.; Kim, H.; Rho, J. Regulation of osteoclast differentiation by cytokine networks. *Immune Netw.* **2018**, *18*, e8. [CrossRef]
23. Marahleh, A.; Kitaura, H.; Ohori, F.; Kishikawa, A.; Ogawa, S.; Shen, W.R.; Qi, J.; Noguchi, T.; Nara, Y.; Mizoguchi, I. TNF-alpha directly enhances osteocyte RANKL expression and promotes osteoclast formation. *Front. Immunol.* **2019**, *10*, 2925. [CrossRef] [PubMed]
24. Agostino, M.; Pohl, S. The structural biology of canonical Wnt signalling. *Biochem. Soc. Trans.* **2020**, *48*, 1765–1780. [CrossRef] [PubMed]
25. Sawakami, K.; Robling, A.G.; Ai, M.; Pitner, N.D.; Liu, D.; Warden, S.J.; Li, J.; Maye, P.; Rowe, D.W.; Duncan, R.L.; et al. The Wnt co-receptor LRP5 is essential for skeletal mechanotransduction but not for the anabolic bone response to parathyroid hormone treatment. *J. Biol. Chem.* **2006**, *281*, 23698–23711. [CrossRef]
26. Komori, T. Molecular mechanism of Runx2-dependent bone development. *Mol. Cells* **2020**, *43*, 168–175.
27. Kanis, J.A.; McCloskey, E.V.; Johansson, H.; Oden, A.; Melton, L.J., 3rd; Khaltaev, N. A reference standard for the description of osteoporosis. *Bone* **2008**, *42*, 467–475. [CrossRef]
28. Osteoporosis Foundation. Epidemiology of Osteoporosis and Fragility Fractures. Available online: https://www.osteoporosis.foundation/facts-statistics/epidemiology-of-osteoporosis-and-fragility-fractures (accessed on 20 April 2025).
29. Patel, D.; Saxena, B. Decoding osteoporosis: Understanding the disease, exploring current and new therapies, and emerging targets. *J. Orthop. Rep.* **2025**, *4*, 100472. [CrossRef]
30. Li, L.; Wang, Z. Ovarian aging and osteoporosis. *Adv. Exp. Med. Biol.* **2018**, *1086*, 199–215.
31. Khosla, S.; Oursler, M.J.; Monroe, D.G. Estrogen and the skeleton. *Trends Endocrinol. Metab.* **2012**, *23*, 576–581. [CrossRef]
32. Cauley, J.A. Estrogen and bone health in men and women. *Steroids* **2015**, *99*, 11–15. [CrossRef] [PubMed]
33. Vilaca, T.; Eastell, R.; Schini, M. Osteoporosis in men. *Lancet Diabetes Endocrinol.* **2022**, *10*, 273–283. [CrossRef]
34. Kassem, M.; Marie, P.J. Senescence-associated intrinsic mechanisms of osteoblast dysfunctions. *Aging Cell* **2011**, *10*, 191–197. [CrossRef]
35. Zhou, S.; Greenberger, J.S.; Epperly, M.W.; Goff, J.P.; Adler, C.; Leboff, M.S.; Glowacki, J. Age-related intrinsic changes in human bone-marrow-derived mesenchymal stem cells and their differentiation to osteoblasts. *Aging Cell* **2008**, *7*, 335–343. [CrossRef] [PubMed]
36. Almeida, M. Aging mechanisms in bone. *Bonekey Rep.* **2012**, *1*, 102. [CrossRef] [PubMed]
37. Jilka, R.L.; Almeida, M.; Ambrogini, E.; Han, L.; Roberson, P.K.; Weinstein, R.S.; Manolagas, S.C. Decreased oxidative stress and greater bone anabolism in the aged, as compared to the young, murine skeleton by parathyroid hormone. *Aging Cell* **2010**, *9*, 851–867. [CrossRef] [PubMed]
38. Manolagas, S.C.; Almeida, M. Gone with the Wnts: Beta-catenin, T-cell factor, forkhead box O, and oxidative stress in age-dependent diseases of bone, lipid, and glucose metabolism. *Mol. Endocrinol.* **2007**, *21*, 2605–2614. [CrossRef]

39. Pfeilschifter, J.; Diel, I.; Pilz, U.; Brunotte, K.; Naumann, A.; Ziegler, R. Mitogenic responsiveness of human bone cells in vitro to hormones and growth factors decreases with age. *J. Bone Miner. Res.* **1993**, *8*, 707–717. [CrossRef]
40. Campisi, J. Senescent cells, tumor suppression, and organismal aging: Good citizens, bad neighbors. *Cell* **2005**, *120*, 513–520. [CrossRef]
41. Han, M.H.; Kwon, H.S.; Hwang, M.; Park, H.H.; Jeong, J.H.; Park, K.W.; Kim, E.J.; Yoon, S.J.; Yoon, B.; Jang, J.W.; et al. Association between osteoporosis and the rate of telomere shortening. *Aging* **2024**, *16*, 11151–11161. [CrossRef]
42. Li, Q.; Cheng, J.C.; Jiang, Q.; Lee, W.Y. Role of sirtuins in bone biology: Potential implications for novel therapeutic strategies for osteoporosis. *Aging Cell* **2021**, *20*, e13301. [CrossRef]
43. Stolzing, A.; Jones, E.; McGonagle, D.; Scutt, A. Age-related changes in human bone marrow-derived mesenchymal stem cells: Consequences for cell therapies. *Mech. Ageing Dev.* **2008**, *129*, 163–173. [CrossRef]
44. Xing, Y.; Liu, C.; Zhou, L.; Li, Y.; Wu, D. Osteogenic effects of rapamycin on bone marrow mesenchymal stem cells via inducing autophagy. *J. Orthop. Surg. Res.* **2023**, *18*, 129. [CrossRef] [PubMed]
45. Suarjana, I.N.; Isbagio, H.; Soewondo, P.; Rachman, I.A.; Sadikin, M.; Prihartono, J.; Malik, S.G.; Soeroso, J. The role of serum expression levels of microRNA-21 on bone mineral density in hypostrogenic postmenopausal women with osteoporosis: Study on level of RANKL, OPG, TGFβ-1, sclerostin, RANKL/OPG ratio, and physical activity. *Acta Med. Indones* **2019**, *51*, 245–252.
46. Zuo, B.; Zhu, J.F.; Li, J.; Wang, C.D.; Zhao, X.Y.; Cai, G.Q.; Li, Z.; Peng, J.; Wang, P.; Shen, C.; et al. MicroRNA-103a functions as a mechanosensitive microRNA to inhibit bone formation through targeting Runx2. *J. Bone Miner. Res.* **2015**, *30*, 330–345. [CrossRef]
47. Liu, H.; Yue, X.; Zhang, G. Downregulation of miR-146a inhibits osteoporosis in the jaws of ovariectomized rats by regulating the Wnt/β-catenin signaling pathway. *Int. J. Mol. Med.* **2021**, *47*, 6. [CrossRef]
48. Trojniak, J.; Sendera, A.; Banaś-Ząbczyk, A.; Kopańska, M. The microRNAs in the pathophysiology of osteoporosis. *Int. J. Mol. Sci.* **2024**, *25*, 6240. [CrossRef]
49. Zamani, A.; Zamani, V.; Heidari, B.; Parsian, H.; Esmaeilnejad-Ganji, S.M. Prevalence of osteoporosis with the World Health Organization diagnostic criteria in the Eastern Mediterranean Region: A systematic review and meta-analysis. *Arch. Osteoporos.* **2018**, *13*, 129. [CrossRef]
50. Salari, N.; Ghasemi, H.; Mohammadi, L.; Behzadi, M.H.; Rabieenia, E.; Shohaimi, S.; Mohammadi, M. The global prevalence of osteoporosis in the world: A comprehensive systematic review and meta-analysis. *J. Orthop. Surg. Res.* **2021**, *16*, 609. [CrossRef]
51. Chen, P.; Li, Z.; Hu, Y. Prevalence of osteoporosis in China: A meta-analysis and systematic review. *BMC Public Health* **2016**, *16*, 1039. [CrossRef]
52. Bliuc, D.; Nguyen, N.D.; Milch, V.E.; Nguyen, T.V.; Eisman, J.A.; Center, J.R. Mortality risk associated with low-trauma osteoporotic fracture and subsequent fracture in men and women. *J. Am. Med. Assoc.* **2009**, *301*, 513–521. [CrossRef] [PubMed]
53. Bandeira, L.; Bilezikian, J.P. Novel Therapies for Postmenopausal Osteoporosis. *Endocrinol. Metab. Clin. N. Am.* **2017**, *46*, 207–219. [CrossRef] [PubMed]
54. Khosla, S.; Riggs, B.L.; Atkinson, E.J.; Oberg, A.L.; McDaniel, L.J.; Holets, M.; Peterson, J.M.; Melton, L.J., 3rd. Effects of sex and age on bone microstructure at the ultradistal radius: A population-based noninvasive in vivo assessment. *J. Bone Miner. Res.* **2006**, *21*, 124–131. [CrossRef] [PubMed]
55. Khosla, S.; Monroe, D.G. Regulation of Bone Metabolism by Sex Steroids. *Cold Spring Harb. Perspect. Med.* **2018**, *8*, a031211. [CrossRef]
56. Szulc, P.; Claustrat, B.; Marchand, F.; Delmas, P.D. Increased risk of falls and increased bone resorption in elderly men with partial androgen deficiency: The MINOS study. *J. Clin. Endocrinol. Metab.* **2003**, *88*, 5240–5247. [CrossRef] [PubMed]
57. Cauley, J.A.; Ewing, S.K.; Taylor, B.C.; Fink, H.A.; Ensrud, K.E.; Bauer, D.C.; Barrett-Connor, E.; Marshall, L.; Orwoll, E.S.; Osteoporotic Fractures in Men Study (MrOS) Research Group. Sex steroid hormones in older men: Longitudinal associations with 4.5-year change in hip bone mineral density—The osteoporotic fractures in men study. *J. Clin. Endocrinol. Metab.* **2010**, *95*, 4314–4323. [CrossRef]
58. LeBlanc, E.S.; Nielson, C.M.; Marshall, L.M.; Lapidus, J.A.; Barrett-Connor, E.; Ensrud, K.E.; Hoffman, A.R.; Laughlin, G.; Ohlsson, C.; Orwoll, E.S.; et al. The effects of serum testosterone, estradiol, and sex hormone binding globulin levels on fracture risk in older men. *J. Clin. Endocrinol. Metab.* **2009**, *94*, 3337–3346. [CrossRef] [PubMed]
59. Kirby, D.J.; Buchalter, D.B.; Anil, U.; Leucht, P. DHEA in bone: The role in osteoporosis and fracture healing. *Arch. Osteoporos.* **2020**, *15*, 84. [CrossRef] [PubMed]
60. D'Amelio, P.; Roato, I.; D'Amico, L.; Veneziano, L.; Suman, E.; Sassi, F.; Bisignano, G.; Ferracini, R.; Gargiulo, G.; Castoldi, F.; et al. Bone and bone marrow pro-osteoclastogenic cytokines are up-regulated in osteoporosis fragility fractures. *Osteoporos. Int.* **2011**, *22*, 2869–2877. [CrossRef]
61. Lips, P. Vitamin D deficiency and secondary hyperparathyroidism in the elderly: Consequences for bone loss and fractures and therapeutic implications. *Endocr. Rev.* **2001**, *22*, 477–501. [CrossRef]
62. Metter, E.J.; Conwit, R.; Tobin, J.; Fozard, J.L. Age-associated loss of power and strength in the upper extremities in women and men. *J. Gerontol. Ser. A Biol. Sci. Med. Sci.* **1997**, *52*, B267–B276. [CrossRef]

63. Izquierdo, M.; Ibañez, J.; Gorostiaga, E.; Garrues, M.; Zúniga, A.; Antón, A.; Larriòn, J.L.; Hakkinen, K. Maximal strength and power characteristics in isometric and dynamic actions of the upper and lower extremities in middle-aged and older men. *Acta Physiol. Scand.* **1999**, *167*, 57–68. [CrossRef]
64. Compston, J. Glucocorticoid-induced osteoporosis: An update. *Endocrine* **2018**, *61*, 7–16. [CrossRef]
65. van Staa, T.P.; Leufkens, H.G.M.; Abenhaim, L.; Zhang, B.; Cooper, C. Oral corticosteroids and fracture risk: Relationship to daily and cumulative doses. *Rheumatology* **2000**, *39*, 1383–1389. [CrossRef]
66. Wu, Z.; Bucher, N.L.R.; Farmer, S.R. Induction of peroxisome proliferator-activated receptor γ during the conversion of 3T3 fibroblasts into adipocytes is mediated by C/EBPβ, C/EBPγ, and glucocorticoids. *Mol. Cell Biol.* **1996**, *16*, 4128–4136. [CrossRef]
67. Weinstein, R.S.; Jilka, R.L.; Parfitt, A.M.; Manolagas, S.C. Inhibition of osteoblastogenesis and promotion of apoptosis of osteoblasts and osteocytes by glucocorticoids—Potential mechanisms of their deleterious effects on bone. *J. Clin. Investig.* **1998**, *102*, 274–282. [CrossRef]
68. Ohnaka, K.; Tanabe, M.; Kawate, H.; Nawata, H.; Takayanagi, R. Glucocorticoid suppresses the canonical Wnt signal in cultured human osteoblasts. *Biochem. Biophys. Res. Commun.* **2005**, *329*, 177–181. [CrossRef]
69. Swanson, C.; Lorentzon, M.; Conaway, H.H.; Lerner, U.H. Glucocorticoid regulation of osteoclast differentiation and expression of receptor activator of nuclear factor-kappaB (NF-kappaB) ligand, osteoprotegerin, and receptor activator of NF-kappaB in mouse calvarial bones. *Endocrinology* **2006**, *147*, 3613–3622. [CrossRef]
70. Hofbauer, L.C.; Gori, F.; Riggs, B.L.; Lacey, D.L.; Dunstan, C.R.; Spelsberg, T.C.; Khosla, S. Stimulation of osteoprotegerin ligand and inhibition of osteoprotegerin production by glucocorticoids in human osteoblasts: Potential paracrine mechanisms of glucocorticoid-induced osteoporosis. *Endocrinology* **1999**, *140*, 4382–4389. [CrossRef]
71. Mazziotti, G.; Formenti, A.M.; Adler, R.A.; Bilezikian, J.P.; Grossman, A.; Sbardella, E.; Minisola, S.; Giustina, A. Glucocorticoid-induced osteoporosis: Pathophysiological role of GH/IGF-I and PTH/vitamin D axes, treatment options and guidelines. *Endocrine* **2016**, *54*, 603–611. [CrossRef]
72. Sato, A.Y.; Richardson, D.; Cregor, M.; Davis, H.M.; Au, E.D.; McAndrews, K.; Zimmers, T.A.; Organ, J.M.; Peacock, M.; Plotkin, L.I.; et al. Glucocorticoids induce bone and muscle atrophy by tissue-specific mechanisms upstream of E3 ubiquitin ligases. *Endocrinology* **2017**, *158*, 664–677. [CrossRef] [PubMed]
73. Skowrońska-Jóźwiak, E.; Lewandowski, K. Editorial: Osteoporosis secondary to endocrine disorders. *Front. Endocrinol.* **2023**, *14*, 1194241. [CrossRef] [PubMed]
74. Xu, Y.; Wu, Q. Trends in osteoporosis and mean bone density among type 2 diabetes patients in the US from 2005 to 2014. *Sci. Rep.* **2021**, *11*, 3693. [CrossRef] [PubMed]
75. Chen, S.C.; Shepherd, S.; McMillan, M.; McNeilly, J.; Foster, J.; Wong, S.C.; Robertson, K.J.; Ahmed, S.F. Skeletal fragility and its clinical determinants in children with type 1 diabetes. *J. Clin. Endocrinol. Metab.* **2019**, *104*, 3585–3594. [CrossRef]
76. Shah, V.N.; Harrall, K.K.; Shah, C.S.; Gallo, T.L.; Joshee, P.; Snell-Bergeon, J.K.; Kohrt, W.M. Bone mineral density at femoral neck and lumbar spine in adults with type 1 diabetes: A meta-analysis and review of the literature. *Osteoporos. Int.* **2017**, *28*, 2601–2610. [CrossRef] [PubMed]
77. Hofbauer, L.C.; Busse, B.; Eastell, R.; Ferrari, S.; Frost, M.; Müller, R.; Burden, A.M.; Rivadeneira, F.; Napoli, N.; Rauner, M. Bone fragility in diabetes: Novel concepts and clinical implications. *Lancet Diabetes Endocrinol.* **2022**, *10*, 207–220. [CrossRef] [PubMed]
78. Holloway-Kew, K.L.; De Abreu, L.L.F.; Kotowicz, M.A.; Sajjad, M.A.; Pasco, J.A. Bone turnover markers in men and women with impaired fasting glucose and diabetes. *Calcif. Tissue Int.* **2019**, *104*, 599–604. [CrossRef]
79. Kitaura, H.; Ogawa, S.; Ohori, F.; Noguchi, T.; Marahleh, A.; Nara, Y.; Pramusita, A.; Kinjo, R.; Ma, J.; Kanou, K.; et al. Effects of incretin-related diabetes drugs on bone formation and bone resorption. *Int. J. Mol. Sci.* **2021**, *22*, 6578. [CrossRef]
80. Schacter, G.I.; Leslie, W.D. Diabetes and osteoporosis: Part I, epidemiology and pathophysiology. *Endocrinol. Metab. Clin. N. Am.* **2021**, *50*, 275–285. [CrossRef]
81. Sheu, A.; Greenfield, J.R.; White, C.P.; Center, J.R. Assessment and treatment of osteoporosis and fractures in type 2 diabetes. *Trends Endocrinol. Metab.* **2022**, *33*, 333–344. [CrossRef]
82. Martínez-Montoro, J.I.; García-Fontana, B.; García-Fontana, C.; Muñoz-Torres, M. Evaluation of quality and bone micro-structure alterations in patients with type 2 diabetes: A narrative review. *J. Clin. Med.* **2022**, *11*, 2206. [CrossRef] [PubMed]
83. Sihota, P.; Yadav, R.N.; Dhaliwal, R.; Bose, J.C.; Dhiman, V.; Neradi, D.; Karn, S.; Sharma, S.; Aggarwal, S.; Goni, V.G.; et al. Investigation of mechanical, material, and compositional determinants of human trabecular bone quality in type 2 diabetes. *J. Clin. Endocrinol. Metab.* **2021**, *106*, e2271–e2289. [CrossRef]
84. Yuan, S.; Wan, Z.H.; Cheng, S.L.; Michaëlsson, K.; Larsson, S.C. Insulin-like growth factor-1, bone mineral density, and fracture: A Mendelian randomization study. *J. Clin. Endocrinol. Metab.* **2021**, *106*, e1552–e1558. [CrossRef]
85. Ouquerke, A.; Blulel, J. Osteoblasts and insulin: An overview. *J. Biol. Regul. Homeost. Agents* **2021**, *35*, 27–35.
86. Govoni, K.E. Insulin-like growth factor-I molecular pathways in osteoblasts: Potential targets for pharmacological manipulation. *Curr. Mol. Pharmacol.* **2012**, *5*, 143–152. [CrossRef]
87. Karim, L.; Bouxsein, M.L. Effect of type 2 diabetes-related non-enzymatic glycation on bone. *Bone* **2016**, *82*, 21–27. [CrossRef]

88. Gortázar, A.R.; Ardura, J.A. Osteocytes and diabetes: Altered function of diabetic osteocytes. *Curr. Osteoporos. Rep.* **2020**, *18*, 796–802. [CrossRef]
89. Kalaitzoglou, E.; Popescu, I.; Bunn, R.C.; Fowlkes, J.L.; Thrailkill, K.M. Effects of type 1 diabetes on osteoblasts, osteocytes, and osteoclasts. *Curr. Osteoporos. Rep.* **2016**, *14*, 310–319. [CrossRef] [PubMed]
90. Palermo, A.; D'Onofrio, L.; Buzzetti, R.; Manfrini, S.; Napoli, N. Pathophysiology of bone fragility in patients with diabetes. *Calcif. Tissue Int.* **2017**, *100*, 122–132. [CrossRef] [PubMed]
91. Skowrońska-Jóźwiak, E.; Krawczyk-Rusiecka, K.; Lewandowski, K.C.; Adamczewski, Z.; Lewiński, A. Successful treatment of thyrotoxicosis is accompanied by a decrease in serum sclerostin levels. *Thyroid Res.* **2012**, *5*, 14. [CrossRef] [PubMed]
92. Skowrońska-Jóźwiak, E.; Lewandowski, K.C.; Adamczewski, Z.; Krawczyk-Rusiecka, K.; Lewiński, A. Mechanisms of normalisation of bone metabolism during recovery from hyperthyroidism: Potential role for sclerostin and parathyroid hormone. *Int. J. Endocrinol.* **2015**, *2015*, 948384. [CrossRef]
93. Yeh, M.W.; Ituarte, P.H.; Zhou, H.C.; Nishimoto, S.; Liu, I.L.A.; Harari, A.; Haigh, P.I.; Adams, A.L. Incidence and prevalence of primary hyperparathyroidism in a racially mixed population. *J. Clin. Endocrinol. Metab.* **2013**, *98*, 1122–1129. [CrossRef]
94. Pallan, S.; Rahman, M.O.; Khan, A.A. Diagnosis and management of primary hyperparathyroidism. *BMJ* **2012**, *344*, e1013. [CrossRef]
95. Thakker, R.V. Genetics of parathyroid tumours. *J. Intern. Med.* **2016**, *280*, 574–583. [CrossRef]
96. Bilezikian, J.P.; Brandi, M.L.; Eastell, R.; Silverberg, S.J.; Udelsman, R.; Marcocci, C.; Potts, J.T., Jr. Guidelines for the management of asymptomatic primary hyperparathyroidism: Summary statement from the Fourth International Workshop. *J. Clin. Endocrinol. Metab.* **2014**, *99*, 3561–3569. [CrossRef]
97. Stein, E.; Shane, E. Secondary osteoporosis. *Endocrinol. Metab. Clin. N. Am.* **2003**, *32*, 115–134. [CrossRef] [PubMed]
98. Lespessailles, E.; Paccou, J.; Javier, R.M.; Thomas, T.; Cortet, B.; GRIO Scientific Committee. Obesity, bariatric surgery, and fractures. *J. Clin. Endocrinol. Metab.* **2019**, *104*, 4756–4768. [CrossRef] [PubMed]
99. Gagnon, C.; Schafer, A.L. Bone health after bariatric surgery. *J. Bone Miner. Res. Plus* **2018**, *2*, 121–133. [CrossRef] [PubMed]
100. Mechanick, J.I.; Apovian, C.; Brethauer, S.; Garvey, W.T.; Joffe, A.M.; Kim, J.; Kushner, R.F.; Lindquist, R.; Pessah-Pollack, R.; Seger, J.; et al. Clinical practice guidelines for the perioperative nutrition, metabolic, and nonsurgical support of patients undergoing bariatric procedures—2019 update: Cosponsored by American Association of Clinical Endocrinologists/American College of Endocrinology, The Obesity Society, American Society for Metabolic & Bariatric Surgery, Obesity Medicine Association, and American Society of Anesthesiologists. *Surg. Obes. Relat. Dis.* **2020**, *16*, 175–247.
101. Schafer, A.L.; Kazakia, G.J.; Vittinghoff, E.; Stewart, L.; Rogers, S.J.; Kim, T.Y.; Carter, J.T.; Posselt, A.M.; Pasco, C.; Shoback, D.M.; et al. Effects of gastric bypass surgery on bone mass and microarchitecture occur early and particularly impact postmenopausal women. *J. Bone Miner. Res.* **2018**, *33*, 975–986. [CrossRef]
102. Lindeman, K.G.; Greenblatt, L.B.; Rourke, C.; Bouxsein, M.L.; Finkelstein, J.S.; Yu, E.W. Longitudinal 5-year evaluation of bone density and microarchitecture after Roux-en-Y gastric bypass surgery. *J. Clin. Endocrinol. Metab.* **2018**, *103*, 4104–4112. [CrossRef]
103. Hansen, S.; Jørgensen, N.R.; Hermann, A.P.; Støving, R.K. Continuous decline in bone mineral density and deterioration of bone microarchitecture 7 years after Roux-en-Y gastric bypass surgery. *Eur. J. Endocrinol.* **2020**, *182*, 303–311. [CrossRef]
104. Diamond, T.; Stiel, D.; Posen, S. Osteoporosis in hemochromatosis: Iron excess, gonadal deficiency, or other factors? *Ann. Intern. Med.* **1989**, *110*, 430–436. [CrossRef]
105. Schneider, A.; Shane, E. Osteoporosis secondary to illnesses and medications. In *Osteoporosis*; Marcus, R., Feldman, D., Kelsey, J., Eds.; Academic Press: San Diego, CA, USA, 2001; pp. 303–327.
106. Bikle, D. Osteoporosis in gastrointestinal, pancreatic, and hepatic diseases. In *Osteoporosis*; Marcus, R., Feldman, D., Kelsey, J., Eds.; Academic Press: San Diego, CA, USA, 2001; pp. 237–258.
107. Sánchez, M.I.; Mohaidle, A.; Baistrocchi, A.; Matoso, D.; Vazquez, H.; Gonzalez, A.; Mazure, R.; Maffei, E.; Ferrari, G.; Smecuol, E.; et al. Risk of fracture in celiac disease: Gender, dietary compliance, or both? *World J. Gastroenterol.* **2011**, *17*, 3035–3042. [CrossRef]
108. Pritchard, L.; Wilson, S.; Griffin, J.; Pearce, G.; Murray, I.A.; Lewis, S. Prevalence of reduced bone mineral density in adults with coeliac disease—Are we missing opportunities for detection in patients below 50 years of age? *Scand. J. Gastroenterol.* **2018**, *53*, 1433–1436. [CrossRef] [PubMed]
109. Meyer, D.; Stavropolous, S.; Diamond, B.; Shane, E.; Green, P.H. Osteoporosis in a North American adult population with celiac disease. *Am. J. Gastroenterol.* **2001**, *96*, 112–119. [CrossRef]
110. Vazquez, H.; Mazure, R.; Gonzalez, D.; Flores, D.; Pedreira, S.; Niveloni, S.; Smecuol, E.; Maurino, E.; Bai, J.C. Risk of fractures in celiac disease patients: A cross-sectional, case-control study. *Am. J. Gastroenterol.* **2000**, *95*, 183–189. [CrossRef]
111. Hernandez, C.J.; Guss, J.D.; Luna, M.; Goldring, S.R. Links between the microbiome and bone. *J. Bone Miner. Res.* **2016**, *31*, 1638–1646. [CrossRef]
112. Bryant, R.V.; Ooi, S.; Schultz, C.G.; Goess, C.; Grafton, R.; Hughes, J.; Lim, A.; Bartholomeusz, F.D.; Andrews, J.M. Low muscle mass and sarcopenia: Common and predictive of osteopenia in inflammatory bowel disease. *Aliment. Pharmacol. Ther.* **2015**, *41*, 895–906. [CrossRef] [PubMed]

113. Haaber, A.B.; Rosenfalck, A.M.; Hansen, B.; Hilsted, J.; Larsen, S. Bone mineral metabolism, bone mineral density, and body composition in patients with chronic pancreatitis and pancreatic exocrine insufficiency. *Int. J. Pancreatol.* **2000**, *27*, 21–27. [CrossRef]
114. Stoltz, D.A.; Meyerholz, D.K.; Welsh, M.J. Origins of cystic fibrosis lung disease. *N. Engl. J. Med.* **2015**, *372*, 351–362. [CrossRef]
115. Paccou, J.; Zeboulon, N.; Combescure, C.; Gossec, L.; Cortet, B. The prevalence of osteoporosis, osteopenia, and fractures among adults with cystic fibrosis: A systematic literature review with meta-analysis. *Calcif. Tissue Int.* **2010**, *86*, 1–7. [CrossRef]
116. Ebeling, P.R. Transplantation osteoporosis. In *Primer on the Metabolic Bone Diseases and Disorders of Mineral Metabolism*, 9th ed.; John, P., Bilezikian, J.W., Eds.; Wiley & Sons, Inc.: Hoboken, NJ, USA, 2019.
117. Stein, E.M.; Ortiz, D.; Jin, Z.; McMahon, D.J.; Shane, E. Prevention of fractures after solid organ transplantation: A meta-analysis. *J. Clin. Endocrinol. Metab.* **2011**, *96*, 3457–3465. [CrossRef]
118. Rajkumar, S.V.; Dimopoulos, M.A.; Palumbo, A.; Blade, J.; Merlini, G.; Mateos, M.V.; Kumar, S.; Hillengass, J.; Kastritis, E.; Richardson, P.; et al. International Myeloma Working Group updated criteria for the diagnosis of multiple myeloma. *Lancet Oncol.* **2014**, *15*, e538–e548. [CrossRef] [PubMed]
119. Thorsteinsdottir, S.; Gislason, G.; Aspelund, T.; Sverrisdottir, I.; Landgren, O.; Turesson, I.; Bjorkholm, M.; Kristinsson, S.Y. Fractures and survival in multiple myeloma: Results from a population-based study. *Haematologica* **2020**, *105*, 1067–1073. [CrossRef]
120. Nador, G.; Ramasamy, K.; Panitsas, F.; Pratt, G.; Sadler, R.; Javaid, M.K. Testing and management for monoclonal gammopathy of uncertain significance and myeloma patients presenting with osteoporosis and fragility fractures. *Rheumatology* **2019**, *58*, 1142–1153. [CrossRef] [PubMed]
121. Kumar, S.K.; Rajkumar, V.; Kyle, R.A.; van Duin, M.; Sonneveld, P.; Mateos, M.V.; Gay, F.; Anderson, K.C. Multiple myeloma. *Nat. Rev. Dis. Primers* **2017**, *3*, 17046. [CrossRef]
122. Hamilton, E.J.; Ghasem-Zadeh, A.; Gianatti, E.; Lim-Joon, D.; Bolton, D.; Zebaze, R.; Seeman, E.; Zajac, J.D.; Grossmann, M. Structural decay of bone microarchitecture in men with prostate cancer treated with androgen deprivation therapy. *J. Clin. Endocrinol. Metab.* **2010**, *95*, E456–E463. [CrossRef]
123. Ng, H.S.; Koczwara, B.; Roder, D.; Vitry, A. Development of comorbidities in men with prostate cancer treated with androgen deprivation therapy: An Australian population-based cohort study. *Prostate Cancer Prostatic Dis.* **2018**, *21*, 403–410. [CrossRef]
124. Lassemillante, A.C.; Doi, S.A.; Hooper, J.D.; Prins, J.B.; Wright, O.R. Prevalence of osteoporosis in prostate cancer survivors: A meta-analysis. *Endocrine* **2014**, *45*, 370–381. [CrossRef] [PubMed]
125. Milat, F.; Vincent, A.J. Management of bone disease in women after breast cancer. *Climacteric* **2015**, *18*, 47–55. [CrossRef]
126. Kim, D.; Oh, J.; Lee, H.S.; Jeon, S.; Park, W.C.; Yoon, C.I. Association between tamoxifen and incidence of osteoporosis in Korean patients with ductal carcinoma in situ. *Front. Oncol.* **2024**, *13*, 1236188. [CrossRef]
127. Coates, A.S.; Keshaviah, A.; Thürlimann, B.; Mouridsen, H.; Mauriac, L.; Forbes, J.F.; Paridaens, R.; Castiglione-Gertsch, M.; Gelber, R.D.; Colleoni, M.; et al. Five years of letrozole compared with tamoxifen as initial adjuvant therapy for postmenopausal women with endocrine-responsive early breast cancer: Update of study BIG 1-98. *J. Clin. Oncol.* **2007**, *25*, 486–492. [CrossRef]
128. Hadji, P.; Colli, E.; Regidor, P.A. Bone health in estrogen-free contraception. *Osteoporos. Int.* **2019**, *30*, 2391–2400. [CrossRef] [PubMed]
129. Diemar, S.S.; Sejling, A.S.; Eiken, P.; Andersen, N.B.; Jørgensen, N.R. An explorative literature review of the multifactorial causes of osteoporosis in epilepsy. *Epilepsy Behav.* **2019**, *100*, 106511. [CrossRef]
130. Ducy, P.; Karsenty, G. The two faces of serotonin in bone biology. *J. Cell Biol.* **2010**, *191*, 7–13. [CrossRef]
131. Poly, T.N.; Islam, M.M.; Yang, H.C.; Wu, C.C.; Li, Y.J. Proton pump inhibitors and risk of hip fracture: A meta-analysis of observational studies. *Osteoporos. Int.* **2019**, *30*, 103–114. [CrossRef]
132. Hosein-Woodley, R.; Hirani, R.; Issani, A.; Hussaini, A.S.; Stala, O.; Smiley, A.; Etienne, M.; Tiwari, R.K. Beyond the surface: Uncovering secondary causes of osteoporosis for optimal management. *Biomedicines* **2024**, *12*, 2558. [CrossRef] [PubMed]
133. Ebeling, P.R.; Nguyen, H.H.; Aleksova, J.; Vincent, A.J.; Wong, P.; Milat, F. Secondary osteoporosis. *Endocr. Rev.* **2022**, *43*, 240–313. [CrossRef]
134. Trajanoska, K.; Morris, J.A.; Oei, L.; Zheng, H.F.; Evans, D.M.; Kiel, D.P.; Ohlsson, C.; Richards, J.B.; Rivadeneira, F. Assessment of the genetic and clinical determinants of fracture risk: Genome-wide association and mendelian randomisation study. *BMJ* **2018**, *362*, k3225. [CrossRef]
135. Formosa, M.M.; Bergen, D.J.M.; Gregson, C.L.; Maurizi, A.; Kämpe, A.; Garcia Giralt, N.; Zhou, W.; Grinberg, D.; Ovejero Crespo, D.; Zillikens, M.C.; et al. A roadmap to gene discoveries and novel therapies in monogenic low and high bone mass disorders. *Front. Endocrinol.* **2021**, *12*, 709711. [CrossRef]

136. Unger, S.; Ferreira, C.R.; Mortier, G.R.; Ali, H.; Bertola, D.R.; Calder, A.; Cohn, D.H.; Cormier Daire, V.; Girisha, K.M.; Hall, C.; et al. Nosology of genetic skeletal disorders: 2023 revision. *Am. J. Med. Genet. A* **2023**, *191*, 1164–1209. [CrossRef]
137. Gistelinck, C.; Weis, M.; Rai, J.; Schwarze, U.; Niyazov, D.; Song, K.M.; Byers, P.H.; Eyre, D.R. Abnormal bone collagen cross-linking in osteogenesis imperfecta/Bruck syndrome caused by compound heterozygous PLOD2 mutations. *J. Bone Miner. Res. Plus* **2021**, *5*, e10454. [CrossRef] [PubMed]
138. Jovanovic, M.; Guterman Ram, G.; Marini, J.C. Osteogenesis Imperfecta: Mechanisms and signaling pathways connecting classical and rare OI types. *Endocr. Rev.* **2022**, *43*, 61–90. [CrossRef] [PubMed]
139. Baron, R.; Kneissel, M. WNT signaling in bone homeostasis and disease: From human mutations to treatments. *Nat. Med.* **2013**, *19*, 179–192. [CrossRef]
140. Gong, Y.; Slee, R.B.; Fukai, N.; Rawadi, G.; Roman Roman, S.; Reginato, A.M.; Wang, H.; Cundy, T.; Glorieux, F.H.; Lev, D.; et al. LDL receptor related protein 5 (LRP5) affects bone accrual and eye development. *Cell* **2001**, *10*, 513–523. [CrossRef]
141. Hu, J.; Lin, X.; Gao, P.; Zhang, Q.; Zhou, B.; Wang, O.; Jiang, Y.; Xia, W.; Xing, X.; Li, M. Genotypic and phenotypic spectrum and pathogenesis of WNT1 variants in a large cohort of patients with OI/osteoporosis. *J. Clin. Endocrinol. Metab.* **2023**, *108*, 1776–1786. [CrossRef]
142. Peris, P.; Monegal, A.; Mäkitie, R.E.; Guañabens, N.; González Roca, E. Osteoporosis related to WNT1 variants: A not infrequent cause of osteoporosis. *Osteoporos. Int.* **2023**, *34*, 405–411. [CrossRef] [PubMed]
143. van Dijk, F.S.; Zillikens, M.C.; Micha, D.; Riessland, M.; Marcelis, C.L.; de Die Smulders, C.E.; Milbradt, J.; Franken, A.A.; Harsevoort, A.J.; Lichtenbelt, K.D.; et al. PLS3 mutations in X-linked osteoporosis with fractures. *N. Engl. J. Med.* **2013**, *369*, 1529–1536. [CrossRef]
144. Kämpe, A.J.; Costantini, A.; Levy Shraga, Y.; Zeitlin, L.; Roschger, P.; Taylan, F.; Lindstrand, A.; Paschalis, E.P.; Gamsjaeger, S.; Raas Rothschild, A.; et al. PLS3 deletions lead to severe spinal osteoporosis and disturbed bone matrix mineralization. *J. Bone Miner. Res.* **2017**, *32*, 2394–2404. [CrossRef] [PubMed]
145. Pekkinen, M.; Terhal, P.A.; Botto, L.D.; Henning, P.; Mäkitie, R.E.; Roschger, P.; Jain, A.; Kol, M.; Kjellberg, M.A.; Paschalis, E.P.; et al. Osteoporosis and skeletal dysplasia caused by pathogenic variants in SGMS2. *JCI Insight* **2019**, *4*, e126180. [CrossRef] [PubMed]
146. Formosa, M.M.; Christou, M.A.; Mäkitie, O. Bone fragility and osteoporosis in children and young adults. *J. Endocrinol. Investig.* **2024**, *47*, 285–298. [CrossRef]
147. Gray, J.R.; Bridges, A.B.; Mole, P.A.; Pringle, T.; Boxer, M.; Paterson, C.R. Osteoporosis and the Marfan syndrome. *Postgrad. Med. J.* **1993**, *69*, 373–375. [CrossRef]
148. Liu, R.; Weng, G.; Zheng, F.; Chen, J.; Wang, K.; Han, J.; Huang, J.; Yan, L.; Jin, J. Generation of an integration-free induced pluripotent stem cell line, FJMAi001-A, from a Marfan syndrome patient with a heterozygous mutation c.2777G > A (p.Cys926Tyr) in FBN1. *Stem Cell Res.* **2024**, *81*, 103591. [CrossRef]
149. Racine, C.; Callier, P.; Touraine, R.; Vitobello, A.; Hanna, N.; Arnaud, P.; Jondeau, G.; Boileau, C.; Thauvin-Robinet, C.; Creveaux, I.; et al. De novo balanced translocations disrupting the *FBN1* gene diagnosed by genome sequencing: An uncommon cause of Marfan syndrome modifying genetic counseling. *Am. J. Med. Genet. Part A* **2025**, *197*, e63923. [CrossRef]
150. Charoenngam, N.; Rittiphairoj, T.; Ponvilawan, B.; Jaroenlapnopparat, A.; Waitayangkoon, P.; Suppakitjanusant, P.; Prasitsumrit, V.; Pongchaiyakul, C.; Holick, M.F. Bone fragility in hereditary connective tissue disorders: A systematic review and meta-analysis. *Endocr. Pract.* **2023**, *29*, 589–600. [CrossRef]
151. Le Parc, J.M.; Plantin, P.; Jondeau, G.; Goldschild, M.; Albert, M.; Boileau, C. Bone mineral density in sixty adult patients with Marfan syndrome. *Osteoporos. Int.* **1999**, *10*, 475–479. [CrossRef] [PubMed]
152. International Osteoporosis Foundation. Modifiable Risk Factors. Available online: https://www.osteoporosis.foundation/health-professionals/about-osteoporosis/risk-factors/modifiable-risks (accessed on 21 April 2025).
153. Reid, I.R.; Billington, E.O. Drug therapy for osteoporosis in older adults. *Lancet* **2022**, *399*, 1080–1092. [CrossRef]
154. Tong, G.; Meng, Y.; Hao, S.; Hu, S.; He, Y.; Yan, W.; Yang, D. Parathyroid hormone activates phospholipase C (PLC)-independent protein kinase C signalling pathway via protein kinase A (PKA)-dependent mechanism: A new defined signalling route would induce alternative consideration to previous conceptions. *Med. Sci. Monit.* **2017**, *23*, 1896–1906. [CrossRef] [PubMed]
155. Vasiliadis, E.S.; Evangelopoulos, D.-S.; Kaspiris, A.; Benetos, I.S.; Vlachos, C.; Pneumaticos, S.G. The role of sclerostin in bone diseases. *J. Clin. Med.* **2022**, *11*, 806. [CrossRef] [PubMed]
156. Liu, B.Y.; Wu, P.W.; Bringhurst, F.R.; Wang, J.T. Estrogen inhibition of PTH-stimulated osteoclast formation and attachment in vitro: Involvement of both PKA and PKC. *Endocrinology* **2002**, *143*, 627–635. [CrossRef]

157. Cosman, F.; Crittenden, D.B.; Adachi, J.D.; Binkley, N.; Czerwinski, E.; Ferrari, S.; Hofbauer, L.C.; Lau, E.; Lewiecki, E.M.; Miyauchi, A.; et al. Romosozumab treatment in postmenopausal women with osteoporosis. *N. Engl. J. Med.* **2016**, *375*, 1532–1543. [CrossRef]
158. Eriksen, E.F.; Boyce, R.W.; Shi, Y.; Brown, J.P.; Betah, D.; Libanati, C.; Oates, M.; Chapurlat, R.; Chavassieux, P. Reconstruction of remodeling units reveals positive effects after 2 and 12 months of romosozumab treatment. *J. Bone Miner. Res.* **2024**, *39*, 729–736. [CrossRef]

BioChem

MDPI

Article

AI-Assisted Identification of the Functional Residues of Ginsenoside Biosynthesis-Associated UGTs

Kisook Jung, Narae Kim, Chaelin Park and Jaewook Kim *

Department of Biology Education, Korea National University of Education, Cheongju 28173, Republic of Korea; sena0109@knue.ac.kr (K.J.); narrr2@knue.ac.kr (N.K.); pch97623@naver.com (C.P.)
* Correspondence: jwkim@knue.ac.kr

Abstract

Background/Objectives: Ginsenosides, one of the most pharmaceutically valuable chemical compounds in *Panax ginseng*, are synthesized with several enzymes, including UGTs. UGTs determine absorbability and physiological function upon consumption. Thus, understanding the functional residues of ginsenoside biosynthesis-associated UGTs is crucial for enhancing the production of valuable ginsenoside varieties. **Methods**: We collected the UGT homologs of high sequence similarity from two rate-limiting steps of the biosynthetic pathway. The 3D structures of these proteins were predicted using the AlphaFold3 model. The ligand-binding interactions of these UGTs were examined using SwissDock and CB-Dock2. Enzyme kinetics were analyzed with MPEK. Using these tools, we performed in silico mutagenic analyses to identify the functional residues of UGTs in detail. **Results**: We elucidated the molecular mechanisms of experimentally verified functional residues in UGTs, many of which were associated with optimal ligand interaction angles that expose target carbons. We also identified putatively important amino acid residues that mediate ligand interactions and modulate reaction kinetics by more than 25%. In this study, residues at positions 62, 224, 397, and 398 were shown to significantly influence enzyme kinetics. **Conclusions**: Our study provides the first structural analysis of the functional residues of ginsenoside biosynthetic UGTs based on their 3D structures. We identified several key amino acid residues essential for proper ginsenoside biosynthesis: (1) residues determining ligand interactions, (2) residues modulating ligand binding angles, and (3) residues affecting reaction kinetics. Our findings demonstrate an effective approach to identifying functional residues in plant enzymes and present valuable UGT candidates for future experimental validation.

Keywords: UGT; ginsenoside biosynthesis; functional residue; artificial intelligence; 3D structure

1. Introduction

Ginsenosides are the triterpenoid compounds accumulated in *Panax* spp., and possess many of the pharmaceutical effects of these plants [1]. In detail, ginsenosides species include protopanaxadiol (PPD), protopanaxatriol (PPT), ocotillol (OCT), and oleanane (OA)-type saponins [2]. Of these, *Panax ginseng* primarily accumulates PPD- and PPT-type ginsenosides. These ginsenoside skeleton structures are formed from squalene through squalene epoxidase (SE) and are subsequently cyclized by dammarendiol-II synthase (DDS) [2]. Next, dammarendiols are converted by CYPs (cytochrome P450) into protopanaxadiol and protopanaxatriol, which are the direct precursors of PPD- and PPT-type ginsenosides [2].

From this step, ginsenoside compounds undergo glycosylation by various UGTs (UDP-dependent glycosyltransferases) [2]. UGTs further catalyze ginsenosides by attaching sugar moieties onto the C3, C6, and C20 positions. PPD can be catalyzed into CK, Rh2, F2, Rg3, Rd, Rb1, Rb2, Rb3, and Rc through a cascade of UGTs [2]. PPT can be catalyzed into Rh1, F1, Rf, and Rg1 through a cascade of UGTs [2].

UGT-based modification mediates the conversion of ginsenosides into various molecules with numerous pharmaceutical benefits. Ginsenoside Rb2 has been reported to possess anti-cancer, anti-diabetic, cardio-protective, anti-inflammatory, anti-oxidative, anti-osteoporosis, antiviral, and anti-photoaging effects [3]. Ginsenoside Rg3 has also been shown to exhibit anti-inflammatory and anti-oxidative effects, thereby contributing to an anti-allergic effect [4]. In addition to these, ginsenoside Re has been reported to possess anti-diabetes mellitus, anti-Parkinson's disease, anti-inflammatory, and anti-cancer effects; ginsenoside Rg1 has demonstrated anti-hyperglycemic and anti-depressive effects; ginsenoside Rc has been reported to have anti-alcoholic liver disease, anti-myocardial necrosis, and anti-endothelial insulin resistance effects; and compound K has shown anti-tumor, anti-inflammatory, and hepatoprotective effects [5]. One previous study indicated that sugar moiety compositions of PPT-type ginsenosides modulate binding affinity to the glucocorticoid receptor (GR) [6]. Another line of evidence showed that sugar moieties attached only to the C3 position possess a potent inhibitory effect on Na^+/K^+-ATPase activity, while this effect is diminished when a monosaccharide is linked to the C6 or C20 position [7]. Thus, modulating the sugar moiety of ginsenoside molecules significantly alters their pharmaceutical activity.

Therefore, many efforts have been dedicated to produce these valuable ginsenosides. Treatment using methyl jasmonate on the root tissue of *P. ginseng* has been shown to enhance the expression of UGT genes, thus increasing ginsenoside Rb1, Rg1, Rh2, and Rg3 (Korean Patent Number 10-2014-0041261). Using a yeast system (*Saccharomyces cerevisiae*) with transgenic expression of UGTs and CYP enzymes, compound K, ginsenoside F1, and ginsenoside Rh1 were successfully synthesized [8,9]. In the case of ginsenoside Rh2, factory-level production was shown to be successfully achieved [10]. Enzymatic assessment was also performed for ginsenoside production. Cellulase KN was shown to produce the ginsenoside F1 [11]. A mutant form of glycosyltransferase BS-YjiC has been reported to catalyze PPD into ginsenoside F12 and Rh2 (US12134789B2). A new UGT named UGT109A1 from *Bacillus subtilis* has been reported to glycosylate the C3 and C12 position, leading to the production of unnatural ginsenoside variants [12].

As reviewed above, the production of ginsenosides has attracted great interest. Thus, assessing the enzymatic activity or specificity of UGTs and understanding them could greatly advance productivity. To do that, we aimed to identify important amino acid residues and reveal their potential roles. Homologs of high sequence similarity (over 99%) were identified as potentially active UGTs in the ginsenoside biosynthetic process. Next, sequence alignment was performed to reveal potentially important amino acids. Through 3D structure modeling with AlphaFold3, we applied a series of AI-based models to assess protein–ligand interaction or catalytic activity, using SwissDock, CB-Dock2, and MPEK with in silico mutated proteins. Through these comprehensive analyses, we identified the first potent amino acid residue in UGTs that may alter their reaction kinetics and several other residues that potentially change ligand interactivity.

2. Materials and Methods

2.1. Identification of Target Sequences

Among many reactions associated with ginsenoside biosynthesis, we focused on two rate-limiting steps: PPD-to-ginsenoside CK and PPT-to-ginsenoside F1. UGTs involved in these reactions were identified using BLASTp with an e-value cutoff of 1×10^{-5} and a similarity of 99% [13,14]. Sequences were aligned with MUSCLE pipeline, revealing several variable regions [15]. To assess sequence conservation, aligned sequences were analyzed with Jalview version 2.11.4.0, and the quality score metric was used [16]. Based on these criteria, we identified eight amino acid residues with potential functional relevance in ginsenoside biosynthesis. To evaluate the applicability of our in-silico approach, we also tested seven amino acid residues previously shown to have functional relevance in ginsenoside biosynthesis [9]. Mutagenesis was performed by substituting residues with naturally occurring counterparts or alanine.

2.2. 3D Structure-Based Protein–Ligand Interactivity and Enzyme Catalytic Activity Assay

3D structures of all proteins were predicted with AlphaFold3, and the most probable structural model for each was used [17]. We utilized AlphaFold3 to predict the 3D structures of the target UGTs, and obtained confidence scores above 0.8 across entire sequences (Figure S1). AlphaFold3 generated highly reliable structural models, as confirmed by comparison between the predicted structure of a representative UGT and the experimentally determined 3D structure of a homologous UGT (Figure S2). To examine UGT–ligand interactivity, we used SwissDock and CB-Dock2 [18–20]. To predict catalytic activity, MPEK was applied to each UGT–ligand pair [21].

2.3. Mutagenic Analysis to Reveal Functional Amino Acid Residue

First, we performed multiple sequence alignment with MUSCLE to identify regions with sequence variation [15]. The 3D structures predicted by AlphaFold3 were then manually analyzed with ChimeraX to identify target sites for mutagenesis [17,22]. Eight amino acid residues were selected for this study. Residue numbering was standardized based on the multiple sequence alignment, and mutant UGT sequences were generated manually. All sequences were then modeled with AlphaFold3, and the most probable structures were used for further analysis [17]. Using these predicted structures or amino acid sequences as input, we performed SwissDock, CB-Dock2, and MPEK with default settings [18–21]. Specifically, we analyzed the AC_score from SwissDock to estimate protein–ligand affinity, inspected CB-Dock2-predicted structures with ChimeraX, and used MPEK to predict kcat values [18–22]. MPEK can predict both kcat and Km values with any protein–ligand pair, and thus was applied in our analysis as well [21]. In applying these models to plant enzymes, we noted the following limitations or considerations for each AI tool: AlphaFold3 showed no discernible limitations; SwissDock was unable to predict the exact binding site; CB-Dock2 predictions of binding affinity were unreliable due to excessive variation; and MPEK reported kcat values and Km values that did not always correspond to kcat results.

2.4. Data Visualization

Heatmaps were generated with TB-tools II [23]. Centered Z-scores for ligand affinity and enzyme activity were calculated, setting the wild-type value to zero. To visualize ligand orientation, we examined the detailed structures to determine whether the ligand exposed the C3, C6, or C20 positions toward the sugar donor. Each prediction was repeated until at least three independent runs yielded consistent results.

3. Results

3.1. Identification of Potential Amino Acids of Functional Relevance in Ginsenoside Biosynthesis

Among the many biochemical reactions involved in the ginsenoside biosynthetic process, we analyzed UGTs associated with two rate-limiting steps: PPD-to-CK and PPT-to-

F1 (Figure 1A; Table 1). Previous studies identified four UGTs that catalyze the conversion of PPD or PPT into CK or F1 (Table 1). To identify amino acids potentially correlated with functionality, we collected homologs with high sequence similarity (over 99%) (Table 2). Sequence similarity analysis identified a total of 27 homologous proteins, including the functional UGTs involved in the target reactions (Table 2). Using these homologs, we sought to identify amino acids with potential functional relevance in ginsenoside biosynthesis.

Table 1. List of target enzymes in this study.

Precursor	Product	Enzyme	UniProtKBID
protopanaxadiol	ginsenoside CK	PgUGT71A27 PnUGT1/UDP-glucosyltransferase PgUGT1/PgUGT71A53	A0A0A7HB61.1 AFO63526.1 AIE12479.1/A0A068J840.1
protopanaxatriol	ginsenoside F1	PnUGT1/UDP-glucosyltransferase PgUGT1/PgUGT71A53 PgUGT101/PgUGT71A55	AFO63526.1 AIE12479.1/A0A068J840.1 A0A0K0PVM5.1

Table 2. List of target UGTs that share high sequence similarity with UGTs in Table 1.

Accession Number	Description
A0A0A6ZFR4.1	PgUGT74AE2 [*Panax ginseng*]
A0A0A7HB61.1	PgUGT71A27 [*Panax ginseng*]
A0A0D5ZDC8.1	PgUGT74AE4 [*Panax ginseng*]
A0A0K0PVL0.1	UGTPg102 [*Panax ginseng*]
A0A0K0PVM5.1	PgUGT71A55 [*Panax ginseng*]
A0A068J840.1	PgUGT71A53 [*Panax ginseng*]
AED99883.1	glycosyltransferase [*Panax notoginseng*]
AFO63526.1	PnUGT1 [*Panax notoginseng*]
AKA44602.1	UGTPg23 [*Panax ginseng*]
AKI06370.1	glycosyltransferase [*Panax quinquefolius*]
ALE15279.1	UDP-glycosyltransferase 3GT1 [*Panax quinquefolius*]
ART66191.1	uridine diphosphate glycosyltransferase1 [*Kalopanax septemlobus*]
ART66192.1	uridine diphosphate glycosyltransferase2 [*Kalopanax septemlobus*]
QEV87497.1	Pq3-O-UGT1 [*Panax quinquefolius*]
QEV87498.1	UDP-glycosyltransferase [*Panax quinquefolius*]
QEV87499.1	UDP-glycosyltransferase [*Panax quinquefolius*]
QOJ43864.1	UDP-glycosyltransferase [*Panax notoginseng*]
QOJ43865.1	UDP-glycosyltransferase [*Panax notoginseng*]
UMX47351.1	UGT73AD2 [*Aralia elata*]
UMX47352.1	UGT74AG6 [*Aralia elata*]
UMX47353.1	UGT73CB3 [*Aralia elata*]
UMX47354.1	UGT73CB2 [*Aralia elata*]
UMX47355.1	UGT73CB4 [*Aralia elata*]

Table 2. *Cont.*

Accession Number	Description
UMX47356.1	UGT73CB5 [*Aralia elata*]
UNO37640.1	UGT7 [*Panax ginseng*]
WLJ58897.1	UDP-glycosyltransferase [*Panax ginseng*]
WPX61740.1	UGT74AG11 [*Hedera helix*]

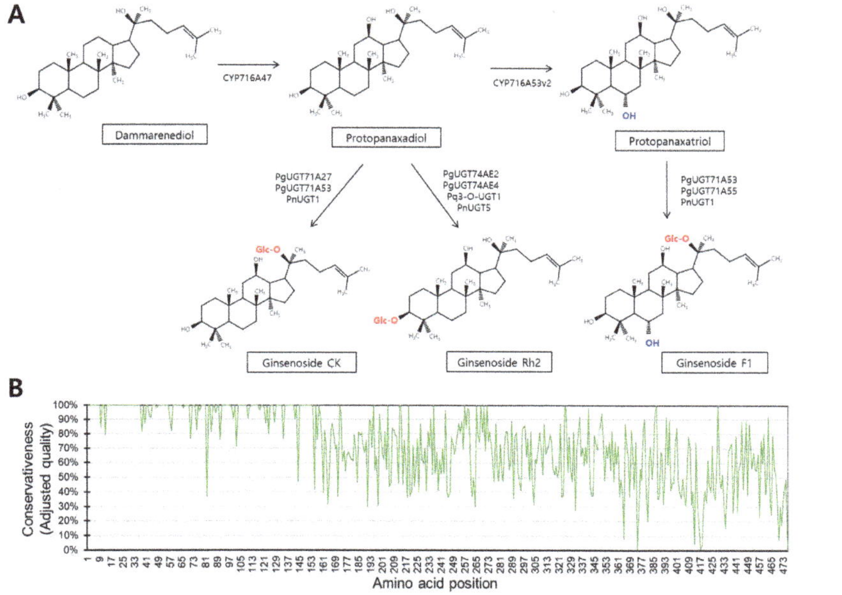

Figure 1. Target reaction and amino acid conservativeness. (**A**) Schematic diagram presenting UGT-mediated reactions catalyzing PPD and PPT into ginsenoside CK, ginsenoside Rh2, and ginsenoside F1. (**B**) Amino acid conservativeness plot made by comparing all close homologs.

Comparison of the sequences of the 27 homologous UGTs revealed variable regions in the amino acid sequences (Figure 1B). Regions with low sequence similarity were located in C-terminus regions (Figure 1B). Previous studies reported functionally relevant amino acids in the N-terminal region, with only one residue identified outside this region (Table 3) [9]. In this study, we aimed to determine how these functional amino acids contribute to catalytic activity and to identify novel residues that affect different aspects of catalytic function, including ligand interaction, the ligand interaction angle, and reaction kinetics.

Table 3. List of functional amino acids that were experimentally shown to alter catalytic activities in the ginsenoside biosynthetic process.

UGT \ Location	10	13	82	142	144	185	339
PgUGT71A27	A	I	H	A	H	L	G
UGTPg1	A	I	H	A	H	L	G
UGTPg101	A	I	H	A	H	L	G
UGTPg102	A	I	H	A	Y	L	G
UGTPg103	V	F	C	T	F	F	K

3.2. Functional Amino Acid Residues from the Previous Study Mainly Alter Ligand Interactivity

We established a strategy to gain detailed functional insights using a mutagenesis approach (Figure 2). In short, we predicted the 3D structures of all UGTs, manually analyzed their structures and sequences, and identified amino acid residues with potential functional relevance (Figure 2). We then performed in silico mutagenesis of these UGTs and analyzed their kinetic constants, ligand interactions, and ligand specificities using different deep-learning models (Figure 2). To validate our strategy, we first examined experimentally proven amino acid sites known to have functional relevance [9].

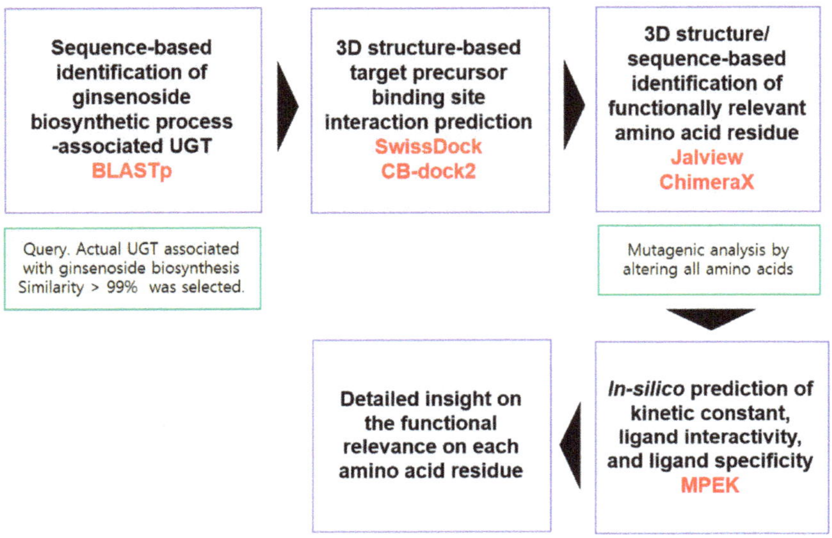

Figure 2. In-silico procedure to predict the specific functionality of each amino acid residue.

A previous study identified seven amino acids that can alter UGT functionality [9]. These residues were reported to affect ligand specificity or interaction, but the available data did not include information on catalytic activity [9]. The majority of these residues were located near the ligand-interacting helix (A10, I13, H82, A142, H144, and L185 (Figure 3A). SwissDock was used to evaluate the relative interaction between each UGT and either PPD or PPT (Figure 3B). Many mutant UGTs showed moderately altered ligand affinity, whereas mutations at H82 or A142 caused significant changes (> 25%) in binding affinity to PPD, and mutations at I13, H82, and A142 significantly affected binding affinity to PPT (Figure 3B).

We analyzed the ligand interaction angle using CB-Dock2 and found that the majority of mutants altered the interacting angle with PPD or PPT (Figure 3C,D). Catalytic activity was then assessed using MPEK, which revealed that these mutations modulated catalytic activity (Figure 3E,F). However, none of these mutations produced a predicted change in catalytic activity greater than 25% (Figure 3E,F). These results are consistent with experimental findings and confirm the validity of our pipeline, allowing us to proceed with the identification of novel amino acid residues important for the detailed functionality of UGTs.

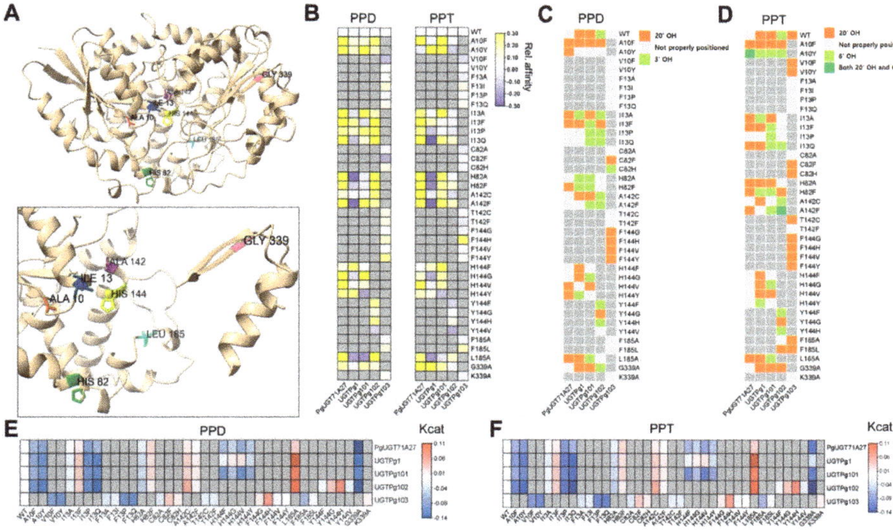

Figure 3. In-silico identification of detailed functional contribution of each amino acid residue. (**A**) Representative visualization of functional amino acid residues in 3D structure. UGTPg1 was selected to visualize physical locations of amino acids. (**B**) Ligand interactivity analysis of each UGT. Relative ligand affinity was visualized by normalizing affinity with Z-score by setting affinity of wild-type UGT as 0. Vertical axis indicates amino acid mutant variants of each UGT annotated in horizontal axis. (**C**,**D**) Ligand interacting angle was visualized with heatmap. By manually analyzing UGT–ligand structure, we determined directionality of ligands. (**C**) Ligand interacting angle of PPD with each UGT. (**D**) Ligand interacting angle of PPT with each UGT. (**E**,**F**) Predicted catalytic activity of all analyzed UGTs. Kcat value was normalized into $\log_2$(wild-type centered Z-score) and visualized with heatmap. (**E**) Catalytic activity of UGTs with PPD were visualized. (**F**) Catalytic activity of UGTs with PPT were visualized.

3.3. Eight Amino Acid Residues with Functional Relevance

We manually analyzed the structures and sequences of all 27 UGTs and curated eight putative amino acid residues based on three criteria: (1) residues showing sequence variation among the 27 UGTs; (2) residues located within ligand-interacting regions; and (3) residues that have not been previously reported (Figure 4A). Many of these residues were located near UDP-sugar binding socket regions (positions 176, 224, 397, and 398), while others were dispersed throughout the structure (Figure 4A). To determine the detailed functional contributions of these amino acids in the ginsenoside biosynthetic process, we generated mutant versions of these residues and analyzed them using our established pipeline (Figures 2 and 4B). First, we assessed the relative affinity of each UGT for PPD or PPT and observed the differential effects of amino acid residues on PPD or PPT affinity (Figure 4B). Notably, almost all amino acid positions altered affinity toward PPD and PPT by at least 25% in at least one enzyme (Figure 4B).

All eight amino acids were capable of altering the ligand interaction orientation to some extent (Figure 5A,B). Specifically, ALE15279 and WPX61740 were unaffected by any of the mutations analyzed in this study (Figure 5A,B). In addition, five UGTs (ART66192, UMX47353, UMX47354, UMX47355, and UMX47356) interacted with PPD and PPT in orientations where sugar attachment was impossible, regardless of mutation—for example, in many cases the ligand was buried inside amino acid regions, as shown by superposition analyses (Figure 5A,B).

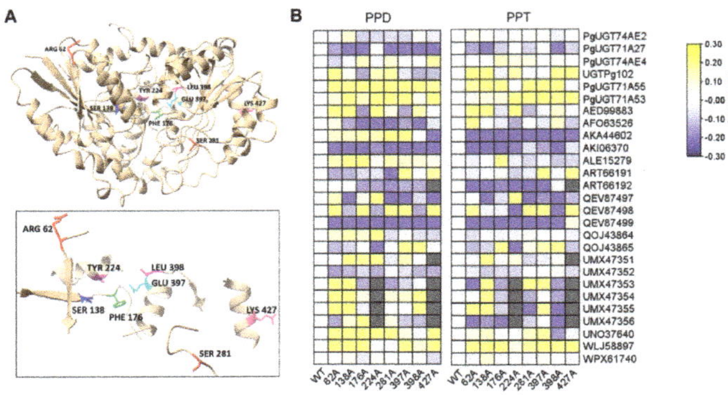

Figure 4. In-silico identification of relative affinity with PPD or PPT of noble amino acid residues. (**A**) Representative visualization of functional amino acid residues in 3D structure. PgUGT71A53 was selected to visualize physical location of amino acids. (**B**) Ligand interactivity analysis of each UGT. Relative ligand affinity was visualized by normalizing affinity with Z-score by setting affinity of wild-type UGT as 0. Horizontal axis indicates amino acid mutant variants of each UGTs annotated in vertical axis.

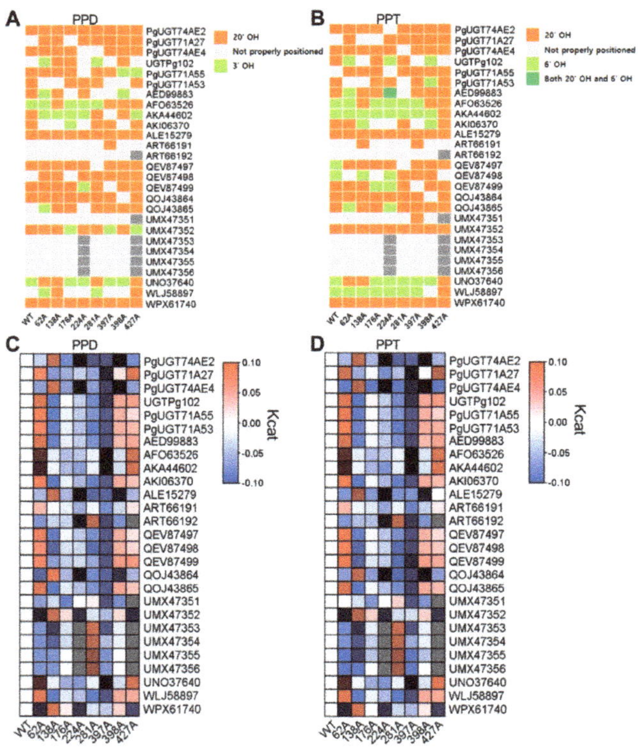

Figure 5. In-silico identification of detailed functional contribution of noble amino acid residues. (**A**,**B**) Ligand interacting angle was visualized with heatmap. By manually analyzing UGT–ligand structure, we determined directionality of ligands. (**A**) Ligand interacting angle of PPD with each UGT. (**B**) Ligand interacting angle of PPT with each UGT. (**C**,**D**) Predicted catalytic activity of all analyzed UGTs. Kcat value was normalized into $\log_2$(wild-type centered Z-score) and visualized with heatmap. (**C**) Catalytic activity of UGTs with PPD were visualized. (**D**) Catalytic activity of UGTs with PPT were visualized.

Next, we analyzed the catalytic activities of all mutant UGTs with PPD or PPT (Figure 5C,D). Catalytic activity varied among mutant variants for the eight amino acid residues, and four positions (62, 224, 397, and 398) were found to modulate activity by more than 25% (Figure 5C,D). Among these, position 62 acted as an inhibitory site for catalytic activity, whereas the others enhanced the catalytic activity of UGTs (Figure 5C,D).

In summary, we developed an easy-to-use pipeline for assessing the functional contributions of amino acid residues in plant enzymes (Figure 2). Notably, the confidence scores in regions near three functional sites and ligand-binding pockets exceeded 0.9 (Figure S1). Thus, we identified three promising amino acid positions that can determine ligand affinity, the ligand interaction angle, and reaction kinetics (Figure 6).

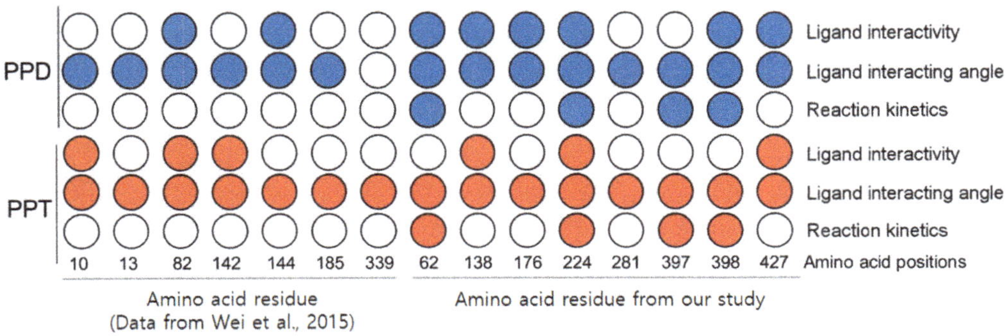

Figure 6. Summary of this study. From left, seven amino acids revealed from previous study [9]; eight amino acids on right side were analyzed in this study.

4. Discussion

Our study took advantage of an in-silico pipeline to analyze the specific functional roles of individual amino acid residues in plant enzymes (Figure 2). Through our approach, we identified functionally relevant amino acid residues in UGTs associated with ginsenoside biosynthesis. Specifically, we confirmed that amino acids determining ligand specificity were also predicted, in our pipeline, to influence ligand interactivity and interaction angles [9] (Figure 6). We then revealed eight putative functionally relevant amino acids and applied our pipeline to them (Figures 2, 4 and 5). As a result, we discovered three amino acid residues that are conserved among 27 UGTs and alter all aspects of UGT function, including kinetics (Figure 6). These three residues have potential value for future industrial applications.

Ginsenoside biosynthetic and modification processes involve not only UGTs, but also other enzyme classes, such as cellulases. For example, cellulase KN was previously shown to catalyze the conversion of ginsenoside Re into F1 [11]. Specifically, the reaction proceeded via ginsenoside Rg1 as an intermediate [11]. This insight suggests an approach similar to ours could be applied to cellulase family members to identify functionally relevant amino acids in cellulase KN (Figure 2). However, the amino acid sequence of cellulase KN is currently unavailable. Therefore, it is necessary to search public databases for cellulase proteins predicted to interact with ginsenosides and exhibit catalytic activity.

A mutant form of glycosyltransferase BS-YjiC was shown to catalyze the conversion of PPD into ginsenoside F12 and Rh2 (US12134789B2). Mutations at positions 125, 178, and 313 were tested, and the most stable variants were double mutants K125I/N178I and K125I/P313W (US12134789B2). The major difference between our analysis and this patent is that catalytic activity in the patented mutants increased due to enhanced stability, whereas our analysis predicted increased kcat values (Figure 6). This study adapted an *insilico* methodology from our ongoing research aimed at identifying novel plant enzymes associated with various enzymatic reactions (Jung et al., in revision). Although the core pipeline is the same, the present study included targeted amino acid selection and mutagenic analysis. The increased enzyme activity predicted in our analysis could also be partially explained by enhanced stability during the reaction process, although other possibilities cannot be excluded. One plausible explanation for increased enzyme activity is that these altered amino acids may improve the flexibility of the catalytic domain, thereby accelerating the reaction. However, the exact reason for these increased kcat values remains to be elucidated, as our pipeline cannot predict it.

UGT109A1 from *Bacillus subtilis* was shown to catalyze glycosylation at the C3, C12, and C20 positions rather than the C3, C6, and C20 positions [12]. UGT109A1 could recognize dammarendiol-II, PPD, and PPT as substrates and convert them [12]. It produced novel ginsenosides such as 3β,12β-Di-O-Glc-PPD from PPD and 3β,12β-Di-O-Glc-PPT from PPT [12]. In cell line experiments, 3β,12β-Di-O-Glc-PPD demonstrated higher anti-lung cancer activity than Rg3 [12]. Considering these findings, our predictions of "not properly located" ligand orientations may need to be revisited (Figures 3C,D and 5A,B). Nevertheless, the majority of these orientations severely disrupted the UDP-sugar binding pockets and failed to expose hydroxyl groups toward the UDP-sugar binding site. However, a few orientations could potentially produce alternative products (Figures 3C,D and 5A,B). Although our primary focus was the two reactions of interest, these additional predictions could be investigated in future studies.

Supplementary Materials: The following supporting information can be downloaded at: https://www.mdpi.com/article/10.3390/biochem5040036/s1, Figure S1: Confidence map of 3D structures predicted from two representative UGTs. Red box indicates the ligand binding amino acid and nearby region. Yellow arrow indicates the identified functional residue sites. A. Confidence map drawn with PgUGT71A27. B. Confidence map drawn with PgUGT74AE2; Figure S2: Structure comparison between alphafold3-predicted 3D structure from PgUGT71A27 and experimentally determined 3D structure of MtUGT71G1 (PDB:2ACW). RMSD was annotated to quantitatively show the similarity. Only monomer structure was aligned and shown.

Author Contributions: Conceptualization, J.K.; methodology, K.J.; software, J.K.; validation, K.J., N.K. and C.P.; formal analysis, K.J.; investigation, J.K.; data curation, K.J.; writing—original draft preparation, J.K.; writing—review and editing, J.K.; visualization, K.J. and J.K.; supervision, J.K. All authors have read and agreed to the published version of the manuscript.

Funding: This research was supported by the Korea National University of Education.

Data Availability Statement: Data are available upon request.

Conflicts of Interest: The authors declare no conflicts of interest.

Abbreviations

The following abbreviations are used in this manuscript:

UGT	UDP-Glycosyltransferase
PPD	Protopanaxadiol
PPT	Protopanaxatriol

References

1. Leung, K.W.; Wong, A.S. Pharmacology of ginsenosides: A literature review. *Chin. Med.* **2010**, *5*, 20. [CrossRef]
2. Hou, M.; Wang, R.; Zhao, S.; Wang, Z. Ginsenosides in Panax genus and their biosynthesis. *Acta Pharm. Sin. B* **2021**, *11*, 1813–1834. [CrossRef]

3. Miao, L.; Yang, Y.; Li, Z.; Fang, Z.; Zhang, Y.; Han, C.-C. Ginsenoside Rb2: A review of pharmacokinetics and pharmacological effects. *J. Ginseng Res.* **2022**, *46*, 206–213. [CrossRef] [PubMed]
4. Huang, W.-C.; Huang, T.-H.; Yeh, K.-W.; Chen, Y.-L.; Shen, S.-C.; Liou, C.-J. Ginsenoside Rg_3 ameliorates allergic airway inflammation and oxidative stress in mice. *J. Ginseng Res.* **2021**, *45*, 654–664. [CrossRef] [PubMed]
5. Wang, Y.; Mou, C.; Hu, Y.; He, Z.; Cho, J.Y.; Kim, J.H. In vivo metabolism, pharmacokinetics, and pharmacological activities of ginsenosides from ginseng. *J. Ginseng Res.* **2025**, *49*, 479–487. [CrossRef]
6. Li, W.; Zheng, L.; Ma, X.; Xia, J.; Sheng, J.; Ge, P.; Yuan, Y.; Fan, Y.; Zhou, Y. The sugar moiety in protopanaxadiol ginsenoside affects its ability to target glucocorticoid receptor to regulate lipid metabolism. *Bioorganic Chem.* **2024**, *153*, 107885. [CrossRef]
7. Chen, R.J.Y.; Chung, T.-Y.; Li, F.-Y.; Lin, N.-H.; Tzen, J.T.C. Effect of sugar positions in ginsenosides and their inhibitory potency on Na+/K+-ATPase activity. *Acta Pharmacol. Sin.* **2009**, *30*, 61–69. [CrossRef]
8. Yan, X.; Fan, Y.; Wei, W.; Wang, P.; Liu, Q.; Wei, Y.; Zhang, L.; Zhao, G.; Yue, J.; Zhou, Z. Production of bioactive ginsenoside compound K in metabolically engineered yeast. *Cell Res.* **2014**, *24*, 770–773. [CrossRef]
9. Wei, W.; Wang, P.; Wei, Y.; Liu, Q.; Yang, C.; Zhao, G.; Yue, J.; Yan, X.; Zhou, Z. Characterization of Panax ginseng UDP-Glycosyltransferases Catalyzing Protopanaxatriol and Biosyntheses of Bioactive Ginsenosides F1 and Rh1 in Metabolically Engineered Yeasts. *Mol. Plant* **2015**, *8*, 1412–1424. [CrossRef]
10. Wang, P.; Wei, W.; Ye, W.; Li, X.; Zhao, W.; Yang, C.; Li, C.; Yan, X.; Zhou, Z. Synthesizing ginsenoside Rh2 in Saccharomyces cerevisiae cell factory at high-efficiency. *Cell Discov.* **2019**, *5*, 5. [CrossRef]
11. Wang, Y.; Choi, K.-D.; Yu, H.; Jin, F.; Im, W.-T. Production of ginsenoside F1 using commercial enzyme Cellulase KN. *J. Ginseng Res.* **2016**, *40*, 121–126. [CrossRef]
12. Liang, H.; Hu, Z.; Zhang, T.; Gong, T.; Chen, J.; Zhu, P.; Li, Y.; Yang, J. Production of a bioactive unnatural ginsenoside by metabolically engineered yeasts based on a new UDP-glycosyltransferase from Bacillus subtilis. *Metab. Eng.* **2017**, *44*, 60–69. [CrossRef] [PubMed]
13. Altschul, S.F.; Gish, W.; Miller, W.; Myers, E.W.; Lipman, D.J. Basic local alignment search tool. *J. Mol. Biol.* **1990**, *215*, 403–410. [CrossRef] [PubMed]
14. Camacho, C.; Coulouris, G.; Avagyan, V.; Ma, N.; Papadopoulos, J.; Bealer, K.; Madden, T.L. BLAST+: Architecture and applications. *BMC Bioinform.* **2009**, *10*, 421. [CrossRef] [PubMed]
15. Edgar, R.C. MUSCLE: Multiple sequence alignment with high accuracy and high throughput. *Nucleic Acids Res.* **2004**, *32*, 1792–1797. [CrossRef]
16. Waterhouse, A.M.; Procter, J.B.; Martin, D.M.A.; Clamp, M.; Barton, G.J. Jalview Version 2—A multiple sequence alignment editor and analysis workbench. *Bioinformatics* **2009**, *25*, 1189–1191. [CrossRef]
17. Abramson, J.; Adler, J.; Dunger, J.; Evans, R.; Green, T.; Pritzel, A.; Ronneberger, O.; Willmore, L.; Ballard, A.J.; Bambrick, J.; et al. Accurate structure prediction of biomolecular interactions with AlphaFold 3. *Nature* **2024**, *630*, 493–500. [CrossRef]
18. Grosdidier, A.; Zoete, V.; Michielin, O. SwissDock, a protein-small molecule docking web service based on EADock DSS. *Nucleic Acids Res.* **2011**, *39*, W270–W277. [CrossRef]
19. Bugnon, M.; Röhrig, U.F.; Goullieux, M.; Perez, M.A.; Daina, A.; Michielin, O.; Zoete, V. SwissDock 2024: Major enhancements for small-molecule docking with Attracting Cavities and AutoDock Vina. *Nucleic Acids Res.* **2024**, *52*, W324–W332. [CrossRef]
20. Liu, Y.; Yang, X.; Gan, J.; Chen, S.; Xiao, Z.-X.; Cao, Y. CB-Dock2: Improved protein–ligand blind docking by integrating cavity detection, docking and homologous template fitting. *Nucleic Acids Res.* **2022**, *50*, W159–W164. [CrossRef]
21. Wang, J.; Yang, Z.; Chen, C.; Yao, G.; Wan, X.; Bao, S.; Ding, J.; Wang, L.; Jiang, H. MPEK: A multitask deep learning framework based on pretrained language models for enzymatic reaction kinetic parameters prediction. *Brief. Bioinform.* **2024**, *25*, bbae387. [CrossRef]
22. Meng, E.C.; Goddard, T.D.; Pettersen, E.F.; Couch, G.S.; Pearson, Z.J.; Morris, J.H.; Ferrin, T.E. UCSF ChimeraX: Tools for structure building and analysis. *Protein Sci.* **2023**, *32*, e4792. [CrossRef]
23. Chen, C.; Chen, H.; Zhang, Y.; Thomas, H.R.; Frank, M.H.; He, Y.; Xia, R. TBtools: An Integrative Toolkit Developed for Interactive Analyses of Big Biological Data. *Mol. Plant* **2020**, *13*, 1194–1202. [CrossRef]

BioChem

MDPI

Review

RASopathy and Sudden Cardiac Death: A Literature Review

Cecilia Salzillo [1,2,*] and Andrea Marzullo [2,*]

[1] Department of Experimental Medicine, PhD Course in Public Health, University of Campania "Luigi Vanvitelli", 80138 Naples, Italy
[2] Department of Precision and Regenerative Medicine and Ionian Area, Pathology Section, University of Bari "Aldo Moro", 70121 Bari, Italy
* Correspondence: cecilia.salzillo@uniba.it (C.S.); andrea.marzullo@uniba.it (A.M.)

Abstract

RASopathies are a heterogeneous group of genetic syndromes caused by germline mutations in genes encoding proteins of the RAS/MAPK pathway, which are essential in the regulation of cell proliferation, differentiation and survival. Although characterized by common phenotypic manifestations such as craniofacial dysmorphism, congenital heart defects, and growth retardation, an aspect of great clinical relevance is the increased risk of sudden cardiac death, especially in relation to hypertrophic cardiomyopathy (HCM) and ventricular arrhythmias. Pathogenic variants in genes such as RAF1, RIT1, PTPN11, BRAF and SHOC2 have been associated with phenotypes with increased incidence of HCM, sometimes with early onset and a rapidly evolving course. The literature highlights the importance of early identification of patients at risk; however, specific surveillance protocols and follow-up strategies are defined in expert guidelines. This literature review aims to provide an updated overview of the main RASopathies with cardiac involvement, highlighting the genotype-phenotype correlations, the pathogenic mechanisms underlying sudden cardiac death, and current diagnosis, monitoring, and prevention strategies. The aim is to promote greater clinical awareness and encourage a multidisciplinary approach aimed at reducing mortality in these rare genetic conditions.

Keywords: RASopathy; RASopathies; sudden cardiac death; sudden death

1. Introduction

RASopathies are rare genetic syndromes caused by germline mutations affecting genes involved in the Ras/mitogen-activated protein kinase (MAPK) pathway, resulting in increased signaling through this pathway [1]. In particular, the RAS/MAPK pathway is involved in the signal transduction cascade of cellular processes such as division, proliferation, differentiation and migration [2].

RASopathies include a group of rare but highly impactful conditions such as Noonan syndrome, Noonan syndrome with multiple lentigines, cardiofaciocutaneous syndrome, Costello syndrome, Mazzanti syndrome, neurofibromatosis type 1 and Legius syndrome [3,4]. These disorders present a wide range of clinical manifestations, including growth abnormalities, musculoskeletal alterations, cardiovascular diseases, craniofacial anomalies, cognitive impairment, renal malformations, coagulation disorders, up to a high susceptibility to tumors [3].

The estimated prevalence of RASopathies ranges from 1 in 1000 to 1 in 2500 live births, these figures are derived from limited cohorts and may underestimate the real prevalence in the general population [5] and are characterized by high genetic heterogeneity with the same condition being caused by several genes belonging to the RAS-MAPK cascade Figure 1. Currently, approximately 20 genes are known whose pathogenic variants are associated with this group of conditions, PTPN11, SOS1, SOS2, NRAS, KRAS, MRAS, RRAS2, RIT1, LZTR1, RAF1, MAP2K1, MAP2K2, SPRED2, BRAF, SPRED1, NF1, SCHOC2, HRAS, PPP1CB and CBL [4,6,7].

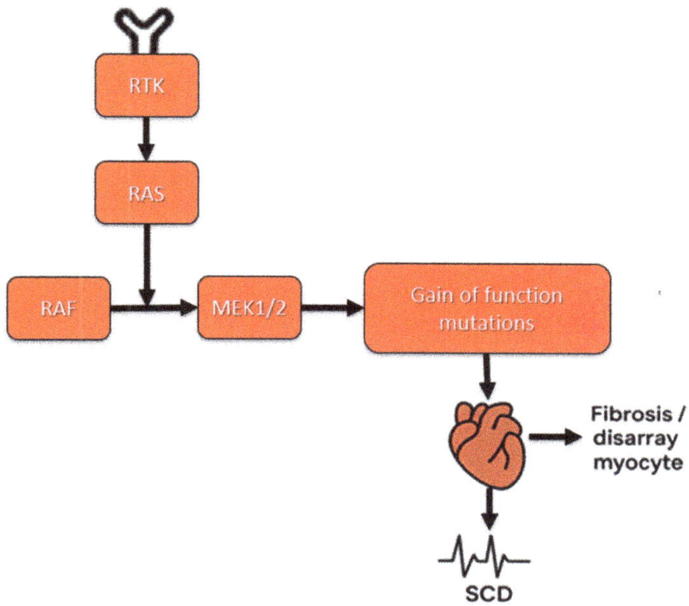

Figure 1. RAS/MAPK Pathway.

The genotype-phenotype correlation is not constant; however, some specific gene associations are described for certain unique clinical features, even though many of the RASopathies have overlapping clinical features such as short stature, facial dysmorphism, congenital heart disease, lymphatic dysfunction and intellectual disability [7].

Specifically, RASopathies-related heart defects include congenital heart disease (CHD), hypertrophic cardiomyopathy (HCM), dilated cardiomyopathy (DCM) [1,2,5] and consequently sudden cardiac death (SCD) [8–13].

Table 1 summarizes the main genes.

Table 1. Genes and cardiac phenotypes in RASopathies.

Gene	Associated Syndrome	Predominant Cardiac Phenotype	Risk of SCD
PTPN11	Noonan syndrome, Noonan syndrome with multiple lentigines	Pulmonary stenosis, less frequent HCM, septal defects	Moderate
RAF1	Noonan syndrome	Early/severe HCM, LVOT obstruction, arrhythmias	High
RIT1	Noonan syndrome	Early HCM, congenital coronary anomalies	High
BRAF	Cardiofaciocutaneous syndrome	HCM + combined defects (ASD, PVS), rapid progression	High
SHOC2	Mazzanti syndrome	Pediatric HCM → arrhythmic risk	Moderate

Table 1. *Cont.*

Gene	Associated Syndrome	Predominant Cardiac Phenotype	Risk of SCD
HRAS	Costello syndrome	HCM (34–61%), arrhythmias (33–48%)	High
NF1	Neurofibromatosis type 1	Congenital malformations, vasculopathy, rare HCM	Low–Moderate
SPRED1	Legius syndrome	Rare valvular anomalies, no documented HCM	Low

Lynch et al., in a multicenter cohort, reported a 10-year cumulative incidence of SCD of 4.7% in patients with RAS-HCM (vs. 4.2% in pediatric controls with non-RAS HCM), with higher non-arrhythmic/transplant mortality in RAS-HCM subjects. This implies an approximate annualized rate of around ~0.4–0.5%/year in observed follow-up, but the incidence varies with age, genotype, and severity [12].

Furthermore, Boleti et al. demonstrated that HCM in RASopathies often develops very early in life, with cases reported in infancy and even the neonatal period, underlining the importance of initiating surveillance as soon as the genetic diagnosis is established [13].

Although many patients present with the disease in early childhood, HCM can also develop later, during adolescence or early adulthood. Therefore, surveillance should continue into young adulthood, as the risk of SCD extends beyond childhood in genetically predisposed individuals [10].

In non-syndromic sarcomeric HCM, the annual incidence of SCD has varied across time periods and populations. Recent reviews estimate rates in the range of ~0.3–0.7% per year in unselected cohorts, while contemporary data from specialized centers suggest lower event rates, largely reflecting advances in risk stratification and the prophylactic use of ICDs [8,9].

Pediatric data from validation studies of the HCM Risk-Kids model and other cohort analyses indicate rates of SCD-equivalent arrhythmic events of approximately 0.7–0.8 per 100 patient-years. These findings underscore that children with HCM experience higher rates of SCD-equivalent events compared with adults [11].

2. Materials and Methods

This article is a literature review. The literature was searched in PubMed/MEDLINE and Scopus using a combination of the following keywords: "RASopathies", "hypertrophic cardiomyopathy", "sudden cardiac death", "pediatric cardiology", and "genetics." Only articles in English and open access were considered.

3. Noonan Syndrome

Noonan syndrome (NS) is an inherited genetic disease of the RASopathies group characterized by heterogeneous phenotypic manifestations that can vary from mild to severe [14].

NS can be inherited with autosomal dominant transmission or arise spontaneously as a de novo mutation, it affects both sexes equally, the prevalence is not entirely known, but the incidence is estimated from 1 in 1000 to 1 in 2500 live births but is probably underestimated due to deaths in the fetal period [15–17].

NS is caused in 50% of cases by pathogenic variants of protein-tyrosine phosphatase non-receptor-like 11 (PTPN11), which encodes SHP-2, and by pathogenic variants in the genes SOS1 in 13%, RAF1 in 10% and RIT1 in 9% [15,18,19]. Other genes have been reported to cause SN in less than 5% of cases, including SHOC2, RASA2, LZTR1, SPRED2, SOS2, CBL, KRAS, NRAS, MRAS, PRAS, BRAF, PPP1CB, A2ML1, MAP2K1, and CDC42 [15,16,18,19].

Some genotype-phenotype correlations are well known, mutations in PTPN11 favor pulmonary stenosis, while those in RAF1 and RIT1 are strongly associated with HCM [19,20].

Clinically, patients with NS frequently present with short stature, facial dysmorphism such as hypertelorism, ptosis, often low-set and rotated ears, skeletal malformations such as pectus carinatum or excavatum, neurological difficulties and coagulation disorders [14].

Cardiac involvement is one of the most relevant manifestations, in fact 50–80% of cases present congenital heart disease, in particular pulmonary valve stenosis in 50–65%, HCM in 20–30% and atrial and ventricular septal defects. NS-related HCM presents early, already in the first months of life, with a higher prevalence of outflow tract obstruction and diastolic dysfunction compared to classic sarcomeric forms [21–23].

Histologically, NS shows a myocellular disarray like idiopathic HCM, suggesting a high predisposition to ventricular dysfunction and arrhythmias [24].

The diagnostic process requires complex clinical and instrumental attention, in particular, an echocardiogram and electrocardiogram (ECG) are recommended from birth with typical signs such as left axis deviation, wide QRS complexes, and abnormal Q waves, even in the absence of cardiac symptoms and must be repeated periodically approximately every 2–5 years. Genetic confirmation is essential, allowing for targeted therapy [18,25].

Cardiovascular treatment ranges from surgical or percutaneous intervention for pulmonary stenosis, to medical therapy with beta-blockers and, in severe cases of symptomatic HCM, to the experimental use of MEK inhibitors [20].

The association of NS with SCD is closely related to the presence of HCM, which has been identified as a significant risk factor for SCD, especially in pediatric patients. The prognosis for these patients can be particularly severe, and the mortality associated with hypertrophic cardiomyopathy in patients with RASopathies is higher than that in the general population [6,12].

Recent case series highlight cases of sudden death associated with main coronary artery atresia in subjects with the RIT1 variant, as well as familial cases discovered with molecular autopsies identifying pathogenic mutations in LZTR1 [22].

Patients with specific genetic variants such as PTPN11 and RAF1 can have greater left ventricular hypertrophy and higher left ventricular outflow tract gradients, factors related to the risk of adverse cardiac events. Predictive factors for SCD include nonsustained ventricular tachycardia, congestive heart failure, and a high LVOT gradient, which indicate a higher likelihood of fatal arrhythmias [26,27].

This evidence underscores the need for very close electrocardiographic monitoring, early consideration of implantable defibrillators in patients with risk factors such as severe obstruction or tachyarrhythmias, and postmortem genetic screening in sudden deaths to prevent family members from dying.

Take-Home Messages
• HCM present in 20–30% of patients, with early onset and a more severe course than sarcomeric forms
• RAF1 and RIT1 mutations are strongly associated with HCM and an increased risk of SCD
• ECG and echocardiogram recommended from birth and repeated regularly
• Risk factors for SCD: syncope, NSVT, LVOT $\geq$ 30 mmHg
• According to international guidelines, ICD implantation should be evaluated in selected high-risk patients within specialized centers.

4. Noonan Syndrome with Multiple Lentigines

Noonan syndrome with multiple lentigines (NSML), formerly known as Leopard syndrome, is a rare autosomal dominant RASopathy, high penetrance, variable expressivity, with approximately 200–300 cases described in the literature and without precise incidence rates [28–31].

In 85% of cases, missense mutations in the PTPN11 gene, specifically exons 7, 12, 13, cause a dominant-negative effect with reduced SHP-2 activity, distinct from the gain-of-function variants typical of NS [30,32].

The most frequent pathogenic variants in 65% of cases are Tyr279Cys [33] and Thr468Met [34], while in 15% in RAF1, BRAF, MAP2K1 and, rarely, in other genes of the RAS/MAPK pathway [29,35].

Clinically, NSML manifests itself with a variable involvement of different organs and systems, making the diagnostic picture complex and multidisciplinary.

One of the most characteristic signs is the appearance of multiple lentigines, which begin to develop in preschool age and then progressively increase throughout childhood and adolescence. Freckles preferentially affect photoexposed areas, but can extend to the oral and genital mucosa, and even to the sclera. They are often associated with café-au-lait spots, and the pigmented lesions can be either epidermal or melanocytic, giving rise to a complex and easily recognizable dermatological phenotype [36].

The facial appearance is like that of NS, but facial dysmorphism tends to be less marked than in other RASopathies, especially in adulthood.

Cardiac involvement is present in approximately 85% of patients and represents the most serious clinical manifestation. Specifically, HCM occurs in 70–80% of patients, often with early onset, with an asymmetric and obstructive pattern, and carries a high risk of ventricular arrhythmias and consequently SCD [11,12]. Less frequently, pulmonary valve stenosis is observed in 25%. Furthermore, up to 75% of patients have electrical conduction abnormalities, including bundle branch blocks, bradyarrhythmias or tachyarrhythmias, detectable on ECG or Holter monitoring [30].

Other systemic manifestations include sensorineural hearing, cryptorchidism, hypospadias, and, more rarely, renal anomalies. Growth may be impaired in less than half of cases, with short stature and postnatal growth retardation, while cognitive involvement is generally mild, with intellectual disability [31].

The clinical diagnosis is based on the acronym LEOPARD, which represents the main characteristics of the syndrome: Lentigines, Electrocardiographic conduction abnormalities, Ocular hypertelorism, Pulmonary stenosis, Abnormalities of genitalia, Retardation of growth, and Deafness. The diagnostic criteria proposed by Voron et al. in 1976 [37] suggest that the presence of lentigines associated with at least two other manifestations or the presence of three clinical signs even in the absence of lentigines can be considered sufficient to raise a diagnostic suspicion, especially in the absence of family history. If there is an affected family member with a confirmed diagnosis, even a smaller number of signs may be considered diagnostically relevant [38,39].

Genetic confirmation is obtained by molecular analysis of the PTPN11 gene, but in subjects with a negative test but a suggestive phenotype, it is indicated to extend the study to other genes [40].

Regular follow-up with echocardiogram, ECG, 24-h Holter monitoring, and, if necessary, exercise testing is recommended for cardiac manifestations, every 6–12 months in high-risk or progressive cases [30].

Therapeutically, management of HCM may include pharmacological treatment with beta-blockers or calcium channel blockers, and in cases of significant obstruction, surgical myoreduction. In patients at high risk of SCD, preventative ICD implantation may be necessary. Pulmonic stenosis may require valvotomy or valve plasty, while sensorineural hearing loss should be treated with hearing aids and careful audiological follow-up [30].

Skin manifestations are treated with laser or depigmenting creams, and careful dermatoscopic surveillance is recommended, given the rare possibility of melanomatous transformation [30].

The risk of SCD is a serious complication of NSML, and is closely associated with HCM, which is frequently asymmetric, obstructive and of early onset, and in a relevant percentage of cases with ventricular arrhythmias, atrioventricular block and conduction alterations, which contribute significantly to the increased arrhythmic risk [11,12].

SCD has been highlighted in patients carrying specific high-risk genetic variants, in particular Tyr279Cys and Thr468Met in PTPN11, as well as in RAF1 mutations, suggesting a clear genotype-phenotype correlation that should guide risk stratification [41].

Risk stratification should be repeated periodically during developmental age, considering the natural progression of the disease and the possible late appearance of clinical risk criteria. Therefore, regular, multidisciplinary cardiac follow-up is essential to prevent avoidable fatal outcomes in patients with NSML.

Take-Home Messages
• HCM present in 70–80% of patients, often asymmetric and obstructive
• High prevalence of conduction abnormalities: AV block, brady/tachyarrhythmias
• High-risk variants: PTPN11 (Tyr279Cys, Thr468Met), RAF1
• High risk of SCD, even at a young age
• Regular follow-up with ECG, Holter monitoring and echocardiogram is advised by expert consensus, with ICD evaluation considered only in high-risk cases.

5. Cardiofaciocutaneous Syndrome

Cardiofaciocutaneous syndrome (CFCS or CFC) is a rare autosomal dominant RASopathy, characterized by a violaceous phenotype and multisystem involvement, and is caused by pathogenic mutations in key genes of the RAS/MAPK pathway, such as BRAF, MAP2K1 (MEK1), MAP2K2 (MEK2) and, more rarely, KRAS [42–44].

Mutations are predominantly missense or small deletions triggering an increase in activity of the respective enzymes of the RAS/MAPK cascade. The most frequent variants include BRAF p.Gln257Arg, MAP2K1 p.Tyr130Cys and MAP2K2 p.Tyr134Cys, and in rare cases a mutation in YWHAZ has also been identified [42].

Approximately 300 cases have been described in the literature, and the real incidence is still unknown [45].

Clinically, CFCS is characterized by multiple congenital anomalies including cardiac defects, ectodermal abnormalities, distinctive craniofacial features, and varying degrees of intellectual disability [46,47].

CHD is present in 75% of cases, with predominantly pulmonary stenosis and HCM, and in 45% of cases with ASD + PVS type defects [48].

The diagnosis of CFCs is based on the integration of clinical and molecular data. However, due to the phenotypic overlap with other RASopathies such as NS and Costello, it is essential to confirm the suspected diagnosis through targeted genetic testing for pathogenic variants in the key genes associated with SCFC [49,50].

Furthermore, regular cardiac follow-up, including echocardiograms, ECGs, and, in case of documented abnormalities, dynamic Holter monitoring and exercise testing, is essential and is particularly important in patients with HCM, given its association with arrhythmias and risk of SCD.

Currently, there is no specific therapy for CFCS, and treatment is essentially symptomatic and multidisciplinary, with beta-blockers or calcium channel blockers to control HCM. When the obstruction is significant, surgical muscle reduction is indicated, and in the most severe and refractory cases, heart transplantation.

HCM associated with CFCS can evolve rapidly, especially in childhood, with disorganized myocytes and possible inherited coronary artery abnormalities. SCD due to ventricular fibrillation is documented, with signs of severe HCM post-mortem [13].

Although there are no specific prognostic models, HCM markers represent valid risk factors for SCD similarly to what is observed in related RASopathies.

Take-Home Messages
• Heart defects in 75%: pulmonic stenosis and HCM
• Most common mutations: BRAF, MAP2K1, MAP2K2
• HCM can progress rapidly and be complicated by ventricular arrhythmias
• Documented risk of SCD; Intensive monitoring is recommended.
• Symptomatic treatment with beta-blockers, calcium channel blockers, and surgical options in severe cases.

6. Mazzanti Syndrome

Mazzanti syndrome (MS), also known as Noonan-like syndrome with loose anagen hair, is a rare autosomal dominant RASopathy caused by gain-of-function mutations in the SHOC2 gene, including the most frequent c.4A > G, p.Ser2Gly, which introduces an aberrant N-myrostylation site, altering the protein localization on the plasma membrane and amplifying the RAS/MAPK signal [51,52].

Other very rare pathogenic variants such as p.Thr411Ala have been described recently, potentially related to more heterogeneous clinical manifestations [53,54].

Clinically it is characterized by suggestive facial features such as hypertelorism, ptosis, low-set ears; reduced growth often with growth hormone deficiency and short stature and slow psychomotor development; sparse and easily extractable hair; skin abnormalities such as keratosis pilaris, hyperpigmentation, and eczema; ligamentous hyperlaxity and HCM [55,56].

The diagnostic-therapeutic process includes a complete clinical/dysmorphological evaluation; SHOC2 gene sequencing, particularly when the most common mutations PTPN11, SOS1, RAF1, KRAS, BRAF, MAP2K1/2 are negative; multidisciplinary management with growth monitoring, often with GH hormone therapy which has shown significant benefits in several clinical cases, even in the absence of GH deficiency; neuropsychological support; dermatological surveillance; rigorous cardiac evaluation with serial ECG and echocardiogram [54,57,58].

In the absence of specific reports of SCD in MS, the presence of HCM at the pediatric level is recognized as a significant risk factor for ventricular arrhythmias and SCD in patients with RASopathies [11]. Therefore, MS requires proactive cardiac follow-up with Holter monitoring and possibly electrophysiological studies in high-risk cases.

Take-Home Messages
• Rare syndrome with a characteristic phenotype, such as loose hair, dysmorphic features, and short stature
• Presence of HCM in childhood → potential risk of SCD
• Typical mutation: SHOC2 p.Ser2Gly
• Regular cardiac surveillance is essential, including ECG, echocardiogram, and Holter monitoring
• Consider electrophysiological studies in high-risk cases

7. Costello Syndrome

Costello syndrome (CS) is a rare autosomal dominant RASopathy caused by heterozygous germline gain-of-function variants in the HRAS gene (11p15.5), caused in 80% by the p.Gly12Ser substitution (c.34G>A) and the second most frequent is the p.Gly12Ala variant [59,60].

CS is extremely rare, with a prevalence in the United Kingdom of 1:380,000 and most cases are de novo with correlation to advanced paternal age [61,62].

Clinically, newborns frequently present with macrosomia, polyhydramnios, hypoglycemia, and severe feeding difficulties; In the following months, growth retardation, psychomotor developmental delay, intellectual disability, coarse facies with relative macrocephaly, epicanthus, prominent ears, broad-tipped nose, full lips, soft skin and joint laxity, cutaneous papillomatosis, and curly or sparse hair are observed [62,63].

Additionally, approximately 63% of patients with CS have cardiac abnormalities, including valvular heart disease such as pulmonary stenosis and septal defects in 30%, HCM in 34–61%, and arrhythmias in 33–48% [64–66].

The diagnostic-therapeutic path includes clinical evaluation with attention to dysmorphic-phenotypic traits; molecular confirmation by HRAS sequencing; multidisciplinary management with nutritional and neuropsychological support, physical and occupational therapy; pediatric oncology surveillance with screening for neuroblastoma, rhabdomyosarcoma and bladder cancer up to 8–10 years of age; rigorous cardiac monitoring, with periodic echocardiogram and ECG/Holter, and medical and/or surgical interventions in cases of obstructive HCM or symptomatic arrhythmias [59].

Regarding the association with SCD, although specific data in CS are limited, the presence of HCM and arrhythmias represents a known risk factor [11,13].

Take-Home Messages
• Mutated HRAS (e.g., p.Gly12Ser) in >80% of cases
• HCM in 34–61%, arrhythmias in 33–48%
• Increased arrhythmic risk → possible SCD, although specific data are rare
• Regular cardiac monitoring: ECG/Holter/Echo
• Multidisciplinary management, including pediatric oncology surveillance

8. Neurofibromatosis Type 1

Neurofibromatosis type 1 (NF1) is an autosomal dominant disease, but in 40% it can arise as a de novo mutation and is characterized by high phenotypic variability even within the same family [67,68].

The NF1 gene, located on chromosome 17q11.2, encodes the protein neurofibromin, a GAP protein that negatively regulates the RAS/MAPK pathway. Pathogenic variants are predominantly intragenic, such as frameshift, nonsense, and splice site, with less than 10% of cases due to whole-gene deletions (WGD), which are associated with a more severe phenotype, increased risk of malignancies and congenital heart disease [69].

NF1 has an incidence of 1 in 2600–3000 individuals [70] and with a prevalence ranging from 1 in 3000 to 6000 [67].

Clinically, NF1 typically manifests in childhood with ≥6 café-au-lait macules, axillary/inguinal freckles, cutaneous or plexiform neurofibromas, Lisch nodules, optic gliomas, bone defects such as sphenoid dysplasia, pseudoarthrosis, and cognitive delay or learning disabilities [67]. The diagnosis is based on the 1987 NIH criteria, confirmed in 1997 and updated in 2021, which require the presence of at least two cardinal features [71].

Cardiovascular manifestations are well documented, with 2% to 27% of patients having congenital heart defects, including pulmonary valve stenosis, septal defects, mitral valve disease, and in some cases HCM [69,72–74]. Recent studies on NF1 subjects without cardiovascular events highlight early alterations in endothelial function, medial carotid thickening and mild impairment of left ventricular contractile function, suggesting the need for careful cardiovascular surveillance starting from childhood [75].

The diagnostic workup includes clinical phenotypic evaluation, molecular confirmation on the NF1 gene, supported by neurological and ophthalmological imaging. Management is multidisciplinary, including neurosurgery for tumor lesions, pediatric oncology, endocrinology, neurology, and medical genetics, with new targeted therapies such as selumetinib, approved for symptomatic, inoperable plexiform neurofibromas in children ≥ 2 years of age [70,76].

Although SCD is rare, at least one case has been documented in a young NF1 patient with intramyocardial coronary vasculopathy, myocardial fibrosis and floppy mitral valve, suggesting a potentially lethal arrhythmic and vascular substrate [77]. Furthermore, cases of myocardial infarction in young NF1 adults due to coronary stenoses, aneurysms or hypertension associated with renal vasculopathy have been described [78,79]. Therefore, although SCD is uncommon, the presence of vascular disease, cardiomyopathy, or arrhythmias requires comprehensive cardiovascular screening, with ECG monitoring, echocardiography, evaluation of hypertension, and vascular imaging in selected patients.

Take-Home Messages
• Cardiac involvement is rare but highly variable
• Congenital malformations, vascular disease, and isolated cases of HCM
• SCD events have been described in young adults with coronary artery disease or myocardial fibrosis
• Cardiac screening with ECG and echocardiogram is recommended in patients with symptoms or vascular abnormalities
• Consider vascular imaging in patients with hypertension or suspected vascular disease

9. Legius Syndrome

Legius syndrome (LS), also known as neurofibromatosis type 1-like syndrome, is an autosomal dominant RASopathy caused by loss-of-function variants in the SPRED1 gene (15q13.2), which interfere with the recruitment of neurofibromin to the membrane and result in excessive activation of the RAS/MAPK pathway [80,81].

The incidence of LS varies between 1:46,000 and 1:75,000 live births, and it is estimated that approximately 1–4% of patients clinically suspected of having NF1 have a SPRED1 variant [82].

Less than 300 cases have been described so far, including approximately 89 heterogeneous mutations such as missense, frameshift, nonsense, deletions/duplications identified in 146 different probands [83].

The phenotype of SL is generally milder than that of NF1 and manifests in childhood with multiple café-au-lait spots, axillary/inguinal freckles, macrocephaly, short stature, and neurobehavioral problems, such as delayed speech, attention deficit or ADHD. Absent are cutaneous or plexiform neurofibromas, Lisch nodules, optic nerve gliomas, and bone lesions specific to NF1 [71,82,84].

The diagnostic process requires a careful clinical evaluation, especially in the presence of café-au-lait and lentigines but in the absence of NF1-specific tumors; SPRED1 sequencing is then performed, identifying approximately 89% of cases, and deletion/duplication analysis. Molecular diagnosis is essential, as up to 50% of subjects with mutated SPRED1 meet clinical criteria for NF1, with consequent implications for medical surveillance [85–87].

There are no specific therapies for LS. The prognosis is favorable with symptomatic management of neurobehavioral delay and more restricted surveillance than that required for NF1, in the absence of the specific tumor risk of NF1 [71,82].

Regarding the risk of SCD, to date no documented cases of SCD associated with LS have been reported in the literature. Reported cardiovascular manifestations are limited to very few isolated clinical cases, such as pulmonary stenosis, mitral valve prolapses or paroxysmal tachycardias, but without evidence of structural cardiomyopathy or substrate for potentially lethal arrhythmias [84]. Therefore, although LS has a RASopathic transmission, intensive cardiac screening like that indicated for other RASopathies with cardiac involvement is not considered necessary. Basic cardiac evaluation remains prudent in the presence of symptoms or family history.

Take-Home Messages
• Caused by SPRED1 mutations with excessive activation of the RAS/MAPK pathway
• Mild phenotype like NF1, with café-au-lait syndrome, macrocephaly, and behavioral disturbances
• No current evidence of HCM or SCD
• Cardiac follow-up only if symptoms or a positive family history is present
• Genetic diagnosis is crucial to distinguish LS from NF1

10. Sudden Cardiac Death

SCD is defined as an unexpected cardiac death that occurs within one hour of the onset of acute symptoms, or within 24 h if the event was not directly observed [88–90]. It is frequently caused by malignant ventricular arrhythmias such as ventricular tachycardia or ventricular fibrillation, often in the presence of predisposing structural heart disease, and is one of the most serious complications of RASopathies with cardiac involvement [11–13,88].

In RASopathies associated with HCM, such as NS, NSML, CFCS and CS, the risk of SCD is significantly increased compared to idiopathic forms of HCM [13,26,27]. The pathological substrate includes myocellular disorganization, fibrosis, marked hypertrophy and, in some cases, congenital coronary anomalies [11,24].

The association between genotype and risk of sudden death has been documented in patients with high-risk mutations in genes such as RAF1, RIT1, PTPN11, BRAF and

LZTR1, which are frequently related to aggressive forms of HCM with early onset, severe obstructive gradient and ventricular tachyarrhythmias [6,19,22,26,27]. The presence of unexplained syncope, non-sustained ventricular tachycardia, left ventricular dysfunction (NSVT), or LVOT gradient $\geq$ 30 mmHg is considered significant predictors of major adverse events, including SCD [12,26].

Recent cohort studies, such as that of Boleti et al., have validated the utility of the HCM Risk-Kids model in pediatric patients with RASopathies, highlighting higher rates of arrhythmic events compared to subjects with sarcomeric HCM [13]. Furthermore, according to the multicenter analysis by Lynch et al., patients with RASopathies present a higher mortality and a greater incidence of fatal events already in childhood, supporting the need for intensive and early cardiac surveillance [12].

Cardiogenetics is central to the management of RASopathies, as molecular confirmation guides cardiac surveillance, prognostic assessment, reproductive counseling, and family screening strategies. Society recommendations emphasize the utility of genetic testing in all inherited cardiomyopathies and in pediatric patients with HCM or a syndromic phenotype [88,89].

In children with RAS-HCM, expert guidelines (PACES, ESC, AHA) emphasize the importance of baseline ECG, periodic Holter monitoring, and multidisciplinary evaluation in specialized centers, with ICD therapy considered only in selected high-risk cases. PACES recommendations and pediatric risk models aid decisions about prophylactic ICD implantation [11,90].

Given the frequent need for chronic surveillance and interventions, patients with RASocardiac diseases should be included in structured pediatric-to-adult transition programs that include an individualized transition plan, readiness assessment, education on warning symptoms, contact information for adult referrals (ACHD/cardiogenetics), and tracking to avoid loss to follow-up. These programs reduce care discontinuity and improve adherence to surveillance recommendations [91].

Family screening through genetic testing is a crucial strategy for identifying at-risk relatives and initiating surveillance and prevention. In cases of RASopathies, due to the potential early onset of HCM, cardiac evaluation (ECG + echocardiogram) should be performed on first-degree relatives already in childhood or at molecular diagnosis; targeted genetic testing is recommended when the variant is known in the proband [12,92].

In cases of SCD, molecular autopsy, defined as postmortem genetic study, is strongly recommended to identify any heritable pathogenic variants and implement genetic surveillance in family members at risk [22,26,93–97].

Finally, even in the absence of overt HCM, some RASopathic phenotypes, such as NSML, may present a high arrhythmic risk linked to conduction abnormalities such as bundle branch blocks and bradyarrhythmias or tachyarrhythmias, suggesting that risk stratification should be dynamic and adapted to the clinical evolution of the individual patient [12,30,42].

11. Conclusions

RASopathies are a heterogeneous group of rare genetic syndromes with relevant cardiovascular implications, particularly due to the increased risk of SCD associated with HCM and ventricular arrhythmia. Genotype-phenotype correlation is crucial in risk stratification and defining personalized diagnostic and therapeutic pathways.

Pathogenic variants in high-risk genes, such as RAF1, RIT1, PTPN11, BRAF, and LZTR1, are frequently associated with aggressive, early-onset forms of HCM with high arrhythmogenic potential. Evidence suggests that early diagnosis, careful cardiac follow-up, and preventive strategies such as ICD in selected patients may reduce the risk of

SCD. Management should be guided by international guidelines and delivered through a multidisciplinary, patient-centered approach.

In conclusion, the management of RASopathies requires a multidisciplinary, patient-centered, and evidence-based approach to significantly reduce mortality related to these rare but potentially lethal conditions.

Author Contributions: Conceptualization, C.S. and A.M.; methodology, A.M.; investigation, C.S.; resources, C.S.; writing—original draft preparation, C.S.; writing—review and editing, A.M.; supervision, A.M. All authors have read and agreed to the published version of the manuscript.

Funding: This research received no external funding.

Data Availability Statement: Not applicable.

Conflicts of Interest: The authors declare no conflicts of interest.

References

1. Chennappan, S.; Kontaridis, M.I. RASopathies in Cardiac Disease. *Annu. Rev. Med.* **2025**, *76*, 301–314. [CrossRef] [PubMed]
2. Garg, K.; Trehan, S.; Fredrick, F.; Singla, A.; Aggarwal, K.; Gupta, A.; Jain, R. RASopathies and Cardiac Complications: Insights into Mechanisms, Diagnosis, and Innovative Treatments. *Curr. Cardiol. Rev.* **2025**, *21*, e1573403X341624. [CrossRef] [PubMed]
3. Pierpont, E.I.; Bennett, A.M.; Schoyer, L.; Stronach, B.; Anschutz, A.; Borrie, S.C.; Briggs, B.; Burkitt-Wright, E.; Castel, P.; Cirstea, I.C.; et al. The 8th International RASopathies Symposium: Expanding research and care practice through global collaboration and advocacy. *Am. J. Med. Genet. Part A* **2024**, *194*, e63477. [CrossRef] [PubMed]
4. Tartaglia, M.; Aoki, Y.; Gelb, B.D. The molecular genetics of RASopathies: An update on novel disease genes and new disorders. *Am. J. Med. Genet. Part C Semin. Med. Genet.* **2022**, *190*, 425–439. [CrossRef]
5. Hilal, N.; Chen, Z.; Chen, M.H.; Choudhury, S. RASopathies and cardiac manifestations. *Front. Cardiovasc. Med.* **2023**, *10*, 1176828. [CrossRef]
6. Faienza, M.F.; Meliota, G.; Mentino, D.; Ficarella, R.; Gentile, M.; Vairo, U.; D'amato, G. Cardiac Phenotype and Gene Mutations in RASopathies. *Genes* **2024**, *15*, 1015. [CrossRef]
7. Yılmaz Uzman, C.; Gürsoy, S.; Özkan, B.; Vuran, G.; Ayyıldız Emecen, D.; Köprülü, Ö.; Bilen, M.M.; Hazan, F. Clinical features and molecular genetics of patients with RASopathies: Expanding the phenotype with rare genes and novel variants. *Eur. J. Pediatr.* **2024**, *184*, 108. [CrossRef]
8. Sclafani, M.; Falasconi, G.; Tini, G.; Musumeci, B.; Penela, D.; Saglietto, A.; Arcari, L.; Bucciarelli-Ducci, C.; Barbato, E.; Berruezo, A.; et al. Substrates of Sudden Cardiac Death in Hypertrophic Cardiomyopathy. *J. Clin. Med.* **2025**, *14*, 1331. [CrossRef]
9. Maron, M.S.; Rowin, E.J.; Maron, B.J. The Paradigm of Sudden Death Prevention in Hypertrophic Cardiomyopathy. *Am. J. Cardiol.* **2024**, *212S*, S64–S76. [CrossRef]
10. Delogu, A.B.; Limongelli, G.; Versacci, P.; Adorisio, R.; Kaski, J.P.; Blandino, R.; Maiolo, S.; Monda, E.; Putotto, C.; De Rosa, G.; et al. The heart in RASopathies. *Am. J. Med. Genet. Part C Semin. Med. Genet.* **2022**, *190*, 440–451. [CrossRef]
11. Boleti, O.D.; Roussos, S.; Norrish, G.; Field, E.; Oates, S.; Tollit, J.; Nepali, G.; Bhole, V.; Uzun, O.; Daubeney, P.E.F.; et al. Sudden cardiac death in childhood RASopathy-associated hypertrophic cardiomyopathy: Validation of the HCM risk-kids model and predictors of events. *Int. J. Cardiol.* **2023**, *393*, 131405. [CrossRef] [PubMed]
12. Lynch, A.; Tatangelo, M.; Ahuja, S.; Steve Fan, C.P.; Min, S.; Lafreniere-Roula, M.; Papaz, T.; Zhou, V.; Armstrong, K.; Aziz, P.F.; et al. Risk of Sudden Death in Patients with RASopathy Hypertrophic Cardiomyopathy. *J. Am. Coll. Cardiol.* **2023**, *81*, 1035–1045. [CrossRef] [PubMed]
13. Boleti, O.; Norrish, G.; Field, E.; Dady, K.; Summers, K.; Nepali, G.; Bhole, V.; Uzun, O.; Wong, A.; Daubeney, P.E.F.; et al. Natural history and outcomes in paediatric RASopathy-associated hypertrophic cardiomyopathy. *ESC Heart Fail.* **2024**, *11*, 923–936. [CrossRef] [PubMed]
14. Allen, M.J.; Sharma, S. Noonan Syndrome. In *StatPearls*; StatPearls Publishing: Treasure Island, FL, USA, 2023.
15. Ouboukss, F.; Adadi, N.; Amasdl, S.; Smaili, W.; Laarabi, F.Z.; Lyahyai, J.; Sefiani, A.; Ratbi, I. High frequency of hotspot mutation in PTPN11 gene among Moroccan patients with Noonan syndrome. *J. Appl. Genet.* **2024**, *65*, 303–308. [CrossRef]
16. Chen, Q.; Hong, D.; Huang, Y.; Zhang, Z.; Wang, S. Phenotypic and genotypic spectrum of noonan syndrome: A retrospective analysis of 46 consecutive pediatric patients presented at a regional cardiac center in China. *Heliyon* **2024**, *10*, e27038. [CrossRef]
17. Selås, M. Language profiles in Noonan Syndrome—A multiple case study. *Clin. Linguist. Phon.* **2024**, *38*, 1025–1039. [CrossRef]
18. Sun, L.; Xie, Y.M.; Wang, S.S.; Zhang, Z.W. Cardiovascular Abnormalities and Gene Mutations in Children with Noonan Syndrome. *Front. Genet.* **2022**, *13*, 915129. [CrossRef]

19. Chan, C.H.; Chu, M.F.; Lam, U.P.; Mok, T.M.; Tam, W.C.; Tomlinson, B.; Coelho, R.; Évora, M. Case report: Distinctive cardiac features and phenotypic characteristics of Noonan syndrome with multiple lentigines among three generations in one family. *Front. Cardiovasc. Med.* **2023**, *10*, 1225667. [CrossRef]
20. Ilic, N.; Krasic, S.; Maric, N.; Gasic, V.; Krstic, J.; Cvetkovic, D.; Miljkovic, V.; Zec, B.; Maver, A.; Vukomanovic, V.; et al. Noonan Syndrome: Relation of Genotype to Cardiovascular Phenotype-A Multi-Center Retrospective Study. *Genes* **2024**, *15*, 1463. [CrossRef]
21. Ko, J.M.; Kim, J.M.; Kim, G.H.; Yoo, H.W. PTPN11, SOS1, KRAS, and RAF1 gene analysis, and genotype-phenotype correlation in Korean patients with Noonan syndrome. *J. Hum. Genet.* **2008**, *53*, 999–1006. [CrossRef]
22. Kraoua, L.; Jaouadi, H.; Allouche, M.; Achour, A.; Kaouther, H.; Ahmed, H.B.; Chaker, L.; Maazoul, F.; Ouarda, F.; Zaffran, S.; et al. Molecular autopsy and clinical family screening in a case of sudden cardiac death reveals ACTN2 mutation related to hypertrophic/dilated cardiomyopathy and a novel LZTR1 variant associated with Noonan syndrome. *Mol. Genet. Genom. Med.* **2022**, *10*, e1954. [CrossRef] [PubMed]
23. Nishikawa, T.; Ishiyama, S.; Shimojo, T.; Takeda, K.; Kasajima, T.; Momma, K. Hypertrophic cardiomyopathy in Noonan syndrome. *Acta Paediatr. Jpn. Overseas Ed.* **1996**, *38*, 91–98. [CrossRef] [PubMed]
24. Burch, M.; Mann, J.M.; Sharland, M.; Shinebourne, E.A.; Patton, M.A.; McKenna, W.J. Myocardial disarray in Noonan syndrome. *Br. Heart J.* **1992**, *68*, 586–588. [CrossRef] [PubMed]
25. Zenker, M.; Wolf, C.M. Cardiovascular aspects of Noonan syndrome and related disorders. *Med. Genet. Mitteilungsblatt Des Berufsverb. Med. Genet. e.V.* **2025**, *37*, 113–124. [CrossRef]
26. Chen, H.; Li, X.; Liu, X.; Wang, J.; Zhang, Z.; Wu, J.; Huang, M.; Guo, Y.; Li, F.; Wang, X.; et al. Clinical and mutation profile of pediatric patients with RASopathy-associated hypertrophic cardiomyopathy: Results from a Chinese cohort. *Orphanet J. Rare Dis.* **2019**, *14*, 29. [CrossRef]
27. Zheng, J.; Peng, L.; Cheng, R.; Li, Z.; Xie, J.; Huang, E.; Cheng, J.; Zhao, Q. RAF1 mutation leading to hypertrophic cardiomyopathy in a Chinese family with a history of sudden cardiac death: A diagnostic insight into Noonan syndrome. *Mol. Genet. Genom. Med.* **2024**, *12*, e2290. [CrossRef]
28. Fan, L.; Jiang, J.; Zhang, Y.; Han, X.; Ding, W.; Xue, X.; Jiang, Y. Case Report: A rare case of Noonan syndrome with multiple lentigines manifesting as cardiac enlargement. *Front. Cardiovasc. Med.* **2025**, *12*, 1490436. [CrossRef]
29. Rivero-García, P.; Campuzano-Estrada, I.D.C.; Hernandez-Felix, J.H. Hypertrophic cardiomyopathy in an adult patient with Noonan syndrome with multiple lentigines. *Clin. Case Rep.* **2023**, *11*, e7607. [CrossRef]
30. Monda, E.; Prosnitz, A.; Aiello, R.; Lioncino, M.; Norrish, G.; Caiazza, M.; Drago, F.; Beattie, M.; Tartaglia, M.; Russo, M.G.; et al. Natural History of Hypertrophic Cardiomyopathy in Noonan Syndrome with Multiple Lentigines. *Circ. Genom. Precis. Med.* **2023**, *16*, 350–358. [CrossRef]
31. Tartaglia, M.; Zampino, G.; Gelb, B.D. Noonan syndrome: Clinical aspects and molecular pathogenesis. *Mol. Syndromol.* **2010**, *1*, 2–26. [CrossRef]
32. Yue, X.; Zhao, X.; Dai, Y.; Yu, L. Leopard syndrome: The potential cardiac defect underlying skin phenotypes. *Hereditas* **2021**, *158*, 34. [CrossRef]
33. Martínez-Quintana, E.; Rodríguez-González, F. LEOPARD Syndrome Caused by Tyr279Cys Mutation in the PTPN11 Gene. *Mol. Syndromol.* **2012**, *2*, 251–253. [CrossRef] [PubMed]
34. Zhang, J.; Shen, J.; Cheng, R.; Ni, C.; Liang, J.; Li, M.; Yao, Z. Identification of a PTPN11 hot spot mutation in a child with atypical LEOPARD syndrome. *Mol. Med. Rep.* **2016**, *14*, 2639–2643. [CrossRef] [PubMed]
35. Gelb, B.D.; Tartaglia, M. Noonan Syndrome with Multiple Lentigines. In *GeneReviews*®; Adam, M.P., Feldman, J., Mirzaa, G.M., Pagon, R.A., Wallace, S.E., Amemiya, A., Eds.; University of Washington: Seattle, WA, USA, 2007.
36. Lin, Y.; Xu, X. LEOPARD Syndrome. *Int. J. Dermatol.* **2024**, *63*, 1601–1602. [CrossRef] [PubMed]
37. Voron, D.A.; Hatfield, H.H.; Kalkhoff, R.K. Multiple lentigines syndrome. Case report and review of the literature. *Am. J. Med.* **1976**, *60*, 447–456. [CrossRef]
38. Martinez-Molina, M.; Fabregat-Pratdepadua, M.; Bielsa Marsol, I. Noonan Syndrome with Multiple Lentigines. *Actas Dermo-Sifiliogr.* **2024**, *115*, T414–T416. [CrossRef]
39. Ghosh, S.K.; Majumdar, B.; Rudra, O.; Chakraborty, S. LEOPARD Syndrome. *Dermatol. Online J.* **2025**, *21*, 13030/qt2d55s0t1. [CrossRef]
40. Lioncino, M.; Monda, E.; Verrillo, F.; Moscarella, E.; Calcagni, G.; Drago, F.; Marino, B.; Digilio, M.C.; Putotto, C.; Calabrò, P.; et al. Hypertrophic Cardiomyopathy in RASopathies: Diagnosis, Clinical Characteristics, Prognostic Implications, and Management. *Heart Fail. Clin.* **2022**, *18*, 19–29. [CrossRef]
41. Bertola, D.R.; Pereira, A.C.; Albano, L.M.; De Oliveira, P.S.; Kim, C.A.; Krieger, J.E. PTPN11 gene analysis in 74 Brazilian patients with Noonan syndrome or Noonan-like phenotype. *Genet. Test.* **2006**, *10*, 186–191. [CrossRef]
42. Rauen, K.A. Cardiofaciocutaneous Syndrome. In *GeneReviews*®; Adam, M.P., Feldman, J., Mirzaa, G.M., Pagon, R.A., Wallace, S.E., Amemiya, A., Eds.; University of Washington: Seattle, WA, USA, 2007.

43. Dentici, M.L.; Sarkozy, A.; Pantaleoni, F.; Carta, C.; Lepri, F.; Ferese, R.; Cordeddu, V.; Martinelli, S.; Briuglia, S.; Digilio, M.C.; et al. Spectrum of MEK1 and MEK2 gene mutations in cardio-facio-cutaneous syndrome and genotype-phenotype correlations. *Eur. J. Hum. Genet. EJHG* **2009**, *17*, 733–740. [CrossRef]
44. Scorrano, G.; David, E.; Calì, E.; Chimenz, R.; La Bella, S.; Di Ludovico, A.; Di Rosa, G.; Gitto, E.; Mankad, K.; Nardello, R.; et al. The Cardiofaciocutaneous Syndrome: From Genetics to Prognostic-Therapeutic Implications. *Genes* **2023**, *14*, 2111. [CrossRef]
45. Jurcă, M.C.; Iuhas, O.A.; Puiu, M.; Chiriţă-Emandi, A.; Andreescu, N.I.; Petcheşi, C.D.; Jurcă, A.D.; Magyar, I.; Jurcă, S.I.; Kozma, K.; et al. Cardiofaciocutaneous syndrome—A longitudinal study of a case over 33 years: Case report and review of the literature. *Rom. J. Morphol. Embryol.* **2021**, *62*, 563–568. [CrossRef]
46. Tzen, E.Y.; Lim, J.Y.; Cheah, S.M.; Choo, J.T.L.; Kam, S.; Ng, Z.M.; Thomas, B.; Jamuar, S.; Koh, A.L.; Tan, E. Diverse Clinical Manifestations of Cardiofaciocutaneous Syndrome Type 3 in Two Patients from South East Asia. *Mol. Syndromol.* **2023**, *14*, 21–29. [CrossRef]
47. Tang, Q.; Gong, D.; Ye, X.M.; Xu, J.R.; Yang, Y.C.; Yan, L.J.; Zou, L.; Wen, X.L. A Case Report of Cardiofaciocutaneous Syndrome with *MAP2K1* Pathogenic Variant. *Pharmacogenom. Pers. Med.* **2023**, *16*, 817–823. [CrossRef] [PubMed]
48. Roberts, A.; Allanson, J.; Jadico, S.K.; Kavamura, M.I.; Noonan, J.; Opitz, J.M.; Young, T.; Neri, G. The cardiofaciocutaneous syndrome. *J. Med. Genet.* **2006**, *43*, 833–842. [CrossRef] [PubMed]
49. Onesimo, R.; Sforza, E.; Giorgio, V.; Viscogliosi, G.; Kuczynska, E.M.; Margiotta, G.; Perri, L.; Limongelli, D.; Proli, F.; De Rose, C.; et al. The "FEEDS (FEeding Eating Deglutition Skills)" over Time Study in Cardiofaciocutaneous Syndrome. *Genes* **2023**, *14*, 1338. [CrossRef] [PubMed]
50. Yi, J.S.; Perla, S.; Bennett, A.M. An Assessment of the Therapeutic Landscape for the Treatment of Heart Disease in the RASopathies. *Cardiovasc. Drugs Ther.* **2023**, *37*, 1193–1204. [CrossRef]
51. Motta, M.; Solman, M.; Bonnard, A.A.; Kuechler, A.; Pantaleoni, F.; Priolo, M.; Chandramouli, B.; Coppola, S.; Pizzi, S.; Zara, E.; et al. Expanding the molecular spectrum of pathogenic SHOC2 variants underlying Mazzanti syndrome. *Hum. Mol. Genet.* **2022**, *31*, 2766–2778. [CrossRef]
52. Cordeddu, V.; Di Schiavi, E.; Pennacchio, L.A.; Ma'ayan, A.; Sarkozy, A.; Fodale, V.; Cecchetti, S.; Cardinale, A.; Martin, J.; Schackwitz, W.; et al. Mutation of SHOC2 promotes aberrant protein N-myristoylation and causes Noonan-like syndrome with loose anagen hair. *Nat. Genet.* **2009**, *41*, 1022–1026. [CrossRef]
53. Wang, Q.; Cheng, S.; Fu, Y.; Yuan, H. Case report: A de novo RASopathy-causing *SHOC2* variant in a Chinese girl with noonan syndrome-like with loose anagen hair. *Front. Genet.* **2022**, *13*, 1040124. [CrossRef]
54. Liu, Z.; Lai, J.; Song, F. Noonan syndrome and Noonan-like syndrome with loose anagen hair: Rare phenotypes may emerge during follow-up. *Transl. Pediatr.* **2024**, *13*, 1161–1168. [CrossRef]
55. Wang, X.O.; Liu, Z.Q.; Shangguan, S.F.; Lai, J.M.; Wang, P.C.; Song, F.Y.; Ye, X.; Du, M.; Huang, S.Y.; Gao, K. Clinical characteristics of children with Noonan-like syndrome and piloerection. *Chin. J. Pediatr.* **2025**, *63*, 405–410.
56. Avery, A.; Metcalf, J.S.; Maize, J.C.; Swanson, L.A. Cutaneous T-cell lymphoma in SHOC2 mutation-associated Noonan-like syndrome with loose anagen hair. *JAAD Case Rep.* **2022**, *24*, 52–55. [CrossRef]
57. Mazzanti, L.; Tamburrino, F.; Scarano, E.; Perri, A.; Vestrucci, B.; Guidetti, M.; Rossi, C.; Tartaglia, M. GH Therapy and first final height data in Noonan-like syndrome with loose anagen hair (Mazzanti syndrome). *Am. J. Med. Genet. Part A* **2013**, *161A*, 2756–2761. [CrossRef]
58. Takasawa, K.; Takishima, S.; Morioka, C.; Nishioka, M.; Ohashi, H.; Aoki, Y.; Shimohira, M.; Kashimada, K.; Morio, T. Improved growth velocity of a patient with Noonan-like syndrome with loose anagen hair (NS/LAH) without growth hormone deficiency by low-dose growth hormone therapy. *Am. J. Med. Genet. Part A* **2015**, *167*, 2425–2429. [CrossRef] [PubMed]
59. Ngulube, M.M.; Sharma, S. Faciocutaneoskeletal Syndrome (Costello Syndrome). In *StatPearls*; StatPearls Publishing: Treasure Island, FL, USA, 2025.
60. Leoni, C.; Viscogliosi, G.; Tartaglia, M.; Aoki, Y.; Zampino, G. Multidisciplinary Management of Costello Syndrome: Current Perspectives. *J. Multidiscip. Healthc.* **2022**, *15*, 1277–1296. [CrossRef] [PubMed]
61. Sol-Church, K.; Stabley, D.L.; Nicholson, L.; Gonzalez, I.L.; Gripp, K.W. Paternal bias in parental origin of HRAS mutations in Costello syndrome. *Hum. Mutat.* **2006**, *27*, 736–741. [CrossRef]
62. Gripp, K.W.; Morse, L.A.; Axelrad, M.; Chatfield, K.C.; Chidekel, A.; Dobyns, W.; Doyle, D.; Kerr, B.; Lin, A.E.; Schwartz, D.D.; et al. Costello syndrome: Clinical phenotype, genotype, and management guidelines. *Am. J. Med. Genet. Part A* **2019**, *179*, 1725–1744. [CrossRef] [PubMed]
63. Ríos-González, B.E.; Rodríguez-Ortiz, J.F.; Castro-Martínez, A.G.; Magaña-Torres, M.T.; Barros-Núñez, P. Clinical and molecular characterization of Costello syndrome in unrelated Mexican patients. *Clin. Dysmorphol.* **2022**, *31*, 55–58. [CrossRef]
64. Lin, A.E.; Grossfeld, P.D.; Hamilton, R.M.; Smoot, L.; Gripp, K.W.; Proud, V.; Weksberg, R.; Wheeler, P.; Picker, J.; Irons, M.; et al. Further delineation of cardiac abnormalities in Costello syndrome. *Am. J. Med. Genet.* **2002**, *111*, 115–129. [CrossRef]

65. Lin, A.E.; Alexander, M.E.; Colan, S.D.; Kerr, B.; Rauen, K.A.; Noonan, J.; Baffa, J.; Hopkins, E.; Sol-Church, K.; Limongelli, G.; et al. Clinical, pathological, and molecular analyses of cardiovascular abnormalities in Costello syndrome: A Ras/MAPK pathway syndrome. *Am. J. Med. Genet. Part A* **2011**, *155A*, 486–507. [CrossRef] [PubMed]
66. Hakim, K.; Boussaada, R.; Hamdi, I.; Msaad, H. Cardiac events in Costello syndrome: One case and a review of the literature. *J. Saudi Heart Assoc.* **2014**, *26*, 105–109. [CrossRef] [PubMed]
67. Moodley, M.; Lopez, K.R. Neurofibromatosis type 1—An update. *Semin. Pediatr. Neurol.* **2024**, *52*, 101172. [CrossRef] [PubMed]
68. Yoshida, Y. Neurofibromatosis 1 (von Recklinghausen Disease). *Keio J. Med.* **2025**, *74*, 37–41. [CrossRef]
69. Pinna, V.; Daniele, P.; Calcagni, G.; Mariniello, L.; Criscione, R.; Giardina, C.; Lepri, F.R.; Hozhabri, H.; Alberico, A.; Cavone, S.; et al. Prevalence, Type, and Molecular Spectrum of *NF1* Mutations in Patients with Neurofibromatosis Type 1 and Congenital Heart Disease. *Genes* **2019**, *10*, 675. [CrossRef]
70. Adil, A.; Koritala, T.; Munakomi, S.; Singh, A.K. Neurofibromatosis Type 1. In *StatPearls*; StatPearls Publishing: Treasure Island, FL, USA, 2023.
71. Legius, E.; Messiaen, L.; Wolkenstein, P.; Pancza, P.; Avery, R.A.; Berman, Y.; Blakeley, J.; Babovic-Vuksanovic, D.; Cunha, K.S.; Ferner, R.; et al. Revised diagnostic criteria for neurofibromatosis type 1 and Legius syndrome: An international consensus recommendation. *Genet. Med. Off. J. Am. Coll. Med. Genet.* **2021**, *23*, 1506–1513. [CrossRef]
72. İncecik, F.; Hergüner, Ö.M.; Alınç Erdem, S.; Altunbaşak, Ş. Neurofibromatosis type 1 and cardiac manifestations. *Turk Kardiyol. Dern. Ars. Turk Kardiyol. Derneginin Yayin. Organidir* **2015**, *43*, 714–716. [CrossRef]
73. Nguyen, R.; Mir, T.S.; Kluwe, L.; Jett, K.; Kentsch, M.; Mueller, G.; Kehrer-Sawatzki, H.; Friedman, J.M.; Mautner, V.F. Cardiac characterization of 16 patients with large NF1 gene deletions. *Clin. Genet.* **2013**, *84*, 344–349. [CrossRef]
74. Kapusta, L.; Beer, G.; Rothschild, E.; Baruch, G.; Barkay, G.; Marom, D.; Grinshpun-Cohen, Y.; Raskind, C.; Constantini, S.; Toledano-Alhadef, H. Cardiac screening in pediatric patients with neurofibromatosis type 1: Similarities with Noonan syndrome? *Int. J. Cardiovasc. Imaging* **2024**, *40*, 1475–1482. [CrossRef]
75. Cutruzzolà, A.; Irace, C.; Frazzetto, M.; Sabatino, J.; Gullace, R.; De Rosa, S.; Spaccarotella, C.; Concolino, D.; Indolfi, C.; Gnasso, A. Functional and morphological cardiovascular alterations associated with neurofibromatosis 1. *Sci. Rep.* **2020**, *10*, 12070. [CrossRef]
76. Kerashvili, N.; Gutmann, D.H. The management of neurofibromatosis type 1 (NF1) in children and adolescents. *Expert Rev. Neurother.* **2024**, *24*, 409–420. [CrossRef]
77. Hamilton, S.J.; Allard, M.F.; Friedman, J.M. Cardiac findings in an individual with neurofibromatosis 1 and sudden death. *Am. J. Med. Genet.* **2001**, *100*, 95–99. [CrossRef] [PubMed]
78. Patel, M.; Patel, D.; Nehme, C.; Ambati, A.; Oostra, C.E. Myocardial Infarction in a 28-Year-Old Male with Neurofibromatosis Type 1. *Cureus* **2020**, *12*, e11254. [CrossRef] [PubMed]
79. Pourirahim, M.; Houshmand, G.; Abdolkarimi, L.; Maleki, M.; Kalayinia, S. Whole-exome sequencing revealed a likely pathogenic variant in *NF1* causing neurofibromatosis type I and Arrhythmogenic Cardiomyopathy. *BMC Cardiovasc. Disord.* **2024**, *24*, 220. [CrossRef] [PubMed]
80. Brems, H.; Legius, E. Legius syndrome, an Update. Molecular pathology of mutations in SPRED1. *Keio J. Med.* **2013**, *62*, 107–112. [CrossRef]
81. Hirata, Y.; Brems, H.; Van der Auweraer, S.; Ohyagi, M.; Iizuka, M.; Mise-Omata, S.; Ito, M.; Messiaen, L.; Mizuno, S.; Takahashi, S.; et al. Legius syndrome mutations in the Ras-regulator SPRED1 abolish its membrane localization and potentially cause neurodegeneration. *J. Biol. Chem.* **2024**, *300*, 107969. [CrossRef]
82. Legius, E.; Stevenson, D. Legius Syndrome. In *GeneReviews®*; Adam, M.P., Feldman, J., Mirzaa, G.M., Pagon, R.A., Wallace, S.E., Amemiya, A., Eds.; University of Washington: Seattle, WA, USA, 2010.
83. Brems, H.; Pasmant, E.; Van Minkelen, R.; Wimmer, K.; Upadhyaya, M.; Legius, E.; Messiaen, L. Review and update of SPRED1 mutations causing Legius syndrome. *Hum. Mutat.* **2012**, *33*, 1538–1546. [CrossRef]
84. Giugliano, T.; Santoro, C.; Torella, A.; Del Vecchio Blanco, F.; Grandone, A.; Onore, M.E.; Melone, M.A.B.; Straccia, G.; Melis, D.; Piccolo, V.; et al. Clinical and Genetic Findings in Children with Neurofibromatosis Type 1, Legius Syndrome, and Other Related Neurocutaneous Disorders. *Genes* **2019**, *10*, 580. [CrossRef]
85. Cammarata-Scalisi, F.; Willoughby, C.E.; Cárdenas Tadich, A.; Ludeña Mendoza, C.; Peralta Aros, C.; Araya Castillo, M.; Callea, M. Legius Syndrome: The importance of molecular differential diagnosis with neurofibromatosis type 1. *Ital. J. Dermatol. Venereol.* **2025**, *160*, 278–279. [CrossRef]
86. Chelleri, C.; Brolatti, N.; De Marco, P.; Ognibene, M.; Diana, M.C.; Madia, F.; Duca, M.D.; Santangelo, A.; Capra, V.; Striano, P.; et al. Novel causative variants in Legius syndrome: SPRED1 Genotype spectrum expansion. *Am. J. Med. Genet. Part A* **2024**, *194*, e63824. [CrossRef]
87. Wang, X.; Zhang, Y.; Du, M.; Yang, H.; Liu, X.; Wang, M.; Chen, J.; Chen, Y.; Wei, H. Clinical and genetic analysis of three cases of Legius syndrome caused by SPRED1 gene mutation. *Chin. J. Med. Genet.* **2024**, *41*, 941–946.

88. Ommen, S.R.; Ho, C.Y.; Asif, I.M.; Balaji, S.; Burke, M.A.; Day, S.M.; Dearani, J.A.; Epps, K.C.; Evanovich, L.; Ferrari, V.A.; et al. 2024 AHA/ACC/AMSSM/HRS/PACES/SCMR Guideline for the Management of Hypertrophic Cardiomyopathy: A Report of the American Heart Association/American College of Cardiology Joint Committee on Clinical Practice Guidelines. *Circulation* **2024**, *149*, e1239–e1311. [CrossRef]
89. Kontorovich, A.R. Approaches to Genetic Screening in Cardiomyopathies: Practical Guidance for Clinicians. *JACC Heart Fail.* **2023**, *11*, 133–142. [CrossRef] [PubMed]
90. Silka, M.J.; Shah, M.J.; Silva, J.N.A.; Balaji, S.; Beach, C.M.; Benjamin, M.N.; Berul, C.I.; Cannon, B.; Cecchin, F.; Cohen, M.I.; et al. 2021 PACES Expert Consensus Statement on the Indications and Management of Cardiovascular Implantable Electronic Devices in Pediatric Patients: Executive Summary. *Ann. Pediatr. Cardiol.* **2022**, *15*, 323–346. [CrossRef] [PubMed]
91. Moons, P.; Bratt, E.-L.; De Backer, J.; Goossens, E.; Hornung, T.; Tutarel, O.; Zühlke, L.; Araujo, J.J.; Callus, E.; Gabriel, H.; et al. Transition to adulthood and transfer to adult care of adolescents with congenital heart disease: A global consensus statement of the ESC Association of Cardiovascular Nursing and Allied Professions (ACNAP), the ESC Working Group on Adult Congenital Heart Disease (WG ACHD), the Association for European Paediatric and Congenital Cardiology (AEPC), the Pan-African Society of Cardiology (PASCAR), the Asia-Pacific Pediatric Cardiac Society (APPCS), the Inter-American Society of Cardiology (IASC), the Cardiac Society of Australia and New Zealand (CSANZ), the International Society for Adult Congenital Heart Disease (ISACHD), the World Heart Federation (WHF), the European Congenital Heart Disease Organisation (ECHDO), and the Global Alliance for Rheumatic and Congenital Hearts (Global ARCH). *Eur. Heart J.* **2021**, *42*, 4213–4223. [PubMed]
92. Aiyer, S.; Kalutskaya, E.; Agdamag, A.C.; Tang, W.H.W. Genetic Evaluation and Screening in Cardiomyopathies: Opportunities and Challenges for Personalized Medicine. *J. Pers. Med.* **2023**, *13*, 887. [CrossRef]
93. Salzillo, C.; Ciccone, M.M.; Introna, F.; Santobuono, V.E.; Solarino, B.; Marzullo, A. Sudden cardiac death in the young: Single-center study of Bari autopsy cases. *Front. Cardiovasc. Med.* **2025**, *12*, 1630511. [CrossRef]
94. Salzillo, C.; Sansone, V.; Napolitano, F. Sudden Cardiac Death in the Young: State-of-the-Art Review in Molecular Autopsy. *Curr. Issues Mol. Biol.* **2024**, *46*, 3313–3327. [CrossRef]
95. Salzillo, C.; Marzullo, A. Hereditary Aortopathies as Cause of Sudden Cardiac Death in the Young: State-of-the-Art Review in Molecular Medicine. *Diseases* **2024**, *12*, 264. [CrossRef]
96. Salzillo, C.; La Verde, M.; Imparato, A.; Molitierno, R.; Lucà, S.; Pagliuca, F.; Marzullo, A. Cardiovascular Diseases in Public Health: Chromosomal Abnormalities in Congenital Heart Disease Causing Sudden Cardiac Death in Children. *Medicina* **2024**, *60*, 1976. [CrossRef]
97. Salzillo, C.; Lucà, S.; Ronchi, A.; Cazzato, G.; Ingravallo, G.; Marzullo, A. Non-atherosclerotic coronary artery disease in molecular autopsy: A literature review. *Vessel Plus* **2025**, *9*, 1. [CrossRef]

BioChem

MDPI

Review

Polyoxometalates' Progress for the Treatment of Alzheimer's Disease

Manuel Aureliano [1,2,*], João Mateus [3] and David Manjua Rijo [3]

1 Faculdade de Ciências e Tecnologia (FCT), Campus de Gambelas, Universidade do Algarve, 8005-139 Faro, Portugal
2 Centro de Ciências do Mar do Algarve (CCMAR/CIMAR LA), Campus de Gambelas, Universidade do Algarve, 8005-139 Faro, Portugal
3 Faculty of Medicine and Biomedical Sciences, University of Algarve, 8005-139 Faro, Portugal; a64508@ualg.pt (J.M.); a71237@ualg.pt (D.M.R.)
* Correspondence: maalves@ualg.pt

Abstract

Alzheimer's disease (AD) signifies a devastating impact on the quality of life of patients and their families. At a biomolecular level, AD is characterized by the deposition of extracellular plaques of β-amyloid (Aβ), affecting language, spatial navigation, recognition abilities and memory. Among the selected 30 articles about polyoxometalates (POMs) and AD published from 2011 to 2025, pure POMs, hybrid POMs and POM nanoparticles can be found. The majority of POMs are polyoxotungstates (62%), the Keggin-type $SiW_{11}O_{39}$ being the most studied in AD. The main effect described is the inhibition of Aβ aggregates. Other effects include reversing the neurotoxicity induced by Aβ aggregates, decreasing ROS production and neuroinflammation, restoring memory and sequestering Zn^{2+} and Cu^{2+}, among others, features that are well known to be associated with the pathology of AD. POMs have also shown the ability to induce the disaggregation of Aβ fibrils, particularly after irradiation, and to inhibit acetylcholinesterase activity at an nM range. Putting it all together, this review highlights a predominant trend in the exploration of POMs to act directly at the level of the formation and/or disaggregation of Aβ aggregates in the treatment of AD.

Keywords: polyoxometalates; Alzheimer's disease; β-amyloid aggregation; neurotoxicity; blood–brain barrier; reactive oxygen species production; β-amyloid disaggregation; acetylcholinesterase inhibitors; transition metal chelators; nanoparticles

1. Introduction

The 21st century poses new challenges with regard to the growing prevalence of diseases associated with aging, particularly neurodegenerative diseases [1]. Neurodegenerative diseases define disorders characterized by the progressive dysfunction of synapses, neurons and glial cells, inducing the decline of the central nervous system (CNS) and, therefore, leading to debilitating conditions that impair the cognitive and physical characteristics of individuals [2]. At a more advanced stage, these disorders culminate in situations of total loss of autonomy and, eventually, death [3]. Studies identify changes in the physiological conformation of certain proteins [4], the formation of protein aggregates (intra- or extracellular) [5], mitochondrial dysfunction [6] and oxidative stress [7] as crucial elements that contribute to neurological diseases such as Parkinson's disease, amyotrophic lateral sclerosis, multiple sclerosis, Huntington's disease and Alzheimer's disease, among others [8].

At the molecular level, AD is characterized, in the majority of the cases, by the deposition of extracellular senile plaques of β-amyloid (Aβ), the formation of intracellular neurofibrillary tangles of hyperphosphorylated tau protein (NFT) and, in rarer cases, the formation of intracellular actin aggregates and actin-associated proteins in neurons, predominantly in the hippocampus, although the cause of AD is not yet known [9]. Approved medications include acetylcholinesterase inhibitors such as donepezil, rivastigmine and galantamine, which increase the concentration of acetylcholine in the synaptic cleft and are indicated for the treatment of mild to moderate symptoms [10]. Normally, these drugs are often associated with side effects, which may include nausea, vomiting and heart problems, limiting their usefulness in certain patients with associated comorbidities [11]. Therefore, the development of new treatments for AD is challenging due to multiple factors, including the need for long follow-up periods to assess the effectiveness of treatments, as well as the difficulty in recruiting and retaining participants [12,13].

In recent decades, POMs have played an increasingly important role as a new age treatment for diseases. The wide range of POM uses in medicine may be due to the modulation of several proteins, as referred to elsewhere [14], possibly being the reason why these metal-based compounds, particularly in cancer treatment, offer unique mechanisms of action and have spurred research into alternative and enhanced therapies [14–16]. Beyond oncology, POMs are also being explored for their potential in treating viral infections, bacterial resistance to antibiotics and neurodegenerative disorders, among other incurable diseases [14–16]. Herein, we provide an overview from an inorganic biochemistry perspective of the evolution of the utilization of POMs in AD studies, focusing on the diverse types of POM structures, the POMs' effects and future directions in the field.

2. Alzheimer's Disease

AD represents one of the main global public health challenges due to its devastating impact on the quality of life of patients and their families, as well as the costs associated with providing healthcare [17]. Approximately 57 million people suffer from dementia, with prospects for this number to triple by 2050 due to population growth and aging [18]. According to data from the World Health Organization (WHO), AD and other types of dementia are the 7th leading cause of deaths worldwide. Women are disproportionately represented, accounting for 65% of deaths [17]. The pathophysiology of AD encompasses both anatomical changes in the brain and biomolecular changes that disrupt neuronal function [19]. AD is characterized by generalized brain atrophy, which begins in regions involved in memory and cognition and progressively covers larger areas of the brain [19]. The hippocampus, entorhinal cortex and other medial temporal lobe structures, essential for forming and retrieving memories, are among the first to exhibit neurodegeneration [20].

Alzheimer's disease was first described in 1906 by Alois Alzheimer as "a severe pathological process peculiar to the cerebral cortex" [21]. AD is currently an irreversible and incurable neurodegenerative disease and the leading cause of dementia in adults over 65 years of age (50–70%). In an initial phase, AD manifests itself as a dysfunction in short-term memory, which progressively evolves into more serious clinical conditions characterized by dementia, aphasia, temporal and spatial disorientation, cognitive deficits, behavioral changes such as apathy, agitation, aggression and irritability, mood changes such as depressive and anxiety syndromes, mutism, insomnia, seizures, paranoia and hallucinations, among others [22].

Although the main risk factor for the development of AD is age, there are a multitude of factors associated with the development of the pathology, namely a family history of dementia, genetic factors such as mutations in the genes coding for amyloid precursor protein (APP), presenilin-1 (PS1) and presenilin-2 (PS2), environmental factors, race, low socioeconomic and/or educational level, hypertension, diabetes, dyslipidemia, obesity and traumatic brain injury, among others [23].

β-amyloid (Aβ) is a peptide, originating from APP, that plays a central role in the pathogenesis of AD. Aβ accumulates in the brain, forming senile plaques, one of the main pathological biomarkers of the disease [24]. APP is a transmembrane protein expressed in various parts of the organism, particularly in neurons [25]. Synapses, fundamental in neurotransmission and neuroplasticity, are particularly sensitive to the presence of Aβ aggregates and NFTs (neurofibrillary tangles), two major pathological hallmarks of AD [26]. The loss of neural connectivity due to Aβ aggregates is reflected in memory decline, changes in executive capabilities and a decrease in the ability to process information [26]. In parallel, the activity of microglia and astrocytes, glial cells with supportive and protective functions in the central nervous system, is also stimulated by Aβ and NFTs [27]. This hyperactive state results in the excessive production of reactive oxygen species (ROS) and pro-inflammatory cytokines, which favor chronic states of neuroinflammation and oxidative stress. These factors, together with the disturbance of calcium homeostasis, contribute to neurotoxicity and cell death [27,28].

Currently, treatments for AD focus primarily on managing the cognitive and behavioral symptoms of the disease [11]. Memantine, an N-methyl-D-aspartate (NMDA) receptor antagonist that inhibits glutamate-induced calcium-mediated excitotoxicity, is prescribed to alleviate moderate to severe symptoms [29]. However, these therapies offer limited benefits and do not prevent disease progression [11]. Acetylcholinesterase inhibitors, for example, may temporarily improve cognitive function or stabilize symptoms in some patients, but do not alter the underlying course of neurodegeneration [10,11]. In short, the complexity of the pathological mechanisms involved (some of which are certainly yet to be discovered) makes the therapeutic target extremely challenging [30]. At the same time, individual variability in the progression and symptoms of the disease makes it difficult to create treatments that are effective across the board [30]. Another significant obstacle is the blood–brain barrier (BBB) [31]. This consists of a highly selective structure that separates the systemic circulation from the brain tissue, protecting the brain from potentially toxic substances present in the blood, allowing the passage of the essential nutrients and metabolites necessary for normal brain functioning [31]. Even so, the blood–brain barrier also prevents many drugs from reaching brain tissue in therapeutic concentrations [13,31]. This panorama demonstrates the urgent need for new therapeutic approaches that not only prevent the symptoms but additionally can act on the original causes of AD (Figure 1). In fact, one of the take-home messages of the present review is to bring together the biochemistry side of AD (left side of Figure 1), illustrated by the synapses, the APP, the Aβ and Aβ aggregates, the neurons and the brain, and AD symptoms (right side of Figure 1), illustrated by the loss of memories of anniversaries and keys, among others. Alzheimer's caregivers, doctors, psychologists, nurses and family members are in the middle between AD biochemistry and symptoms, and the current review is also aimed at them (Figure 1).

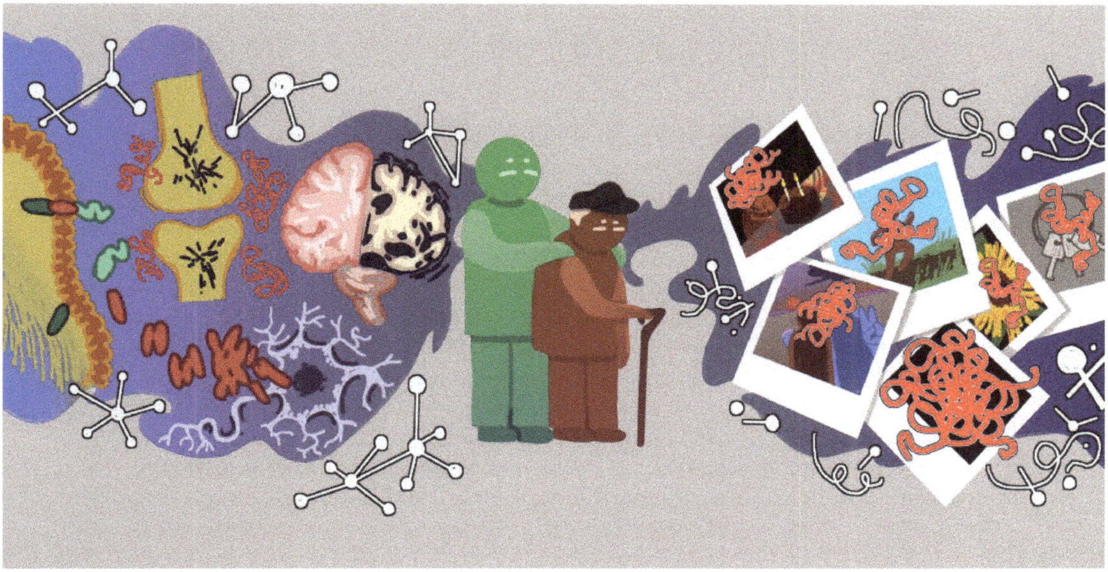

Figure 1. Alzheimer's disease, "a severe pathological process peculiar to the cerebral cortex". In an initial phase, it manifests itself as a dysfunction in short-term memory, which progressively evolves into more detrimental clinical conditions characterized by dementia, aphasia, temporal and spatial disorientation, cognitive deficits and behavioral changes such as apathy and agitation, among others. β-amyloid becomes a central figure in the pathogenesis of AD. Alzheimer's caregivers, doctors, psychologists, nurses and family members are in the middle between AD biochemistry and symptoms, and the current review is also aimed at them.

3. Polyoxometalates

POMs represent a fascinating and diverse class of inorganic compounds that have captured the attention of researchers due to their unique structures, chemical versatility and wide range of potential applications [32]. These compounds consist of molecular clusters formed by transition metal oxoanions, such as tungsten (W), molybdenum (Mo) and vanadium (V), linked by oxygen atoms [14,15,32]. The concept of POMs as a distinct class of compounds dates back to 1826 through the studies of Swedish chemist Jöns Jacob Berzelius, who extensively studied metallic acids and their properties [33]. Since then, the practically unlimited ability to chemically modify, adjust properties and structures and form hybrid systems with other particles, compounds or molecules has turned the study of POMs into a vibrant and constantly evolving area of research [14,15]. The structure of POMs is remarkably diverse, varying in size, shape and composition, which is due to the highly modular nature of oxoanions [32]. The basic skeleton of a POM generally consists of a central cluster of metal atoms surrounded by oxygen atoms, forming polyhedral units such as tetrahedra and/or octahedra, among others [32,34]. These units can be linked in different ways, removed, added or even replaced by other elements, creating structures that range from simple rings or chains to complex three-dimensional clusters [32,34].

The applications of POMs are as diverse as their structures [14,15,34]. Due to their catalytic action, POMs can be used to facilitate chemical reactions such as the oxidation of organic compounds and the degradation of environmental pollutants [35–37]. In the energy industry, POMs have been explored as components of energy storage devices, such as batteries and supercapacitors, due to their ability to reversibly accept and donate electrons [38,39]. In the field of medicine, the antiviral, antibacterial, antifungal, anticancer and antidiabetic properties (among many others), as well as the inhibition of the formation of Aβ aggregates, of certain POMs pave the way for the development of new pharmaceutical treatments [14–16,40–47].

4. Polyoxometalate Studies in Alzheimer's Disease

On 18 August 2025, the number of articles published with the keyword "polyoxometalate" was 15,894, determined after a literature search on the Web of Science. The number of articles found with the keywords "polyoxometalate" and "Alzheimer's disease" was 33, also considering review articles. From the articles analyzed, about 65 POMs were found (Table S1). Most POMs found in AD research studies are polyoxotungstates, POTs (62%), followed by polyoxomolybdates, POMos (32%), with minor contributions from polyoxoniobates, PONbs (5%), and polyoxovanadates, POVs (2%). From these, several POM types and structures that were studied in AD were selected for this study (Figure 2). As can be observed in Figure 2, not all were just pure POMs—for example, the one described in the first studies (2011) [48]—but also POM nanoparticles with a peptide with specificity for Aβ [49] (2013); Wells–Dawson POMds with histidine-chelating metals [50] (2014); systems combining cerium nanoparticles with POMs [51] (2016); POM gold nanorods (AuNRs) [52] (2017); hybrid POMs [53,54] (2018, 2019) and those functionalized with a platinum-substituted POT [55] (2019); other pure POMs [56] (2020); and a nanohybrid system consisting of gold nanoparticles covered with POMs and polyethylene glycol [57] (2023), among others (Figure 2). Not all are pure POMs (46%), and, as described below, hybrid POMs (27%) and hybrid nanoparticle POM systems (27%) can also be found.

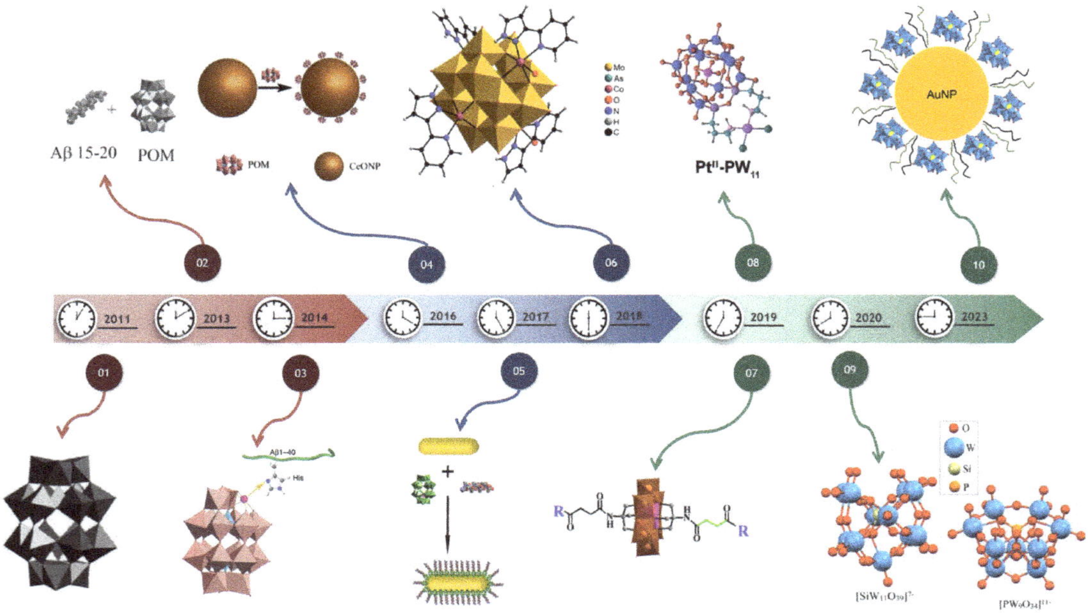

Figure 2. Timeline of selected POM structures included in this review. POMs alone (2011, 2013, 2014 and 2020) or combined in nanoparticles and/or nanorods (2016, 2017), and hybrid POMs (2018, 2019) with a platinum-substituted (2019) and nanohybrid system (2023) showed promising applications in AD.

4.1. Pure Polyoxometalate Studies in Alzheimer's Disease

In the first study described, in 2011, Geng et al. explored the inhibition of Aβ aggregate formation between different POMs [48]. The results suggest that POTs such as

$K_8[P_2CoW_{17}O_{61}]$, $K_7[PTi_2W_{10}O_{40}]$, α-$Na_9H[SiW_9O_{34}]$ and K_8[β-$SiW_{11}O_{39}$] prevent Aβ fibrillation by establishing electrostatic interactions with the peptide, with IC_{50} values of inhibition between 10 and 39 μM. On the other hand, polyoxomolybdates (POMos) such as $H_3[PMo_{12}O_{40}]$ and $Na_5[IMo_6O_{24}]$ did not demonstrate an inhibitory activity. The study highlighted that POMs appear to specifically bind to charged recognition motifs in the structure of the Aβ peptide, suggesting that this interaction is crucial for preventing fibril formation. In the same year, Zhou et al. [58] showed that two representative POMs, the Keggin-type $[W_{12}O_{42}]^{12-}$ (abbreviated W_{12}) and the donut-shaped $[P_5W_{30}O_{110}]^{14-}$ (abbreviated P_5W_{30}), interact with Aβ1–40, affecting the degree of aggregation, being W_{12} more potent than P_5W_{30}, and inhibiting its fibrillization upon 5 days of incubation in PBS medium at pH 7.4 [58]. A thioflavin T (ThT) fluorescence assay was used. Upon the binding of ThT, a fluorescent dye, to the β sheet conformation of Aβ fibrils, a fluorescence emission at 480 nm is observed that is proportional to the number of fibrils formed [58]. Most relevant were the findings that both POMs dramatically prevent the Aβ1–40 random coil changes to the β sheet secondary structure and therefore prevent the formation of aggregates. In order to understand the POM binding patterns of interaction with Aβ1–40, a complete characterization was performed using dynamic light scattering (DLS) and isothermal titration calorimetry (ITC). Different binding patterns were found for the two POMs (Figure 3), and the mode of interaction was mainly electrostatic [58].

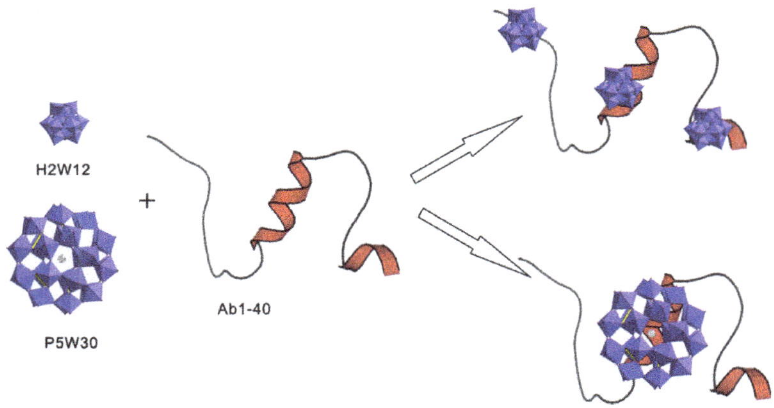

Figure 3. Different binding patterns upon W_{12} and P_5W_{30} interaction with Aβ1–40 [58]. Reproduced from Ref. [58] with permission from Elsevier.

Two years later, Li et al. [59] used a phosphotungstate ($K_8[P_2CoW_{17}O_{61}]$) to test a new therapeutic approach specifically focused on Aβ aggregation. The innovative aspect of this study lies in the application of the photocatalytic properties of phosphotungstate to inhibit Aβ aggregation. Moreover, it induces the degradation of preformed Aβ monomers and oligomers, while also decreasing ROS production [59]. The phosphotungstate compounds used in this study demonstrated the ability not only to inhibit the aggregation of Aβ peptides, but also to degrade both Aβ monomers and oligomers when exposed to UV light radiation. In the same year, Iqbal et al. described several POMs as inhibitors of acetylcholinesterase and butyrylcholinesterase activities [60]. They state that $[H_2W_{12}O_{42}]^{10-}$ and $[TeW_6O_{24}]^{6-}$ are the most potent acetylcholinesterase inhibitors, with IC_{50} values of 0.29 and 0.31 μM, respectively, while $[(O_3PCH_2PO_3)_4W_{12}O_{36}]^{16-}$ was the most potent and selective inhibitor of butyrylcholinesterase, with an IC_{50} value of 0.18 μM, pointing to POMs as a new class of acetyl and butyrylcholinesterase inhibitors [60].

In 2014, Chen et al. [61] described the inhibition of Aβ aggregation by three spherically shaped anionic POMo nanoclusters, based in a cluster skeleton with a basic Mo (Mo_5) unit and built up of 12 Mo_{11} fragments. All three POMo nanoclusters prevent Aβ aggregation, induced by Zn^{2+} and by Cu^{2+}. Circular dicroism (CD) spectroscopy studies also showed that these POMos prevent the conformational conversion of Aβ40 from random coils to β sheets. The PC12 cell neurotoxicity induced by Aβ40 (20 μM) with or without Zn^{2+} or Cu^{2+} is partially reverted by the POMs (10 μg/mL), from about 50% to almost 80% for the most potent POM (the POM cluster 3). It was also observed that Zn^{2+} induces Aβ40 aggregates, which in turn also promote H_2O_2 production, both these effects being inhibited by these POMos. It was suggested that Aβ40 induces PC12 cell death through apoptosis. Moreover, the depolarization of mitochondria was reduced from 74% to 6–12% (depending on the POMo) when the cells were incubated with these POMos [61]. In the same year, Gao et al. [50] designed a series of transition metal-functionalized POM derivatives, Wells–Dawson POMds such as $K_8[P_2NiW_{17}O_{61}]$ and $K_8[P_2CuW_{17}O_{61}]$ with defined histidine-chelating metals (such as Ni, Co, Cu, Fe and Mn) (Figure 4), which have much better Aβ aggregation inhibition and peroxidase-like activity inhibition effects than the parent POM [50] and the ones found previously [48].

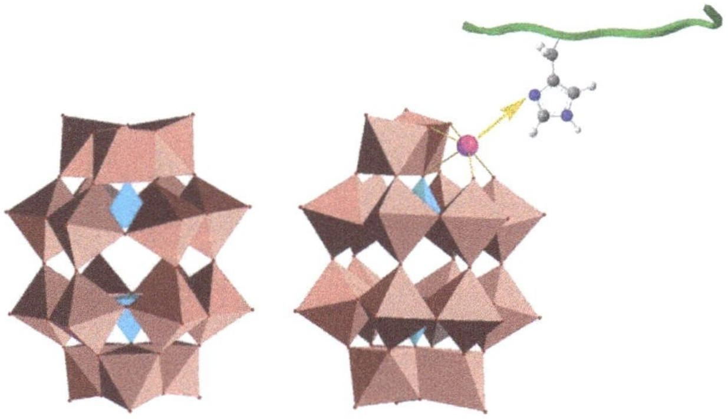

Figure 4. Structures of POM and POMds. (**Left**) Wells–Dawson structure; (**Right**) Wells–Dawson POMds with histidine-chelating metals (such as Ni, Co, Cu, Fe and Mn). The PO_4 and WO_6 polyhedra are shown in blue and light pink, respectively. The histidine-chelating metal is shown as a purple ball. The O, C, N and H atoms are shown as red, gray, dark blue and white balls, respectively [50]. Reproduced from Ref. [50] with permission from Springer Nature.

In 2017, two Keggin-type heteropolytungstates, $K_7[Ti_2PW_{10}O_{40}]\cdot 6H_2O$ and $K_6H[SiV_3W_9O_{40}]\cdot 3H_2O$, with different inhibitory potencies toward acetylcholinesterase activity (IC_{50} values of 1.04 μM and 4.80 mM, respectively) were evaluated regarding their toxicity [62]. Although no severe toxicity was verified, further toxicological studies are needed [62]. Two years later, Li et al. used nanoclusters of MoO_7 in neurons to test its protective properties against oxidative stress in a cerebral ischemia/reperfusion injury (I/R) model [63]. These POMo clusters were intrathecally administrated and effectively scavenged ROS. It was also found that the treatment with these POMos recovered superoxide dismutase (SOD) and lipid peroxidation (LPO) levels, in addition to inflammatory parameters such as TNFά and IL6. Moreover, edema volume in the brain was also reduced upon POM administration [63]. The authors point out that POMs play a role in ROS and LPO.

Recently, several features of ROS and LPOs induced in vivo by POMs were reviewed, particularly polyoxovanadates (POVs), where associations with cancer, diabetes and neurological diseases were also analyzed [64]. It was suggested that, because POMs can directly and indirectly induce ROS formation, the study of LPOs arising from increased ROS should analyze the contribution of both processes [64]. Therefore, by inhibiting amyloid aggregation, POMs might also prevent, indirectly, the neurotoxicity induced by Aβ aggregates, decreasing ROS production and neuroinflammation, among other effects.

In 2020, a POT ($H_4[SiW_{12}O_{40}]$ (WSiA) with an nM potency (72.3 nM) of achetylcolinesterase (AChE) inhibition was described [56], at least 2.5× and 13× more potent than was described before [60,62]. A new allosteric binding site, termed the β-allosteric site (β-AS), which was considered responsible for the inhibition of AChE by POMs, was deduced through a molecular docking approach [56]. No POM genotoxic effects in human lymphocytes were observed, which indicates their potential to be used as medicinal drugs. In fact, as discussed above, acetylcholinesterase inhibitors may temporarily improve cognitive function or stabilize symptoms in some patients [10,11].

However, at pH 8, both studied POTs ($H_4[SiW_{12}O_{40}]$ (WSiA) and $H_3[PW_{12}O_{40}]$ (WPA)) suffered transformations in the experimental medium solution. That is, WSiA was converted into $[SiW_{11}O_{39}]^{7-}$, while WPA was converted into $PW_9O_{34}]^{11-}$. The studies were performed in a phosphate buffer at pH 8. A speciation analysis deduced that $[SiW_{11}O_{39}]^{7-}$, a derivative of WSiA, and $[PW_9O_{34}]^{11-}$, a derivative of WPA, were considered as the active species present in the solution [56]. Thus, care must be taken to deduce which POM chemical species are responsible for the effects observed [43,65]. In fact, the stability of the POM compounds at all experimental conditions and times upon exposition should be ascertained, highlighting the importance of POMs' speciation for deducing their biological effects [43,65,66]. Determining POMs' speciation and stability under physiological conditions is critical for understanding their biological effects. Without this knowledge, POM biomedical applications could be compromised [66,67].

One year later, Chandry et al. studied the effects of decaniobate $[Nb_{10}O_{28}]^{6-}$ (abbreviated Nb_{10}) and monotitanoniobate $[TiNb_9O_{28}]$ (abbreviated $TiNb_{10}$) on S100A9 protein amyloide assembly (Figure 5) [68]. S100A9 is pro-inflammatory and amyloidogenic protein central to the amyloid-neuroinflammatory cascade in neurodegenerative diseases. The inhibition of fibril formation from 37.5 to 700 µM for Nb_{10} and $TiNb_9$, respectively, was verified. These polyoxoniobates (PONbs) inhibit amyloid aggregation at an early stage of the aggregation process of the S100A9 protein, and not when added to preformed amyloid fibrils. In fact, these POMs do not disaggregate preformed fibrils [68].

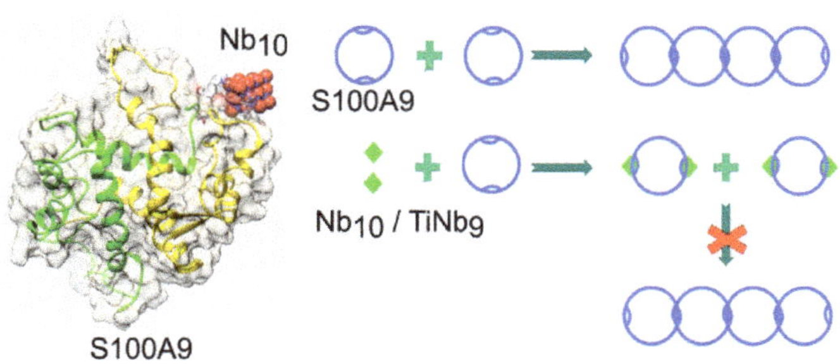

Figure 5. Niobium POMs, Nb_{10} and $TiNb_9$, can act as potent inhibitors of S100A9 amyloid assembly [68]. Reproduced from Ref. [68] with permission from American Chemical Society.

It was suggested that these PONb interactions are favored by a Lys-rich cluster in the S100A9 surface. The S100A9-Nb_{10} complex interacts with Lys50, Lys51 and Lys54, while the S100A9-$TiNb_9$ complex forms interactions additionally with Lys106 [68] (Figure 5). Aureliano's group and collaborators extensively studied the interactions of POMs with proteins [14,43,44,64,65] such as Nb_{10} and, particularly, the isostructural V_{10} [69]. In several cases, these POM–protein interactions are mainly electrostatic in nature between the negatively charged metal clusters and the positively charged regions of proteins [43,65]. For specific proteins such as actin, it was determined that V_{10} prevents G-actin aggregation into F-actin [43,65], a feature that can explain, at least in part, many of the POMs' associated biological activities, such as the ones preventing amyloid assembly. However, in the case of G-actin, it was found that the POMs' interaction is prevented by the native protein ligand ATP [69].

In 2022, Blasco et al. [70] focused on the dual functionality of POMs $K_4[\alpha\text{-}SiW_{12}O_{40}]$ and $K_8[\alpha\text{-}SiW_{11}O_{39}]$ as copper-chelating agents and inhibitors of Aβ aggregate formation [70]. Their results suggest that these POTs can prevent the deleterious effects of Cu(II) (Aβ) interactions through Cu(II) sequestration, giving rise to redox-inert complexes, thus inhibiting the production of ROS associated with Cu(II)-bound Aβ. Furthermore, they demonstrated that POMs can induce the formation of less toxic Aβ aggregates [70]. One year later, Wang et al. [71] described a $K_8\{[Co(H_2O)_4][HP_2Mo_5O_{23}]_2\}_2$ Strandberg-type POM (abbreviated CoPOM) with the ability to inhibit Aβ self-induced misfolding [71]. It is suggested that the CoPOM prevents Aβ40 from misfolding into itself. Thus, it interferes with the formation of β sheets. While CoPOM was able to reduce Aβ40 misfolding into fibrils, induced by Zn^{2+}, it has also reduced ROS formation in Aβ40 induced by Cu^{2+}. However, it is suggested that CoPOM works as an interfering agent, rather than a metal ion chelator [71].

In the same year, Lei et al. studied the antioxidant properties of Keggin-type POMos [72]. They analyzed the effects of 11 POMos for the scavenging capacities of ROS, particularly the anion superoxide and hydroxyl radical, in addition to measuring the total antioxidant capacity. As discussed above, POMos have been used for preventing oxidative stress in the brains of mice suffering from I/R [63]. Thus, it was verified that POMos present in vitro antioxidant activity, PMo_{12} being the one with the most potent scavenger activity [72]. On the contrary, as discussed above, POMs can induce ROS formation both directly and indirectly [64]. However, as previously reviewed, POMs' effects on oxidative stress may differ between in vitro and in vivo studies [43,64].

Recently (2025), molecular docking was used to predict the binding pose and affinities of POTs such as $ZrPW_{11}O_{39}{}^{3-}$, $SiW_{11}O_{39}{}^{7-}$ and $TeW_6O_{24}{}^{6-}$ to the top 10 targets associated with AD, such as Aβ, acetylcholinesterase (AChE) and butyryl acetylcholinesterase (BChE), among others [73]. POMs can synergistically address multiple hallmarks of AD by acting as multi-target agents. This includes their established role in inhibiting Aβ aggregation and simultaneously inhibiting acetylcholinesterase (AChE) by binding to its catalytic site. Through such binding, POMs offer the potential to mitigate APP cleavage, Aβ oligomer neurotoxicity and Aβ aggregation, thereby attenuating disease progression [74].

As mentioned above, molecular docking studies represent a powerful tool facilitating the development of novel treatments for AD [73]. In order to elucidate specific POM–peptide interactions, molecular docking and multiscale simulations can be used as a strategy to accelerate the rational design and optimization of next-generation POMs against AD [75].

The major side effects of clinically approved acetylcholinesterase inhibitors for Alzheimer's include a loss of appetite and weight loss, followed by agitation, insomnia and depression [76]. Although, as mentioned above, the toxicity of POMs with anti-acetylcholinesterase activity cannot be considered as severe, it was suggested that the potential clinical application would require a more complex toxicological study [62]. These studies reinforce the necessity for rigorous toxicological and pharmacokinetic POM studies in AD to be conducted in parallel with efficacy assessments in preclinical models. Therefore, establishing a robust safety profile is essential for any potential clinical development.

Moreover, a translational strategy for POM-based therapeutics and the development of POM informatics should promote the developed of appropriate drugs for fighting against AD, among others diseases and POMs applications [77,78].

4.2. Hybrid Polyoxometalate Studies in Alzheimer's Disease

Modified POMs, such as those with peptide functionalization, show an enhanced ability to target Aβ and block metal-induced aggregation, as described elsewhere [40]. In 2018, Ma et al. introduced a modified POV "CAM" $[CoL(H_2O)]_2[CoL]_2[HAsVMoV_6MoV_{I6}O_{40}]$, where L = 2-(1H-pyrazol-3-yl)pyridine and functions as a group targeting Aβ, giving it specificity [53]. This hybrid POV acts as a conformational modulator, converting the Aβ structure, rich in β sheets, into other less toxic conformers and potentially disrupting the aggregation pathway. It has been demonstrated that CAM prevents the formation of self-induced and metal-induced Aβ aggregates (Cu^{2+} and Zn^{2+}) through the formation of hydrogen bonds with β sheets.

One year later, Gao et al. [54] showed that D/L-amino acid-functionalized POMo derivatives bind to the Aβ13–23 segment and exhibit a strong enantioselectivity in inhibiting Aβ aggregation. A series of D- and L-amino acid-modified POMo ($MnMo_8O_{26}$) derivatives were synthesized, including positively charged amino acids (D-His and L-His) and negatively charged (D-Glu and L-Glu) and hydrophobic amino acids (D-Leu, L-Leu, D-Phe and L-Phe), to modulate Aβ aggregation [54]. It was suggested that chiral POT binds to the cationic cluster from His13 to Lys16 [54]. In the same year, Zhao et al. [55] explored the use of a platinum-substituted POT $(Me4N)_3[PW_{11}O_{40}(SiC_3H_6NH_2)_2PtCl_2)$ that prevents the oligomerization and fibril formation of Aβ [55]. The study highlights the interaction of Pt^{2+} platinum with an amine group of Aβ42 through electrostatic interactions, hydrogen bonds and Van der Waals forces, which not only inhibited Aβ deposition but also rescued memory loss in APP/PS1 transgenic mice (used as AD models).

In 2022, Hua et al. [79] studied the effects of a nanoscale POT base on a tetra-Cd cluster, sandwiched by trivacant Keggin-type tungstoarsenate POT, $(H_2dap)_6[Cd_4\text{-}Cl_2(B\text{-}\alpha\text{-}AsW_9O_{34})_2]\cdot 8H_2O$ (dap = 1,2-diaminopropane, abbreviated as CdAW), in the modulation of beta sheet-rich fibrils of Aβ peptides. CD was used to verify that CdAW can block the beta sheet conformation with or without Cu^{2+}/Zn^{2+} (Figure 6). Moreover, using the ThT assay, it was verified that, after the incubation of the Aβ40 solution (20 μM) with CdAW, a gradual decrease in fluorescence was observed from 15 to 120 min. As mentioned before, the ThT assay has been widely used to detect the β sheet content in Aβ aggregates [80]. These results point out that CdAW not only inhibits the formation of β sheet-rich aggregates but also has an ability to reverse the formed β sheet conformation [79].

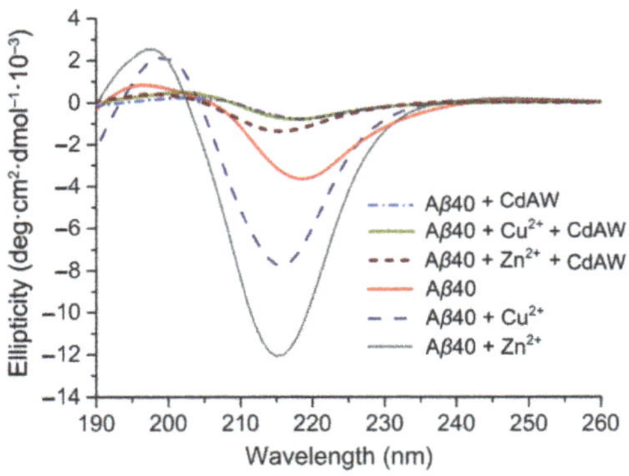

Figure 6. CD spectra of Aβ40 (20 μM) and Aβ40 in the presence of Cu^{2+} or Zn^{2+} (20 μM), with or without CdAW (20 μM), after incubation at 37 °C for 24 h [79]. Reproduced from Ref. [79] with permission from Royal Society of Chemistry.

In the same year, Gao et al. [81] followed previous studies but now presented a new approach to inhibiting amyloid aggregation through the chemical post-translational modification (PTM) of protein modification using POMs as inhibitors of Aβ aggregates [81]. Thus, after the POMs were modified with thiazolidinethione (TZ), the resulting POMD-TZ (POMD is α-$K_6P_2W_{18}O_{62}\cdot 14H_2O$) covalently modified Aβ site-selectively at Lys16. Thus, a new way of regulating amyloid aggregation by preventing inhibitor loss was suggested (Figure 7).

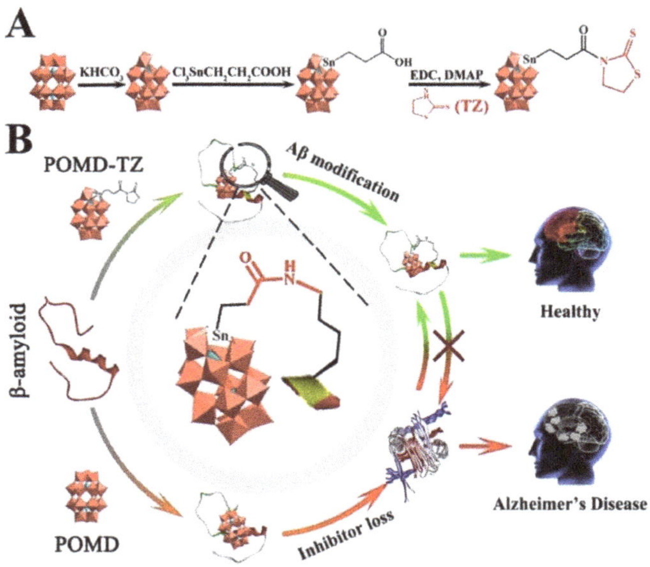

Figure 7. (**A**) Synthetic route to POMD-TZ (the chemical composition of POMD is α-$K_6P_2W_{18}O_{62}\cdot 14H_2O$). (**B**) POMD-TZ can specifically modify Aβ at the Lys16 site and inhibit Aβ aggregation [81]. Reproduced from Ref. [81] with permission from Royal Society of Chemistry.

Also in 2022, Hu et al. [82] suggested novel chiral Aβ inhibitors based on POTs $[(CH_3)_2NH_2]_{15}\{[\alpha\text{-}P_2W_{15}O_{55}(H_2O)]Zr_3(\mu 3\text{-}O)(H_2O)(L\text{-}tartH)[\alpha\text{-}P_2W_{16}O_{59}]\}$ (L-POM) and $[(CH_3)_2NH_2]_{15}\{[\alpha\text{-}P_2W_{15}O_{55}(H_2O)]Zr_3(\mu 3\text{-}O)(H_2O)(D\text{-}tartH)[\alpha\text{-}P_2W_{16}O_{59}]\}$ (D-POM). It was found that, in contrast to L-POT, D-POT displayed a higher Aβ40-binding affinity, greater brain biodistribution and the ability to reduce Aβ40 accumulation, scavenge ROS and better rescue memory deficits in a mouse model of AD. Thus, chiral POMs can improve the enantioselectivity and specificity for the Aβ protein and for AD treatment [81].

One year earlier, Díaz et al. [83] focused on the effects of decavanadate (V_{10}) and metforminium decavanadate on cognitive decline associated with pathophysiological processes of aging, such as increased ROS and neuroinflammation (also present in AD) [83]. As mentioned above, V_{10}, belonging to the class of POVs, is perhaps the most studied POM in biological systems, as well in biomedical applications [14,43,44,64,67], although studies in AD, to our knowledge, were not yet described. Their results indicated that these POMs modulate Nrf2 activity, a transcription factor that regulates the expression of antioxidant enzymes promoting redox balance and reducing oxidative stress and neuroinflammation [84]. POMs mitigate neurodegeneration by inhibiting microglial activation and modulating the Nrf2 signaling pathway by preventing ROS production and iron overload, thus reducing the formation of aggregates. Moreover, POMs' Nrf2 pathway activation leads to an increase in antioxidant enzymes, reducing the oxidative stress and inflammation that contribute to neurodegenerative diseases [85].

In 2025, it was shown that a nano-linear nickel-substituted Strandberg-type phosphomolybdate $(H_2en)_6\{[Ni(H_2O)_4](P_2Mo_5O_{23})\}_3$ (abbreviated as NiPM) (en = ethanediamine) can modulate the β sheet-rich conformation of Aβ aggregates [86]. The interaction between NiPM and Aβ involves metal ions in NiPM coordinating with the imidazole group of histidine residues present in the Aβ peptide. This interaction can influence Aβ's aggregation process, potentially affecting the formation of amyloid plaques associated with AD [86]. In another recent study, in order to further explore the inhibitory mechanisms of action of POMs in AD, a multiscale computational strategy integrating active-learning Bayesian Optimization (BO) and density functional theory (DFT) was employed to explore low-energy configurations of isolated amino acids, $[PMo_{12}O_{40}]^{3-}$–amino acid complexes and $[PMo_{12}O_{40}]^{3-}$–peptide systems [87]. The simulations predict that the cluster preferentially targets hydrophobic amino acids with alkyl chains (valine, lysine, leucine, isoleucine) located in peptide regions critical for Aβ aggregation, pointing out the role of multivalent weak interactions in POM-mediated inhibition [87].

4.3. Nanoparticle-Based Polyoxometalate Studies in Alzheimer's Disease

Besides using pure POMs [48], Li et al. [49] reported in another study the putative AD applications of a self-assembly of Aβ15–20 peptides and a POT ($K_8[P_2CoW_{17}O_{61}]$) formation of hybrid colloidal spheres in water. These POM@P hybrid particles were tested as bifunctional Aβ aggregation inhibitors (Figure 8). As evidenced by AFM and spectral analysis and demonstrated in PC12 cells, the POT@P was effective in inhibiting the aggregation of Aβ1–40 [49].

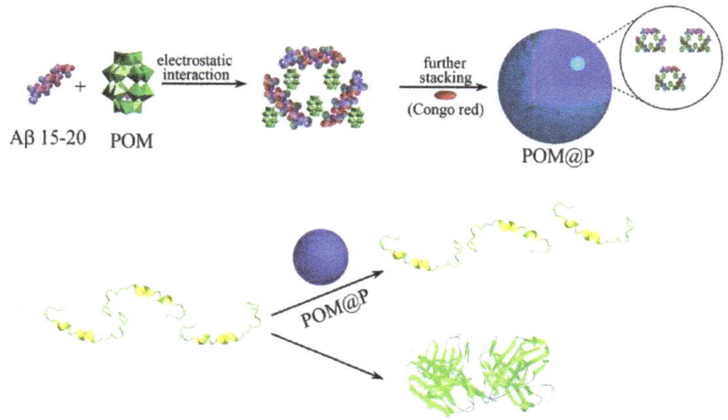

Figure 8. Schematic representation of the peptide–POT conjugates used for AD treatment. Upper panel: The schematic illustration of self-assembly of Aβ15–20 and POMs to hybrid spheres. Lower panel: The assembled peptide and POM nanoparticles can effectively inhibit Aβ1–40 aggregation [49]. Reproduced from Ref. [49] with permission from Wiley.

One year later (2014), Gao et al. studied a hybrid system that acts as a multifunctional inhibitor of Aβ [88]. This system combines POMs, gold nanorods (AuNRs) and a peptide with a specificity for Aβ. A synergistic effect was observed between the constituents of the system, culminating in the inhibition of Aβ aggregation, the dissociation of preformed Aβ fibrils and the decrease in Aβ-mediated peroxidase activity and Aβ-induced cytotoxicity. Furthermore, AuNRs functioned as a vehicle for crossing the blood–brain barrier [88]. In 2016, another study from Gao et al. also involved a hybrid system, this time Cerium/POMs, describing proteolytic and antioxidant effects, the latter similar to superoxide dismutase (SOD) [51]. This hybrid system was shown to effectively degrade Aβ aggregates and decrease oxidative stress (due to SOD activity), attenuating Aβ-mediated toxicity. Notably, it has also demonstrated the ability to inhibit microglial activation, thereby reducing neuroinflammation, and also to cross the blood–brain barrier (BBB) [51].

Two years later, also with proteolytic and antioxidant abilities as described previously [51], but now with a completely different approach, Gao et al. described a POM-based nanozyme, AuNPs@POMD, and octa-peptides, AuNPs@POMD-8pep, both with a protease-like activity for depleting Aβ aggregates (Figure 9). As before, an SOD-like activity for scavenging Aβ-mediated ROS was described. Furthermore, this nanozyme acts as a metal chelator to remove Cu from Cu-induced Aβ oligomers [89].

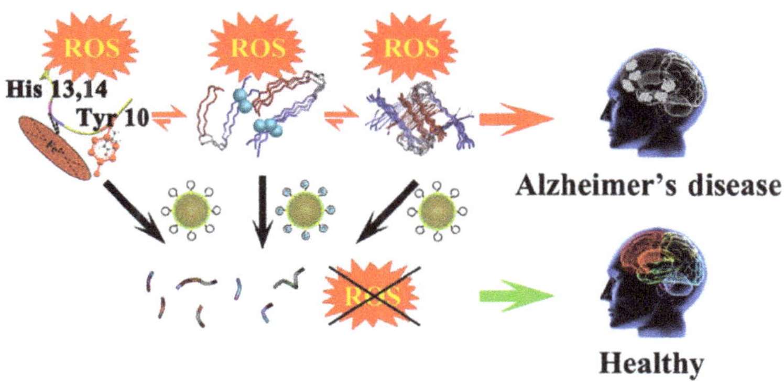

Figure 9. Proteolytic degradation of amyloid-β (Aβ) aggregates and clearance of Aβ-induced reactive oxygen species (ROS) by a POM based nanozyme, AuNPs@POMD, for the treatment of AD [89]. Reproduced from Ref. [89] with permission from Springer Nature.

In 2017, four years after the study described above by Li et al. [49], the same research group published a new study where they again studied photothermal therapy, while also using a hybrid system (AuP; Figure 10A) that combines the benefits of gold nanoparticles (AuNRs), POT ($K_8[P_2CoW_{17}O_{61}]$) and a specific Aβ inhibitor, Aβ15–20 [52]. This system relies on the strong NIR (near-infrared) optical absorption of AuNRs to generate localized hyperthermia, which effectively disaggregates the preformed Aβ fibrils (Figure 10B).

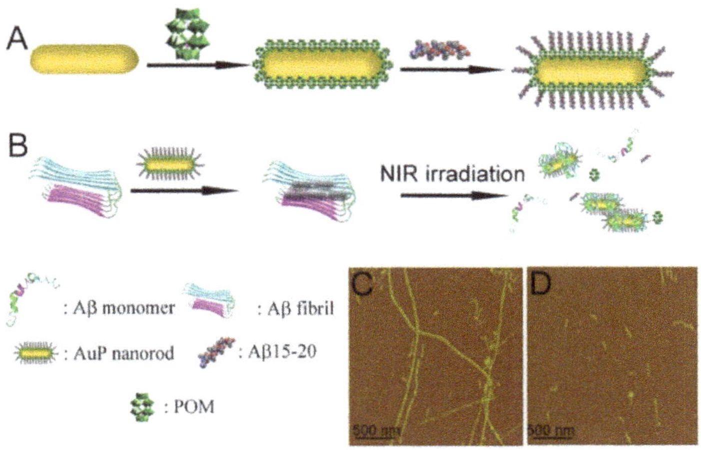

Figure 10. Degradation of Aβ fibrils by AuP. (**A**) Representative scheme of the AuP hybrid system; (**B**) representative scheme of the degradation mechanism of Aβ fibrils through NIR irradiation; (**C**) morphology of Aβ1–40 fibrils in the presence of AuP, without NIR irradiation. (**D**) Morphology of Aβ1–40 fibrils in the presence of AuP, with NIR irradiation. Reproduced from Ref. [52] with permission from Ivyspring International.

The results demonstrated the ability of AuP to disaggregate preformed Aβ fibrils after NIR irradiation (Figure 10D), reduce Aβ-induced cytotoxicity by about 45% and also cross the blood–brain barrier. The Aβ1–40 fibrils were incubated in an aggregation buffer (10 mM Tris, 150 mM NaCl, pH 7.3) with AuP for 20 min at 37 °C. Next, the morphology of Aβ fibrils was observed through atomic force microscopy (AFM). It was observed that Aβ15–20 cannot establish stable bonds with AuNRs. Thus, the POT ($P_2CoW_{17}O_{61}$) has the dual function of a link between AuNRs and Aβ15–20, and also functions as an inhibitor of fibril formation [52]. It was possible to observe long well-defined fibrils on the left side of the image, as well as clusters of fibrils near the right margin (Figure 10C). Dispersed Aβ segments of reduced size are observed in comparison to the fibrils present in C, without it being possible to observe evident aggregates, suggestive of the photocatalytic activity of AuP after NIR irradiation (Figure 10D).

One year later, Ma et al. [90] developed a redox-activated POM-based nanoplatform sensitive to NIR radiation [90]. The nanoplatform, (rPOMs@MSNs@copolymer), silica nanoparticles (MSNs), rPOMs (rPOMs with various structures, which included Wells–Dawson structure (rPOMDs), Keggin structure (rPOMKs) and Anderson structure (rPOMAs)) and the thermal responsive copolymer poly(*N*-isopropylacrylamide-*co*-acrylamide) are activated by NIR radiation, which induces a thermal response capable of disaggregating the Aβ fibrils. This response is particularly advantageous, as it allows non-invasive activation and a localized effect, reducing off-target effects. Furthermore, the nanoplatform demonstrated antioxidant properties that give it the ability to inhibit the production of ROS induced by Aβ, as well as the ability to cross the blood–brain barrier [90].

Compared with unmodified POMs, POM-based inorganic–organic hybrids and POM-based nanocomposite structures show a significantly enhanced bioactivity and reduced side effects in AD [40]. Moreover, POM-based nanomedicine developed hybrid and nanoparticle systems that enhance BBB penetration, increase efficiency and reduce toxicity. In fact, nanotechnology has emerged as a transformative strategy for precise brain-targeted treatment [91,92].

In 2019, Liu et al. verified that Zn induces Aβ aggregation, but this is prevented by peptide@Mo-POM nanoparticles [93]. Only with the peptide–POM is a strong prevention of beta aggregation observed. By preventing ROS production induced by Aβ, the peptide@Mo-POM nanoparticles block neuron shrinkage. It was also observed that the peptide enhanced BBB penetration and the affinity for Aβ species. No morphological changes and no toxicity effects were evaluated in several tissues. Moreover, it was suggested that Mo-POMs can also sequester Zn^{2+} [93].

In the year 2023, Perich et al. [57] developed a nanohybrid system (Figure 11A) consisting of gold nanoparticles covered with $[\beta 2\text{-}SiW_{11}O_{39}]^{8-}$ (Figure 11B) and polyethylene glycol (PEG), AuNPs@POM@PEG, with the aim of reducing the fibrillization of Aβ [61]. In order to evaluate the inhibitory capacity of AuNPs@POM@PEG in vitro, Aβ1–42 peptides were incubated with increasing concentrations of AuNPs@POM@PEG (1, 2.5 and 5 nM) for 2 h at 37 °C. The results demonstrated a strong inhibitory effect of AuNPs@POM@PEG, with an inhibition of approximately 75% for the highest concentration (5 nM) compared to the untreated Aβ sample (Figure 10C).

The above recent insights into the use of POMs in the treatment of AD has demonstrated a growing potential in different forms of action, particularly with regard to the inhibition of the formation of aggregates, one of the main pathophysiological hallmarks of the disease. Reflecting the increasing interest of the putative applications of POMs in AD treatment and research, several review papers have been published recently pointing out different points of view [40,45,75,94]. Moreover, POM nanoparticle-based systems have also emerged as promising tools for AD treatment not only because they improve drug delivery across the blood–brain barrier, but also because they improve efficiency, stability and bioavailability [92,95,96].

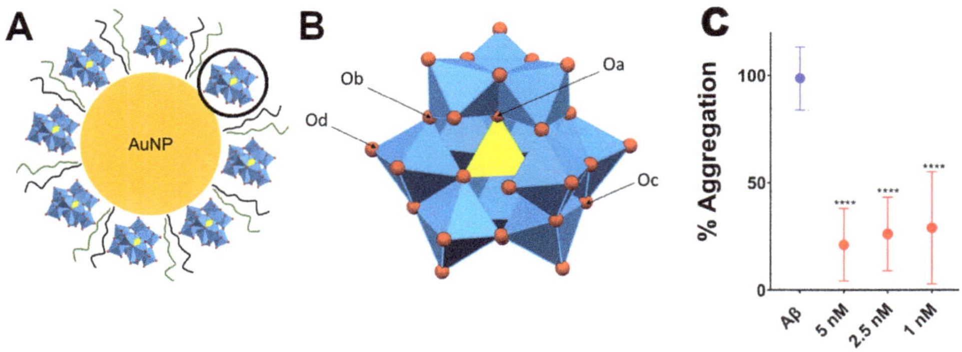

Figure 11. Inhibition of Aβ aggregation by AuNPs@POM@PEG. (**A**) Representative scheme of the AuNPs@POM@PEG system, where AuNP is the central gold nanoparticle, the surrounding blue structures are the POM and the intercalated filaments are the PEG. (**B**) Schematic representation of the POM structure. In yellow you can see the silicon core (Si) and in blue the tungstates. The red spheres correspond to oxygen atoms. (**C**) The % aggregation of Aβ fibrils compared to untreated Aβ. **** $p < 0.0001$. Reproduced from Ref. [57] with permission from MDPI.

In summary, from the selected papers described above in Figure 2, in addition to Aβ aggregation, we can find other effects induced by POMs, such as the following: (1) the disaggregation of Aβ fibrils; (2) acetylcholinesterase inhibition activity; (3) SOD activity; (4) reduced neuroinflammation; (5) reduced ROS; (6) reduced Aβ citotoxicity; (7) reduced peroxidase-like activity; (8) the prevention of Aβ conformational shift (Figure 12).

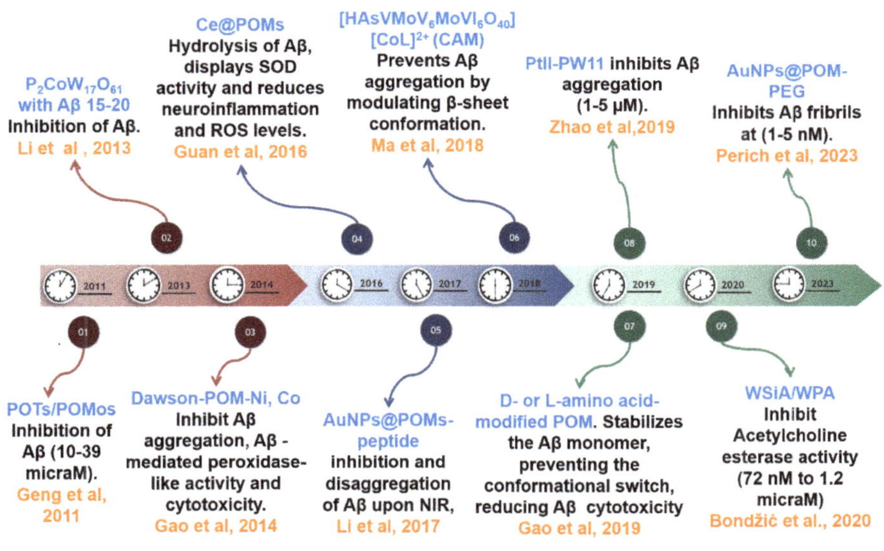

Figure 12. Timeline of POMs' effects in Alzheimer's disease [48–57].

However, several other effects were described, such as the chelation of transition metals (4 articles) and the improvement of memory (2 articles), among others (Figure 13). The main effect observed in the selected articles was exactly the inhibition of Aβ aggregation (76%), with the majority of the studies (23 articles, from 30) describing this result [48–57]. Other major effects found to be promoted by POMs included the following: (i) the decrease in neurotoxicity induced by Aβ (66%); (ii) the crossing of the blood–brain barrier (53%); (iii) the decrease in ROS (33%); and (iv) Aβ degradation (26%). Other effects, such as the inhibition of AChE (13%) and transition metal chelators (13%) among other effects, were described only in a few studies (Figure 13).

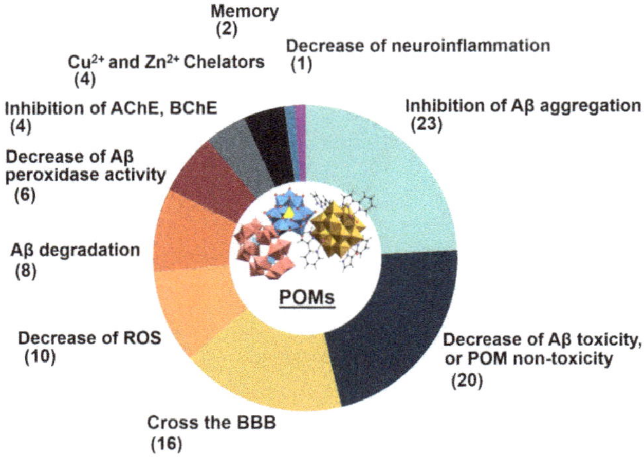

Figure 13. POMs' effects in Alzheimer's disease studies from 2011 to 2025. Note that the numbers in parentheses refer to the number of articles, from a total of 30.

Globally, the majority of the POM/AD studies used POTs (62%), but also POMos (32%), PONbs (5%) and POVs (2%) (Table S1). However, in the first seven years (2011–2018), we found that the majority of the studies were performed with POTs (80%), with minor contributions from POMos (13%) and POVs (7%), whereas, in the last seven years (2019–2025), POMos (44%) increased but were still below POTs (50%), while PONbs (6%) also had a fresh contribution. Among the POMs described above, the ones most referred to in the AD studies were W_{11} (six articles) followed by W_{17} and W_{12} (three articles). The first one (W_{11}) was said to (1) prevent Aβ fibrillation (2011); (2) be used in a platinum-substituted POT $(Me4N)_3[PW_{11}O_{40}(SiC_3H_6NH_2)_2PtCl_2)$ that prevents the oligomerization and fibril formation of Aβ [55] (2018); (3) induce achetylcolinesterase inhibition (2020); (4) be a copper-chelating agent (2022) [58]; (5) be also present in a nanohybrid system consisting of gold nanoparticles covered with $[\beta 2\text{-}SiW_{11}O_{39}]^{8-}$ and polyethylene glycol (AuNPs@POM@PEG), with the aim of reducing the fibrillization of Aβ [61] (2023); and (6) be analyzed in docking studies as a POM target for AD. Conversely, the Keggin-type W_{12} also interacts with Aβ1–40, affecting the degree of aggregation (2011) [50], and also is a potent acetylcholinesterase inhibitor (2011) [60] and a copper-chelating agent and inhibitor of Aβ aggregate formation (2020) [68]. Regarding the POMs present in the nanohybrid systems, the most common POM used is the Wells–Dawson-type P_2W_{17} (Table S1).

While much AD research has focused on amyloid-beta and tau proteins, the role of POMs in modulating calcium signaling offers a promising but underexplored alternative therapeutic strategy [97]. In fact, calcium dysregulation also plays a key role in AD progression [98]. As described above, AD chronic states of neuroinflammation and oxidative stress, together with the disturbance of calcium homeostasis, contribute to neurotoxicity and cell death [27]. A growing number of studies identify the dysregulation of calcium (Ca^{2+}) homeostasis, particularly the anomalous increase in cytosolic Ca^{2+} concentrations, as being implicated in the cascade of pathophysiological events that trigger the clinical picture of AD [28]. Recent studies suggest that calcium dysregulation is not just a consequence of the pathological processes of AD, but rather a central player in the pathogenesis of the disease [99]. It is becoming increasingly clear that strategies aimed at correcting or modulating calcium signaling pathways constitute promising approaches for the study of therapeutic alternatives for AD [100]. Recently, polyoxometalates (POMs) were described as presenting agonistic properties on purinergic P2 receptors from neuron cells [47]. Thus, POMs inhibited the P-type ATPases [46] and also modulated the cytosolic calcium concentrations in neurons, proving to be a useful tool in the studies of pathological processes of AD [46,47], as well as in others neurodegenerative diseases [101].

Taking into consideration the articles described in the present review, interdisciplinary collaborations are fundamental for bringing together experienced and also young researchers for the evolution of the applications of POMs in the treatment of AD. Moreover, we believe that inorganic biochemistry also represents an extraordinary platform for promoting communication between Alzheimer's caregivers, doctors, psychologists, nurses and family members, just like a key cog in the "clock of the knowledge", dynamically moving chemistry, biology and medicine, among others, to promote innovation and urgent developments in the treatment of AD [102,103].

As summarized in Figure 12, the POM studies in AD started in 2011, with POMs such as W_{11} and W_{17} being pure or in nanoparticle (NP) systems and finished also with W_{11} that was pure and/or in NPs, but now with a higher potency of inhibition, that is, with ranging nM concentrations as inhibitors of AChE activity and Aβ aggregation, well-known hallmarks of AD. Therefore, POMs' progress in AD shows an increased efficiency that, together with a decreased toxicity, potentiate the development of more effective and sustainable applications in this challenging 21st-century neurodegenerative disease (Figure 14). In fact, Figure 14 was based on the observation of a relative with AD who, upon arriving at the beach, sat with her back to the sea, as if disoriented. It is suggested that POMs can reverse this disorientation and allow a return to the sun (Figure 14). However, it should be realized that POMs' putative applications in the treatment of AD are yet in the early stages. Collaborations from several fields and points of view are absolutely essential for the progress of the usage of new drugs against this disease, which is devastating to the lives of patients and their families.

Figure 14. Sunflower on the beach: Alzheimer's disease and polyoxometalates.

5. Conclusions and Outlook

Polyoxometalates are evolving in AD treatment through an increased understanding of their roles in disease and the development of POM-based therapies, including hybrids and POM nanoparticles (NPs@POMs) for preventing Aβ aggregates and the subsequent neurotoxicity induced by β-amyloid aggregates and decreasing ROS production and neuroinflammation, in addition to restoring memory and sequestering transition metals ions such as Zn^{2+} and Cu^{2+}, features that are well known to be associated with the pathology of Alzheimer's disease.

Herein, we investigated the potential of POMs in modulating the pathophysiology of Alzheimer's disease, with a special focus on the possible associated neuroprotective effects. From an in-depth analysis of the studies, relevant data emerged on how polyoxometalates interact with the pathogenic mechanisms of Alzheimer's disease. The majority of the AD studies use POTs (62%), but also POMos (32%), PONbs (5%) and POVs (2%), W_{11} being the most studied. Not all are pure POMs (46%), and we also found hybrid POMs (27%) and hybrid nanoparticle POM systems (27%).

Notably, most studies have focused on interactions with Aβ aggregates, which play a central role in the cascade of pathological events present in Alzheimer's disease. POMs have the ability to reverse the formed β sheet conformation and also to induce the disaggregation of Aβ fibrils, particularly after NIR and/or UV/Vis irradiation. POMs were also described as a novel class of acetylcholinesterase inhibitors known to improve cognitive function in AD. Thus, by inhibiting acetylcholinesterase and the formation of protein aggregates, or by removing the formed aggregates, POMs represent a promising therapeutic strategy for AD.

In the pathology of Alzheimer's disease, other processes can be prevented by POMs, pointing to putative applications in AD treatment. Future studies will point out POMs' ability to inhibit microglial activation, thereby reducing neuroinflammation, and also to cross the blood–brain barrier. Still, applications of POMs in AD could be associated with many other mechanisms of action. For instance, we cannot exclude the possibility that POMs can be used as the modulators of neurons for calcium homeostasis, since it is also known that neurodegenerative diseases are associated with calcium dysregulation.

Putting it all together, the studies performed so far suggested that, in several ways, POMs can be used in AD treatment, though it should be stated that putative applications in the treatment of the disease are yet in their early stages. In fact, while the applications of POMs in cancer and in bacterial and/or viral infections are clearly in progress, POM studies in neurodegenerative diseases are still very few. Moreover, a current limitation, besides the paucity of in vivo validation, is the need for greater interdisciplinary collaboration between chemistry, neuroscience and pharmacology. We believe that the present review, with an inorganic biochemistry perspective on the evolution of the utilization of POMs in AD studies, will push forward those interdisciplinary collaborations, particularly with neurosciences, in order to promote new insights into POM applications in AD, among other neurodegenerative diseases with overlapping proteinopathies.

Supplementary Materials: The following supporting information can be downloaded at https://www.mdpi.com/article/10.3390/biochem5040041/s1. Table S1: POM studies associated with Alzheimer´s disease (2011–2025). Year, POM information and POM composition, referred to as pure (P), combined with a organic ligand (H) or POM nanoparticles-based compounds (NP), and the classes of POMs, namely polyoxotungstates (POTs), polyoxomolybdates (POMos), polyoxobiobates (PONbs) and polyoxovanadates (POVs) [48–63,68,70–73,79,81,82,86–90,93].

Author Contributions: Conceptualization, M.A.; methodology, M.A., D.M.R. and J.M.; validation, M.A.; formal analysis, M.A.; investigation, M.A., D.M.R. and J.M.; resources, M.A.; data curation, M.A.; writing—M.A.; writing—review and editing, M.A., D.M.R. and J.M.; visualization, M.A., D.M.R. and J.M.; supervision, M.A.; project administration, M.A.; funding acquisition, M.A. All authors have read and agreed to the published version of the manuscript.

Funding: Thanks to Algarve University and to Portuguese national funds from FCT—Foundation for Science and Technology, through contracts UID/04326/2025, UID/PRR/04326/2025 and LA/P/0101/2020 (DOI:10.54499/LA/P/0101/2020).

Data Availability Statement: The original contributions presented in this study are included in the article/Supplementary Material. Further inquiries can be directed to the corresponding author.

Acknowledgments: M.A. thanks Algarve University and Portuguese national funds from FCT—Foundation for Science and Technology through contracts UID/04326/2025, UID/PRR/04326/2025 and LA/P/0101/2020.

Conflicts of Interest: The authors declare no conflicts of interest.

Abbreviations

The following abbreviations are used in this manuscript:

AD	Alzheimer's disease
AChE	Achetylcolinesterase
AFM	Atomic force microscopy
APP	Amyloid precursor protein
AuNRs	Gold nanorods
Aβ	β-amyloid
BBB	Blood–brain barrier
BO	Bayesian Optimization
CD	Circular dicroism
CNS	Central nervous system
DFT	Density functional theory
DLS	Dynamic light scattering
IC_{50}	Half maximal inhibitory concentration
IL6	Interleukin-6
ITC	Isothermal titration calorimetry
LPO	Lipid peroxidation
NFT	Neurofibrillary tangles
NIR	Near-infrared radiation
NMDA	N-methyl-D-aspartate
NPs	Nanoparticles
Nrf2	Transcription factor
P2	Purinergic P2 receptors
PBS	Phosphate-buffered solution
PC12	Cell line from rat pheochromocytoma
PEG	Polyethylene glycol
POMo	Polyoxmolybdates
POMs	Polyoxometalates
POVs	Polyoxovanadates
PS1	Presenilin-1
PS2	Presenilin-2
PTM	Post-translational modification
ROS	Reactive oxygen species
SOD	Superoxide dismutase
TNF	Tumor necrosis factor
TZ	Thiazolidinethione
WHO	World Health Organization

References

1. Morris, J.C. Neurodegenerative Disorders of Aging: The Down Side of Rising Longevity. *Mo Med.* **2013**, *110*, 393–3944.
2. Dugger, B.N.; Dickson, D.W. Pathology of Neurodegenerative Diseases. *Cold Spring Harb. Perspect. Biol.* **2017**, *9*, a028035. [CrossRef]
3. Heemels, M.T. Neurodegenerative diseases. *Nature* **2016**, *539*, 179. [CrossRef] [PubMed]
4. Wakabayashi, K. Cellular pathology of neurodegenerative disorders. *Clin. Neurol.* **2013**, *53*, 609–617.
5. Davis, A.A.; Leyns, C.E.G.; Holtzman, D.M. Intercellular Spread of Protein Aggregates in Neurodegenerative Disease. *Annu. Rev. Cell Dev. Biol.* **2018**, *34*, 545–568. [CrossRef]

6. Lin, M.T.; Beal, M.F. Mitochondrial dysfunction and oxidative stress in neurodegenerative diseases. *Nature* **2006**, *443*, 787–795. [CrossRef]
7. Chen, X.; Guo, C.; Kong, J. Oxidative stress in neurodegenerative diseases. *Neural Regen. Res.* **2012**, *7*, 376.
8. Lamptey, R.N.L.; Chaulagain, B.; Trivedi, R.; Gothwal, A.; Layek, B.; Singh, J. A Review of the Common Neurodegenerative Disorders: Current Therapeutic Approaches and the Potential Role of Nanotherapeutics. *Int. J. Mol. Sci.* **2022**, *23*, 1851. [CrossRef]
9. Breijyeh, Z.; Karaman, R. Comprehensive Review on Alzheimer's Disease: Causes and Treatment. *Molecules* **2020**, *25*, 5789. [CrossRef] [PubMed]
10. Anand, P.; Singh, B. A review on cholinesterase inhibitors for Alzheimer's disease. *Arch. Pharm. Res.* **2013**, *36*, 375–399. [CrossRef] [PubMed]
11. Passeri, E.; Elkhoury, K.; Morsink, M.; Broersen, K.; Linder, M.; Tamayol, A.; Malaplate, C.; Yen, F.T.; Arab-Tehrany, E. Alzheimer's Disease: Treatment Strategies and Their Limitations. *Int. J. Mol. Sci.* **2022**, *23*, 13954. [CrossRef]
12. Frozza, R.L.; Lourenco, M.V.; de Felice, F.G. Challenges for Alzheimer's Disease Therapy: Insights from Novel Mechanisms Beyond Memory Defects. *Front. Neurosci.* **2018**, *12*, 37. [CrossRef] [PubMed]
13. Pardridge, W.M. Treatment of Alzheimer's Disease and Blood–Brain Barrier Drug Delivery. *Pharmaceuticals* **2020**, *13*, 394. [CrossRef]
14. Aureliano, M.; Mitchell, S.G.; Yin, P. Editorial: Emerging polyoxometalates with biological, biomedical, and health applications. *Front. Chem.* **2022**, *10*, 977317. [CrossRef]
15. Čolović, M.B.; Lacković, M.; Lalatović, J.; Mougharbel, A.S.; Kortz, U.; Krstić, D.Z. Polyoxometalates in Biomedicine: Update and Overview. *Curr. Med. Chem.* **2020**, *27*, 362–379. [CrossRef] [PubMed]
16. Woźniak-Budych, M.J.; Staszak, K.; Bajek, A.; Pniewski, F.; Jastrząb, R.; Staszak, M.; Tylkowski, B.; Wieszczycka, K. The future of polyoxymetalates for biological and chemical apllications. *Coord. Chem. Rev.* **2023**, *493*, 215306. [CrossRef]
17. Georges, J.; Bintener, C.; Miller, O. *Dementia in Europe Yearbook 2019: Estimating the prevalence of dementia in Europe*; Alzheimer Europe: Sandweiler, Luxembourg, 2020; pp. 1–108.
18. GBD 2019 Dementia Forecasting Collaborators. Estimation of the global prevalence of dementia in 2019 and forecasted prevalence in 2050: An analysis for the Global Burden of Disease Study 2019. *Lancet Public Health* **2022**, *7*, e105–e125. [CrossRef]
19. DeTure, M.A.; Dickson, D.W. The neuropathological diagnosis of Alzheimer's disease. *Mol. Neurodegener.* **2019**, *14*, 32. [CrossRef]
20. Igarashi, K.M. Entorhinal cortex dysfunction in Alzheimer's disease. *Trends Neurosci.* **2023**, *46*, 124. [CrossRef]
21. Hippius, H.; Neundörfer, G. The discovery of Alzheimer's disease. *Dialogues Clin. Neurosci.* **2003**, *5*, 101–108. [CrossRef]
22. Madnani, R.S. Alzheimer's disease: A mini-review for the clinician. *Front. Neurol.* **2023**, *14*, 1178588. [CrossRef] [PubMed]
23. Silva, M.V.F.; Loures, C.d.M.G.; Alves, L.C.V.; De Souza, L.C.; Borges, K.B.; Carvalho, M.G. Alzheimer's disease: Risk factors and potentially protective measures. *J. Biomed. Sci.* **2019**, *26*, 33. [CrossRef]
24. Chen, G.F.; Xu, T.H.; Yan, Y.; Zhou, Y.R.; Jiang, Y.; Melcher, K.; Xu, H.E. Amyloid beta: Structure, biology and structure-based therapeutic development. *Acta Pharmacol. Sin.* **2017**, *38*, 1205–1235. [CrossRef]
25. Coronel, R.; Bernabeu-Zornoza, A.; Palmer, C.; Muñiz-Moreno, M.; Zambrano, A.; Cano, E.; Liste, I. Role of Amyloid Precursor Protein (APP) and Its Derivatives in the Biology and Cell Fate Specification of Neural Stem Cells. *Mol. Neurobiol.* **2018**, *55*, 7107–7117. [CrossRef]
26. Griffiths, J.; Grant, S.G.N. Synapse pathology in Alzheimer's disease. *Semin. Cell. Dev. Biol.* **2023**, *139*, 13–23. [CrossRef]
27. Woodburn, S.C.; Bollinger, J.L.; Wohleb, E.S. The semantics of microglia activation: Neuroinflammation, homeostasis, and stress. *J. Neuroinflamm.* **2021**, *18*, 258. [CrossRef] [PubMed]
28. Joshi, M.; Joshi, S.; Khambete, M.; Degani, M. Role of calcium dysregulation in Alzheimer's disease and its therapeutic implications. *Chem. Biol. Drug Des.* **2023**, *101*, 453–468. [CrossRef] [PubMed]
29. Liu, J.; Chang, L.; Song, Y.; Li, H.; Wu, Y. The Role of NMDA Receptors in Alzheimer's Disease. *Front. Neurosci.* **2019**, *13*, 43. [CrossRef]
30. Tatulian, S.A. Challenges and hopes for Alzheimer's disease. *Drug Discov. Today* **2022**, *27*, 1027–1043. [CrossRef]
31. Pardridge, W.M. Alzheimer's disease drug development and the problem of the blood-brain barrier. *Alzheimer's Dement.* **2009**, *5*, 427. [CrossRef]
32. Gao, Y.; Choudhari, M.; Such, G.K.; Ritchie, C. Polyoxometalates as chemically and structurally versatile components in self-assembled materials. *Chem. Sci.* **2022**, *13*, 2510–2527. [CrossRef]
33. Hill, C.L. Introduction: Polyoxometalates Multicomponent Molecular Vehicles To Probe Fundamental Issues and Practical Problems. *Chem. Rev.* **1998**, *98*, 1–2. [CrossRef] [PubMed]
34. Gumerova, N.I.; Rompel, A. Synthesis, structures and applications of electron-rich polyoxometalates. *Nat. Rev. Chem.* **2018**, *2*, 112. [CrossRef]
35. Chen, Y.; Li, F.; Li, S.; Zhang, L.; Sun, M. A review of application and prospect for polyoxometalate-based composites in electrochemical sensor. *Inorg. Chem. Commun.* **2022**, *135*, 109084. [CrossRef]

36. Liu, R.; Streb, C. Polyoxometalate-Single Atom Catalysts (POM-SACs) in Energy Research and Catalysis. *Adv. Energy Mater.* **2021**, *11*, 2101120. [CrossRef]
37. Samaraj, E.; Balaraman, E.; Manickam, S. Functional POM-catalyst for selective oxidative dehydrogenative couplings under aerobic conditions. *Mol. Catal.* **2021**, *502*, 111396. [CrossRef]
38. Asif, H.M.; Shakoor, Z.; Ibraheem, S.; Salama, A.M.; Khan, M.A.; Nguyen, T.A.; Yasin, G. Chapter 12—Polyoxometalate-based metal organic frameworks (POMOFs) for lithium-ion batteries. In *Micro and Nano Technologies; Metal-Organic Framework-Based Nanomaterials for Energy Conversion and Storage*; Gupta, R.K., Nguyen, T.A., Yasin, G., Eds.; Elsevier: Amsterdam, The Netherlands, 2022; pp. 245–268. [CrossRef]
39. Zhang, Y.; Li, Y.; Guo, H.; Guo, Y.; Song, R. Recent advances in polyoxometalate-based materials and their derivatives for electrocatalysis and energy storage. *Mater. Chem. Front.* **2024**, *8*, 732–768. [CrossRef]
40. Wang, X.; Wei, S.; Zhao, C.; Li, X.; Jin, J.; Shi, X.; Su, Z.; Li, J.; Wang, J. Promising application of polyoxometalates in the treatment of cancer, infectious diseases and Alzheimer's disease. *J. Biol. Inorg. Chem.* **2022**, *27*, 405–419. [CrossRef]
41. Li, B.; Xu, X.; Lv, Y.; Wu, Z.; He, L.; Song, Y.F. Polyoxometalates as Potential Artificial Enzymes toward Biological Applications. *Small* **2024**, *20*, e2305539. [CrossRef]
42. Bijelic, A.; Aureliano, M.; Rompel, A. Polyoxometalates as Potential Next-Generation Metallodrugs in the Combat Against Cancer. *Angew. Chem. Int. Ed.* **2019**, *58*, 2980–2999. [CrossRef]
43. Aureliano, M.; Gumerova, N.I.; Sciortino, G.; Garribba, E.; Rompel, A.; Crans, D.C. Polyoxovanadates with Emerging Biomedical Activities. *Coord. Chem. Rev.* **2021**, *447*, 214143. [CrossRef]
44. De Sousa-Coelho, A.L.; Fraqueza, G.; Aureliano, M. Repurposing Therapeutic Drugs Complexed to Vanadium in Cancer. *Pharmaceuticals* **2024**, *17*, 12. [CrossRef]
45. Gonzalez-Cano, S.I.; Flores, G.; Guevara, J.; Morales-Medina, J.C.; Treviño, S.; Diaz, A. Polyoxidovanadates a new therapeutic alternative for neurodegenerative and aging diseases. *Neural Regen. Res.* **2024**, *19*, 571–577. [CrossRef]
46. Aureliano, M.; Fraqueza, G.; Berrocal, M.; Cordoba-Granados, J.J.; Gumerova, N.I.; Rompel, A.; Gutierrez-Merino, C.; Mata, A.M. Inhibition of SERCA and PMCA Ca^{2+}-ATPase Activities by Polyoxotungstates. *J. Inorg. Biochem.* **2022**, *236*, 111952. [CrossRef] [PubMed]
47. Poejo, J.; Gumerova, N.I.; Rompel, A.; Mata, A.M.; Aureliano, M.; Gutierrez-Merino, C. Unveiling the Agonistic Properties of Preyssler-Type Polyoxotungstates on Purinergic P2 Receptors. *J. Inorg. Biochem.* **2024**, *259*, 112640. [CrossRef] [PubMed]
48. Geng, J.; Li, M.; Ren, J.; Wang, E.; Qu, X. Polyoxometalates as Inhibitors of the Aggregation of Amyloid β Peptides Associated with Alzheimer's Disease. *Angew. Chem. Int. Ed.* **2011**, *50*, 4184–4188. [CrossRef]
49. Li, M.; Xu, C.; Wu, L.; Ren, J.; Wang, E.; Qu, X. Self-assembled peptide-polyoxometalate hybrid nanospheres: Two in one enhances targeted inhibition of amyloid B-peptide aggregation associated with Alzheimer's disease. *Small* **2013**, *9*, 3455–3461. [CrossRef] [PubMed]
50. Gao, N.; Sun, H.; Dong, K.; Ren, J.; Duan, T.; Xu, C.; Qu, X. Transition-metal-substituted polyoxometalate derivatives as functional anti-amyloid agents for Alzheimer's disease. *Nat Commun.* **2014**, *5*, 3422. [CrossRef] [PubMed]
51. Guan, Y.; Li, M.; Dong, K.; Gao, N.; Ren, J.; Zheng, Y.; Qu, X. Ceria/POMs hybrid nanoparticles as a mimicking metallopeptidase for treatment of neurotoxicity of amyloid-β peptide. *Biomaterials* **2016**, *98*, 92–102. [CrossRef]
52. Li, M.; Guan, Y.; Zhao, A.; Ren, J.; Qu, X. Using Multifunctional Peptide Conjugated Au Nanorods for Monitoring β-amyloid Aggregation and Chemo-Photothermal Treatment of Alzheimer's Disease. *Theranostics* **2017**, *7*, 2996–3006. [CrossRef]
53. Ma, X.; Hua, J.; Wang, K.; Zhang, H.; Zhang, C.; He, Y.; Guo, Z.; Wang, X. Modulating Conformation of Aβ-Peptide: An Effective Way to Prevent Protein-Misfolding Disease. *Inorg. Chem.* **2018**, *57*, 13533–13543. [CrossRef]
54. Gao, N.; Du, Z.; Guan, Y.; Dong, K.; Ren, J.; Qu, X. Chirality-Selected Chemical Modulation of Amyloid Aggregation. *J. Am. Chem. Soc.* **2019**, *141*, 6915–6921. [CrossRef]
55. Zhao, J.; Li, K.; Wan, K.; Sun, T.; Zheng, N.; Zhu, F.; Ma, J.; Jiao, J.; Li, T.; Ni, J.; et al. Organoplatinum-Substituted Polyoxometalate Inhibits β-amyloid Aggregation for Alzheimer's Therapy. *Angew. Chem. Int. Ed. Engl.* **2019**, *58*, 18032–18039. [CrossRef]
56. Bondžić, A.M.; Lazarević-Pašti, T.D.; Leskovac, A.R.; Petrović, S.Ž.; Čolović, M.B.; Parac-Vogt, T.N.; Janjić, G.V. A new acetylcholinesterase allosteric site responsible for binding voluminous negatively charged molecules—The role in the mechanism of AChE inhibition. *Eur. J. Pharm. Sci.* **2020**, *151*, 105376. [CrossRef]
57. Perxés Perich, M.; Palma-Florez, S.; Solé, C.; Goberna-Ferrón, S.; Samitier, J.; Gómez-Romero, P.; Mir, M.; Lagunas, A. Polyoxometalate-Decorated Gold Nanoparticles Inhibit β-Amyloid Aggregation and Cross the Blood–Brain Barrier in a µphysiological Model. *Nanomaterials* **2023**, *13*, 2697. [CrossRef]
58. Zhou, Y.; Zheng, L.; Han, F.; Zhang, G.; Ma, Y.; Yao, J.; Keita, B.; Oliveira, P.; Nadjo, L. Inhibition of amyloid-β protein fibrillization upon interaction with polyoxometalates nanoclusters, *Colloids Surf. A Physicochem. Eng. Asp.* **2011**, *375*, 97–101. [CrossRef]
59. Li, M.; Xu, C.; Ren, J.; Wang, E.; Qu, X. Photodegradation of β-sheet amyloid fibrils associated with Alzheimer's disease by using polyoxometalates as photocatalysts. *Chem. Commun.* **2013**, *49*, 11394–11396. [CrossRef] [PubMed]

60. Iqbal, J.; Barsukova-Stuckart, M.; Ibrahim, M.; Ali, S.U.; Khan, A.A.; Kortz, U. Polyoxometalates as potent inhibitors for acetyl and butyrylcholinesterases and as potential drugs for the treatment of Alzheimer's disease. *Med. Chem. Res.* **2013**, *22*, 1224–1228. [CrossRef]
61. Chen, Q.; Yang, L.; Zheng, C.; Zheng, W.; Zhang, J.; Zhou, Y.; Liu, J. Mo polyoxometalate nanoclusters capable of inhibiting the aggregation of Aβ-peptide associated with Alzheimer's disease. *Nanoscale* **2014**, *6*, 6886–6897. [CrossRef]
62. Čolović, M.B.; Medić, B.; Ćetković, M.; Kravić Stevović, T.; Stojanović, M.; Ayass, W.W.; Mougharbel, A.S.; Radenković, M.; Prostran, M.; Kortz, U.; et al. Toxicity evaluation of two polyoxotungstates with anti-acetylcholinesterase activity. *Toxicol. Appl. Pharmacol.* **2017**, *333*, 68–75. [CrossRef] [PubMed]
63. Li, S.; Jiang, D.; Ehlerding, E.B.; Rosenkrans, Z.T.; Engle, J.W.; Wang, Y.; Liu, H.; Ni, D.; Cai, W. Intrathecal Administration of Nanoclusters for Protecting Neurons against Oxidative Stress in Cerebral Ischemia/Reperfusion Injury. *ACS Nano.* **2019**, *13*, 13382–13389. [CrossRef]
64. Aureliano, M.; De Sousa-Coelho, A.L.; Dolan, C.C.; Roess, D.A.; Crans, D.C. Biological Consequences of Vanadium Effects on Formation of Reactive Oxygen Species and Lipid Peroxidation. *Int. J. Mol. Sci.* **2023**, *24*, 5382. [CrossRef]
65. Aureliano, M.; Gumerova, N.I.; Sciortino, G.; Garribba, E.; McLauchlan, C.C.; Rompel, A.; Crans, D.C. Polyoxidovanadates' Interactions with Proteins: An Overview. *Coord. Chem. Rev.* **2022**, *454*, 214344. [CrossRef]
66. Gumerova, N.I.; Rompel, A. Polyoxometalates in Solution: Speciation under Spotlight. *Chem. Soc. Rev.* **2020**, *49*, 7568–7601. [CrossRef]
67. Gumerova, N.I.; Rompel, A. Speciation atlas of polyoxometalates in aqueous solutions. *Sci Adv.* **2023**, *9*, eadi0814. [CrossRef] [PubMed] [PubMed Central]
68. Chaudhary, H.; Iashchishyn, I.A.; Romanova, N.V.; Rambaran, M.A.; Musteikyte, G.; Smirnovas, V.; Holmboe, M.; Ohlin, C.A.; Svedružić, Z.M.; Morozova-Roche, L.A. Polyoxometalates as effective nano-inhibitors of amyloid aggregation of pro-inflammatory S100A9 protein involved in neurodegenerative diseases. *ACS Appl. Mater. Interfaces* **2021**, *13*, 26721–26734. [CrossRef]
69. Sciortino, G.; Aureliano, M.; Garribba, E. Rationalizing the Decavanadate(V) and Oxidovanadium(IV) Binding to G-Actin and the Competition with Decaniobate(V) and ATP. *Inorg. Chem.* **2021**, *60*, 334–344. [CrossRef] [PubMed]
70. Atrian-Blasco, E.; de Cremoux, L.; Lin, X.; Mitchell-Heggs, R.; Sabater, L.; Blanchard, S.; Hureau, C. Keggin-type polyoxometalates as Cu(II) chelators in the context of Alzheimer's disease. *Chem. Commun.* **2022**, *58*, 2367–2370. [CrossRef]
71. Wang, M.; Hua, J.; Zheng, P.; Tian, Y.; Kang, S.; Chen, J.; Duan, Y.; Ma, X. A Nanoscale Cobalt Functionalized Strandberg-Type Phosphomolybdate with *b*-Sheet Conformation Modulation Ability in Anti-Amyloid Protein Misfolding. *Inorganics* **2023**, *11*, 442. [CrossRef]
72. Lei, S.; Yang, H.; Li, J.X.; Li, Y.; Wang, L.; Chen, B.N.; Li, J. A study of the antioxidant properties of Keggin-type polyoxometalates. *Dalton Trans.* **2023**, *52*, 9673–9683. [CrossRef]
73. Ben Zaken, K.; Bouhnik, R.; Omer, N.; Bloch, N.; Samson, A.O. Polyoxometalates bind multiple targets involved in Alzheimer's disease. *J. Biol. Inorg. Chem.* **2025**, *30*, 299–309. [CrossRef]
74. Ramesh, M.; Balachandra, C.; Andhare, P.; Govindaraju, T. Rationally Designed Molecules Synergistically Modulate Multifaceted Aβ Toxicity, Microglial Activation, and Neuroinflammation. *ACS Chem. Neurosci.* **2022**, *13*, 2209–2221. [CrossRef] [PubMed]
75. Ma, M.; Liu, Z.; Zhao, H.; Zhang, H.; Ren, J.; Qu, X. Polyoxometalates: Metallodrug agents for combating amyloid aggregation. *Natl. Sci. Rev.* **2024**, *11*, nwae226. [CrossRef] [PubMed] [PubMed Central]
76. Bittner, N.; Funk, C.S.M.; Schmidt, A.; Bermpohl, F.; Brandl, E.J.; Algharably, E.E.A.; Kreutz, R.; Riemer, T.G. Psychiatric Adverse Events of Acetylcholinesterase Inhibitors in Alzheimer's Disease and Parkinson's Dementia: Systematic Review and Meta-Analysis. *Drugs Aging* **2023**, *40*, 953–964. [CrossRef]
77. Kondinski, A.; Gumerova, N.I.; Rompel, A.; Falcaro, P.; Schreck, T. Data-Driven Polyoxometalate Chemistry. Chem. *Eur. J.* **2025**, *31*, e01528. [CrossRef] [PubMed]
78. Hardman, T.C.; Aitchison, R.; Scaife, R.; Edwards, J.; Slater, G. The future of clinical trials and drug development: 2050. *Drugs Context.* **2023**, *12*, 2023-2-2. [CrossRef] [PubMed] [PubMed Central]
79. Hua, J.A.; Wei, X.M.; Ma, X.; Jiao, J.Z.; Chai, B.H.; Wu, C.B.; Zhang, C.L.; Niu, Y.L. A $\{Cd_4Cl_2O_{14}\}$ cluster functionalized sandwich-type tungstoarsenate as a conformation modulator for misfolding Aβ peptides. *Cryst. Eng. Comm.* **2022**, *24*, 1171–1176. [CrossRef]
80. Xue, C.; Lin, T.Y.; Chang, D.; Guo, Z. Thioflavin T as an amyloid dye: Fibril quantification, optimal concentration and effect on aggregation. *R. Soc. Open Sci.* **2017**, *4*, 160696. [CrossRef]
81. Gao, N.; Liu, Z.; Zhang, H.; Liu, C.; Yu, D.; Ren, J.; Qu, X. Site-Directed Chemical Modification of Amyloid by Polyoxometalates for Inhibition of Protein Misfolding and Aggregation. *Angew. Chem. Int. Ed. Engl.* **2022**, *61*, e202115336. [CrossRef] [PubMed]
82. Hu, S.; Ning, X.; Lv, J.; Wei, Y.; Zhang, H.; Li, M. Enantioselective modulation of amyloid burden and memory deficits by chiralpolyoxometalates for Alzheimer's disease treatment. *Inorg. Chem. Front.* **2023**, *10*, 5347–5356. [CrossRef]

83. Diaz, A.; Muñoz-Arenas, G.; Venegas, B.; Vázquez-Roque, R.; Flores, G.; Guevara, J.; Gonzalez-Vergara, E.; Treviño, S. Metforminium Decavanadate (MetfDeca) Treatment Ameliorates Hippocampal Neurodegeneration and Recognition Memory in a Metabolic Syndrome Model. *Neurochem. Res.* **2021**, *46*, 1151–1165. [CrossRef] [PubMed]
84. Copple, I.M. The Keap1-Nrf2 Cell Defense Pathway—A Promising Therapeutic Target? *Adv. Pharmacol.* **2012**, *63*, 43–79.
85. la Torre, A.; Lo Vecchio, F.; Angelillis, V.S.; Gravina, C.; D'Onofrio, G.; Greco, A. Reinforcing Nrf2 Signaling: Help in the Alzheimer's Disease Context. *Int. J. Mol. Sci.* **2025**, *26*, 1130. [CrossRef]
86. Ma, X.; Li, H.; Zhao, Z.; Zhang, L.; Su, H.; Zhao, Y.; Sun, Y.; Li, C.; Wang, Y.; Hua, J. An interesting nano-linear nickel-modified Strandberg-type phosphomolybdate acting as a conformation modulator anti-misfolding peptide. *J. Mol. Struct.* **2025**, *1329*, 141498. [CrossRef]
87. Fang, L.; Peng, R.; Xia, L.; Zhuang, G.-L. Active Learning-Assisted Exploration of $[PO_{40}Mo_{12}]^{3-}$ for Alzheimer's Therapy Insights. *Adv. Sci.* **2025**, *12*, e08702. [CrossRef] [PubMed]
88. Gao, N.; Sun, H.; Dong, K.; Ren, J.; Qu, X. Gold-Nanoparticle-Based Multifunctional Amyloid-β Inhibitor against Alzheimer's Disease. *Chem. A Eur. J.* **2015**, *21*, 829–835. [CrossRef]
89. Gao, N.; Dong, K.; Zhao, A.; Sun, H.; Wang, Y.; Ren, J.; Qu, X. Polyoxometalate-based nanozyme: Design of a multifunctional enzyme for multi-faceted treatment of Alzheimer's disease. *Nano Res.* **2016**, *9*, 1079–1090. [CrossRef]
90. Ma, M.; Gao, N.; Sun, Y.; Du, X.; Ren, J.; Qu, X. Redox-Activated Near-Infrared-Responsive Polyoxometalates Used for Photothermal Treatment of Alzheimer's Disease. *Adv. Healthc. Mater.* **2018**, *7*, e1800320. [CrossRef]
91. Gao, L.; Wang, J.; Bi, Y. Nanotechnology for Neurodegenerative Diseases: Recent Progress in Brain-Targeted Delivery, Stimuli-Responsive Platforms, and Organelle-Specific Therapeutics. *Int. J. Nanomed.* **2025**, *20*, 11015–11044. [CrossRef] [PubMed] [PubMed Central]
92. Liu, L.; He, H.; Du, B.; He, Y. Nanoscale drug formulations for the treatment of Alzheimer's disease progression. *RSC Adv.* **2025**, *15*, 4031–4078. [CrossRef] [PubMed]
93. Liu, Y.; Sun, J.; Gong, Y.; Zhou, H.; Chen, X.; Zhu, X.; Zhao, Y.; Wen, Y.; Qin, X.; Liu, J. Peptide-modified Mo polyoxometalate nanoparticles suppress Zn^{2+}-induced Aβ aggregation. *ChemNanoMat* **2019**, *5*, 897–910. [CrossRef]
94. Díaz, A.; Vázquez-Roque, R.; Carreto-Meneses, K.; Moroni-González, D.; Moreno-Rodríguez, J.A.; Treviño, S. Polyoxidovanadates as a pharmacological option against brain aging. Polyoxidovanadates as a pharmacological option against brain aging. *J. Chem. Neuroanat.* **2023**, *129*, 102256. [CrossRef]
95. Huang, Y.; Chang, Y.; Liu, L.; Wang, J. Nanomaterials for Modulating the Aggregation of β-Amyloid Peptides. *Molecules* **2021**, *26*, 4301. [CrossRef]
96. Lo, C.H.; Cheong, L.Y.T.; Zeng, J. Nanoplatforms Targeting Intrinsically Disordered Protein Aggregation for Translational Neuroscience Applications. *Nanomaterials* **2025**, *15*, 704. [CrossRef] [PubMed]
97. Baracaldo-Santamaría, D.; Avendaño-Lopez, S.S.; Ariza-Salamanca, D.F.; Rodriguez-Giraldo, M.; Calderon-Ospina, C.A.; González-Reyes, R.E.; Nava-Mesa, M.O. Role of Calcium Modulation in the Pathophysiology and Treatment of Alzheimer's Disease. *Int. J. Mol. Sci.* **2023**, *24*, 9067. [CrossRef] [PubMed]
98. Chaudhary, B.; Kumari, S.; Dhapola, R.; Sharma, P.; Paidlewar, M.; Vellingiri, B.; Medhi, B.; HariKrishnaReddy, D. Calcium dysregulation in Alzheimer's disease: Unraveling the molecular nexus of neuronal dysfunction and therapeutic opportunities. *Biochem Pharmacol.* **2025**, *242 Pt 2*, 117211. [CrossRef]
99. Alzheimer's Association Calcium Hypothesis Workgroup. Calcium Hypothesis of Alzheimer's disease and brain aging: A framework for integrating new evidence into a comprehensive theory of pathogenesis. *Alzheimer's Dement.* **2017**, *13*, 178–182.e17. [CrossRef]
100. Popugaeva, E.; Bezprozvanny, I.; Chernyuk, D. Reversal of calcium dysregulation as potential approach for treating Alzheimer's disease. *Curr. Alzheimer Res.* **2020**, *17*, 344. [CrossRef]
101. Chen, J.; Yang, W.-Z.; Chen, H.; Ding, X.; Chen, H.; Zhan, C.-H.; Jin, Z. Targeting protein aggregation: The promising application of polyoxometalates in neurodegenerative diseases. *Inorg. Chem. Front.* **2024**, *11*, 7238–7255. [CrossRef]
102. Maret, W. The quintessence of metallomics: A harbinger of a different life science based on the periodic table of the bioelements. *Metallomics* **2022**, *14*, mfac051. [CrossRef] [PubMed]
103. Aureliano, M.; Ma, B. Feature Papers in *BioChem*. *BioChem* **2025**, *5*, 17. [CrossRef]

BioChem

MDPI

Article

Drug Allergy in Hospitalized Patients: Three Years of Consultation Experience in a Tertiary Care Setting

Christian P. Ratti [1], Alessandra Chiei Gallo [1,*], Francesca Barei [2], Alice Botta [1], Matteo Cavara [1], Eleonora Bono [1], Lea Caron [1], Valeria G. R. Ortolani [1] and Enrico Iemoli [1]

[1] Allergy and Clinical Immunology Unit, Sacco Hospital, 20157 Milan, Italy; christian.ratti@unimi.it (C.P.R.); alice.botta@unimi.it (A.B.); matteo.cavara@unimi.it (M.C.); eleonora.bono@unimi.it (E.B.); caron.lea@asst-fbf-sacco.it (L.C.); ortolani.valeria@asst-fbf-sacco.it (V.G.R.O.); iemoli.enrico@asst-fbf-sacco.it (E.I.)

[2] Dermatology Unit, IRCCS Foundation Ca' Granda Ospedale Maggiore Policlinico, 20122 Milan, Italy

* Correspondence: alessandra.chiei@unimi.it

Abstract

Background/Objectives: Drug hypersensitivity reactions (DHRs) are an important cause of morbidity in hospitalized patients, but their epidemiology and management in the inpatient setting are not well defined. Mislabeling of drug allergies may lead to inappropriate treatment and reduced antimicrobial stewardship. This study aimed to characterize the clinical profile, diagnostics, and management of inpatients referred for suspected drug allergy in a tertiary care hospital. **Methods**: We retrospectively reviewed all adult inpatients (≥18 years) at Luigi Sacco Hospital (Milan, Italy) who received allergology consultation between 1 June 2022 and 31 May 2025. Data on demographics, reaction type, culprit drugs, investigations, and management were collected. Immediate reaction severity was graded using the United States Drug Allergy Registry (USDAR) scale; delayed reactions were classified as severe cutaneous adverse reactions (SCARs) or non-SCARs. Logistic regression identified predictors of severity. **Results**: Among 35,438 admissions, 334 patients (0.9%) were evaluated; median age was 65 years, 51.2% were female, 67.4% had atopic comorbidities, and 55.1% reported prior drug allergy. Immediate reactions occurred in 49.1%, delayed in 43.7%. Cutaneous involvement was present in 86.8%, anaphylaxis in 6.6%, and SCARs in 3.9%. Antibiotics—particularly β-lactams—were most often implicated. In multivariate analysis, antibiotic exposure and older age were linked to more severe immediate reactions, while the absence of atopy predicted SCARs. Desensitization was successfully performed in 16.2% of patients. **Conclusions**: DHRs in inpatients are frequent and often involve high-risk drugs. Structured inpatient allergology services and an "allergy stewardship" approach may reduce DHR-related risks, support optimal therapy, and improve antimicrobial use strategies in tertiary care settings.

Keywords: allergy consultation; drug allergy; adverse drug reaction; epidemiology; inpatient allergology; anaphylaxis; scar; drug allergy diagnosis; desensitization; allergy stewardship

1. Introduction

The World Health Organization's (WHO) stated definition of an adverse drug reaction (ADR) is a response to a drug that is noxious and unintended and occurs at doses normally used in man for prophylaxis, diagnosis or therapy of disease, or for the modification of physiological function [1].

In hospitalized patients, ADRs—encompassing both allergic and non-allergic mechanisms—are reported in 10–20% of cases [2]. Among these, drug hypersensitivity reactions (DHRs) represent a clinically significant subset due to their association with severe manifestations, including anaphylaxis and Severe Cutaneous Adverse Reactions (SCARs), prolonged hospital stays, and increased reliance on broader-spectrum or costlier medications [2,3]. Mislabeling of drug allergies, particularly antibiotics, exacerbates these issues by limiting appropriate antimicrobial use and contributing to antimicrobial resistance [4].

The true burden of drug allergy remains poorly defined, despite being frequently reported in clinical settings. Approximately 10% of patients accessing healthcare services report a drug allergy, most commonly to penicillin [5]. However, the majority of these self-reported allergies lack confirmatory evaluation, resulting in the unnecessary avoidance of first-line therapies and contributing to suboptimal treatment strategies.

While Allergy and Immunology (A/I) is often considered an outpatient subspecialty, A/I specialists also provide crucial support in inpatient and emergency settings, where their expertise is essential for the accurate diagnosis and management of suspected DHRs. Nevertheless, studies specifically addressing inpatient allergy consultations remain limited [6–12], and available data frequently focus predominantly on pediatric cohorts or do not distinguish between general A/I referrals and those related to drug hypersensitivity.

Given the growing global emphasis on antimicrobial stewardship and the clinical consequences of inaccurate drug allergy labeling [13], a deeper understanding of inpatient allergy referrals is warranted. The aim of this study is to characterize the clinical features of inpatients referred for allergology consultation due to suspected drug allergy at Luigi Sacco Hospital, a tertiary care center in Milan, over a three-year period, and to detail the allergist's approach to reaction management, diagnostic evaluation, and guidance for subsequent therapeutic planning.

2. Materials and Methods

2.1. Patients and Data Collection

Luigi Sacco Hospital is a 558-bed facility, ranking among the five largest hospitals in Milan. It also functions as a major teaching institution affiliated with a medical school. This retrospective descriptive study included all patients aged 18 years or older who were hospitalized at Luigi Sacco Hospital between 1 June 2022 and 31 May 2025, and who received an A/I consultation for a suspected drug allergy.

Data were retrospectively extracted from electronic medical records using a standardized collection form. For each patient, the following variables were recorded: demographic characteristics, reason for admission, requesting department, referral question, type of reaction, implicated drugs, and inpatient management. Patients with an inconclusive diagnosis and/or an urgent need for home therapy were promptly re-evaluated after discharge in an outpatient setting, where they underwent diagnostic procedures including skin testing and drug provocation tests (DPTs), performed either with the suspected drug or with suitable therapeutic alternatives.

Although latex is not a pharmaceutical agent, consultations related to suspected hypersensitivity to latex were also included in the study, as latex is commonly used in medical devices and these reactions frequently fall under the scope of hospital-based drug allergy consultations [14].

DHRs were categorized as either immediate (onset within 1–6 h of drug exposure) or delayed (onset after more than 1 h), following a pragmatic binary classification [15]. Reactions occurring between 1 and 6 h were classified as immediate or delayed based on their clinical phenotype (e.g., urticaria, angioedema, bronchospasm for immediate-type; maculopapular rash for delayed-type), acknowledging the overlapping nature of reaction timing in some cases. Reactions for which latency could not be determined were classified in the "latency undetermined" group.

The severity of immediate reactions was graded using the United States Drug Allergy Registry (USDAR) Grading [16]. For statistical analysis, reactions of grades 1–2 (mild) and 3–4 (severe) were combined.

Anaphylaxis was defined according to the criteria outlined in the 2021 position paper by the European Academy of Allergy and Clinical Immunology (EAACI) [17].

Delayed reactions were further classified into *SCARs* and *non-SCARs*. SCARs were defined in accordance with EAACI criteria [18] and included Stevens-Johnson Syndrome/Toxic Epidermal Necrolysis (SJS/TEN), Drug Reaction with Eosinophilia and Systemic Symptoms (DRESS), and Acute Generalized Exanthematous Pustulosis (AGEP). Other delayed reactions such as Symmetrical Drug-Related Intertriginous and Flexural Exanthema (SDRIFE) and Erythema Multiforme Majus (EMM) were also classified using the same EAACI definitions.

For each suspected ADR, the likelihood of a causal relationship between the drug and the reaction was assessed using the Naranjo Adverse Drug Reaction Probability Scale [19].

Hospital-wide statistics, including the total number of annual admissions during the study period, were obtained from institutional records and used to contextualize consultation incidence.

2.2. Statistical Analyses

Descriptive statistics are reported as mean and standard deviation (SD) or median and quartiles (Q1–Q3) for quantitative variables based on the distribution of the population. Absolute numbers (n) and frequencies (%) are used for categorical variables. Categorical variables were analyzed using a chi-square test or Fisher's exact test as appropriate. We performed univariate regression and multivariate regression analyses to evaluate a potential predominant factor affecting the determination of the USDAR score of 3–4 in immediate reactions and the determination of SCAR in the delayed reactions. As covariates, we considered age, sex, current outpatient therapy, previous drug reaction, atopic comorbidities, receiving antibiotics, beta-lactams, NSAIDs, iodine-based contrast media, or opioids. Multicollinearity was assessed using the Variance Inflation Factor (VIF), variance proportions, and the condition index in SPSS. Thresholds of VIF < 10, condition index < 30, and absence of high shared variance proportions were considered acceptable. Considering a chi-square test, with a significance level of 0.05 and 80% power, a degree of freedom of 3, and assuming an effect size of 0.5, we determined that a minimum sample size of 44 patients was necessary. All statistical analyses were two-tailed with an alpha error set at 0.05, considering a p-value < 0.05 as significant. Statistical analyses were carried out using SPSS-IBM software (version 29.0; IBM Corp., Armonk, NY, USA). To calculate the sample size, G*Power software (version 3.1, Heinrich Heine University Düsseldorf, Düsseldorf, Germany) was used.

3. Results

3.1. Study Population

During the study period, 35,438 patients were admitted to Luigi Sacco Hospital. Among them, 334 (0.9%) underwent an allergology consultation for suspected drug allergy and were included in this study. Of these patients, 171 (51.2%) were female and 163 (48.8%) male, with a median age of 65 years (18–96 years). A history of atopic comorbidities was reported in 225/334 patients (67.4%), and previous drug allergy was documented in 184/334 cases (55.1%). Additional atopic conditions are listed in Table 1. Most patients (268/334; 80.2%) were on polypharmacy at home, defined as the use of more than three concomitant medications. The most frequent reason for hospitalization was infectious disease (105/334; 31.4%), followed by surgical procedures (92/334; 27.5%) and cardiovascular conditions (84/334; 25.1%). Consequently, the departments most frequently requesting allergology consultations were Cardiology (93/334; 27.8%), Infectious Diseases (82/334; 24.6%), and General Surgery (54/334; 16.2%).

Table 1. Baseline characteristics of the overall population and of the three subgroups (immediate, delayed, and latency undetermined).

Baseline Characteristics	Overall Population	*n* [a]	Immediate Reactions	*n* [a]	Delayed Reactions	*n* [a]	Latency Undetermined	*n* [a]
Sex, *n* (%)								
- Female	171 (51.2)	334	89 (54.3)	164	72 (49.3)	146	10 (41.7)	24
- Male	163 (48.8)		75 (45.7)		74 (50.7)		14 (58.3)	
Age, mean (SD)	63.8 (16.7)	334	61.1 (16.9)	164	66.7 (16.4)	146	64.2 (14.8)	24
Age, median	65		62		68		64	
Age, min	18		18		18		30	
Age, max	96		91		96		89	
Age distribution, *n* (%)								
- 18–44-year-old	42 (12.6)	334	28 (17.1)	164	12 (8.2)	146	2 (8.3)	24
- 45–64-year-old	122 (36.5)		61 (37.2)		51 (34.9)		10 (41.7)	
- 65–74-year-old	59 (17.7)		30 (18.3)		23 (15.8)		6 (25.0)	
- ≥75-year-old	111 (33.2)		45 (27.4)		60 (41.1)		6 (25.0)	
Atopic comorbidities, *n* (%)	225 (67.4)	334	148 (90.3)	164	62 (42.5)	146	15 (62.5)	24
- Oculorhinitis	38 (32.3)		25 (15.2)		8 (5.5)		5 (20.8)	
- Asthma	21 (6.3)		17 (10.4)		1 (0.7)		3 (12.5)	
- Food allergy	23 (6.9)		12 (7.3)		5 (3.4)		6 (25.0)	
- Drug allergy (previous)	184 (55.1)		127 (77.4)		49 (33.6)		8 (33.3)	
- Chronic spontaneous urticaria	6 (1.8)		2 (1.2)		2 (1.4)		2 (8.3)	
- Atopic dermatitis	4 (1.2)		2 (1.2)		1 (0.7)		1 (4.2)	
- Allergic contact dermatitis	39 (11.7)		26 (15.9)		13 (8.9)		0 (0.0)	
- Hymenoptera venom allergy	2 (0.6)		2 (1.2)		0 (0.0)		0 (0.0)	
- Nasal polyposis	3 (0.9)		2 (1.2)		1 (0.7)		0 (0.0)	
Reason for hospitalization, *n* (%)								
- Infectious diseases	105 (31.4)	334	30 (18.3)	164	65 (44.5)	146	10 (41.7)	24
- Surgical procedures	92 (27.5)		62 (37.8)		25 (17.1)		5 (20.8)	
- Cardiological diseases	84 (25.1)		55 (33.5)		24 (16.4)		5 (20.8)	
- Renal failure	11 (3.3)		4 (2.4)		6 (4.1)		1 (4.2)	
- Respiratory failure	7 (2.1)		2 (1.2)		4 (2.7)		1 (4.2)	
- Malignancy	11 (3.3)		5 (3.0)		6 (4.1)		0 (0.0)	
- Other	24 (7.2)		6 (3.7)		16 (11.0)		2 (8.3)	
Current outpatient therapy, *n* (%)								
- None	43 (12.9)	334	26 (15.9)	164	13 (8.9)	146	4 (16.7)	24
- <3 drugs	23 (6.9)		19 (6.1)		12 (8.2)		1 (4.2)	
- ≥3 drugs	268 (80.2)		128 (78.0)		121 (82.9)		19 (79.2)	

Note: Valid data [a].

3.2. Allergological Characteristics of Patients

Consultations were categorized based on the reason for referral as follows: ongoing reactions (163/334; 48.8%), previous reactions (139/334; 41.6%), and suspected latex allergy (35/334; 10.5%) (Table 2). In our population, immediate reactions were identified in 164/334 cases (49.1%), delayed reactions in 146/334 (43.7%), while latency could not be determined in 24/334 patients (7.2%).

Cutaneous involvement was observed in the majority of cases (290/334; 86.8%), with erythematous rash in 39.0%, urticaria in 19.7%, maculopapular exanthema in 18.7%, angioedema in 15.5%, and enanthem in 9.0% (Table 3). Anaphylaxis occurred in 22/334 (6.6%) patients, and 5.7% experienced hypotension or loss of consciousness. One case of Kounis syndrome with cardiac involvement was documented.

The severity of immediate reactions was assessed using the USDAR grading scale in 131/334 (39.2%) patients: 95/131 (72.5%) of them were grade 1, 4/131 (3.1%) grade 2, 20/131 (15.3%) grade 3, and 12/131 (9.2%) grade 4.

Among the delayed reactions, 13/334 (3.9%) cases were classified as SCARs, including 5 SJS (38.5%), 3 DRESS (23.1%), and 5 AGEP (38.5%). Additionally, 1 case of SDRIFE and 6 cases of likely drug-induced EMM were recorded. ACE inhibitor-induced angioedema was diagnosed in 4/334 (1.2%) patients.

Blood tests were performed in 164/334 patients, revealing eosinophilia (eosinophil count in the peripheral blood > 500/µL) in 48 cases (29.3%).

Table 2. Features of consultations in the overall population and in the three subgroups (immediate, delayed, and latency undetermined).

Consultation	Overall Population	n [a]	Immediate Reactions	n [a]	Delayed Reactions	n [a]	Latency Undetermined	n [a]
Reason for request, n (%)								
- Ongoing reaction	163 (48.8)		29 (17.7)		117 (80.1)		17 (70.8%)	
- Previous reaction	139 (41.6)	334	114 (68.2)	164	21 (14.4)	146	4 (16.7)	24
- Suspected latex allergy	35 (10.5)		24 (14.6)		8 (5.5)		3 (12.5)	
Timing of previous reaction, n (%)								
- Less than 1 year	18 (10.3)		13 (9.4)		4 (13.8)		1 (14.2)	
- Between 1 and 5 years	28 (16.1)		24 (17.4)		4 (13.8)		0 (0.0)	
- Between 5 and 10 years	15 (8.6)	174	7 (5.1)	138	8 (27.6)	29	0 (0.0)	7
- More than 10 years	68 (39.1)		58 (42.0)		7 (24.1)		3 (42.9)	
- Unknown	42 (24.1)		33 (23.9)		6 (20.7)		3 (42.9)	

Note: Valid data [a].

Table 3. Drug reaction characteristics in the overall population and in the three subgroups (immediate, delayed, and latency undetermined).

	Overall Population	n [a]	Immediate Reactions	n [a]	Delayed Reactions	n [a]	Latency Undetermined	n [a]
Drug involved								
Type of drugs, n (%)								
- Antibiotics	149 (44.6)		60 (36.6)		82 (56.2)		7 (29.2)	
- Beta-lactam	122 (36.5)		54 (32.9)		61 (41.8)		7 (29.2)	
- Non-Beta-lactam	44 (13.2)	334	15 (9.1)	164	27 (18.5)	146	2 (8.4)	24
- NSAIDs	80 (24.0)		57 (34.7)		20 (13.7)		3 (12.5)	
- Opioids	18 (5.4)		13 (7.9)		5 (3.4)		0 (0.0)	

Table 3. *Cont.*

- Local anesthetics	7 (2.1)	334	7 (4.3)	164	0 (0.0)	146	0 (0.0)	24
- General anesthetics	14 (4.2)		13 (7.9)		1 (0.7)		0 (0.0)	
- Iodinate contrast media	31 (9.3)		15 (9.1)		16 (11.0)		0 (0.0)	
- Gadolinium-based medium contrast	2 (0.6)		2 (1.2)		0 (0.0)		0 (0.0)	
- Heparins	17 (5.1)		2 (1.2)		14 (9.6)		1 (4.2)	
- Uric acid synthesis inhibitors	7 (2.1)		0 (0.0)		7 (4.8)		0 (0.0)	
- Others	67 (20.0)		17 (10.3)		46 (31.5)		4 (16.7)	
- Unknown	17 (5.1)		7 (4.3)		5 (3.4)		5 (20.8)	
Administration route, *n* (%)		334		164		146		24
- Intravenous	137 (41.0)		55 (33.5)		75 (51.4)		7 (29.2)	
- Oral	157 (47.0)		82 (50.0)		69 (47.3)		6 (25.0)	
- Subcutaneous	24 (7.2)		9 (5.5)		14 (9.6)		1 (4.2)	
- Others	36 (10.4)		23 (14.0)		11 (7.6)		2 (8.3)	
- Unknown	21 (6.3)		6 (3.7)		6 (4.1)		9 (37.5)	
Concomitant medications during the reaction, *n* (%)	291 (87.1)	334	138 (84.1)	164	133 (91.1)	146	20 (83.3)	24
Reaction type								
Reaction type, *n* (%)		334		164		146		24
- Cutaneous reaction	290 (86.8)		134 (81.7)		142 (97.3)		14 (58.3)	
- Urticaria	57 (17.1)		40 (24.4)		13 (8.9)		4 (16.7)	
- Angioedema	45 (13.5)		35 (21.3)		9 (6.2)		1 (4.2)	
- Erythematous rash	113 (33.8)		61 (37.2)		50 (34.2)		2 (8.3)	
- Maculopapular exanthema	54 (16.2)		0 (0.0)		54 (37.0)		0 (0.0)	
- Enanthem	26 (7.8)		7 (4.3)		19 (13.0)		0 (0.0)	
- Pruritus sine materia	7 (2.1)		3 (1.8)		4 (2.7)		0 (0.0)	
- SCAR	13 (3.9)		0 (0.0)		13 (8.9)		0 (0.0)	
- Throat tightness	13 (3.9)		12 (7.3)		1 (0.7)		0 (0.0)	
- Dyspnea/Asthma	17 (5.1)		16 (9.8)		1 (0.7)		0 (0.0)	
- Gastrointestinal symptoms	8 (2.4)		2 (1.2)		0 (0.0)		6 (25.0)	
- Loss of consciousness/hypotension	19 (5.7)		19 (11.7)		0 (0.0)		0 (0.0)	
- Anaphylaxis	22 (6.6)		22 (13.4)		0 (0.0)		0 (0.0)	
- Kounis syndrome	1 (0.3)		1 (0.6)		0 (0.0)		0 (0.0)	
- Others	16 (4.8)		5 (3.0)		4 (2.7)		7 (29.2)	
- Unknown	5 (1.5)		0 (0.0)		0 (0.0)		5 (20.8)	
Eosinophilia	48 (14.4)	164	1 (0.6)	32	41 (28.1)	116	6 (25.0)	16
Naranjo Score								
Naranjo Score, *n* (%)		288		141		136		11
- <0 doubtful	5 (1.5)		1 (0.6)		1 (0.7)		3 (12.25)	
- 1–4 possible	120 (35.9)		52 (31.7)		61 (41.8)		7 (29.2)	
- 5–8 probable	155 (46.4)		85 (51.8)		69 (47.3)		1 (4.2)	
- ≥9 definite	8 (2.4)		3 (1.8)		5 (3.4)		0 (0.0)	

Table 3. *Cont.*

Patient management								
Continuation of ongoing therapy, *n* (%)	41 (12.3)	334	3 (1.8)	164	27 (18.5)	146	11 (45.8)	24
Suggested alternatives without testing, *n* (%)	171 (51.2)	334	72 (43.8)	164	93 (63.6)	146	6 (25.0)	24
Desensitization, *n* (%)	54 (16.2)	334	45 (27.4)	164	9 (6.2)	146	0 (0.0)	24
Allergological work-up, *n* (%)		334		164		146		24
- Post-discharge evaluation	41 (12.3)		17 (10.4)		22 (15.1)		2 (8.3)	
- Specific IgE testing	39 (11.7)		26 (15.9)		11 (7.5)		2 (8.3)	
- Skin testing	40 (12.0)		28 (17.1)		10 (6.8)		2 (8.3)	
- Provocation test (alternative drug)	21 (6.3)		12 (7.3)		8 (5.5)		1 (4.2)	
- Provocation test (implicated drug)	14 (4.2)		10 (6.1)		4 (2.7)		0 (0.0)	

Note: Valid data [a]. Abbreviation: IgE, immunoglobulin E; SCAR, Severe Cutaneous Adverse Reactions; NSAIDs, non-steroidal Anti-inflammatory Drugs.

3.3. Drugs Involved

The data regarding the drug involved are reported in Table 3. At the time of the reaction, 291/334 (87.1%) patients were receiving multiple medications. Antibiotics were the most frequently implicated drug class (149/334; 44.6%), particularly beta-lactams (122/334; 36.5%) (Table 3). The most common route of administration was oral (157/334; 47%), followed by intravenous (137/334; 41%) and subcutaneous (24/334; 7.2%).

To assess the likelihood of drug causality, the Naranjo algorithm was applied in 288/344 (86.2%) cases (Table 3). The median score was 5, and most patients (275/288; 95.5%) fell into the "possible" (score 1–4) or "probable" (score 5–8) categories.

3.4. Reaction Management

The data regarding the reaction management are reported in Table 3. The majority of reactions were managed with corticosteroids (171/334; 51.2%) and antihistamines (133/334; 39.8%). Epinephrine was administered in 11/334 (3.3%) cases, while inhalation therapy or oxygen supplementation was used in 10/334 (3.0%) patients. In 75/334 (22.5%) patients, the reaction resolved without pharmacological intervention.

Regarding therapeutic decisions (Table 3), the ongoing treatment was maintained in 41/334 (12.3%) patients, and 15/334 (4.5%) received premedication and comedication with antihistamines. These were mainly patients who had experienced mild, predominantly cutaneous reactions, and for whom there was an urgent need to continue the ongoing therapy. Alternative therapies were recommended in 171/334 (51.2%) patients to ensure treatment continuation. In 25/334 (7.5%) cases, the alternative drug was administered using an incremental dosing protocol (1/1000, 1/100, 1/10, and full dose at 30 min intervals), while 9/334 (2.7%) patients received premedication. The incremental protocol was mainly used when the alternative drug was potentially cross-reactive with the culprit agent, particularly in patients with moderate to severe reactions—for example, cephalosporins in those with a history of penicillin hypersensitivity. Only 2/334 (0.6%) patients failed to tolerate the alternative drug.

Desensitization to the suspected culprit drug was performed in 54/334 (16.2%) patients. The majority of these involved acetylsalicylic acid (ASA) (49/54; 90.7%) for antiplatelet therapy, followed by beta-lactam antibiotics (4/54; 7.4%) and prasugrel (1/54; 1.9%). Seven patients (13%) had previously undergone ASA desensitization but required a repeat procedure due to treatment discontinuation. In 4/54 (7.4%) cases, ASA desensitization was not tolerated, and an alternative treatment with indobufen was required.

Most patients (274/334; 82.0%) required a single consultation; however, 60 (18.0%) needed two or more follow-up evaluations during hospitalization. After discharge, re-evaluation in an outpatient setting to complete the diagnostic process was recommended for 95/334 (28.4%) patients, but follow-up assessment was performed in only 41 (12.3%).

3.5. Allergological Work-Up

Specific IgE testing for beta-lactams (penicillin G, penicillin V, amoxicillin, ampicillin, and cefaclor) was performed in 39/334 (11.7%) patients, with positive results in 2 (0.6%). Latex-specific IgE was measured in 22/334 (6.6%) patients, with 3 positive cases (0.9%).

Skin testing was carried out in 40/334 (12.0%) patients across various drug classes: beta-lactams (22/334; 6.6%), with 1 patient testing positive for aminopenicillins and 4 for cephalosporins; iodinated contrast media (9/334; 2.7%) with 2 positivities; local anesthetics (5/334; 1.5%); general anesthetics (9/334; 2.7%) with 3 positivities; heparins (3/334; 0.9%); and vancomycin (3/334; 0.9%) with 1 positive case. For suspected latex allergy, 33/334 (9.9%) patients underwent skin testing, with 2 positive results. It should be noted that in many cases it was not possible to perform skin testing within the optimal time window for its execution, further underscoring the crucial role of specialist consultation in the diagnostic pathway.

Following diagnostic evaluation, 21/334 (6.3%) patients underwent drug provocation testing with alternative medications in a controlled setting, all of which yielded negative results. Additionally, 14/334 (4.2%) patients were re-exposed to the suspected drug, and only 1/334 (0.3%) patient developed a mild cutaneous reaction.

Among the 35 patients with suspected latex allergy, a latex-free surgical protocol was recommended in only 2/35 (5.7%) cases based on clinical history and testing.

3.6. Final Diagnosis and Follow-Up

Following allergology consultation, an ADR was confirmed in 240/334 (78%) patients. In 78/334 (23.4%) cases, a drug-related cause was excluded, and in 16/334 (4.8%) cases, the diagnosis remained uncertain. Only 5/334 (1.5%) patients experienced new reactions after the conclusion of the allergological assessment and required subsequent evaluation.

3.7. USDAR Grading Scale and Naranjo Score in Immediate Hypersensitivity Reactions

An USDAR scale of 3–4 was significantly associated with antibiotic use ($p = 0.029$; $\chi^2 = 4.783$) and with beta-lactam antibiotic use ($p = 0.045$; $\chi^2 = 4.014$) (Table 4). No significant differences were found for NSAIDs, opioids, general anesthetics, and iodine MC. This result was confirmed by the univariate logistic regression analysis (Table 5), where antibiotics and beta-lactams were associated with an increased risk of having an USDAR score of 3–4 (antibiotics: OR = 2.449, 95% CI 1.085–5.530, $p = 0.031$; beta-lactams: OR = 2.267, 95% CI 1.008–5.097, $p = 0.048$). In the multivariate analysis, the antibiotic use remained significant (OR = 3.635, 95% CI 1.471–8.985, $p = 0.005$) together with age (OR = 1.039, 95% CI 1.007–1.071, $p = 0.017$). As for the Naranjo score, most patients receiving beta-lactams had a score of 1–4 or 5–8 ($p = 0.021$).

Table 4. Severity and Causality of Immediate Hypersensitivity Reactions by Drug Class.

Immediate Reactions	USDAR Grading Scale (N = 131)			Naranjo Score (N = 141)				
	Grade 1–2 [g]	Grade 3–4 [g]	*p*-Value	Grade < 0 [h]	Grade 1–4 [h]	Grade 5–8 [h]	Grade ≥ 9 [h]	*p*-Value
Antibiotics, *n* (%)	37/99 (37.4) [a]	19/32 (59.4) [a]	**0.029** **(χ^2 = 4.783)**	0/60 (0.0) [i]	28/60 (46.7) [i]	32/60 (53.3) [i]	0/60 (0.0) [i]	0.050
Beta-lactam, *n* (%)	33/99 (33.3) [b]	17/32 (53.1) [b]	**0.045** **(χ^2 = 4.014)**	0/54 (0.0) [j]	27/54 (50.0) [j]	27/54 (50.0) [j]	0/54 (0.0) [j]	**0.021** **(χ^2 = 8.103)**
NSAIDs, *n* (%)	44/99 (44.4) [c]	9/32 (28.1) [c]	0.102	0/56 (0.0) [k]	19/56 (33.9) [k]	35/56 (62.5) [k]	2/56 (3.6) [k]	0.714
Opioids, *n* (%)	7/99 (7.1) [d]	5/32 (15.6) [d]	0.165	0/13 (0.0) [l]	6/13 (46.2) [l]	7/13 (53.8) [l]	0/13 (0.0) [l]	0.701
General anesthetics, *n* (%)	7/99 (7.1) [e]	6/32 (18.8) [e]	0.840	0/13 (0.0) [m]	6/13 (46.2) [m]	7/13 (53.8) [m]	0/13 (0.0) [m]	0.701
Iodinate contrast media, *n* (%)	13/99 (13.1) [f]	2/32 (6.3) [f]	0.358	0/15 (0.0) [n]	5/15 (33.3) [n]	9/15 (60.0) [n]	1/15 (6.7) [n]	0.530

Note: 56/131 patients received antibiotics [a]; 50/131 patients received beta-lactams [b]; 53/131 patients received NSAIDs [c]; 12/131 patients received opioids [d]; 13/131 received general anesthetics [e]; 15/131 received iodinate contrast media [f]; 99/131 patients had a USDAR Grading Scale of 1–2, 32/131 patients had a USDAR Grading Scale of 3–4 [g]; 1/141 patient had a Naranjo score of <0, 52/141 patients had a Naranjo score of 1–4, 85/141 patients had a Naranjo score of 5–8 and 3/141 patients had a Naranjo score of ≥9 [h]; 60/141 patients received antibiotics [i]; 54/141 patients received beta-lactams [j]; 56/141 patients received NSAIDs [k]; 13/141 patients received opioids [l]; 13/141 received general anesthetics [m]; 15/151 received iodinate contrast media [n]. Abbreviations: NSAIDs, non-steroidal Anti-inflammatory Drugs; USDAR, United States Drug Allergy Registry.

Table 5. Univariate and multivariate regression analyses to assess the potential predominant factor influencing the achievement of an USDAR Score of 3–4 in the immediate reactions.

	Univariate Analysis (USDAR Score 3–4)			Multivariate Analysis (USDAR Score 3–4)		
Characteristics	**OR**	**95% CI**	***p*-Value**	**OR**	**95% CI**	***p*-Value**
Age	1.018	0.992–1.046	0.179	**1.039**	**1.007–1.071**	**0.017**
Sex	0.938	0.422–2.083	0.874	0.713	0.300–1.694	0.443
Atopic comorbidities	0.875	0.258–2.966	0.830	/	/	/
Current outpatient therapy	0.889	0.293–2.696	0.836	0.462	0.124–1.716	0.249
Previous drug reactions	0.655	0.226–1.895	0.435	0.565	0.182–1.754	0.323
Antibiotics	**2.449**	**1.085–5.530**	**0.031**	**3.635**	**1.471–8.985**	**0.005**
Beta-lactams	**2.267**	**1.008–5.097**	**0.048**	/	/	/
NSAIDs	0.489	0.206–1.164	0.106	/	/	/
Iodinate contrast media	0.441	0.094–2.069	0.299	/	/	/
Opioids	2.434	0.715–8.287	0.155	/	/	/

Note: As covariates, we considered age (as continuous variable), sex (female vs. male), current outpatient therapy (≥3 drugs vs. <3 drugs), previous drug reaction (yes vs. no), atopic comorbidities (at least one vs. none), receiving antibiotics (yes vs. no), receiving beta-lactams (yes vs. no), receiving NSAIDs (yes vs. no), receiving iodinate contrast media (yes vs. no), and receiving opioids (yes vs. no). The drugs were analyzed together only in the univariate analysis, whereas in the multivariate analysis, only antibiotics were included alongside baseline characteristics. The variable "atopic comorbidities" was excluded from the multivariate analysis due to collinearity with the variable "previous drug reaction". Variables that were significant in the univariable analyses were not all included in the multivariable logistic regression model because their inclusion led to model instability and convergence issues. Abbreviations: NSAIDs, non-steroidal Anti-inflammatory Drugs; USDAR, United States Drug Allergy Registry; CI, confidence interval; OR, odds ratio.

3.8. SCARs and Naranjo Score in Delayed Hypersensitivity Reactions

For delayed reactions (Table 6), with severity defined by the occurrence of SCAR, the only significant association was between antibiotic use and causality (p = 0.019). No other statistically significant results were observed for the remaining drug classes.

Table 6. Severity and causality of delayed hypersensitivity reactions by drug class.

Delayed Reactions	SCAR (N = 142)			Naranjo Score (N = 136)				
	No [a]	Yes [a]	*p*-Value	Grade < 0 [b]	Grade 1–4 [b]	Grade 5–8 [b]	Grade ≥ 9 [b]	*p*-Value
Antibiotics, *n* (%)	73/129 (56.6) [c]	9/13 (69.2) [c]	0.379	0/82 (0.0) [g]	30/82 (37.0) [g]	49/82 (60.5) [g]	2/82 (2.5) [g]	**0.019 (χ^2 = 8.754)**
Beta-lactams, *n* (%)	58/129 (45.0) [d]	3/13 (23.1) [d]	0.129	0/60 (0.0) [h]	25/60 (41.7) [h]	33/60 (55.0) [h]	2/60 (3.3) [h]	0.816
NSAIDs, *n* (%)	18/129 (14.0) [e]	1/13 (7.7) [e]	0.697	0/19 (0.0) [i]	8/19 (42.1) [i]	10/19 (52.6) [i]	1/19 (5.3) [i]	1.000
Iodinate contrast media, *n* (%)	14/129 (10.9) [f]	2/13 (15.4) [f]	0.642	0/16 (0.0) [j]	7/16 (43.8) [j]	8/16 (50.0) [j]	1/16 (6.3) [j]	1.000

Note: 129/142 did not have a SCAR, 13/142 had a SCAR [a]; 1/136 patient had a Naranjo score of <0, 61/136 patients had a Naranjo score of 1–4, 69/136 patients had a Naranjo score of 5–8 and 5/136 patients had a Naranjo score of ≥9 [b]; 82/142 patients received antibiotics [c]; 61/142 patients received beta-lactams [d]; 19/142 received NSAIDs [e]; 16 patients received iodinate contrast media [f]. 81/136 patients received antibiotics [g]; 60/136 patients received beta-lactams [h]; 19/136 received NSAIDs [i]; 16/136 received iodinate contrast media [j]. Abbreviations: NSAIDs, non-steroidal Anti-inflammatory Drugs; SCAR, Severe Cutaneous Adverse Reactions.

The univariate logistic regression analysis (Table 7) showed that the absence of atopic comorbidities was significantly associated with a higher risk of developing SCARs (OR = 9.500, 95% CI 1.200–75.239, p = 0.033), while no other variables reached statistical significance. In the multivariate model, the absence of atopic comorbidities remained significantly associated with the development of SCARs (OR = 10.400, 95% CI 1.273–84.999, p = 0.029).

Table 7. Univariate and multivariate regression analyses to assess the potential predominant factor influencing the development of SCAR in the delayed reactions.

	Univariate Analysis (SCAR)			Multivariate Analysis (SCAR)		
Characteristics	OR	95% CI	*p*-Value	OR	95% CI	*p*-Value
Age	0.989	0.956–1.022	0.499	0.994	0.958–1.031	0.740
Sex	0.898	0.286–2.818	0.854	1.275	0.383–4.244	0.692
Atopic comorbidities	**9.500**	**1.200–75.239**	**0.033**	**10.400**	**1.273–84.999**	**0.029**
Current outpatient therapy	0.541	0.136–2.155	0.383	0.558	0.111–2.813	0.480
Previous drug reactions	0.127	0.016–1.010	0.051	/	/	/
Antibiotics	1.726	0.505–5.894	0.384	/	/	/
NSAIDs	0.514	0.063–4.196	0.534	/	/	/
Iodinate contrast media	1.494	0.300–7.438	0.624	/	/	/
Opioids	2.604	0.269–25.206	0.409	/	/	/

Note: As covariates, we considered age (as continuous variable), sex (female vs. male), current outpatient therapy (≥3 drugs vs. <3 drugs), previous drug reaction (yes vs. no), atopic comorbidities (none vs. at least one), receiving antibiotics (yes vs. no), receiving NSAIDs (yes vs. no), receiving iodinate contrast media (yes vs. no), and receiving opioids (yes vs. no). Variables that were significant in the univariable analyses were not all included in the multivariable logistic regression model because their inclusion led to model instability and convergence issues. Abbreviations: NSAIDs, non-steroidal Anti-inflammatory Drugs; SCAR, Severe Cutaneous Adverse Reactions; CI, confidence interval; OR, odds ratio.

4. Discussion

This study presents a comprehensive analysis of inpatient allergology consultations for suspected drug allergy at Luigi Sacco Hospital over a three-year period. Our findings underscore the relevance and complexity of DHRs in hospitalized patients, with an incidence of 0.9% among all admissions, comparable to previous reports from tertiary care centers worldwide [8–10].

The demographic profile of our cohort (median age 65 years; 51.2% female) reflects a predominance of elderly, polymedicated individuals—a well-recognized risk group for adverse drug reactions (ADRs) [5]. Notably, for immediate reactions, increasing patient age was significantly associated with a greater risk of reaction severity in multivariate analysis, with an OR of 1.039 for each additional year of age. The high prevalence of atopic comorbidities (67.4%) and prior drug allergy history (55.1%) further underscores the clinical complexity of this population. Our findings are consistent with previous reports indicating that inpatient allergology services frequently manage patients with multiple comorbidities and a high baseline drug exposure [9,11].

Immediate-type reactions accounted for 49.1% of cases, with cutaneous manifestations being the most frequent clinical presentation (86.8%), echoing the results of Thong et al., who observed similar trends in Singapore [8]. This also aligns with the 94.3% skin involvement reported by Büyük Yaytokgil et al. in children [10], confirming the skin as the primary target organ of DHRs regardless of age. Anaphylaxis occurred in 6.6% of our cohort—higher than in some previous inpatient studies (e.g., 0% in Thong et al.) but consistent with reports from European centers [5,8]. This likely reflects the overall rise in anaphylaxis incidence, both drug-induced and from other causes, since the publication of the Thong et al. study in 2003 [5].

Notably, 13 patients developed SCARs, including SJS, DRESS, and AGEP—conditions associated with significant morbidity and mortality, especially in older adults [5,20]. In both univariate and multivariate analyses, the absence of atopic comorbidities was significantly associated with an increased likelihood of developing a SCAR. To the best of our knowledge, this finding has not been previously reported in the literature. A plausible explanation lies in the distinct immunopathogenesis of atopic comorbidities, which are typically IgE-mediated and therefore classified as type I hypersensitivity reactions, compared with SCARs, which are type IV reactions [20,21]. Accordingly, a predisposition to type I reactions may act as a protective factor against the development of type IV reactions such as SCARs.

Eosinophilia, a known marker of delayed hypersensitivity, was detected in nearly 30% of tested patients in our study, further corroborating the immune-mediated nature of many of these reactions.

In our cohort, antibiotics were the most frequently implicated drug class (44.6%), with β-lactams predominating (36.5%). This distribution is consistent with global epidemiological trends, in which penicillin allergy labels remain disproportionately common and β-lactams—particularly aminopenicillins and aminocephalosporins—account for a substantial proportion of severe antimicrobial hypersensitivity reactions [4,22]. Notably, antibiotic use was significantly associated with greater reaction severity, a finding confirmed in both univariate and multivariate analyses. This association was also observed for β-lactams alone in univariate analysis. These findings align with prior multicenter data indicating that penicillins and cephalosporins are among the leading culprits in antimicrobial-associated anaphylaxis, and that β-lactam exposure is linked to a significant proportion of ICU admissions for severe reactions [22]. The predominance of β-lactams in our setting may reflect both their widespread use in tertiary care—particularly in the management of severe infections and in perioperative prophylaxis—and their intrinsic allergenic potential.

A statistically significant association was also identified between immediate reactions to β-lactam antibiotics and Naranjo score categories ranging from 1 to 8. This observation may be explained by the difficulty of determining definitive causality for a specific drug in a polytreated inpatient population. A similar association was also observed for delayed reactions when considering antibiotics as a whole, suggesting that this challenge is not limited to immediate hypersensitivity events. Consistent with these findings, the causality assessment using the Naranjo algorithm classified most reactions as "possible" or "probable" (median score: 5), reflecting the diagnostic uncertainty frequently encountered in the inpatient setting, where multiple concurrent therapies and complex comorbidities often confound causal attribution.

In agreement with the findings reported by Doña et al. regarding the general population [5], nonsteroidal anti-inflammatory drugs (NSAIDs) also represent the second most frequently implicated drug class in our cohort, followed by iodinated contrast media, whose incidence has been increasing in recent years [5].

Pharmacological treatment (mostly corticosteroids and antihistamines) was necessary in about 80% of patients, while epinephrine use was limited to severe cases. Interestingly, although 5.7% of patients experienced hypotension, only 3.3% received epinephrine, suggesting a possible underuse or hesitation in its administration. This finding aligns with previously reported trends and underscores the need to strengthen awareness and adherence to guidelines recommending early epinephrine use in anaphylaxis [23]. Notably, 16.2% of patients underwent desensitization, primarily to aspirin and beta-lactams—a proportion significantly higher than in studies from Kuwait or pediatric centers [9–11], likely reflecting the infrequent use of antiplatelet-dose aspirin in pediatric practice.

Interestingly, the success rate of desensitization in our adult cohort was high, reinforcing the safety of this strategy when performed under specialist supervision—a position strongly supported by Doña et al. and recent EAACI recommendations [5], and the consensus statement by Cernadas et al. [24].

Our approach to maintaining essential therapy through alternative drugs or graded challenge protocols proved largely successful, with only two patients (0.6%) unable to tolerate the alternative. These results underscore the importance of allergist-guided interventions to avoid unnecessary discontinuation of first-line therapies [5].

Notably, in only two out of the 35 patients evaluated for suspected latex allergy was the diagnosis confirmed, and a latex-free perioperative pathway was recommended. This suggests that the perception of latex allergy in the general population may be greater than its actual prevalence, which is approximately 1% [25].

Compared to other international experiences (e.g., 0.008% incidence in Kuwait, 0.42–0.87% in broader literature), our higher referral rate likely reflects institutional awareness and the availability of an in-hospital allergology service [9,12]. As Weber et al. highlighted, centers with well-integrated A/I services see more frequent and timely consultations [12].

The variety of clinical departments requesting consultations—most notably Cardiology, Infectious Diseases, and Surgery—also mirrors patterns seen in academic centers, as described by Kempe et al. [11]. The demand for rapid diagnostic clarification in patients needing critical medications (e.g., antibiotics, antiplatelets) emphasizes the indispensable role of allergology in inpatient care.

5. Limitations

This was a retrospective single-center study, and despite the structured data collection, underreporting of minor or self-resolving reactions may have occurred. Moreover, the availability of in vivo tests (e.g., drug provocation testing) was limited during hospitalization, and a complete diagnostic work-up was not always feasible. Additional limitations include the fact that many patients did not undergo follow-up evaluation after discharge, and the need for future multicenter studies to confirm and expand upon these findings.

6. Conclusions

Our findings underscore the importance of establishing structured inpatient allergology services in tertiary hospitals, given the complexity of polypharmacy and comorbidities in the admitted population. Where in-house expertise is lacking, the involvement of an external allergist with specific experience in drug hypersensitivity reactions may be crucial to ensuring optimal patient management. A substantial proportion of patients benefited from specific interventions, including desensitization and diagnostic workup, enabling the safe continuation of essential therapies.

In this context, and drawing on the model of antimicrobial stewardship [13], the development of an allergy stewardship approach could enhance the identification and management of DHR. This would promote timely allergist involvement, reduce inappropriate drug avoidance, and support access to diagnostic procedures such as skin testing, graded challenges, and desensitization.

Efforts should focus on standardizing referral criteria, implementing prospective reporting systems, and increasing post-discharge follow-up to complete allergologic evaluations, as suggested in other models [8,10,11]. With an aging population, rising rates of chronic infections and antimicrobial resistance, increasingly complex drug regimens, and inappropriate medication use, the demand for inpatient allergology services is likely to continue increasing. Future studies should assess the impact of structured allergology interventions on patient outcomes, resource utilization, and healthcare system sustainability.

Author Contributions: Conceptualization, C.P.R.; Methodology, C.P.R., F.B., V.G.R.O. and E.I.; Software, F.B.; Validation, V.G.R.O. and E.I.; Formal analysis, F.B.; Investigation, A.C.G., A.B., M.C., E.B. and L.C.; Resources, L.C.; Data curation, C.P.R., A.C.G., F.B., A.B., M.C., E.B., L.C., V.G.R.O. and E.I.; Writing—original draft, C.P.R. and A.C.G.; Writing—review and editing, F.B., L.C., V.G.R.O. and E.I.; Visualization, A.C.G., A.B., M.C. and E.B.; Supervision, C.P.R. and A.C.G.; Project administration, C.P.R. All authors have read and agreed to the published version of the manuscript.

Funding: This research received no external funding.

Institutional Review Board Statement: Since the study was exclusively observational and based only on routine analyses, formal approval by an external ethical committee was not warranted or requested. The study was conducted following the ethical principles of the 2013 revision of the Declaration of Helsinki and the code of Good Clinical Practice.

Informed Consent Statement: The study was conducted in accordance with the ethical principles for medical research involving human subjects (WMA declaration of Helsinki and its later amendments). Written informed consent was obtained from all patients.

Data Availability Statement: Data supporting the findings of this study are available upon request from the corresponding author due to privacy.

Conflicts of Interest: The authors declare no conflicts of interest.

References

1. World Health Organization (WHO). *International Drug Monitoring: The Role of National Centres*; WHO Technical Report Series 498; World Health Organization: Geneva, Switzerland, 1972.
2. Gomes, E.R.; Demoly, P. Epidemiology of Hypersensitivity Drug Reactions. *Curr. Opin. Allergy Clin. Immunol.* **2005**, *5*, 309–316. [CrossRef]
3. Sousa-Pinto, B.; Cardoso-Fernandes, A.; Araújo, L.; Fonseca, J.A.; Freitas, A.; Delgado, L. Clinical and Economic Burden of Hospitalizations with Registration of Penicillin Allergy. *Ann. Allergy Asthma Immunol.* **2018**, *120*, 190–194.e2. [CrossRef] [PubMed]
4. Blumenthal, K.G.; Peter, J.G.; Trubiano, J.A.; Phillips, E.J. Antibiotic Allergy. *Lancet* **2019**, *393*, 183–198. [CrossRef]
5. Doña, I.; Torres, M.J.; Celik, G.; Phillips, E.; Tanno, L.K.; Castells, M. Changing Patterns in the Epidemiology of Drug Allergy. *Allergy* **2024**, *79*, 613–628. [CrossRef] [PubMed]
6. England, R.W.; Ho, T.C.; Napoli, D.C.; Quinn, J.M. Inpatient Consultation of Allergy/Immunology in a Tertiary Care Setting. *Ann. Allergy Asthma Immunol.* **2003**, *90*, 393–397. [CrossRef] [PubMed]
7. Otto, H.F.; England, R.W.; Quinn, J.M. Inpatient Allergy/Immunology Consultations in a Tertiary Care Setting. *Allergy Asthma Proc.* **2010**, *31*, 244–251. [CrossRef]
8. Thong, B.Y.; Leong, K.P.; Tang, C.Y.; Chng, H.H. Drug Allergy in a General Hospital: Results of a Novel Prospective Inpatient Reporting System. *Ann. Allergy Asthma Immunol.* **2003**, *90*, 342–347. [CrossRef]
9. Al-Ahmad, M.; Bouza, T.R. Pattern of Inpatient Referrals to a Drug Allergy Unit in Kuwait. *Eur. Ann. Allergy Clin. Immunol.* **2017**, *49*, 276–280. [CrossRef]

10. Yaytokgil, Ş.B.; Selmanoğlu, A.; Kulhas Çelik, I.; Emeksiz, Z.Ş.; Ginis, T.; Karaatmaca, B.; Toyran, M.; Civelek, E.; Misirlioğlu, E.D. Evaluation of the Frequency and Characteristics of Drug Hypersensitivity Reactions in Hospitalized Children: Real Life-Cohort Study. *World Allergy Organ. J.* **2024**, *17*, 100893. [CrossRef]
11. Kempe, E.; Stukus, D.R.; Strothman, K.; Scherzer, R. Allergy and Immunology Inpatient and Emergency Department Consultations at a Pediatric Academic Medical Center. *Ann. Allergy Asthma Immunol.* **2016**, *116*, 373–374. [CrossRef]
12. Weber, R.W. The Allergist as Consultant. *Ann. Allergy Asthma Immunol.* **2003**, *90*, 369–370. [CrossRef] [PubMed]
13. Lanckohr, C.; Bracht, H. Antimicrobial Stewardship. *Curr. Opin. Crit. Care* **2022**, *28*, 551–556. [CrossRef] [PubMed]
14. Preheim, E.A.; Phillips, C.M.; Weant, K.A. Latex Allergy Alerts in the Age of CPOE: Assessing the Risk from Injectable Medications. *Hosp. Pharm.* **2022**, *57*, 413–415. [CrossRef]
15. Brockow, K.; Przybilla, B.; Aberer, W.; Bircher, A.J.; Brehler, R.; Dickel, H.; Fuchs, T.; Jakob, T.; Lange, L.; Pfützner, W.; et al. Guideline for the Diagnosis of Drug Hypersensitivity Reactions: S2K-Guideline of the German Society for Allergology and Clinical Immunology (DGAKI) and the German Dermatological Society (DDG) in Collaboration with the Association of German Allergologists (AeDA), the German Society for Pediatric Allergology and Environmental Medicine (GPA), the German Contact Dermatitis Research Group (DKG), the Swiss Society for Allergy and Immunology (SGAI), the Austrian Society for Allergology and Immunology (ÖGAI), the German Academy of Allergology and Environmental Medicine (DAAU), the German Center for Documentation of Severe Skin Reactions and the German Federal Institute for Drugs and Medical Products (BfArM). *Allergo J. Int.* **2015**, *24*, 94–105. [CrossRef] [PubMed]
16. Khan, D.A.; Phillips, E.J.; Accarino, J.J.; Gonzalez-Estrada, A.; Otani, I.M.; Ramsey, A.; Arroyo, A.C.; Banerji, A.; Chow, T.; Liu, A.Y.; et al. United States Drug Allergy Registry (USDAR) Grading Scale for Immediate Drug Reactions. *J. Allergy Clin. Immunol.* **2023**, *152*, 1581–1586. [CrossRef]
17. Muraro, A.; Worm, M.; Alviani, C.; Cardona, V.; DunnGalvin, A.; Garvey, L.H.; Riggioni, C.; de Silva, D.; Angier, E.; Arasi, S.; et al. EAACI Guidelines: Anaphylaxis (2021 Update). *Allergy* **2022**, *77*, 357–377. [CrossRef]
18. Barbaud, A.; Garvey, L.H.; Torres, M.; Laguna, J.J.; Arcolaci, A.; Bonadonna, P.; Hofmeier, K.S.; Chiriac, A.M.; Cernadas, J.; Caubet, J.C.; et al. EAACI/ENDA Position Paper on Drug Provocation Testing. *Allergy* **2024**, *79*, 565–579. [CrossRef]
19. Naranjo, C.A.; Busto, U.; Sellers, E.M.; Sandor, P.; Ruiz, I.; Roberts, E.A.; Janecek, E.; Domecq, C.; Greenblatt, D.J. A Method for Estimating the Probability of Adverse Drug Reactions. *Clin. Pharmacol. Ther.* **1981**, *30*, 239–245. [CrossRef]
20. Hung, S.I.; Mockenhaupt, M.; Blumenthal, K.G.; Abe, R.; Ueta, M.; Ingen-Housz-Oro, S.; Phillips, E.J.; Chung, W.-H. Severe Cutaneous Adverse Reactions. *Nat. Rev. Dis. Primers* **2024**, *10*, 30. [CrossRef]
21. Jutel, M.; Agache, I.; Zemelka-Wiacek, M.; Akdis, M.; Chivato, T.; del Giacco, S.; Gajdanowicz, P.; Gracia, I.E.; Klimek, L.; Lauerma, A.; et al. Nomenclature of Allergic Diseases and Hypersensitivity Reactions: Adapted to Modern Needs: An EAACI Position Paper. *Allergy* **2023**, *78*, 2851–2874. [CrossRef]
22. Hall, V.; Wong, M.; Munsif, M.; Stevenson, B.R.; Elliott, K.; Lucas, M.; Baird, A.J.; Athan, E.; Young, M.; Pickles, R.; et al. Antimicrobial Anaphylaxis: The Changing Face of Severe Antimicrobial Allergy. *J. Antimicrob. Chemother.* **2020**, *75*, 229–235. [CrossRef]
23. Kamath, P.; Rao, R.R.; Kamath, A. Study of Epinephrine-Induced Cardiovascular Adverse Events in Patients with Anaphylaxis Using Two Spontaneously Reported Adverse Event Databases from 2004 to 2024. *Sci. Rep.* **2025**, *15*, 37088. [CrossRef]
24. Cernadas, J.R.; Brockow, K.; Romano, A.; Aberer, W.; Torres, M.J.; Bircher, A.; Campi, P.; Sanz, M.L.; Castells, M.; Demoly, P.; et al. General Considerations on Rapid Desensitization for Drug Hypersensitivity—A Consensus Statement. *Allergy* **2010**, *65*, 1357–1366. [CrossRef]
25. Charous, B.L.; Blanco, C.; Tarlo, S.; Hamilton, R.G.; Baur, X.; Beezhold, D.; Sussman, G.; Yunginger, J.W. Natural Rubber Latex Allergy After 12 Years: Recommendations and Perspectives. *J. Allergy Clin. Immunol.* **2002**, *109*, 31–34. [CrossRef] [PubMed]

BioChem

MDPI

Review

Lactic Acid Bacteria: From Bioprocessing to Nanomedicine

Maryam Rezvani [1,2,*], Maria Manconi [1] and Nejat Düzgüneş [3]

1 Department of Life and Environmental Sciences, University of Cagliari, Cittadella Universitaria di Monserrato, Monserrato, 09124 Cagliari, Italy; manconi@unica.it
2 Advanced Nanobiotechnology and Nanomedicine Research Group (ANNRG), Iran University of Medical Sciences, Tehran 14496-4535, Iran
3 Department of Biomedical Sciences, Arthur A. Dugoni School of Dentistry, University of the Pacific, San Francisco, CA 94103, USA; nduzgunes@pacific.edu
* Correspondence: rezvani.mary@yahoo.com

Abstract

Background/Objectives: Lactic acid bacteria have long been recognized as pivotal microorganisms in food fermentation and health promotion. However, their significance has recently grown due to innovative applications in various fields, particularly at the intersection of biotechnology and nanotechnology. This study aimed to provide a comprehensive overview of these emerging applications. **Methods**: The latest scientific literature was drawn from online databases and thoroughly reviewed. The new nomenclature system based on the post-2020 reclassification was used for reports. **Results**: The current study highlighted the evolving role of lactic acid bacteria, beyond their traditional use as starter cultures for food fermentation, in newer challenges, including the production of high-value bioactive compounds through bioprocessing under optimal conditions to enhance the yield, underlining the involved genes and pathways. Furthermore, this review addressed the beneficial effects of lactic acid bacteria as probiotics, postbiotics, and paraprobiotics in the treatment of various diseases and disorders, their application in the production of functional foods, and the encapsulation of their bioproducts to produce advanced health-promoting functional ingredients. The potential use of lactic acid bacteria to synthesize metallic nanoparticles, minicells, and carbon dots was also explored, promising significant advancements in nanomedicine. **Conclusions**: This review could open a new horizon for leveraging the potential of lactic acid bacteria in biotechnology, food science, and nanomedicine. The multilateral perspective offered here would provide a foundation for future research and development to exploit the capabilities of lactic acid bacteria across these innovative fields.

Keywords: lactic acid bacteria; nanomedicine; *Lactobacillus*; postbiotics; paraprobiotics; probiotics; encapsulation; functional foods; nanoparticles

1. Introduction

Lactic acid bacteria (LAB) are Gram-positive, catalase-negative, acid-tolerant, non-motile, non-sporulating, aerotolerant, and non-respiring microorganisms that can exist in rod-shaped or cocci forms [1]. Their primary characteristic is the production of lactic acid via the fermentation of carbohydrates, which serve as their main carbon source. LAB comprise the genera of *Lactobacillus*, *Lactococcus*, *Pediococcus*, *Streptococcus*, *Leuconostoc*, *Aerococcus*, *Carnobacterium*, *Alloiococcus*, *Enterococcus*, *Dolosigranulum*, *Oenococcus*, *Weissella*, *Vagococcus*, and *Tetragenococcus* [2].

Traditionally, the classification of LAB was based on biochemical and physiological characteristics. Recently, techniques such as 16S rRNA gene sequencing, random amplified polymorphic DNA profiling, soluble protein patterns, PCR-based fingerprinting, and other molecular characterization methods have been employed to identify and classify LAB [3]. Advances in bacterial taxonomy have been driven by the development of whole-genome sequencing and improved computational capacities, enabling the rapid comparison of entire bacterial genomes [4]. In 1901, Beijerinck proposed the genus *Lactobacillus*, which includes Gram-positive, facultatively anaerobic, non-spore-forming, and fermentative bacteria [5]. For decades, the most newly discovered bacteria of this type were routinely added to this genus. By March 2020, *Lactobacillus* encompassed 261 species that exhibited diversity at genotypic, ecological, and phenotypic levels. In April 2020, Zheng et al. [5] reclassified the *Lactobacillus* genus and proposed an emendation to describe the Lactobacillaceae family, incorporating all genera previously classified under the Lactobacillaceae and Leuconostocaceae families. Through this taxonomic reorganization of LAB, over 300 species, formerly categorized within seven genera and two families, were reclassified into the Lactobacillaceae family, comprising 31 genera including *Lactobacillus, Pediococcus, Paralactobacillus, Leuconostoc, Fructobacillus, Weissella, Oenococcus,* and *Convivina*, as well as 23 newly established genera [5,6]. The proposed novel genera include *Bombilactobacillus, Amylolactobacillus, Holzapfelia, Companilactobacillus, Lapidilactobacillus, Agrilactobacillus, Loigolactobacillus, Schleiferilactobacillus, Latilactobacillus, Lacticaseibacillus, Dellaglioa, Liquorilactobacillus, Lactiplantibacillus, Ligilactobacillus, Paucilactobacillus, Furfurilactobacillus, Fructilactobacillus, Limosilactobacillus, Acetilactobacillus, Apilactobacillus, Lentilactobacillus, Secundilactobacillus,* and *Levilactobacillus* [5]. This revised taxonomy has streamlined regulatory approval processes and opened new avenues for scientific exploration. Since 2020, several new species have been described, aligning with the metabolic and ecological traits of the proposed genera [6]. The key *Lactobacillus* species studied in this review and their updated names (post-2020 revision) are presented in Table 1.

Table 1. Updated *Lactobacillus* species nomenclature following the 2020 reclassification.

Basionym	New Name	Nomenclature Status
Lactobacillus casei	*Lacticaseibacillus casei*	Changed
Lactobacillus paracasei	*Lacticaseibacillus paracasei*	Changed
Lactobacillus fermentum	*Limosilactobacillus fermentum*	Changed
Lactobacillus brevis	*Levilactobacillus brevis*	Changed
Lactobacillus rhamnosus	*Lacticaseibacillus rhamnosus*	Changed
Lactobacillus kefiri	*Lentilactobacillus kefiri*	Changed
Latilactobacillus sakei	*Latilactobacillus sakei*	Changed
Lactobacillus reuteri	*Limosilactobacillus reuteri*	Changed
Lactobacillus pentosus	*Lactiplantibacillus pentosus*	Changed
Lactobacillus plantarum	*Lactiplantibacillus plantarum*	Changed
Lactobacillus curvatus	*Latilactobacillus curvatus*	Changed
Lactobacillus kimchicus	*Secundilactobacillus kimchicus*	Changed
Lactobacillus paraplantarum	*Lactiplantibacillus paraplantarum*	Changed
Lactobacillus buchneri	*Lentilactobacillus buchneri*	Changed
Lactobacillus delbrueckii	*Lactobacillus delbrueckii*	Unchanged
Lactobacillus helveticus	*Lactobacillus helveticus*	Unchanged
Lactobacillus acidophilus	*Lactobacillus acidophilus*	Unchanged

LAB have been extensively used in the food industry for centuries due to their vital role in fermentation. They ferment food carbohydrates, promote the decomposition of lipids and proteins, and contribute to product flavor development by producing various

https://doi.org/10.3390/biochem6010003

organic acids, ketones, aldehydes, alcohols, and esters. Additionally, they can improve the product texture and serve as bioprotective cultures by producing antimicrobial compounds. In recent decades, the importance of LAB has expanded beyond their traditional role in producing various fermented foods. They have garnered considerable attention for their crucial contributions to biotechnology, nanotechnology, and the development of functional foods and health-promoting products, such as probiotic- and symbiotic-based products. Probiotics are live microorganisms that confer health benefits to the host when administered in adequate amounts. A product containing at least 10^6 CFU/g of these viable microorganisms may be considered a probiotic-containing product [7]. In symbiotic-based products, probiotic LAB are combined with prebiotics—nondigestible dietary carbohydrates that promote the growth or metabolic activity of beneficial intestinal microorganisms. Combining probiotics with prebiotics, such as fructooligosaccharides, inulin, and galactooligosaccharides, can enhance their effectiveness [8]. The primary mechanism behind the health benefits of probiotics is their ability to colonize the human gut and interact with the complex ecosystem of the gastrointestinal tract. Therefore, probiotics speed up gastrointestinal transit, modulate the gastrointestinal microbiota, and inhibit the growth of pathogenic microorganisms by competing for essential nutrients and limiting their colonization and adhesion to the gastrointestinal mucosa. Moreover, probiotics regulate lipid metabolism, synthesize vitamins within the gastrointestinal tract, improve electrolyte absorption in the intestine, neutralize toxins, decrease the production of pro-inflammatory cytokines, and promote immune system function [9,10]. Recently, postbiotics and paraprobiotics have emerged as promising alternatives to traditional probiotics, particularly for immunocompromised individuals [11].

Paraprobiotics, also known as inactivated or ghost probiotics, are nonviable microbial cells or crude cell extracts that still confer health-promoting effects when administered in adequate amounts [12]. By contrast, postbiotics are soluble products or metabolites secreted by live bacteria or released during bacterial cell lysis, which exert health benefits [13]. However, in many cases, the term postbiotics has been used to describe both paraprobiotics and postbiotics [14].

Functional foods can be defined as foods containing bioactive compounds, offering health benefits beyond basic nutrition [15]. A recently proposed definition for functional foods refers to formulated foods that contain live microorganisms or substances that provide health benefits at safe and adequately high concentrations for the intended advantage [16]. Optimizing culture conditions to maximize yield is critical when producing functional bioproducts by LAB. These bioactive compounds can be incorporated directly into functional products during bioprocessing or added afterward. Encapsulation techniques are frequently employed to prevent undesirable interactions between these bioactive compounds and other components of functional products while improving their bioavailability [17].

LAB's ability to synthesize nanoparticles (NPs) opens new avenues for applications in drug delivery, as well as biosensing and therapeutic interventions, positioning these bacteria as critical players in emerging nanomedicine.

This review will focus on these exciting and innovative aspects, particularly highlighting recent achievements.

2. Methodology

A comprehensive search was conducted using the online databases, PubMed, Scopus, MEDLINE, and Google Scholar to select relevant English language articles published between 2016 and 2025. The articles were then independently screened for quality and validity, with the scientific findings from the highest-quality studies incorporated into the

https://doi.org/10.3390/biochem6010003

review. In this process, expert opinions, compendiums, symposia, conference proceedings, and letters to the editor were excluded. The inclusion criteria focused on observational, longitudinal, in vitro and in vivo studies, review articles, systematic reviews, and book chapters. The new nomenclature system, following the 2020 revision, was used for reports.

3. The Role of LAB in Bioprocessing

3.1. LAB-Mediated Bioprocess for Lactic Acid Production

LAB function as biofactories for various metabolites. Through the fermentation of carbohydrates, LAB generate lactic acid as the primary product. Based on their carbohydrate fermentation capabilities, LAB can be categorized as homofermentative or heterofermentative bacteria. Homolactic fermentation, carried out by genera such as *Streptococcus* and *Lactococcus*, produces two lactate molecules for every glucose molecule. By contrast, heterofermentative genera, such as *Weissella* and *Leuconostoc*, convert one glucose molecule into lactate, ethanol, and carbon dioxide (Figure 1) [3].

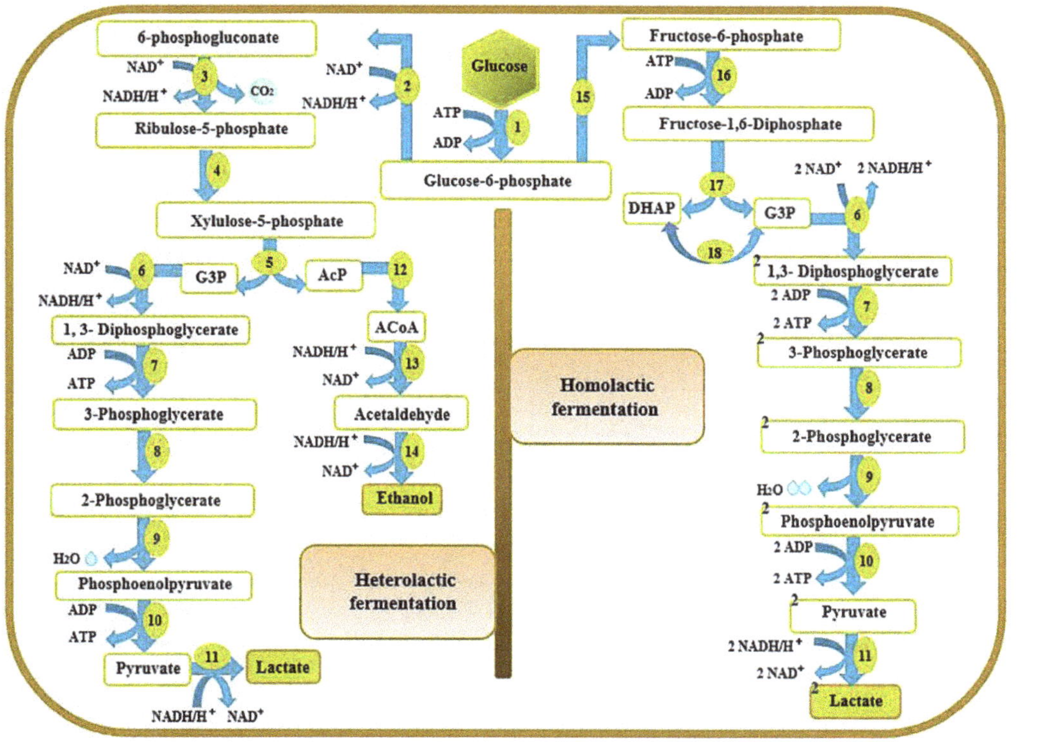

Figure 1. Schematic representation of metabolic pathways of heterofermentative and homofermentative lactic acid bacteria. 1: Hexokinase; 2: Glucose-6-phosphate dehydrogenase; 3: 6-Phosphogluconate dehydrogenase; 4: Ribulose-5-phosphate epimerase; 5: Phosphoketolase; 6: Glyceraldehyde-3-phosphate dehydrogenase; 7: Phosphoglycerate kinase; 8: Phosphoglycerate mutase; 9: Enolase; 10: Pyruvate kinase; 11: Lactate dehydrogenase; 12: Phosphate acetyltransferase; 13: Acetaldehyde dehydrogenase; 14: Alcohol dehydrogenase; 15: Phosphoglucose isomerase; 16: Phosphofructokinase; 17: Aldolase; 18: Triosephosphate isomerase; AcP: Acetyl-phosphate; G3P: Glyceraldehyde-3- phosphate; ACoA: Acetyl-Coenzyme A; DHAP: Dihydroxyacetone-phosphate.

The conversion of glucose into end-products in homofermentative LAB is mediated by the Embden–Meyerhof pathway (EMP). By contrast, heterofermentative LAB use the

https://doi.org/10.3390/biochem6010003

phosphoketolase pathway (PKP) to generate lactic acid and by-products [18]. The enzymes involved in these pathways are illustrated in Figure 1. Lactate dehydrogenase (LDH), the key enzyme in lactic acid production, catalyzes the conversion of pyruvate into lactate [19]. Genomic analyses have revealed that the lactate dehydrogenase gene (*ldh*) plays a crucial role in lactic acid production, and its expression can be increased through genetic engineering to achieve high industrial lactate production [20].

3.2. LAB-Mediated Bioprocess for Fatty Acid Production

LAB can produce short-chain fatty acids (SCFAs), carboxylic acids with fewer than six carbon atoms, primarily acetate, propionate, and butyrate [21]. The production of these SCFAs is closely linked to the activity of genes such as *pdhA* (pyruvate dehydrogenase A), *pdhB*, *pdhC*, *pdhD*, *ldh*, *acyP* (acylphosphatase P), and *adhE* (alcohol/acetaldehyde dehydrogenase) within the glycolysis/gluconeogenesis pathway. Specifically, the *acyP*, *pdhA*, and *pdhB* genes were reported to be associated with acetate formation from pyruvate, while *adhE* was shown to be implicated in butyrate production [22].

Moreover, conjugated linoleic acid (CLA) can be synthesized during fermentation from precursors such as linoleic, ricinoleic, and vaccenic acids or from oils rich in these fatty acids. CLA, a polyunsaturated fatty acid, is well-known for its health-promoting properties, including anticarcinogenic, anti-obesity, and anti-osteoporotic effects, as well as its ability to reduce blood lipid levels and lower the risk of cardiovascular diseases [23]. Although the precise mechanism of CLA production has remained unclear, it is hypothesized to be a bacterial detoxification strategy to counter the inhibitory effects of elevated free linoleic acid on bacterial growth [24]. The biosynthesis of CLA is conducted by multiple enzymes, including CLA hydratase (CLA-HY), CLA short-chain dehydrogenase (CLA-DH), and CLA acetoacetate decarboxylase (CLA-DC). These enzymes catalyze hydration, dehydration, and isomerization reactions, respectively [25]. Some studies suggest that activation of multiple genes in a multiple-step reaction as a stress response triggers CLA production [26]. The genes lp_0139 (*cla-hy*), lp_0060 (*cla-dh*), and lp_0061 (*cla-dc*) encode the enzymes CLA-HY, CLA-DH, and CLA-DC, respectively [25]. A correlation was reported between the CLA yield in *Lactiplantibacillus plantarum* and the transcriptional maintenance ability of the involved genes [27].

3.3. LAB-Mediated Bioprocess for Bacteriocin Production

Bacteriocins are active, ribosomally synthesized peptides known for their antimicrobial properties. These compounds can be categorized into three main classes: class I, small (<5 kDa), post-translationally modified peptides or lantibiotics; class II, small (<10 kDa), post-translationally unmodified peptides; and class III, large (>30 kDa) macromolecules [28]. LAB produce a variety of protease-sensitive, heat-stable bacteriocins [29]. Ongoing efforts aim to develop new LAB-induced bacteriocins and antimicrobial peptides. Commercialization of bacteriocins requires process optimization and purification. Some of the commercially available bacteriocins include nisin (class I), marketed under names such as Nisaplin® (Danisco, Copenhagen, Denmark), Delvo® Nis (DSM, Delft, The Netherlands), and Nisin Z® (Handary, Brussel, Belgium), as well as pediocin PA-1/AcH (class II), available under ALTA™ 2341 (Quest International, Sarasota, FL, USA) and MicroGARD™ (Gillco, San Marcos, CA, USA) [30,31]. Class I lantibiotics undergo post-translational modifications and involve supplementary enzyme-encoding genes, making them more complex than class II bacteriocins. In bacteriocin biosynthesis, structural genes play a pivotal role. Nisin is the most well-characterized class I bacteriocin. The nisin gene cluster comprises 11 genes *nisABTCIPRKFEG*, which are involved in regulation, biosynthesis, maturation, and immunity. During nisin biosynthesis, *nisA* encodes pre-nisin, while *nisB* and *nisC* encode the enzymes required for post-translational modification. The modified pre-nisin is transported out of the cell via a transport protein encoded by *nisT* and the cleavage of pre-nisin to release active nisin A is carried out by a membrane-bound serine protease encoded by *nisP*. The genes *nisI*, *nisF*, *nisE*, and *nisG* encode immunity proteins, while *nisK* and *nisR* encode the involved proteins in the regulation of nisin synthesis. The pediocin *AcH* operon includes *papA*, which encodes pre-pediocin; *papB*, which encodes the immunity protein; and *papC* and *papD*, which encode the transport protein [32,33].

3.4. LAB-Mediated Bioprocess for γ-Aminobutyric Acid (GABA) Production

LAB can also produce GABA in response to abiotic stress, enabling them to survive in harsh environments [34]. GABA is a bioactive compound that promotes sleep and exhibits physiological functions such as antidepressant and anti-hypertensive effects. Several LAB species, including *Lactobacillus delbrueckii* subsp. *Bulgaricus*, *Lactobacillus helveticus*, *Levilactobacillus brevis*, *Limosilactobacillus fermentum*, *Lacticaseibacillus paracasei*, *Lentilactobacillus buchneri*, *Lactiplantibacillus plantarum*, *Lactococcus lactis*, and *Streptococcus thermophilus* are known to produce GABA. They can be utilized in the production of functional foods. Recently, some species from the genera *Leuconostoc*, *Enterococcus*, *Weissella*, and *Pediococcus* have also been reported to produce GABA [35]. This compound is synthesized through the decarboxylation of glutamic acid by the enzyme glutamate decarboxylase (GAD). In LAB, the *gad* operon consists of *gadR* (encoding the transcriptional regulator), *gadB* (encoding GAD), and *gadC* (encoding the glutamate/GABA antiporter) [36].

3.5. LAB-Mediated Bioprocess for Exopolysaccharide (EPS) Production

EPSs synthesized by LAB have demonstrated various biological activities, including antimicrobial, antioxidant, anticarcinogenic, immunomodulatory, gut microbiome modulating, prebiotic, and cholesterol-lowering effects. These properties make EPS a promising ingredient for functional foods. Additionally, EPSs are critical in improving the sensorial and rheological properties and thermal stability of food products. Their unique characteristics, such as film- and gel-forming abilities, as well as biodegradability and biocompatibility, have made them valuable for encapsulating bioactive substances in both functional foods and drug delivery systems in biomedicine. The encapsulation process can be optimized by tailoring these EPSs to achieve specific charge, size, and porosity [37–39].

The *eps* gene cluster responsible for EPS biosynthesis typically includes the genes *epsA*, *epsB*, *epsC*, *epsD* and *epsE*, as well as the flippase *wzx*, polymerase *wzy*, and genes encoding glucosyltransferase or other polymer-modifying enzymes. The genes *epsE*, *wzy*, *wzx*, and *epsA* are involved in the polysaccharide assembly machinery, while *epsBCD* are modulatory genes [40]. Many LAB, including certain *Lactococcus* and *Streptococcus* species, biosynthesize heteropolysaccharide-type EPSs through the Wzx/Wzy-dependent pathway [41]. This pathway involves several key steps, including the transportation and phosphorylation of monosaccharides and disaccharides, the formation of sugar nucleotides, the synthesis of repeating subunits, and the translocation from the cytoplasmic to the extracellular side of the membrane, facilitated by flippase (Wzx). The final step, polymerization of the repeating subunits, is catalyzed by the polymerization protein (Wzy), leading to the release of long-chain polymers into the extracellular space. However, some LAB genera, such as *Leuconostoc*, *Weissella*, and *Pediococcus*, produce EPSs through an extracellular synthesis pathway. They generate homopolysaccharides through enzymatic polymerization of monosaccharides derived from the hydrolysis of substrates like sucrose. The resulting polymers are directly released into the extracellular environment [41,42].

https://doi.org/10.3390/biochem6010003

3.6. LAB-Mediated Bioprocess for Vitamin Production

LAB commonly exhibit vitamin auxotrophy. However, some LAB strains have the genetic capacity to biosynthesize B-group vitamins, such as riboflavin (vitamin B_2) and folates (vitamin B_9). Enhancing the production of these vitamins can be achieved through genetic modification of LAB strains. B-group vitamins are essential for maintaining energy levels, cellular metabolism, and brain function. Since the body cannot store these vitamins, they must be replenished daily [43].

Riboflavin is crucial as a precursor to the redox-active coenzymes, FAD (flavin adenine dinucleotide) and FMN (flavin mononucleotide). The biosynthesis of riboflavin from ribulose-5-phosphate and guanosine triphosphate (GTP) is catalyzed by enzymes encoded by the *ribG*, *ribB*, *ribA*, and *ribH* genes, which are organized in an operon. However, the order of genes and enzymatic reactions can vary. The *ribA*, the third gene in this operon, encodes GTP cyclohydrolase II, the enzyme that catalyzes the first step in the riboflavin biosynthesis pathway. Additionally, this gene encodes the enzyme that converts ribulose-5-phosphate into a four-carbon unit in a later step of the biosynthesis pathway. The second and third steps of riboflavin biosynthesis are controlled by pyrimidine deaminase and pyrimidine reductase, encoded by the *ribG* gene, the first gene in the operon. The final two steps are catalyzed by lumazine synthase and riboflavin synthase, encoded by *ribH* and *ribB*, respectively. In certain LAB strains with disrupted operons, riboflavin must be present in the culture medium for the strain to grow [44].

LAB's ability to produce folate is strain-dependent and affected by fermentation conditions, as well as the availability of folate or its precursors in the medium. In general, folate-producing LAB show a tendency for consumption of the available folate rather than its synthesis. Some LAB strains can produce tetrahydrofolate and methyl-tetrahydrofolate as the predominant forms of folate [45]. 6-hydroxymethyl-7,8-dihydropterin pyrophosphate (DHPPP), p-aminobenzoic acid (PABA), and glutamate are essential for folate biosynthesis. Since most LAB cannot synthesize glutamate and PABA, these compounds must be added to the growth medium. The folate synthetic pathway in LAB is conducted by genes *folE*, *folQ*, *folB*, *folK*, *folP*, *folC1*, and *folA*, which encode GTP cyclohydrolase I, dITP/XTP pyrophosphatase, dihydroneopterin aldolase, hydroxymethyl dihydropterin pyrophosphokinase, dihydropteroate synthase, dihydrofolate synthase, and dihydrofolate reductase, respectively. These enzymes catalyze the conversion of GTP into tetrahydrofolate. The *folC2* gene encodes folylpolyglutamate synthase, which adds glutamate residues to the monoglutamate form of tetrahydrofolate, resulting in the formation of tetrahydrofolate polyglutamate [46].

Some LAB species have also been shown to produce other B-group vitamins, such as cobalamin (vitamin B12). Genomic analysis of *Pediococcus pentosaceus* L51, *Levilactobacillus brevis* G31, and *Lactococcus lactis* E32, isolated from *Chlorella vulgaris* grown in tubular photobioreactors, revealed the presence of key genes involved in cobalamin biosynthesis, including *cobC*, *cobD*, *cbiT*, and *hemL*. Among these species, *Pediococcus pentosaceus* L51, harboring additional genes such as *pduV*, *pduU*, and *cobA*, provided evidence for the presence of the cobalamin operon [47].

3.7. Optimization of LAB-Mediated Bioprocesses

The LAB-induced multiple-product process can be optimized to enhance the production of lactic acid and other by-products. For example, the dark fermentation of raw ovine whey, rich in LAB, conducted in a 2 L bioreactor at 39 ± 1 °C, 150 rpm and a controlled pH of 5.5, resulted in the highest lactic acid concentration (47.9 g L^{-1}) after 53 h of fermentation. At this point, SCFAs, including acetic, propionic, and butyric acids, were undetectable. Following this peak, a second fermentation phase began, characterized by

https://doi.org/10.3390/biochem6010003

lactic acid degradation and biogas (H_2 and CO_2) and SCFA production, continuing for up to 80 h [48]. Another example of such a bioprocess optimization was reported for a fed-batch fermentation using free/immobilized *Lentilactobacillus buchneri* ATCC 4005 with xylose as the substrate, at pH 4 and under anoxic conditions. The process yielded 32 g L^{-1} of acetic acid and 67 g L^{-1} lactic acid (as high concentration levels) without inhibiting cell growth. Notably, fermentation with immobilized cells resulted in higher acetic and lactic acid production rates than free-cell fermentation [49]. The study on the large-scale fermentation using *Limosilactobacillus reuteri* KMP-P4-S03 in 5 L and 50 L bioreactors at 37 °C with sucrose as the carbon source (without requiring beef extract as a nitrogen source) achieved a lactate/acetate concentration ratio of 6.9:1.1 after 12 h. Generally, *Limosilactobacillus reuteri* KMP-P4-S03 produced more lactic acid than acetic acid. Other SCFAs, such as propionic acid and butyric acid, were undetectable [50]. In a study, optimum fermentation conditions for improved SCFA production by *Limosilactobacillus fermentum* ATCC9338, *Limosilactobacillus fermentum* 19 SH, and *Lactobacillus acidophilus* ATCC4356 were reported as 30 °C, pH 6, and a low salt concentration of 2% [51].

In a separate study, fed-batch fermentation using a metabolically engineered *Lactococcus lactis* F44 in a 5 L bioreactor at 30 °C and 100 rpm for 30 h was optimized for the co-production of nisin and GABA. A two-stage pH control strategy was employed: the pH was maintained at 6.0 for the first 16 h to maximize nisin production, followed by adjustment to pH 4.8 to enhance GABA production. This approach resulted in a GABA yield of 9.12 g L^{-1}, which is 2.2 times higher than in constant pH fermentation [52].

The maximum biomass of *Lactobacillus acidophilus* ATCC 4356 and cost-effective generation of Acidocin 4356, a bacteriocin used to combat *Pseudomonas aeruginosa* infections, were achieved in a lab-scale fermenter. This study was performed under optimal conditions of 40 °C and the concentrations of 4 g L^{-1} yeast extract and 8 g L^{-1} peptone. A basal medium consisting of 40 g L^{-1} whey and 5 g L^{-1} sodium acetate was used. Bacteriocin production was successfully scaled up under optimized conditions at a constant pH 5 in a 3 L fermenter [53]. Another study examined bacteriocin-like inhibitory substances produced by *Pediococcus pentosaceus* ATCC 43200 in a bioreactor under anaerobic conditions at 30 °C and 200 rpm for 24 h. These substances exhibited potential as bio-preservative agents in the food industry, demonstrating 50% and 100% inhibition of *Listeria seeligeri* and *Listeria innocua*, respectively [54]. Similarly, leucocin produced by *Leuconostoc lactis* SM2 in shake flasks at 37 °C and pH 7.0, as the optimum conditions, effectively inhibited the growth of *Escherichia coli*, *Bacillus subtilis*, *Staphylococcus aureus*, *Serratia*, *Klebsiella*, and *Pseudomonas putida* [55].

Lactococcus lactis NCDO 2118, cultivated in a 2 L bioreactor with a complex medium containing 0.3 M NaCl at 30 °C and pH 6.6 for 11 h, followed by adjustment to pH 4.6, produced 413 mM of GABA after 56 h. The findings indicated that GABA production was influenced by the nature of salts added to the medium, with no direct relationship between GABA production and osmolarity. The data showed that chloride ions were the most significant factor in enhancing GABA production under acidic stress, whereas sulfate ions had no stimulatory effect. Chloride ions caused an increase in *gadBC* expression and GAD synthesis [34].

Tarhana is a cereal-based traditional Turkish fermented food with yogurt as an ingredient, rich in LAB species [56]. In a study, strains of *Lactiplantibacillus plantarum* (PFC308, PFC309, PFC310, PFC311, PFC312, PFC313), isolated from Tarhana, were cultivated in a 2 L bioreactor at 30 °C, pH 6, and 100 rpm and were found to produce ropy EPSs [57]. Research revealed that co-culturing LAB with yeast could influence metabolism and enhance EPS generation in a strain-dependent manner. A co-culture fermentation using *Lacticaseibacillus rhamnosus* and *Saccharomyces cerevisiae* was performed in 1 L bioreactors containing supplemented whey permeate medium and was inoculated at 1% (*v*/*v*) for 48 h at 37 °C

https://doi.org/10.3390/biochem6010003

with an agitation rate of 200 rpm. EPS production increased by 49%, 42%, and 39% for *Lacticaseibacillus rhamnosus* ATCC 9595, *Lacticaseibacillus rhamnosus* RW-9595M, and *Lacticaseibacillus rhamnosus* R0011, respectively, compared to their monoculture fermentation. The results indicated that co-culture fermentation of *Lacticaseibacillus rhamnosus* RW-9595M with yeast led to overexpression of key genes compared to monoculture, thereby increasing EPS production under acid stress. The gene co-expression network analysis revealed direct correlations between the expression of the EPS-associated genes and stress response, as well as sugar and lipid metabolism and amino acid biosynthesis [58].

The optimal conditions for CLA production by *Lactiplantibacillus plantarum* through the conversion of linoleic acid were pH 6.5 and 40 °C, which resulted in a yield of 37.5 μg mL^{-1} after 48 h. Supplementing the MRS broth with a mixture of yeast extract and glucose increased the production rate to 76.6 μg mL^{-1} [59]. Furthermore, it was reported that the biomass and CLA production yields of *Lactiplantibacillus plantarum* were influenced by the carbon source used for bacterial growth. The highest CLA concentration (37.08 μg mL^{-1}) and biomass (3.66 g L^{-1}) were achieved when a combined lactose–glucose mixture (1:1, *v:v*) was used as the carbon source in biofilm reactors. Supplementing the fermentation broth with the combined glucose–lactose mixture and introducing thin perforated and wavy metal sheets into the bioreactor as metal support for bacterial self-immobilization led to biofilm accumulation, producing a high amount of CLA isomers. The CLA isomers identified in the fermentation broth were *cis*-9, *trans*-11 C18:2 and *trans*-9, *trans*-11 C18:2, at concentrations of 22.12 μg mL^{-1} and 14.96 μg mL^{-1}, respectively [60].

Various LAB have been detected in a traditional Korean fermented vegetable product, most often made of seasoned and salted cabbage or radish—kimchi [61]. Amongst these bacteria, *Lactiplantibacillus plantarum* HY7715 demonstrated an ability to overproduce riboflavin due to the high expression of the *ribA*, *ribB*, *ribC*, *ribH*, and *ribG* operon genes. HY7715 was cultivated in a 3 L bioreactor using four different media at various temperatures (25 °C to 37 °C) and pH values (4.5, 5, 5.5 and 6). The highest riboflavin production was achieved using a medium containing 1% (*v*/*w*) soy peptone and 3% (*v*/*w*) yeast extract as nitrogen sources. The optimal temperature and pH for riboflavin production by HY7715 were 35 °C and 5.5, respectively. Under these optimal conditions, mass cultivation of HY7715 led to the production of 34.5 ± 2.41 mg L^{-1} of riboflavin within 24 h [62]. The results of another study revealed that *fol* gene expression and folate production by *Streptococcus gallolyticus* subsp. *Macedonicus* CRL415 increased when glucose was used as the primary carbon source under optimal conditions of 42 °C and a controlled pH of six in a 2 L fermentor with 100 rpm agitation [63].

4. The Role of LAB in the Production of Functional Products

4.1. LAB-Derived Functional Products

Growing consumer awareness of the relationship between diet and health has driven specialists to design and develop various functional products. LAB are the most important probiotics and numerous studies have demonstrated their potential as ingredients for functional foods with therapeutic and bioprotective properties [64]. For example, *Lactobacillus delbrueckii* subsp. *lactis* LL10 and *Limosilactobacillus reuteri* LR12, isolated from pineapple puree, showed promising potential for functional food production. These strains exhibited acid and bile tolerance, the ability to adhere to the intestinal epithelial cells, antioxidant activity, and potent antibiofilm effect against *Bacillus cereus*, *Staphylococcus aureus* and *Enterococcus faecalis*. Adding pineapple puree (up to 3%) to goat milk as a starter culture for yogurt production did not negatively affect the sensorial or physical properties of the yogurt in comparison to the control. These probiotic strains survived well during cold storage [65]. Among the LAB strains isolated from microalga *Chlorella vulgaris* cultures,

https://doi.org/10.3390/biochem6010003

Pediococcus pentosaceus L51 produced the highest cobalamin levels (28.2 ± 2.3 pg mL^{-1}). It also exhibited the highest resilience against harsh gastrointestinal conditions, suggesting its potential for developing functional products with high cobalamin content and probiotic activity [47]. *Weissella confusa*, isolated from pozol, a nixtamalized fermented maize product, showed promising traits for developing novel functional foods with new strains [66]. Another study on *Weissella confusa* (KR780676), isolated from Idli batter, an Indian rice-based food, revealed its tolerance to simulated gastrointestinal conditions and potential for adhesion to intestinal cells. This strain also demonstrated antioxidant and cholesterol removal properties, as well as an antibiofilm effect against pathogenic *Pseudomonas aeruginosa* KT266804. Furthermore, it showed no gelatinase, DNase, or hemolytic activity. *Weissella confusa* (KR780676) exhibited good probiotic and safety characteristics and technological traits such as thermostability and the ability to produce β-galactosidase and proteolytic enzymes, making it a promising candidate for functional food development [67].

Fermentation of cashew apple juice by *Lacticaseibacillus casei*, *Lactobacillus acidophilus*, *Lactiplantibacillus plantarum*, and *Leuconostoc mesenteroides* was conducted at 37 °C for 48 h to enhance the production of B-group vitamins and prebiotic oligosaccharides. Fermentation with *Lacticaseibacillus casei* and *Lactobacillus acidophilus* produced higher levels of B vitamins, by 23.1% and 19.3%, respectively. All fermented products experienced a decrease in thiamine (vitamin B1) levels during fermentation; however, *Lactobacillus acidophilus* was able to maintain a higher level of thiamine by the end of fermentation, suggesting that this strain may not require thiamine for growth. Additionally, all products showed an upward trend in nicotinamide (vitamin B3) content during fermentation. Fermentation with *Lacticaseibacillus casei* increased pyridoxine (vitamin B6) content in the product, while fermentations by *Leuconostoc mesenteroides* and *Lactobacillus acidophilus* led to decreases, possibly due to the requirement of these LAB for pyridoxine during growth. The highest cobalamin levels were observed in fermentation with *Lactiplantibacillus plantarum* (41.6 mg L^{-1}), followed by *Lacticaseibacillus casei* (36.1 mg L^{-1}), and *Lactobacillus acidophilus* (35.7 mg L^{-1}), which were achieved after 24, 48, and 24 h, respectively. Riboflavin levels did not differ significantly among the various probiotic-fermented products. Higher levels of prebiotic oligosaccharides were found in cashew apple juice fermented by *Lactobacillus acidophilus* and *Lactiplantibacillus plantarum*, with the former showing the highest level of fructooligosaccharides and the latter showing the highest level of raffinose family oligosaccharides [68].

Given the human body's inability to produce folate—essential for DNA replication, repair and methylation, and amino acid metabolism—the fortification of food products with LAB-generated folate has been proposed to prevent disorders arising from insufficient intake of this vitamin. For instance, *Streptococcus gallolyticus* subsp. *macedonicus* CRL415 produced folate-enriched fermented milk when inoculated (at 2% v/v) in reconstituted non-fat powdered milk and incubated at 42 °C. This strain showed potential as a probiotic and functional starter culture in the dairy industry due to its resistance to simulated gastrointestinal conditions, proper adhesion to intestinal cells, and antibiotic susceptibility [63].

Among the LAB isolated from Mexican milk kefir grains, *Lentilactobacillus kefiri* BIOTEC014, *Leuconostoc pseudomesenteroides* BIOTEC012, *Lactococcus lactis* BIOTEC008, BIOTEC007, and BIOTEC006 demonstrated GABA production capacity, which was classified as medium [69]. Furthermore, incorporating *Lactobacillus acidophilus* (ATCC 4356) as a probiotic into ice cream resulted in viable bacterial counts exceeding 10^6 CFU/g (the target level) after 90 days of cold storage in the samples that experienced a kind of fermentation before freezing. This approach was proposed to develop low-fat ice cream with health-promoting properties [70].

Table 2 summarizes the culture conditions for producing some of the LAB's bioactive compounds and their use in some functional/model products.

https://doi.org/10.3390/biochem6010003

Table 2. Cultural conditions to produce LAB's bioproducts, their production yield, and their use in some functional/model products.

Lactic Acid Bacteria	Bioproduct	Conditions	Production Yield	Functional/Model Product	Ref.
Lactiplantibacillus plantarum AB20-961	CLA	pH: 7.9, T: 37 °C, t: 79 h, 8 log CFU/g initial count and 5% added safflower free fatty acids	7.9 mg g^{-1} fat	Fermented ground beef	[71]
Lactiplantibacillus plantarum DSM 2601		pH: 7.7, T: 37 °C, t: 73 h, 8 CFU/g initial count and 5% added safflower free fatty acids	log 38.3 mg g^{-1} fat		
Lactococcus lactis UTMC 109	Nisin	pH: 7–7.2, T: 37 °C, t: 24 h, agitation rate: 200 rpm	27.5 IU mL^{-1} h^{-1}	-	[72]
Co-cultivation of *Lactococcus lactis* ATCC 11454 and *Lacticaseibacillus rhamnosus* ATCC 7469	Nisin	Sequential fed-batch fermentation, pH: 6.5, T: 37 °C, flow rate of MRS: 10 mL/min, pump speed: 30 rpm, pump duration: 25 min	38.4 ng mL^{-1} at 10 h	-	[73]
	Riboflavin		9 mm and 10 mm inhibition zone: against *Listeria monocytogenes* and *Bacillus subtilis*, respectively	-	
Lactiplantibacillus plantarum LP-9	GABA	pH: 6.0, T: 35 °C; t: 96 h in MRS broth + 1% *w*/*v* MSG (without sugar)	1.14–1.30 g L^{-1}	Saccharified agro residues	[74]
			1.14–1.39 g L^{-1}	Saccharified cassava	
	Lactic acid		22.83–29.82 g L^{-1}	Saccharified agro residues	
			22.83–31.76 g L^{-1}	Saccharified cassava	

CLA: conjugated linoleic acid; GABA: γ-aminobutyric acid; MRS: De Man, Rogosa and Sharpe; MSG: monosodium glutamate.

4.2. Encapsulated LAB as Functional Ingredients

Despite the well-documented health benefits of probiotic LAB, several challenges remain regarding their use in producing functional products. These bacteria should not negatively affect the food products' sensorial, technological, or physicochemical properties during processing and storage. To fully harness the health-promoting effects of LAB, these bacteria must survive throughout the food processing stages, storage, and their transit through the gastrointestinal tract. To address these challenges, encapsulation has been proposed to protect LAB from environmental stresses and prevent any undesirable effects from directly adding probiotics to food matrices.

Encapsulation of LAB serves multiple purposes. For example, *Lactococcus lactis* QMF 11 and *Lactiplantibacillus paraplantarum* FT-259, when encapsulated in casein/pectin microparticles, were used as bioprotective cultures to control pathogens in fresh Minas cheese, showing significant antilisterial activity. Micro-encapsulation improved the viability of these bacterial strains, maintaining counts of more than 6.2 log CFU/g after 3 months of refrigerated storage. Encapsulating bacterial strains with simultaneous probiotic and antagonistic properties can facilitate the development of bio-preserved green functional products [75]. In a study evaluating the effect of alginate-based encapsulation on the survivability of *Lactiplantibacillus plantarum* during storage, food processing, and passage through the gastrointestinal tract and also on the production of bioactive compounds, various combinations of alginate with denatured whey protein, skim milk, dextrin, and chitosan were

https://doi.org/10.3390/biochem6010003

tested. The results showed that micro-encapsulation with alginate–skim milk preserved the viability of *Lactiplantibacillus plantarum* above the minimum recommended dose of probiotic bacteria under conditions such as pasteurization, frozen storage, simulated gastrointestinal conditions, and varying NaCl concentrations. Microencapsulated *Lactiplantibacillus plantarum* did not exhibit differences in the generation of EPSs and antibacterial compounds compared to unencapsulated ones [76]. The viability of *Limosilactobacillus reuteri* in the gastrointestinal tract of mice was maintained through encapsulation in poly-γ-glutamic acid, which had a positive effect on gut microbiota and SCFA content [77]. Similarly, the micro-encapsulation of *Lactobacillus acidophilus* in carrageenan and sodium alginate to develop probiotic yogurt resulted in improved bacterial viability in yogurt and simulated gastrointestinal conditions, demonstrating its potential for producing functional foods [78].

4.3. Encapsulated LAB-Derived Bioproducts as Functional Ingredients

Directly adding bioproducts generated by LAB to food matrices as functional ingredients often encounters several challenges, such as the loss of bioactivity, interactions with other components of the formulation, low physicochemical stability, and undesirable effects on the properties of the food product [79]. Additionally, various factors, including oxygen, light, and heat, can negatively affect the functionality of LAB-generated bioactive ingredients during industrial processing or storage of functional foods. These compounds may also be susceptible to degradation during gastrointestinal digestion. Micro- and nano-encapsulation have been proposed to address these challenges and protect these bioactive compounds, ensuring their stability and preventing breakdown before reaching their target site. For instance, the encapsulation of the pediocin, extracted from *Pediococcus pentosaceus* KC692718, in nanoliposomes significantly improved its stability, enhanced its controlled release, and exhibited an inhibitory effect against *Listeria monocytogenes*, a foodborne pathogen, during 48 h of incubation in tryptic soy broth with yeast extract. This encapsulation process achieved 89% encapsulation efficiency and retained 50% of pediocin activity [80]. Similarly, incorporating plantaricin, an extracellular bacteriocin produced by *Lactiplantibacillus plantarum*, into silver NPs reduced the minimum inhibitory concentration (MIC) and increased antibacterial activity against *Listeria monocytogenes*. It extended the refrigerated stability period from 5 days for the nonencapsulated bacteriocin to 60 days for the encapsulated version [81]. Nisin-loaded multi-component colloidosomes, which consisted of an alginate–chitosan multi-layer around the nanoliposomes, exhibited higher encapsulation efficiency, stability, and antibacterial activity against *Enterococcus faecalis*, *Listeria monocytogenes*, and *Staphylococcus aureus* compared to both nisin-loaded nanoliposomes and free nisin. These colloidosomes demonstrated biocompatibility, promoted controlled release, and offered promising potential for incorporating bacteriocins into functional foods [82].

The application of EPS-LN60, an EPS produced by *Lactiplantibacillus plantarum* DF60Mi, for coating alginate capsules containing *Lactobacillus acidophilus* LA-3, led to enhanced protection and improved probiotic survival during refrigerated storage and under simulated gastrointestinal conditions. EPS-LN60 also demonstrated antioxidant activity and exhibited an inhibitory effect on tumor cell growth. Additionally, its prebiotic capacity was higher than the commercial fructooligosaccharide. These results revealed the bioactivity of this EPS and its potential as a functional ingredient for probiotic encapsulation in the food industry [83]. KM01 EPS, produced by *Leuconostoc holzapfelii* KM01, isolated from the Thai fermented dessert, had a molecular weight of 500 kDa and could form aggregates with molecular weights exceeding 2000 kDa in an aqueous solution. This EPS could form hydrogels and NPs without exhibiting cytotoxicity. Quercetin, a potent antioxidant with limited applications due to poor solubility, was effectively encapsulated by KM01 EPS,

https://doi.org/10.3390/biochem6010003

with an efficiency of 7.8 mg quercetin/g EPS. The solubility of quercetin was enhanced 2.5-fold using 1% (*w*/*v*) KM01 EPS [84].

CLA's oily taste, low oxidative stability, and hydrophobic nature hinder the direct fortification of yogurt with this bioactive compound. To address this challenge, reassembled casein micelles were used to encapsulate CLA. The resulting NPs effectively protected CLA against UV-induced oxidation. Fortification of yogurt with encapsulated CLA did not negatively affect sensory properties or acid development during storage. The fortified yogurt maintained acceptable quality with an even distribution of encapsulated CLA [85].

Biotechnological hyalonutriosomes, which encapsulated fermented whey enriched in lactic acid, showed excellent stability in a simulated gastric environment. These hyalonutriosomes promoted the proliferation of the human commensal bacterium *Streptococcus salivarius* and were effectively internalized by Caco-2 cells, protecting them against oxidative stress. This demonstrated their significant potential for promoting intestinal health [48].

Table 3 highlights the advantages of encapsulating probiotic LAB and their key bacteriocins.

Table 3. The effect of encapsulation on the functionality of probiotics and bacteriocins.

Encapsulated Probiotic/Bacteriocin	Encapsulant System	Advantages	Ref.
Lacticaseibacillus rhamnosus	Hyaluronate-adipic dihydrazide/aldehyde-terminated pluronic F127/fucoidan hydrogel	Probiotics-loaded hydrogel promoted antibacterial activity with a significant inhibitory effect on inflammation and pseudomonas aeruginosa infection, improved collagen formation and re-epithelialization, and accelerated the healing of full-thickness superbug-infected wounds.	[86]
Lactobacillus acidophilus	Gelatin–chitosan polyelectrolyte-coated nanoliposomes	Encapsulation significantly promoted the viability of probiotics when exposed to simulated gastrointestinal conditions. Nanoliposomes were proposed as effective delivery systems for probiotics to produce functional foods.	[87]
Lacticaseibacillus paracasei KS-199	Alginate-based electrospun nanofiber mats	Encapsulation enhanced the survival of the bacterial strain in kefir and simulated gastric juice.	[88]
Lacticaseibacillus rhamnosus GG	Multi-layer PLGA-pullulan-PLGA electrospun nanofibers	Encapsulation facilitated the intestine delivery of viable probiotic strain and its colonization in the cecum and jejunum.	[89]
Pediocin	Nanoliposome	Pediocin produced by *Pediococcus acidilactici* ITV26 remained stable when encapsulated in liposomes and was released in a controlled manner across all tested pH values (4.0, 5.0, 6.0, and 6.8). The best release occurred at pH 4, resulting in a 2.5 log reduction in Listeria *innocua* AST-062.	[90]
Nisin	Pectin–chitosan polyelectrolyte complex nanoparticles	The antibacterial activity of encapsulated nisin was more effective against *Staphylococcus aureus* than free nisin. The release rate of the nisin from nanoparticles was higher at pH 3 compared to pH 6.	[91]
	Alginate–chitosan nanoparticles	Nanoparticles released nisin in a sustained manner, extending its antilisterial activity in vacuum-sealed refrigerated beef.	[92]
	Pathogen-responsive polyion complex nanoparticles	Nisin release from nanoparticles was selective, occurring preferentially in the presence of *Staphylococcus aureus* (a hyaluronidase-producing pathogen) while showing minimal inhibitory effect on *Bacillus cereus* (a non-hyaluronidase-producing bacterium) in culture and milk.	[93]

PLGA: polylactic-co-glycolic acid.

https://doi.org/10.3390/biochem6010003

5. Paraprobiotics and Postbiotics of LAB

While probiotic bacteria offer numerous health benefits, their administration has raised several safety concerns and limitations. These include the potential for developing antibiotic resistance, risks of systemic infections, unknown molecular mechanisms, short action duration, and the need for maintaining stability and viability during the production process and storage. These challenges have led to the exploration of probiotic derivatives—paraprobiotics and postbiotics—as alternatives for harnessing health benefits. These derivatives address many of the risks associated with probiotics and can extend the shelf-life of products without requiring cold storage to maintain probiotic viability. Additionally, paraprobiotics and postbiotics have demonstrated various therapeutic effects, including wound healing and antitumor, antimicrobial, anti-inflammatory, and immunomodulatory properties (Table 4).

Table 4. Advantages of using paraprobiotics and postbiotics of lactic acid bacteria in various models.

Lactic Acid Bacteria	Postbiotics/ Paraprobiotics	Model	Advantages	Ref.
Latilactobacillus sakei CNS001WB and *Lactiplantibacillus pentosus* WB693	Heat-killed bacteria	LPS-stimulated RAW 264.7 cells	Exhibiting anti-inflammatory and antioxidant effects, reduction in nitric oxide production, and expression of cyclooxygenase-2 and pro-inflammatory cytokines, inhibition of the NF-κB and MAPK pathways and ROS production	[94]
Lacticaseibacillus casei	Heat-killed bacteria	CD1 male mice infected with tachyzoites of *Toxoplasma gondii* RH strain	Remote immunostimulation, increase in generation of MCP-1 and T $CD4^+$ $CD44^+$ lymphocytes, macrophage activation and tachyzoites destruction by intracellular nitric oxide production, reduction in parasitic load of tachyzoites, and increase in animal survival	[95]
Lacticaseibacillus caseii	Heat-killed bacteria and cell-free supernatant *	Caco-2 and MRC5 cell lines	Cytotoxic effects on colorectal cancer cells, greater effectiveness of live cells and cell-free supernatant of *Lacticaseibacillus casei* on Caco-2 cells compared to heat-killed bacterial cells	[96]
Lactobacillus delbrueckii CIDCA 133	Heat-killed bacteria and cell-free supernatant *	Mouse model of 5-Fluorouracil drug-induced mucositis	Reduction in neutrophil infiltration into the small intestinal mucosa, downregulation of inflammatory markers, upregulation of immunoregulatory and epithelial barrier markers, reduction in inflammatory damage	[97]
Latilactobacillus curvatus	Curvatcin LHM (Bacteriocin)	*Streptococcus mutans* and *Streptococcus sanguinis*	Eradication of *streptococci* from the biofilm, exhibiting antibacterial, antibiofilm, and anti-cariogenic activities	[98]
Lactiplantibacillus plantarum PBS067	Plantaricin P1053 (Bacteriocin)	*Staphylococcus aureus* ATCC 6538, *Escherichia coli* ATCC 25922, *Enterococcus faecalis* ATCC 2922, *Pseudomonas aeruginosa* ATCC 9027, Healthy human cell line CCD 841, colon cancer E705 cell line	Exhibiting antimicrobial activity against Gram-negative and Gram-positive bacteria, enhancing the viability of healthy cells, reducing carcinogenic intestinal cell proliferation	[99]

https://doi.org/10.3390/biochem6010003

Table 4. *Cont.*

Lactic Acid Bacteria	Postbiotics/ Paraprobiotics	Model	Advantages	Ref.
Pediococcus acidilactici, *Latilactobacillus sakei*/*Streptococcus xylosus*	Postbiotics containing 19 different phenolic and flavonoids	*Listeria monocytogenes*, *Salmonella* Typhimurium, total mesophilic aerobic bacteria	Demonstrating vigorous antioxidant activity, exhibiting bacteriostatic effect against *Salmonella* Typhimurium and bactericidal effect against *Listeria monocytogenes*, showing potential for extending shelf-life of meat and poultry meat products	[100]
Leuconostoc mesenteroides J27	Postbiotic (soluble by-products)	*Vibrio parahaemolyticus*, *Pseudomonas aeruginosa*, and *Escherichia coli*	Exhibiting antibiofilm activity on seafood and relevant processing surfaces (combined with essential oils), retaining antibacterial activity at high temperatures (100 and 121 °C) and low pH (pH 1–6), and during storage (30 days), showing good potential for use as a preservative in the seafood industry	[101]
Lacticaseibacillus rhamnosus, *Lacticaseibacillus paracasei*	Postbiotics containing different antibacterial and antifungal substances	*Listeria monocytogenes*, *Staphylococcus aureus*, *Escherichia coli*, *Salmonella* Typhimurium, *Aspergillus flavus*, *Penicillium citrinum*	Successfully incorporating in bacterial nanocellulose, showing antimicrobial activity with the highest effect on *Listeria monocytogenes* and the least antifungal effect on *Penicillium citrinum*	[102]
Levilactobacillus brevis KB290	Heat-killed bacteria	C57BL/6J mice fed with high-fat diet	Reduction in diet-induced visceral fat accumulation, improving intestinal microbiota modifications and diet-induced metabolic symptoms, suggesting potential as a novel paraprobiotic for developing functional foods targeting visceral fat reduction	[103]
Latilactobacillus curvatus B67	Postbiotic (soluble by-products containing organic acids, lactic acid, and acetic acid)	*Listeria monocytogenes*, *Salmonella enterica* serovar Typhimurium	Exhibiting antibacterial and antibiofilm activity on processed pork sausage and meat-processing surfaces in combination with quercetin, maintaining stability across diverse pH (1–6) and temperature (40–121 °C) ranges, proposing potential as bioprotective agents in the meat industry	[104]

NF-κB: nuclear factor kappa-light-chain-enhancer of activated B cells; MAPK: mitogen-activated protein kinase; ROS: reactive oxygen species; MCP-1: monocyte chemoattractant protein. *: Representing combined experimental approaches rather than conceptual overlap.

The cell-free supernatant of *Limosilactobacillus fermentum* demonstrated potential as an antiaging agent by alleviating senescence markers such as P38 mitogen-activated protein kinase (p38MAPK), senescence-associated β-galactosidase (SA-β-gal), cyclooxygenase-2 (COX-2), inducible nitric oxide synthase (iNOS), Nuclear factor kappa-light-chain-enhancer of activated B cells (NF-κB), $p21^{WAF1}$, p53, reactive oxygen species (ROS), and DNA damage response [105].

Peptidoglycan produced by *Lacticaseibacillus rhamnosus* CRL1505 retained the unique immunomodulatory characteristics of viable probiotic bacteria. Nasal administration of this peptidoglycan to malnourished mice improved the innate and adaptive immune response against *Streptococcus pneumoniae* infection [106].

The free cultural supernatant of dietary *Lacticaseibacillus casei*, containing varying levels of CLA, demonstrated potential gastrointestinal health-promoting and significant antitumor effects on colorectal cancer cells. Daily administration of dietary *Lacticaseibacillus casei* to mice for one week led to the modulation of gut microflora composition and a reduction in sulfidogenic bacteria abundance. Both these probiotic bacteria and their free cultural supernatant exhibited antioxidant and anti-inflammatory properties [107].

Heat-inactivated *Lacticaseibacillus paracasei* and *Levilactobacillus brevis* induced apoptosis in human colon adenocarcinoma cells and inhibited their growth, with *Levilactobacillus brevis* showing greater cytotoxic and antiproliferative effects than *Lacticaseibacillus paracasei* [108].

Administration of both probiotic *Lacticaseibacillus rhamnosus* GG and their paraprobiotics (heat-inactivated bacteria) to rats with diet-induced nonalcoholic fatty liver disease partially prevented oxidative stress and inflammation. However, hepatic triglyceride content was more effectively reduced by probiotics than paraprobiotics [109].

6. LAB-Mediated Synthesis of NPs

6.1. Metallic NPs

In addition to the previously discussed nanotechnology applications, LAB can be a microbial cell factory for generating metallic NPs. The unique chemical, electronic, and optical properties of these microbial metallic NPs and their cost-effective and eco-friendly production methods compared to conventional chemical and physical techniques have garnered significant scientific interest. The metallic NPs have promising applications in the food industry, cosmetics, and nanomedicine. They exhibit antioxidant, anticancer, antibacterial, neurotherapeutic, biosensing, and wound-healing effects. Metallic NPs can be biosynthesized through either extracellular or intracellular mechanisms [110]. In the extracellular mechanism, biosynthesis is induced by reducing metal ions into their atomic forms by secreted molecules or by enzymatic reduction on the bacterial cell surface. The intracellular mechanism involves the transport of ions into the bacterial cell, where they are reduced into their elemental forms through enzymatic activity, electrostatic interactions, and subsequent NP formation within the bacterial cell (Figure 2).

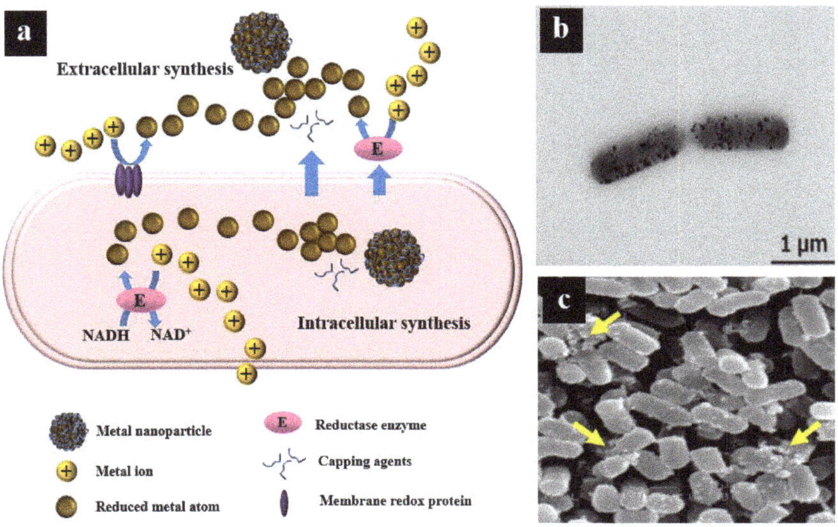

Figure 2. Schematic illustration of the intracellular and extracellular synthesis of metal nanoparticles (**a**), TEM image of gold nanoparticles intracellularly synthesized by *Lacticaseibacillus casei* (**b**), and SEM micrograph of silver nanoparticles extracellularly synthesized by *Lactiplantibacillus plantarum* demonstrated by the yellow arrows (**c**). Parts **b** and **c** are reproduced under the CC BY 4.0 license from Kikuchi et al., 2016 [111] and Mohd Yusof et al., 2020 [112], respectively.

https://doi.org/10.3390/biochem6010003

Bacterial biomolecules act as reducing, chelating, stabilizing, and capping agents during NP biosynthesis. Proteins, peptides, amino acids, enzymes, polysaccharides, and carboxylic acids play key roles as bioreducing agents. Furthermore, they prevent the aggregation of NPs and help their long-term stability by providing natural capping [113]. Al-Asbahi and colleagues [114] demonstrated the contribution of the capping proteins produced by *Lactobacillus* species (sensu lato, including species now classified into new genera) isolated from milk to the synthesis and stability of silver NPs. El. Fadly and colleagues [115] reported that the electrostatic interactions between the biosynthetic silver NPs produced by *Lactobacillus acidophilus* and *Lactobacillus delbrueckii* and the carbonyl groups of peptides help stabilize the NPs by forming a coating layer. Amino acids, as well, can serve as capping agents via interaction of their side chain residues with other compounds [116]. Histidine was reported as the best reducing and capping agent for the synthesis of gold NPs [117]. The microbial enzymes, such as reductases, participate in the synthesis of metal NPs by bioreduction of metal ions [118]. Glutathione reductase in some LAB strains is involved in the synthesis of selenoNPs [119]. Polysaccharides, including bacterial EPSs, are capable of reducing metal ions to form NPs and act as capping agents to stabilize them [120]. Various monomers of LAB-produced EPSs, including galactose, mannose, glucose and fructose, are involved in redox reactions to synthesize silver NPs [121]. Furthermore, the carboxylic acid molecules participate in the biosynthesis of the metal NPs as reducing and stabilizing agents [122]. However, the exact components and mechanism of this biosynthesis are not completely elucidated and require more investigation for clarification in the near future.

Several studies have focused on optimizing the biosynthesis of metallic NPs to produce uniform and stable NPs with small sizes and controlled morphology [123]. LAB have been reported to biosynthesize gold, silver, and gold–silver alloy nanocrystals, within their cells [124]. Silver NPs biosynthesized using the culture filtrate of the LAB strain LCM5, isolated from brined cucumbers, were small and spherical particles with proper dispersion. These NPs exhibited antifungal activity, showing a larger growth inhibition zone for *Penicillium expansum* and smaller zones for *Aspergillus ochraceus* and *Aspergillus flavus*. Their highest antimicrobial activity was observed against *Chromobacterium violaceum*, a Gram-negative bacterium [125]. Silver NPs biosynthesized by mixing the culture supernatant of *Lacticaseibacillus rhamnosus* with silver nitrate solution were used to target biofilm formation and virulence factors mediated by quorum sensing, offering a novel approach to combat antimicrobial resistance. These NPs significantly inhibited the generation of prodigiosin by *Serratia marcescens*, violacein by *Chromobacterium violaceum* and pyocyanin, LasA protease, LasB elastase, rhamnolipid and pyoverdine by *Pseudomonas aeruginosa*. Additionally, the ability of tested Gram-negative bacteria to form biofilms, produce EPS, and exhibit swimming motility was notably reduced in the presence of these silver NPs [126].

Exposing the cell-free culture supernatant of *Lactiplantibacillus plantarum*, *Limosilactobacillus fermentum*, *Lacticaseibacillus paracasei*, and *Lacticaseibacillus casei* to chloroauric acid ($HAuCl_4$) led to the biosynthesis of gold NPs [127]. Korean kimchi-isolated *Secundilactobacillus kimchicus* DCY51^T was used for the biosynthesis of gold NPs through the intracellular membrane-bound mechanism. The obtained spherical NPs showed biocompatibility, antioxidant activity, and long-term stability in biological media and physiological buffers [128].

The cell-biomass and cell-free supernatant of *Lactiplantibacillus plantarum* TA4 were used as reducing agents for the biosynthesis of zinc oxide NPs, resulting in the formation of NPs with an irregular shape and flowerlike pattern, respectively. These zinc oxide NPs exhibited concentration-dependent antibacterial effects on pathogenic *Escherichia coli*, *Salmonella*, *Staphylococcus aureus*, and *Staphylococcus epidermidis* [129]. Zinc oxide NPs synthesized by LAB isolated from cow milk displayed biocompatibility and effective antifungal

https://doi.org/10.3390/biochem6010003

and antibacterial activities, with potent bactericidal effects on *Clostridium perfringens* and *Escherichia coli* [130]. *Limosilactobacillus fermentum*-synthesized zinc oxide NPs exhibited the highest antimicrobial activity against *Vibrio harveyi* at 20 mM concentration [131].

Iron oxide NPs produced using *Lactiplantibacillus plantarum* extracellular product solution as a reducing and stabilizing agent showed potent antibacterial activity against *Pseudomonas aeruginosa* [132]. Furthermore, *Lacticaseibacillus casei* cytoplasmic extract was used to green synthesize iron oxide nanocrystals through an efficient, simple, cost-effective, and biologically safe method with potential applications in targeted drug delivery [133].

Metallic NPs synthesized by bacteria have shown a wide range of biomedical applications [134]. Silver and gold NPs produced in this manner have demonstrated promising potential for photothermal therapy, an effective cancer treatment. These NPs induce localized hyperthermia under near-infrared (NIR) light, effectively ablating cancer cells. For instance, DCY51^{T}–gold NPs, generated by *Secundilactobacillus kimchicus* DCY51^{T} through the intracellular membrane-bound mechanism, were used for the delivery of ginsenoside compound K. These gold NPs, non-covalently bioconjugated to the drug, showed in vitro photothermal effect and anticancer activity against stomach, colon, and lung adenocarcinoma cells. Upon NIR laser irradiation, these ginsenoside-loaded gold NPs enhanced apoptosis in cancer cells more effectively than cells treated with drug-gold NPs alone [135].

LAB-synthesized metal NPs also contributed to the photocatalytic processes, wherein electron excitation from the valence band to the conduction band occurs under light irradiation with energy exceeding the band gap of the photocatalytic substance. This electron movement creates a positive electron hole on the valence band. The reaction of these positive holes and negative electrons with dissolved oxygen generates ROS, which in turn inactivate pathogens and degrade pollutant compounds [136]. Gold NPs synthesized using *Lactobacillus acidophilus* supernatant exhibited antimicrobial activity against multi-drug-resistant strains of enteroaggregative *Escherichia coli* and methicillin-resistant *Staphylococcus aureus*. These probiosynthesized NPs showed a photocatalytic disinfection effect (within 120 min) on multi-drug-resistant bacterial strains and photocatalytic degradation of cationic methylene blue (within 60 min), indicating potential use in antibiotic removal and environmental bioremediation [137]. Similarly, silver NPs produced by *Lactobacillus* sensu lato species isolated from cow milk caused photocatalytic degradation of methylene red dye. These biosynthesized NPs exhibited cytotoxic effects on human colon cancer cells and antimicrobial activity against pathogens [138]. Moreover, silver NPs have shown the ability to dye degradation through redox potential techniques. Silver NPs stabilized by *Leuconostoc lactis* EPS exhibited excellent thermal stability and facilitated the degradation of Congo red and Methyl orange (azo dyes), offering a promising potential for treating industrial textile effluent. The dye degradation was attributed to the destruction of the chromophore structure through electron transfer from the reducing agent to the dye molecule via NPs [121].

Furthermore, metallic NPs have shown great potential in managing infected wounds, utilizing the controlled release of metal ions with antimicrobial effects. For instance, silver NPs biosynthesized using *Lactiplantibacillus plantarum* supernatant demonstrated the ability to promote wound healing and exhibited significant antibacterial activity against human infectious pathogens [139].

6.2. Minicells

Minicells, emerging as nanometric cells, are generated through abnormal bacterial division, primarily due to mutations in associated genes [140]. Although minicells retain some components of their parent cells—such as RNA, protein, ribosomes, and peptidoglycan—they lack chromosomal DNA (Figure 3).

https://doi.org/10.3390/biochem6010003

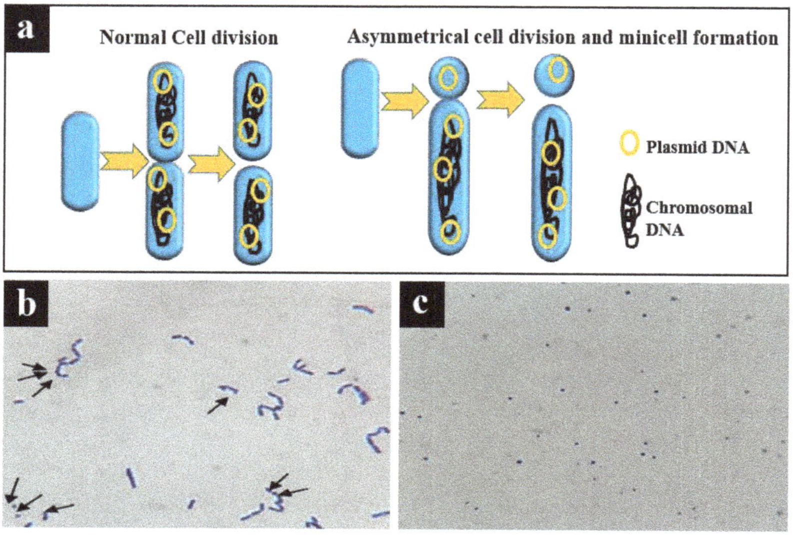

Figure 3. Schematic representation of minicell generation: asymmetrical cell division and minicell formation versus normal cell division (**a**); Gram staining of minicell formation (black arrows) by *Lactobacillus acidophilus* (**b**); and isolated minicells of *Lactobacillus acidophilus* (**c**) observed under a light microscope at 100× magnification. Parts **b** and **c** are reproduced under the CC BY 3.0 license from Ho et al., 2024 [141].

Minicells derived from bacteria offer several advantages, including biocompatibility, biodegradability, safety, high drug-loading capacity, surface modifiability, reduced immunogenicity, and site-specific targeting. They can be employed to deliver therapeutics to treat different diseases [142]. For example, minicells generated by *Lactobacillus acidophilus* can facilitate the delivery of both hydrophobic and hydrophilic therapeutics without undergoing phagocytosis [143]. In a study, *Lactobacillus acidophilus* VTCC-B-871 produced a significant number of minicells in modified MRS broth with 10 g L^{-1} of fructose as the carbon source, highlighting the effect of the carbon source on minicell production by this strain. These minicells demonstrated the ability to encapsulate paclitaxel and cephalosporin after varying incubation times [144]. *Lactobacillus acidophilus* ATCC 4356 minicells, generated using sugar stress, demonstrated containing ability of a significant amount of chloroquine, an anti-malaria and anticancer agent. This drug could be bound to the cell wall of the minicells and could be released in a prolonged manner. These minicells were proposed as a novel targeted drug delivery system to mitigate the toxicity of long-term use and high doses of chloroquine [141]. Additionally, nanosized minicells produced by *Leuconostoc mesenteroides* VTCC B-871 encapsulated tinidazole to 90% efficiency and led to faster absorption than free tinidazole in fed mice models without interference from food intake. The rate of tinidazole release from these minicells was higher in acidic pH than in basic medium, possibly due to the involvement of the acidic environment in hydrolyzing the bond between tinidazole and the minicells [145].

6.3. Carbon Dots

Carbon dots are another type of NP synthesized from LAB, showing great potential for effectively inactivating resistant pathogens. Hydrothermally synthesized carbon dots from *Lactobacillus acidophilus* supernatant demonstrated antibacterial effects against *Listeria monocytogenes* and *Escherichia coli*. An antimicrobial/UV protective nanopaper was developed by embedding photoluminescent carbon dots into nanocellulose, resulting in a fluores-

https://doi.org/10.3390/biochem6010003

cent appearance with potential application in forgery-proof packaging. These nanopapers were more effective in inhibiting *Listeria monocytogenes* than *Escherichia coli* and exhibited superior ultraviolet-blocking properties, flexibility, and stretchability compared to nanocellulose [146]. Carbon dots synthesized from *Lactiplantibacillus plantarum* biomass through one-step hydrothermal carbonization showed antibiofilm activity against *Escherichia coli* and exhibited biocompatibility, addressing the cytotoxicity concerns of current antibiofilm agents [147]. Similarly, carbon dots synthesized from *Lactobacillus acidophilus* via the hydrothermal method displayed antibacterial activity against carbapenem-resistant *Klebsiella pneumonia*, with an MIC of 50 mg mL^{-1} [148].

Carbon dots can also be used in photodynamic therapy as novel photosensitizers or as carriers for other photosensitizers [149]. Under NIR irradiation, photosensitizers can generate ROS, leading to the inactivation of bacterial or cancer cells through mechanisms such as DNA damage, lipid peroxidation, protein denaturation, or cell membrane destruction [150]. Additionally, the ability of carbon dots to convert light into heat has made them valuable as photothermal agents for innovative antibacterial and antitumor applications [151].

7. Conclusions

LAB have demonstrated tremendous potential in bioprocessing, producing a wide range of valuable bioactive compounds, including lactic acid, SCFAs, CLA, bacteriocins, vitamins, and EPSs. The optimized production of these bioactive compounds has significantly expanded the market of functional products. Using LAB as probiotics, postbiotics and paraprobiotics have established these bacterial strains as unparalleled microorganisms across diverse fields. Additionally, the future of nanomedicine is poised to be significantly influenced by LAB-biosynthesized NPs. Integrating biotechnology and nanotechnology through the exploitation of LAB offers numerous advantages and has garnered significant interest. Ongoing efforts in this area are expected to yield significant advances across various fields, ranging from food science to nanomedicine. The biological origin of biosynthesized NPs may confer greater biocompatibility and biodegradability than physically or chemically synthesized NPs, thereby reducing the risk of in vivo toxicity. However, the safety of nanomaterials remains a crucial challenge, especially concerning their long-term use. Despite using safe substances and synthesis methods, the bioaccumulation of NPs in tissues, along with their potential cytotoxicity and genotoxicity, are among the critical concerns, which can arise from repeated administration and the interactions between NPs and cells. Unintended immunogenicity is also a challenge that can lead to immune dysregulation and chronic inflammation. However, completely assessing these risks for new nanodelivery platforms is not feasible due to the limitations in systematic long-term data. Addressing these concerns will require more robust legislation, testing standards, and inspections to ensure their safe application in industrial and medical contexts.

Author Contributions: M.R.: conceptualization, investigation, writing—original draft, writing—review and editing; M.M.: writing—review and editing; N.D.: writing—review and editing. All authors have read and agreed to the published version of the manuscript.

Funding: This research received no external funding.

Data Availability Statement: Data sharing is not applicable.

Conflicts of Interest: The authors declare no conflicts of interest.

References

1. Coelho, M.C.; Malcata, F.X.; Silva, C.C. Lactic acid bacteria in raw-milk cheeses: From starter cultures to probiotic functions. *Foods* **2022**, *11*, 2276. [CrossRef]

https://doi.org/10.3390/biochem6010003

2. Tang, H.; Huang, W.; Yao, Y.F. The metabolites of lactic acid bacteria: Classification, biosynthesis and modulation of gut microbiota. *Microb. Cell* **2023**, *10*, 49–62. [CrossRef]
3. Ayivi, R.D.; Gyawali, R.; Krastanov, A.; Aljaloud, S.O.; Worku, M.; Tahergorabi, R.; Silva, R.C.; Ibrahim, S.A. Lactic acid bacteria: Food safety and human health applications. *Dairy* **2020**, *1*, 202–232. [CrossRef]
4. Oberg, T.S.; McMahon, D.J.; Culumber, M.D.; McAuliffe, O.; Oberg, C.J. Invited review: Review of taxonomic changes in dairy-related lactobacilli. *J. Dairy Sci.* **2022**, *105*, 2750–2770. [CrossRef]
5. Zheng, J.; Wittouck, S.; Salvetti, E.; Franz, C.M.; Harris, H.M.; Mattarelli, P.; O'Toole, P.W.; Pot, B.; Vandamme, P.; Walter, J.; et al. A taxonomic note on the genus *Lactobacillus*: Description of 23 novel genera, emended description of the genus *Lactobacillus* Beijerinck 1901 and union of *Lactobacillaceae* and *Leuconostocaceae*. *Int. J. Syst. Evol. Microbiol.* **2020**, *70*, 2782–2858. [CrossRef]
6. Qiao, N.; Wittouck, S.; Mattarelli, P.; Zheng, J.; Lebeer, S.; Felis, G.E.; Gänzle, M.G. After the storm—Perspectives on the taxonomy of *Lactobacillaceae*. *JDS Commun.* **2022**, *3*, 222–227. [CrossRef]
7. Seyirt, S.; Tezel, B.U.; Şanlıbaba, P. Shelf-life dynamics of key probiotic populations in dairy formulations: A temporal stability assessment. *Food Biol.* **2025**, *14*, 6–11. [CrossRef]
8. Martínez-Miranda, J.G.; Chairez, I.; Durán-Páramo, E. Mannitol production by heterofermentative lactic acid bacteria: A review. *Appl. Biochem. Biotechnol.* **2022**, *194*, 2762–2795. [CrossRef]
9. Gomes, I.A.; Venâncio, A.; Lima, J.P.; Freitas-Silva, O. Fruit-based non-dairy beverage: A new approach for probiotics. *Adv. Biol. Chem.* **2021**, *11*, 302–330. [CrossRef]
10. Maftei, N.M.; Raileanu, C.R.; Balta, A.A.; Ambrose, L.; Boev, M.; Marin, D.B.; Lisa, E.L. The potential impact of probiotics on human health: An update on their health-promoting properties. *Microorganisms* **2024**, *12*, 234. [CrossRef]
11. Mehta, J.P.; Ayakar, S.; Singhal, R.S. The potential of paraprobiotics and postbiotics to modulate the immune system. *Rev. Microbiol. Res.* **2023**, *275*, 127449. [CrossRef]
12. Siciliano, R.A.; Reale, A.; Mazzeo, M.F.; Morandi, S.; Silvetti, T.; Brasca, M. Paraprobiotics: A new perspective for functional foods and nutraceuticals. *Nutrients* **2021**, *13*, 1225. [CrossRef]
13. Aguilar-Toalá, J.E.; Garcia-Varela, R.; Garcia, H.S.; Mata-Haro, V.; González-Córdova, A.F.; Vallejo-Cordoba, B.; Hernández-Mendoza, A. Postbiotics: An evolving term within the functional foods field. *Trends Food Sci. Technol.* **2018**, *75*, 105–114. [CrossRef]
14. Nataraj, B.H.; Ali, S.A.; Behare, P.V.; Yadav, H. Postbiotics-parabiotics: The new horizons in microbial biotherapy and functional foods. *Microb. Cell Factories* **2020**, *19*, 168. [CrossRef]
15. Sik, B.; Székelyhidi, R.; Lakatos, E.; Kapcsándi, V.; Ajtony, Z. Analytical procedures for determination of phenolics active herbal ingredients in fortified functional foods: An overview. *Eur. Food Res. Technol.* **2022**, *248*, 329–344. [CrossRef]
16. Temple, N.J. A rational definition for functional foods: A perspective. *Front. Nutr.* **2022**, *9*, 957516. [CrossRef]
17. Rezvani, M.; Hesari, J.; Peighambardoust, S.H.; Manconi, M.; Hamishehkar, H.; Escribano-Ferrer, E. Potential application of nanovesicles (niosomes and liposomes) for fortification of functional beverages with Isoleucine-Proline-Proline: A comparative study with central composite design approach. *Food Chem.* **2019**, *293*, 368–377. [CrossRef]
18. Vivek, N.; Hazeena, S.H.; Rajesh, R.O.; Godan, T.K.; Anjali, K.B.; Nair, L.M.; Mohan, B.; Nair, S.C.; Sindhu, R.; Pandey, A.; et al. Genomics of lactic acid bacteria for glycerol dissimilation. *Mol. Biotechnol.* **2019**, *61*, 562–578. [CrossRef]
19. Bleckwedel, J.; Mohamed, F.; Mozzi, F.; Raya, R.R. Major role of lactate dehydrogenase D-LDH1 for the synthesis of lactic acid in *Fructobacillus tropaeoli* CRL 2034. *Appl. Microbiol. Biotechnol.* **2020**, *104*, 7409–7426. [CrossRef]
20. Fang, S.; Song, X.; Cui, L.; Bai, J.; Lu, H.; Wang, S. The lactate dehydrogenase gene is involved in the growth and metabolism of *Lacticaseibacillus paracasei* and the production of fermented milk flavor substances. *Front. Microbiol.* **2023**, *14*, 1195360. [CrossRef]
21. Fusco, W.; Lorenzo, M.B.; Cintoni, M.; Porcari, S.; Rinninella, E.; Kaitsas, F.; Lener, E.; Mele, M.C.; Gasbarrini, A.; Collado, M.C.; et al. Short-chain fatty-acid-producing bacteria: Key components of the human gut microbiota. *Nutrients* **2023**, *15*, 2211. [CrossRef]
22. Yang, M.; Meng, F.; Gu, W.; Fu, L.; Zhang, F.; Li, F.; Tao, Y.; Zhang, Z.; Wang, X.; Yang, X.; et al. Influence of polysaccharides from Polygonatum kingianum on short-chain fatty acid production and quorum sensing in *Lactobacillus faecis*. *Front. Microbiol.* **2021**, *12*, 758870. [CrossRef]
23. Badawy, S.; Liu, Y.; Guo, M.; Liu, Z.; Xie, C.; Marawan, M.A.; Ares, I.; López-Torres, B.; Martínez, M.; Maximiliano, J.E.; et al. Conjugated linoleic acid (CLA) as a functional food: Is it beneficial or not? *Food Res. Int.* **2023**, *172*, 113158. [CrossRef]
24. Yang, B.; Gao, H.; Stanton, C.; Ross, R.P.; Zhang, H.; Chen, Y.Q.; Chen, H.; Chen, W. Bacterial conjugated linoleic acid production and their applications. *Prog. Lipid Res.* **2017**, *68*, 26–36. [CrossRef]
25. Iorizzo, M.; Di Martino, C.; Letizia, F.; Crawford, T.W., Jr.; Paventi, G. Production of conjugated linoleic acid (CLA) by *Lactiplantibacillus plantarum*: A review with emphasis on fermented foods. *Foods* **2024**, *13*, 975. [CrossRef]
26. Salsinha, A.S.; Pimentel, L.L.; Fontes, A.L.; Gomes, A.M.; Rodríguez-Alcalá, L.M. Microbial production of conjugated linoleic acid and conjugated linolenic acid relies on a multienzymatic system. *Microbiol. Mol. Biol. Rev.* **2018**, *82*, e00019-18. [CrossRef]

https://doi.org/10.3390/biochem6010003

27. Liu, X.X.; Zhang, H.Y.; Song, X.; Yang, Y.; Xiong, Z.Q.; Xia, Y.J.; Ai, L.Z. Reasons for the differences in biotransformation of conjugated linoleic acid by *Lactobacillus plantarum*. *J. Dairy Sci.* **2021**, *104*, 11466–11473. [CrossRef]
28. Darbandi, A.; Asadi, A.; Mahdizade, A.M.; Ohadi, E.; Talebi, M.; Zadeh, M.H.; Emamie, A.D.; Ghanavati, R.; Kakanj, M. Bacteriocins: Properties and potential use as antimicrobials. *J. Clin. Lab. Anal.* **2022**, *36*, e24093. [CrossRef]
29. Bhattacharya, D.; Nanda, P.K.; Pateiro, M.; Lorenzo, J.M.; Dhar, P.; Das, A.K. Lactic acid bacteria and bacteriocins: Novel biotechnological approach for biopreservation of meat and meat products. *Microorganisms* **2022**, *10*, 2058. [CrossRef]
30. Guitián, M.V.; Ibarguren, C.; Soria, M.C.; Hovanyecz, P.; Banchio, C.; Audisio, M.C. Anti-Listeria monocytogenes effect of bacteriocin-incorporated agar edible coatings applied on cheese. *Int. Dairy J.* **2019**, *97*, 92–98. [CrossRef]
31. Soltani, S.; Zirah, S.; Rebuffat, S.; Couture, F.; Boutin, Y.; Biron, E.; Subirade, M.; Fliss, I. Gastrointestinal stability and cytotoxicity of bacteriocins from Gram-positive and Gram-negative bacteria: A comparative in vitro study. *Front. Microbiol.* **2022**, *12*, 780355. [CrossRef]
32. Sidhu, P.K.; Nehra, K. Bacteriocins of lactic acid bacteria as potent antimicrobial peptides against food pathogens. In *Biomimetics*; Wiley: Hoboken, NJ, USA, 2021; p. 10. [CrossRef]
33. Wu, J.; Zang, M.; Wang, S.; Zhao, B.; Bai, J.; Xu, C.; Shi, Y.; Qiao, X. Nisin: From a structural and meat preservation perspective. *Food Microbiol.* **2023**, *111*, 104207. [CrossRef]
34. Laroute, V.; Mazzoli, R.; Loubière, P.; Pessione, E.; Cocaign-Bousquet, M. Environmental conditions affecting GABA production in Lactococcus lactis NCDO 2118. *Microorganisms* **2021**, *9*, 122. [CrossRef]
35. Cui, Y.; Miao, K.; Niyaphorn, S.; Qu, X. Production of gamma-aminobutyric acid from lactic acid bacteria: A systematic review. *Int. J. Mol. Sci.* **2020**, *21*, 995. [CrossRef]
36. Valenzuela, J.A.; Flórez, A.B.; Vázquez, L.; Vasek, O.M.; Mayo, B. Production of γ-aminobutyric acid (GABA) by lactic acid bacteria strains isolated from traditional, starter-free dairy products made of raw milk. *Benef. Microbes* **2019**, *10*, 579–587. [CrossRef]
37. Jurášková, D.; Ribeiro, S.C.; Silva, C.C. Exopolysaccharides produced by lactic acid bacteria: From biosynthesis to health-promoting properties. *Foods* **2022**, *11*, 156. [CrossRef]
38. Liu, W.; Wei, Y.; Xiang, R.; Dong, B.; Yang, X. Lactic Acid Bacteria Exopolysaccharides Unveiling Multifaceted Insights from Structure to Application in Foods and Health Promotion. *Foods* **2025**, *14*, 823. [CrossRef]
39. Saadat, Y.R.; Khosroushahi, A.Y.; Gargari, B.P. A comprehensive review of anticancer, immunomodulatory and health beneficial effects of the lactic acid bacteria exopolysaccharides. *Carbohydr. Polym.* **2019**, *217*, 79–89. [CrossRef]
40. Zeidan, A.A.; Poulsen, V.K.; Janzen, T.; Buldo, P.; Derkx, P.M.; Øregaard, G.; Neves, A.R. Polysaccharide production by lactic acid bacteria: From genes to industrial applications. *FEMS Microbiol. Rev.* **2017**, *41*, 168–200. [CrossRef]
41. Zhang, J.; Xiao, Y.; Wang, H.; Zhang, H.; Chen, W.; Lu, W. Lactic acid bacteria-derived exopolysaccharide: Formation, immunomodulatory ability, health effects and structure-function relationship. *Microbiol. Res.* **2023**, *274*, 127432. [CrossRef]
42. Zhou, Y.; Cui, Y.; Qu, X. Exopolysaccharides of lactic acid bacteria: Structure, bioactivity and associations: A review. *Carbohydr. Polym.* **2019**, *207*, 317–332. [CrossRef]
43. Hanna, M.; Jaqua, E.; Nguyen, V.; Clay, J. B Vitamins: Functions and Uses in Medicine. *Perm. J.* **2022**, *26*, 89–97. [CrossRef]
44. Thakur, K.; Tomar, S.K.; De, S. Lactic acid bacteria as a cell factory for riboflavin production. *Microb. Biotechnol.* **2016**, *9*, 441–451. [CrossRef]
45. Mahara, F.A.; Nuraida, L.; Lioe, H.N. Folate in milk fermented by lactic acid bacteria from different food sources. *Prev. Nutr. Food Sci.* **2021**, *26*, 230. [CrossRef]
46. Mahara, F.A.; Nuraida, L.; Lioe, H.N.; Nurjanah, S. The occurrence of folate biosynthesis genes in lactic acid bacteria from different sources. *Food Technol. Biotechnol.* **2023**, *61*, 226–237. [CrossRef]
47. Ribeiro, M.; Maciel, C.; Cruz, P.; Darmancier, H.; Nogueira, T.; Costa, M.; Laranjeira, J.; Morais, R.M.S.C.; Teixeira, P. Exploiting Potential Probiotic Lactic Acid Bacteria Isolated from *Chlorella vulgaris* Photobioreactors as Promising Vitamin B12 Producers. *Foods* **2023**, *12*, 3277. [CrossRef]
48. Manconi, M.; Rezvani, M.; Manca, M.L.; Escribano-Ferrer, E.; Fais, S.; Orrù, G.; Lammers, T.; Asunis, F.; Muntoni, A.; Spiga, D.; et al. Bridging biotechnology and nanomedicine to produce biogreen whey-nanovesicles for intestinal health promotion. *Int. J. Pharm.* **2023**, *633*, 122631. [CrossRef]
49. Veeravalli, S.S.; Mathews, A.P. Continuous fermentation of xylose to short chain fatty acids by *Lactobacillus buchneri* under low pH conditions. *Chem. Eng. J.* **2018**, *337*, 764–771. [CrossRef]
50. Boontun, C.; Vatanyoopaisarn, S.; Hankla, S.; Kuraya, E.; Tamaki, Y. Modification of media using food-grade components for the fermentation of Bifidobacterium and *Lactobacillus* strains in large-scale bioreactors. *Prep. Biochem. Biotechnol.* **2021**, *54*, 1017–1027. [CrossRef]
51. Hadinia, N.; Dovom, M.R.; Yavarmanesh, M. The effect of fermentation conditions (temperature, salt concentration and pH) with *Lactobacillus* strains for producing short chain fatty acids. *LWT* **2022**, *165*, 113709. [CrossRef]

https://doi.org/10.3390/biochem6010003

52. Liu, J.; Meng, F.; Du, Y.; Nelson, E.; Zhao, G.; Zhu, H.; Caiyin, Q.; Zhang, Z.; Qiao, J. Co-production of nisin and γ-aminobutyric acid by engineered Lactococcus lactis for potential application in food preservation. *Front. Microbiol.* **2020**, *11*, 49. [CrossRef]
53. Modiri, S.; Kermanshahi, R.K.; Soudi, M.R.; Dad, N.; Ebadi, M.; Zahiri, H.S.; Akbari Noghabi, K. Growth optimization of *Lactobacillus acidophilus* for production of antimicrobial peptide acidocin 4356: Scale up from flask to lab-scale fermenter. *Iran. J. Biotechnol.* **2021**, *19*, 2686. [CrossRef]
54. Azevedo, P.O.D.S.; Converti, A.; Gierus, M.; Oliveira, R.P.D.S. Antimicrobial activity of bacteriocin-like inhibitory substance produced by Pediococcus pentosaceus: From shake flasks to bioreactor. *Mol. Biol. Rep.* **2019**, *46*, 461–469. [CrossRef]
55. Lahiri, D.; Chakraborti, S.; Jasu, A.; Nag, M.; Dutta, B.; Dash, S.; Ray, R.R. Production and purification of bacteriocin from *Leuconostoc lactis* SM 2 strain. *Biocatal. Agric. Biotechnol.* **2020**, *30*, 101845. [CrossRef]
56. Mitaf, N.A.; Kirca, D.S.; Simsek, O. The metagenomic properties of Uşak tarhana dough. *Fermentation* **2024**, *10*, 620. [CrossRef]
57. Şentürk, D.Z.; Dertli, E.; Erten, H.; Şimşek, Ö. Structural technological characterization of ropy exopolysaccharides produced by *Lactobacillus plantarum* strains isolated from Tarhana. *Food Sci. Biotechnol.* **2020**, *29*, 121–129. [CrossRef]
58. Bertsch, A.; Roy, D.; LaPointe, G. Enhanced exopolysaccharide production by *Lactobacillus rhamnosus* in co-culture with Saccharomyces cerevisiae. *Appl. Sci.* **2019**, *9*, 4026. [CrossRef]
59. Devi, P.U.M.; Rashmi, H.K. Effect of physico-chemical factors on the production of conjugated linoleic acid by *Lactobacillus plantarum*. *Asian J. Microbiol. Biotechnol. Environ. Sci.* **2017**, *19*, 412–418.
60. Razmjooei, M.; Shad, E.; Nejadmansouri, M.; Safdarianghomsheh, R.; Delvigne, F.; Khalesi, M. Effect of metal support and different carbon sources on CLA production using *Lactobacillus plantarum*. *Biochem. Eng. J.* **2020**, *162*, 107715. [CrossRef]
61. Park, J.; Heo, S.; Lee, G.; Hong, S.W.; Jeong, D.W. Bacterial diversity of baechu-kimchi with seafood based on culture-independent investigations. *Food Sci. Biotechnol.* **2024**, *33*, 1661–1670. [CrossRef]
62. Kim, J.Y.; Choi, E.J.; Lee, J.H.; Yoo, M.S.; Heo, K.; Shim, J.-J.; Lee, J.-L. Probiotic potential of a novel vitamin B2-overproducing *Lactobacillus plantarum* strain, HY7715, isolated from Kimchi. *Appl. Sci.* **2021**, *11*, 5765. [CrossRef]
63. Laiño, J.E.; Levit, R.; LeBlanc, A.D.M.; Giori, G.S.; LeBlanc, J.G. Characterization of folate production and probiotic potential of *Streptococcus gallolyticus* subsp. macedonicus CRL415. *Food Microbiol.* **2019**, *79*, 20–26. [CrossRef]
64. Erginkaya, Z.; Konuray-Altun, G. Potential biotherapeutic properties of lactic acid bacteria in foods. *Food Biosci.* **2022**, *46*, 101544. [CrossRef]
65. Al-Dhabi, N.A.; Arasu, M.V.; Vijayaraghavan, P.; Esmail, G.A.; Duraipandiyan, V.; Kim, Y.O.; Kim, H.; Kim, H.-J. Probiotic and antioxidant potential of *Lactobacillus reuteri* LR12 and *Lactobacillus lactis* LL10 isolated from pineapple puree and quality analysis of pineapple-flavored goat milk yoghurt during storage. *Microorganisms* **2020**, *8*, 1461. [CrossRef]
66. Hernández-Oaxaca, D.; López-Sánchez, R.; Lozano, L.; Wacher-Rodarte, C.; Segovia, L.; Munguía, A.L. Diversity of *Weissella confusa* in pozol and its carbohydrate metabolism. *Front. Microbiol.* **2021**, *12*, 629449. [CrossRef]
67. Sharma, S.; Kandasamy, S.; Kavitake, D.; Shetty, P.H. Probiotic characterization and antioxidant properties of *Weissella confusa* KR780676, isolated from an Indian fermented food. *LWT Food Sci. Technol.* **2018**, *97*, 53–60. [CrossRef]
68. Kaprasob, R.; Kerdchoechuen, O.; Laohakunjit, N.; Somboonpanyakul, P. B vitamins and prebiotic fructooligosaccharides of cashew apple fermented with probiotic strains *Lactobacillus* spp., *Leuconostoc mesenteroides* and *Bifidobacterium longum*. *Process. Biochem.* **2018**, *70*, 9–19. [CrossRef]
69. Hurtado-Romero, A.; Toro-Barbosa, M.; Gradilla-Hernández, M.S.; Garcia-Amezquita, L.E.; García-Cayuela, T. Probiotic properties, prebiotic fermentability and GABA-producing capacity of microorganisms isolated from Mexican milk kefir grains: A clustering evaluation for functional dairy food applications. *Foods* **2021**, *10*, 2275. [CrossRef]
70. Arslan, A.A.; Gocer, E.M.C.; Demir, M.; Atamer, Z.; Hinrichs, J.; Kücükcetin, A. Viability of *Lactobacillus acidophilus* ATCC 4356 incorporated into ice cream using three different methods. *Dairy Sci. Technol.* **2016**, *96*, 477–487. [CrossRef]
71. Özer, C.O.; Kılıç, B. Optimization of pH, time, temperature, variety and concentration of the added fatty acid and the initial count of added lactic acid Bacteria strains to improve microbial conjugated linoleic acid production in fermented ground beef. *Meat Sci.* **2021**, *171*, 108303. [CrossRef]
72. Papiran, R.; Hamedi, J. Adaptive evolution of *Lactococcus lactis* to thermal and oxidative stress increases biomass and nisin production. *Appl. Biochem. Biotechnol.* **2021**, *193*, 3425–3441. [CrossRef]
73. Halid, N.A.; Ibrahim, M.F.; Haryani, Y.; Zaini, N.S.; Hasan, H. Scaling Up Co-cultivation Fermentation of *Lactococcus lactis* ATCC 11454 and *Lactobacillus rhamnosus* ATCC 7469 with Various Feeding Strategies. *J. Biochem. Microbiol. Biotechnol.* **2024**, *12*, 96–99. [CrossRef]
74. Sharma, P.; Sharma, A.; Singh, J.; Singh, N.; Singh, S.; Tomar, G.S.; Nain, P.K.S.; Khare, S.K.; Nain, L. Co-production of gamma amino butyric acid (GABA) and lactic acid using *Lactobacillus plantarum* LP-9 from agro-residues. *Environ. Technol. Innov.* **2021**, *23*, 101650. [CrossRef]
75. Ribeiro, L.L.S.M.; Araújo, G.P.; Ribeiro, K.O.; Torres, I.M.S.; Martinis, E.C.P.; Marreto, R.N.; Alves, V.F. Use of encapsulated lactic acid bacteria as bioprotective cultures in fresh Brazilian cheese. *Braz. J. Microbiol.* **2021**, *52*, 2247–2256. [CrossRef]

https://doi.org/10.3390/biochem6010003

76. Mahmoud, M.; Abdallah, N.A.; El-Shafei, K.; Tawfik, N.F.; El-Sayed, H.S. Survivability of alginate-microencapsulated *Lactobacillus plantarum* during storage, simulated food processing and gastrointestinal conditions. *Heliyon* **2020**, *6*, e03541. [CrossRef]
77. Jang, W.J.; Oh, D.N.; Lee, E.J.; Lee, J.M. Encapsulation of probiotics with poly-γ-glutamic acid alters gut microbiota and short-chain fatty acid content by maintaining cell viability in the gastrointestinal tract. *Food Hydrocoll.* **2024**, *147*, 109328. [CrossRef]
78. Afzaal, M.; Khan, A.U.; Saeed, F.; Ahmed, A.; Ahmad, M.H.; Maan, A.A.; Tufail, T.; Anjum, F.M.; Hussain, S. Functional exploration of free and encapsulated probiotic bacteria in yogurt and simulated gastrointestinal conditions. *Food Sci. Nutr.* **2019**, *7*, 3931–3940. [CrossRef]
79. Rezvani, M.; Hesari, J.; Peighambardoust, S.H.; Manconi, M.; Hamishehkar, H. Development and characterization of nanostructured pharmacosomal mesophases: An innovative delivery system for bioactive peptides. *Adv. Pharm. Bull.* **2018**, *8*, 609. [CrossRef]
80. Suganthi, V.; Ethiraj, S.; Anbalagan, N.; Siddique, J.F.; Vaithilingam, M. Encapsulation of purified pediocin of *Pediococcus pentosaceus* into liposome based nanovesicles and its antilisterial effect. *Protein Pept. Lett.* **2021**, *28*, 1115–1126. [CrossRef]
81. Amer, S.A.; Abushady, H.M.; Refay, R.M.; Mailam, M.A. Enhancement of the antibacterial potential of plantaricin by incorporation into silver nanoparticles. *J. Genet. Eng. Biotechnol.* **2021**, *19*, 13–16. [CrossRef]
82. Niaz, T.; Shabbir, S.; Noor, T.; Abbasi, R.; Raza, Z.A.; Imran, M. Polyelectrolyte multicomponent colloidosomes loaded with nisin Z for enhanced antimicrobial activity against foodborne resistant pathogens. *Front. Microbiol.* **2018**, *8*, 2700. [CrossRef]
83. Neto, J.H.P.L.; Soares, S.; Leite, K.S.; Campos, M.I.F.; Grisi, C.V.B.; Ribeiro, J.E.S.; Sobral, M.V.; Amorim, G.M.W.; Luciano, W.A.; Gadelha, T.S.; et al. Bioactivity evaluation of microbial exopolysaccharide and its use as an encapsulating agent for *Lactobacillus acidophilus* (LA-3). *Food Biosci.* **2023**, *55*, 102946. [CrossRef]
84. Charoenwongpaiboon, T.; Wangpaiboon, K.; Pichyangkura, R.; Nepogodiev, S.A.; Wonganan, P.; Mahalapbutr, P.; Field, R.A. Characterization of a nanoparticulate exopolysaccharide from *Leuconostoc holzapfelii* KM01 and its potential application in drug encapsulation. *Int. J. Biol. Macromol.* **2021**, *187*, 690–698. [CrossRef]
85. Soliman, T.N.; Elaaser, M.M. Fortification of Yoghurt with Conjugated Linoleic Acid (CLA) Encapsulated in Reassembled Casein. *Egypt. J. Chem.* **2022**, *65*, 1559–1565. [CrossRef]
86. Mei, L.; Zhang, D.; Shao, H.; Hao, Y.; Zhang, T. Injectable and Self-Healing Probiotics-Loaded Hydrogel for Promoting Superbacteria-Infected Wound Healing. *ACS Appl. Mater. Interfaces* **2022**, *14*, 20538–20550. [CrossRef]
87. Adeel, M.; Afzaal, M.; Saeed, F.; Ahmed, A.; Mahmood, K.; Abbas shah, Y.; Ateeq, H.; Sibat, A.; Khan, M.R.; Busquets, R. Encapsulation of probiotic bacteria using polyelectrolytes stabilized nanoliposomes for improved viability under hostile conditions. *J. Food Sci.* **2023**, *88*, 3839–3849. [CrossRef]
88. Yilmaz, M.T.; Taylan, O.; Karakas, C.Y.; Dertli, E. An alternative way to encapsulate probiotics within electrospun alginate nanofibers as monitored under simulated gastrointestinal conditions and in kefir. *Carbohydr. Polym.* **2020**, *244*, 116447. [CrossRef]
89. Ajalloueian, F.; Guerra, P.R.; Bahl, M.I.; Torp, A.M.; Hwu, E.T.; Licht, T.R.; Boisen, A. Multi-layer PLGA-pullulan-PLGA electrospun nanofibers for probiotic delivery. *Food Hydrocoll.* **2022**, *123*, 107112. [CrossRef]
90. García-Toledo, J.A.; Torrestiana-Sánchez, B.; Martínez-Sánchez, C.E.; Tejero-Andrade, J.M.; García-Bórquez, A.; Mendoza-García, P.G. Nanoencapsulation of a bacteriocin from *Pediococcus acidilactici* ITV26 by microfluidization. *Food Bioprocess Technol.* **2019**, *12*, 88–97. [CrossRef]
91. Wang, H.; Yang, B.; Sun, H. Pectin-Chitosan polyelectrolyte complex nanoparticles for encapsulation and controlled release of nisin. *Am. J. Polym. Sci. Technol.* **2017**, *3*, 82. [CrossRef]
92. Zimet, P.; Mombrú, Á.W.; Faccio, R.; Brugnini, G.; Miraballes, I.; Rufo, C.; Pardo, H. Optimization and characterization of nisin-loaded alginate-chitosan nanoparticles with antimicrobial activity in lean beef. *LWT* **2018**, *91*, 107–116. [CrossRef]
93. Çelen, T.; Anumudu, C.; Miri, T.; Onyeaka, H.; Fernandez-Trillo, P. Pathogen-responsive delivery of Nisin. *Food Hydrocoll.* **2024**, *154*, 110076. [CrossRef]
94. Jung, H.S.; Lee, N.K.; Paik, H.D. Heat-killed *LatiLactobacillus sakei* CNSC001WB and *Lactobacillus pentosus* WB693 have an anti-inflammatory effect on LPS-stimulated RAW 264.7 cells. *Probiotics Antimicrob. Proteins* **2023**, *16*, 1875–1885. [CrossRef]
95. Salas-Lais, A.G.; Robles-Contreras, A.; Balderas-López, J.A.; Lucio, V.M.B. Immunobiotic and paraprobiotic potential effect of *Lactobacillus casei* in a systemic toxoplasmosis murine model. *Microorganisms* **2020**, *8*, 113. [CrossRef] [PubMed]
96. Elham, N.; Naheed, M.; Elahe, M.; Hossein, M.M.; Majid, T. Selective cytotoxic effect of probiotic, paraprobiotic and postbiotics of *L. casei* strains against colorectal cancer cells: Invitro studies. *Braz. J. Pharm. Sci.* **2022**, *58*, e19400. [CrossRef]
97. Batista, V.L.; De Jesus, L.C.; Tavares, L.M.; Barroso, F.L.; Fernandes, L.J.; Freitas, A.D.; Americo, M.F.; Drumond, M.M.; Mancha-Agresti, P.; Ferreira, E.; et al. Paraprobiotics and postbiotics of *Lactobacillus delbrueckii* CIDCA 133 mitigate 5-FU-induced intestinal inflammation. *Microorganisms* **2022**, *10*, 1418. [CrossRef]
98. Mahdi, L.H.; Mahdi, N.Z.; Sajet, R.M.; Auda, I.G.; Mater, H.N.; Zwain, L.A.H.; Kadhem, B.M.; Alsaadi, L.G. Anticariogenic and antibiofilm of purified bacteriocin of *Lactobacillus curvatus* and immunomodulatory effect of *L. curvatus* in str eptococcal bacteremia. *Rev. Med. Microbiol.* **2019**, *30*, 26–35. [CrossRef]

https://doi.org/10.3390/biochem6010003

99. Giani, A.D.; Bovio, F.; Forcella, M.; Fusi, P.; Sello, G.; Di Gennaro, P. Identification of a bacteriocin-like compound from *Lactobacillus plantarum* with antimicrobial activity and effects on normal and cancerogenic human intestinal cells. *AMB Express* **2019**, *9*, 88. [CrossRef]
100. İncili, G.K.; Karatepe, P.; Akgöl, M.; Güngören, A.; Koluman, A.; İlhak, O.İ.; Kanmaz, H.; Kaya, B.; Hayaloğlu, A.A. Characterization of lactic acid bacteria postbiotics, evaluation in-vitro antibacterial effect, microbial and chemical quality on chicken drumsticks. *Food Microbiol.* **2022**, *104*, 104001. [CrossRef]
101. Toushik, S.H.; Park, J.H.; Kim, K.; Ashrafudoulla, M.; Ulrich, M.S.I.; Mizan, M.F.R.; Roy, P.K.; Shim, W.-B.; Kim, Y.-M.; Park, S.H.; et al. Antibiofilm efficacy of *Leuconostoc mesenteroides* J. 27-derived postbiotic and food-grade essential oils against Vibrio parahaemolyticus, Pseudomonas aeruginosa and Escherichia coli alone and in combination and their application as a green preservative in the seafood industry. *Food Res. Int.* **2022**, *156*, 111163. [CrossRef]
102. Mohammadi, R.; Moradi, M.; Tajik, H.; Molaei, R. Potential application of postbiotics metabolites from bioprotective culture to fabricate bacterial nanocellulose based antimicrobial packaging material. *Int. J. Biol. Macromol.* **2022**, *220*, 528–536. [CrossRef] [PubMed]
103. Watanabe, J.; Hashimoto, N.; Yin, T.; Sandagdorj, B.; Arakawa, C.; Inoue, T.; Suzuki, S. Heat-killed *Lactobacillus* brevis KB290 attenuates visceral fat accumulation induced by high-fat diet in mice. *J. Appl. Microbiol.* **2021**, *131*, 1998–2009. [CrossRef]
104. Toushik, S.H.; Kim, K.; Park, S.H.; Park, J.H.; Ashrafudoulla, M.; Ulrich, M.S.I.; Mizan, M.F.R.; Hossain, M.I.; Shim, W.-B.; Kang, I.; et al. Prophylactic efficacy of *Lactobacillus curvatus* B67-derived postbiotic and quercetin, separately and combined, against Listeria monocytogenes and Salmonella enterica ser. Typhimurium on processed meat sausage. *Meat Sci.* **2023**, *197*, 109065. [CrossRef] [PubMed]
105. Kumar, R.; Sharma, A.; Gupta, M.; Padwad, Y.; Sharma, R. Cell-free culture supernatant of probiotic *Lactobacillus fermentum* protects against H_2O_2-induced premature senescence by suppressing ROS-Akt-mTOR axis in murine preadipocytes. *Probiotics Antimicrob.* **2020**, *12*, 563–576. [CrossRef] [PubMed]
106. Kolling, Y.; Salva, S.; Villena, J.; Alvarez, S. Are the immunomodulatory properties of *Lactobacillus rhamnosus* CRL1505 peptidoglycan common for all *Lactobacilli* during respiratory infection in malnourished mice? *PLoS ONE* **2018**, *13*, e0194034. [CrossRef]
107. Peng, M.; Lee, S.H.; Rahaman, S.O.; Biswas, D. Dietary probiotic and metabolites improve intestinal homeostasis and prevent colorectal cancer. *Food Funct.* **2020**, *11*, 10724–10735. [CrossRef]
108. Ardestani, S.K.; Tafvizi, F.; Ebrahimi, M.T. Heat-killed probiotic bacteria induce apoptosis of HT-29 human colon adenocarcinoma cell line via the regulation of Bax/Bcl2 and caspases pathway. *Hum. Exp. Toxicol.* **2019**, *38*, 1069–1081. [CrossRef]
109. Arellano-García, L.; Trepiana, J.; Martínez, J.A.; Portillo, M.P.; Milton-Laskibar, I. Beneficial effects of viable and heat-inactivated *Lactobacillus rhamnosus* GG administration on oxidative stress and inflammation in diet-induced NAFLD in rats. *Antioxidants* **2023**, *12*, 717. [CrossRef]
110. Kato, Y.; Suzuki, M. Synthesis of metal nanoparticles by microorganisms. *Crystals* **2020**, *10*, 589. [CrossRef]
111. Kikuchi, F.; Kato, Y.; Furihata, K.; Kogure, T.; Imura, Y.; Yoshimura, E.; Suzuki, M. Formation of gold nanoparticles by glycolipids of *Lactobacillus casei*. *Sci. Rep.* **2016**, *6*, 34626. [CrossRef]
112. Yusof, H.M.; Rahman, N.A.A.; Mohamad, R.; Zaidan, U.H. Microbial mediated synthesis of silver nanoparticles by *Lactobacillus Plantarum* TA4 and its antibacterial and antioxidant activity. *Appl. Sci.* **2020**, *10*, 6973. [CrossRef]
113. Singh, P.; Kim, Y.J.; Zhang, D.; Yang, D.C. Biological synthesis of nanoparticles from plants and microorganisms. *Trends Biotechnol.* **2016**, *34*, 588–599. [CrossRef]
114. Al-Asbahi, M.G.; Al-Ofiry, B.A.; Saad, F.A.; Alnehia, A.; Al-Gunaid, M.Q. Silver nanoparticles biosynthesis using mixture of *Lactobacillus* sp. and *Bacillus* sp. growth and their antibacterial activity. *Sci. Rep.* **2024**, *14*, 10224. [CrossRef] [PubMed]
115. El Fadly, E.B.; Salah, A.S.; Abdella, B.; Al Ali, A.; AlShmrany, H.; ElBaz, A.M.; Abdelatty, N.S.; Khamis, E.F.; Maagouz, O.F.; Salamah, M.A.; et al. Mapping a sustainable approach: Biosynthesis of lactobacilli-silver nanocomposites using whey-based medium for antimicrobial and bioactivity applications. *Microb. Cell Factories* **2024**, *23*, 195. [CrossRef]
116. Báez-Cruz, R.; Baptista, L.A.; Ntim, S.; Manidurai, P.; Espinoza, S.; Ramanan, C.; Cortes-Huerto, R.; Sulpizi, M. Role of pH in the Synthesis and Growth of Gold Nanoparticles Using L-Asparagine: A Combined Experimental and Simulation Study. *J. Physics Condens. Matter* **2021**, *33*, 254005. [CrossRef]
117. Sidhu, A.K.; Verma, N.; Kaushal, P. Role of biogenic capping agents in the synthesis of metallic nanoparticles and evaluation of their therapeutic potential. *Front. Nanotechnol.* **2022**, *3*, 801620. [CrossRef]
118. Ovais, M.; Khalil, A.T.; Ayaz, M.; Ahmad, I.; Nethi, S.K.; Mukherjee, S. Biosynthesis of metal nanoparticles via microbial enzymes: A mechanistic approach. *Int. J. Mol. Sci.* **2018**, *19*, 4100. [CrossRef]
119. Martínez, F.G.; Moreno-Martin, G.; Pescuma, M.; Madrid-Albarrán, Y.; Mozzi, F. Biotransformation of selenium by lactic acid bacteria: Formation of seleno-nanoparticles and seleno-amino acids. *Front. Bioeng. Biotechnol.* **2020**, *8*, 506. [CrossRef]
120. Escárcega-González, C.E.; Garza-Cervantes, J.A.; Vázquez-Rodríguez, A.; Morones-Ramírez, J.R. Bacterial exopolysaccharides as reducing and/or stabilizing agents during synthesis of metal nanoparticles with biomedical applications. *Int. J. Polym. Sci.* **2018**, *2018*, 7045852. [CrossRef]

https://doi.org/10.3390/biochem6010003

121. Saravanan, C.; Rajesh, R.; Kaviarasan, T.; Muthukumar, K.; Kavitake, D.; Shetty, P.H. Synthesis of silver nanoparticles using bacterial exopolysaccharide and its application for degradation of azo-dyes. *Biotechnol. Rep.* **2017**, *15*, 33–40. [CrossRef]
122. Hazbar, A.M.; Mohammed Noori Jassim, A.; Taha Mohammed, M. Green Synthesis of Gold Nanoparticles using Eruca Sativa Plant Extracts. *J. Nanostruct.* **2025**, *15*, 158–167. [CrossRef]
123. Bahrulolum, H.; Nooraei, S.; Javanshir, N.; Tarrahimofrad, H.; Mirbagheri, V.S.; Easton, A.J.; Ahmadian, G. Green synthesis of metal nanoparticles using microorganisms and their application in the agrifood sector. *J. Nanobiotechnol.* **2021**, *19*, 86–112. [CrossRef]
124. Ghosh, S.; Ahmad, R.; Zeyaullah, M.; Khare, S.K. Microbial nano-factories: Synthesis and biomedical applications. *Front. Chem.* **2021**, *9*, 626834. [CrossRef]
125. Matei, A.; Matei, S.; Matei, G.M.; Cogălniceanu, G.; Cornea, C.P. Biosynthesis of silver nanoparticles mediated by culture filtrate of lactic acid bacteria, characterization and antifungal activity. *EuroBiotech J.* **2020**, *4*, 97–103. [CrossRef]
126. Awadelkareem, A.M.; Siddiqui, A.J.; Noumi, E.; Ashraf, S.A.; Hadi, S.; Snoussi, M.; Badraoui, R.; Bardakci, F.; Ashraf, M.S.; Danciu, C.; et al. Biosynthesized silver nanoparticles derived from probiotic *Lactobacillus rhamnosus* (AgNPs-LR) targeting biofilm formation and quorum sensing-mediated virulence factors. *Antibiotics* **2023**, *12*, 986. [CrossRef] [PubMed]
127. Miran, S.B.; Ali, F.A. Biosynthesis of gold nanoparticles with four different *Lactobacillus* species. *Eurasian J. Sci. Eng.* **2023**, *9*, 47–56. [CrossRef]
128. Markus, J.; Mathiyalagan, R.; Kim, Y.J.; Abbai, R.; Singh, P.; Ahn, S.; Perez, Z.E.J.; Hurh, J.; Yang, D.C. Intracellular synthesis of gold nanoparticles with antioxidant activity by probiotic *Lactobacillus kimchicus* DCY51T isolated from Korean kimchi. *Enzym. Microb. Technol.* **2016**, *95*, 85–93. [CrossRef]
129. Yusof, H.M.; Rahman, N.A.A.; Mohamad, R.; Zaidan, U.H.; Samsudin, A.A. Biosynthesis of zinc oxide nanoparticles by cell-biomass and supernatant of *Lactobacillus plantarum* TA4 and its antibacterial and biocompatibility properties. *Sci. Rep.* **2020**, *10*, 19996. [CrossRef]
130. Suba, S.; Vijayakumar, S.; Vidhya, E.; Punitha, V.N.; Nilavukkarasi, M. Microbial mediated synthesis of ZnO nanoparticles derived from *Lactobacillus* spp.: Characterizations, antimicrobial and biocompatibility efficiencies. *Sens. Int.* **2021**, *2*, 100104. [CrossRef]
131. Shanmugam, R.; Munusamy, T.; Jayakodi, S.; Al-Ghanim, K.A.; Nicoletti, M.; Sachivkina, N.; Govindarajan, M. Probiotic-bacteria (*Lactobacillus fermentum*)-wrapped zinc oxide nanoparticles: Biosynthesis, characterization and antibacterial activity. *Fermentation* **2023**, *9*, 413. [CrossRef]
132. Qasim, M.A.; Yaaqoob, L.A. Evaluation of antibacterial activity of iron oxide nanoparticles synthesis by extracellular *Lactobacillus* against pseudomonas aeruginosa. *J. Med. Chem. Sci.* **2023**, *6*, 1100. [CrossRef]
133. Torabian, P.; Ghandehari, F.; Fatemi, M. Biosynthesis of iron oxide nanoparticles by cytoplasmic extracts of bacteria *Lactobacillus casei*. *Asian J. Green. Chem.* **2018**, *2*, 171–280. [CrossRef]
134. Shao, M.; Xiong, G.; He, G. Bacterial synthesis of metallic nanoparticles for biomedical applications. *Mater. Chem. Horiz.* **2023**, *2*, 303–313. [CrossRef]
135. Kim, Y.J.; Perumalsamy, H.; Markus, J.; Balusamy, S.R.; Wang, C.; Kang, S.H.; Lee, S.; Park, S.Y.; Kim, S.; Castro-Aceituno, V.; et al. Development of *Lactobacillus kimchicus* DCY51T-mediated gold nanoparticles for delivery of ginsenoside compound K: In vitro photothermal effects and apoptosis detection in cancer cells. *Artif. Cells Nanomedicine Biotechnol.* **2019**, *47*, 30–44. [CrossRef]
136. Ganguly, P.; Byrne, C.; Breen, A.; Pillai, S.C. Antimicrobial activity of photocatalysts: Fundamentals, mechanisms, kinetics and recent advances. *Appl. Catal. B* **2018**, *225*, 51–75. [CrossRef]
137. Abishad, P.; Vergis, J.; Arya, P.R.; Unni, V.; Vinod, V.K.; Juliet, S.; Kurkure, N.V.; Barbuddhe, S.B.; Byrappa, K.; Rawool, D.B. Facile one-pot synthesis of gold nanoparticles using *Lactobacillus acidophilus* as a potential photocatalytic agent against multi-drug-resistant pathogens of public health importance. *Biomass Convers. Biorefinery* **2023**, *14*, 29787–29794. [CrossRef]
138. Suba, S.; Vijayakumar, S.; Nilavukkarasi, M.; Vidhya, E.; Punitha, V.N. Eco synthesized silver nanoparticles as a next generation of nanoproduct in multidisciplinary applications. *Environ. Toxicol. Chem.* **2022**, *4*, 13–19. [CrossRef]
139. Vijayakumar, G.; Kim, H.J.; Rangarajulu, S.K. In Vitro antibacterial and wound healing activities evoked by silver nanoparticles synthesized through probiotic bacteria. *Antibiotics* **2023**, *12*, 141. [CrossRef] [PubMed]
140. Feng, Z.; Wang, Y.; Xu, H.; Guo, Y.; Xia, W.; Zhao, C.; Zhao, X.; Wu, J. Recent advances in bacterial therapeutics based on sense and response. *Acta Pharm. Sin. B* **2023**, *13*, 1014–1027. [CrossRef]
141. Ho, T.M.; Tran, Y.T.H.; Nguyen, T.H.K. Study on chloroquine containing ability of minicells derived from *Lactobacillus acidophilus* ATCC 4356. *J. Appl. Pharm. Sci.* **2024**, *14*, 91–96. [CrossRef]
142. Ali, M.K.; Liu, Q.; Liang, K.; Li, P.; Kong, Q. Bacteria-derived minicells for cancer therapy. *Cancer Lett.* **2020**, *491*, 11–21. [CrossRef]
143. Nguyen, H.N.; Jovel, S.R.; Nguyen, T.H. Nanosized minicells generated by lactic acid bacteria for drug delivery. *J. Nanomater.* **2017**, *2017*, 6847297. [CrossRef]
144. Vinh, D.T.; Khue, N.T. Study on Minicell Generation of *Lactobacillus acidophilus* VTCC-B-871 for Drug Delivery. *J. Appl. Pharm. Sci.* **2013**, *3*, 33–36. [CrossRef]

https://doi.org/10.3390/biochem6010003

145. Kinh, L.; Nguyen, D.; Jovel, S.R.; Khue, T.N.H. Tinidazole Delivery Improved by Nanosized Minicells Originated from *Leuconostoc mesenteroides*. *J. Nanomater.* **2019**, *2019*, 7684795. [CrossRef]
146. Kousheh, S.A.; Moradi, M.; Tajik, H.; Molaei, R. Preparation of antimicrobial/ultraviolet protective bacterial nanocellulose film with carbon dots synthesized from lactic acid bacteria. *Int. J. Biol. Macromol.* **2020**, *155*, 216–225. [CrossRef]
147. Lin, F.; Li, C.; Chen, Z. Bacteria-derived carbon dots inhibit biofilm formation of Escherichia coli without affecting cell growth. *Front. Microbiol.* **2018**, *9*, 259. [CrossRef]
148. Shamkhali, L.; Mobarez, A.M.; Siadat, S.D.; Pajavand, H. Comparing the production of carbon dots synthesized from *Lactobacillus acidophilus* and *Bifidobacterium bifidum* and investigating their antibacterial effects. *J. Nanomed. Res.* **2024**, *9*, 80–89. [CrossRef]
149. Sun, L.; Zhao, Y.; Peng, H.; Zhou, J.; Zhang, Q.; Yan, J.; Liu, Y.; Guo, S.; Wu, X.; Li, B. Carbon dots as a novel photosensitizer for photodynamic therapy of cancer and bacterial infectious diseases: Recent advances. *J. Nanobiotechnol.* **2024**, *22*, 210. [CrossRef] [PubMed]
150. Rezvani, M. Oxidative Stress-Induced Gastrointestinal Diseases: Biology and Nanomedicines—A Review. *BioChem* **2024**, *4*, 189–216. [CrossRef]
151. Balou, S.; Shandilya, P.; Priye, A. Carbon dots for photothermal applications. *Front. Chem.* **2022**, *10*, 1023602. [CrossRef]

https://doi.org/10.3390/biochem6010003

MDPI AG
Grosspeteranlage 5
4052 Basel
Switzerland
Tel.: +41 61 683 77 34

BioChem Editorial Office
E-mail: biochem@mdpi.com
www.mdpi.com/journal/biochem

www.ingramcontent.com/pod-product-compliance
Lightning Source LLC
Chambersburg PA
CBHW061610220926
51850CB00005B/472

* 9 7 8 3 7 2 5 8 7 7 3 9 3 *